Cognitive Radio Communications and Networks
Principles and Practice

Edited by

Alexander M. Wyglinski, Ph.D.,
Worcester Polytechnic Institute

Maziar Nekovee, Ph.D.,
BT Research and University College London

Y. Thomas Hou, Ph.D.,
Virginia Polytechnic Institute and State University

AMSTERDAM • BOSTON • HEIDELBERG • LONDON
NEW YORK • OXFORD • PARIS • SAN DIEGO
SAN FRANCISCO • SINGAPORE • SYDNEY • TOKYO
Academic Press is an imprint of Elsevier

ELSEVIER

Academic Press is an imprint of Elsevier
30 Corporate Drive, Suite 400, Burlington, MA 01803, USA
525 B Street, Suite 1900, San Diego, California 92101-4495, USA
The Boulevard, Langford Lane, Kidlington, Oxford, OX5 1GB, UK

Notices
Knowledge and best practice in this field are constantly changing. As new research and experience
broaden our understanding, changes in research methods, professional practices, or medical treatment
may become necessary. Practitioners and researchers must always rely on their own experience and
knowledge in evaluating and using any information, methods, compounds, or experiments described
herein. In using such information or methods they should be mindful of their own safety and the safety
of others, including parties for whom they have a professional responsibility.
To the fullest extent of the law, neither the Publisher nor the authors, contributors, or editors assume
any liability for any injury and/or damage to persons or property as a matter of products liability,
negligence or otherwise, or from any use or operation of any methods, products, instructions, or ideas
contained in the material herein.

Library of Congress Cataloging-in-Publication Data
Cognitive radio communications and networks: principles and practice/edited by Alexander M.
Wyglinski, Maziar Nekovee, and Y. Thomas Hou.
    p. cm.
Includes bibliographical references and index.
ISBN 978-0-12-374715-0 (alk. paper)
1. Cognitive radio networks. I. Wyglinski, Alexander M. II. Nekovee, Maziar. III. Hou, Y. Thomas.
TK5103.4815.C63 2010
621.384-dc22   2009040908

British Library Cataloguing-in-Publication Data
A catalogue record for this book is available from the British Library.
For information on all Academic Press publications
visit our Web site at www.elsevierdirect.com

Printed and bound by CPI Group (UK) Ltd, Croydon, CR0 4YY

Transferred to digital print 2013

Contents

CHAPTER 3 Digital communication fundamentals for cognitive radio   41

Si Chen and Alexander M. Wyglinski

CHAPTER 4   Spectrum sensing and identification   85
Qing Zhao and Ananthram Swami

CHAPTER 5   Spectrum access and sharing   113
Alireza Attar, Oliver Holland, and Hamid Aghvami

CHAPTER 6 Agile transmission techniques   149
Srikanth Pagadarai, Rakesh Rajbanshi, Gary J. Minden,
and Alexander M. Wyglinski

CHAPTER 7 Reconfiguration, adaptation, and optimization   177
Timothy R. Newman, Joseph B. Evans,
and Alexander M. Wyglinski

CHAPTER 12 Cross-layer optimization for multihop cognitive radio networks 335

Yi Shi and Y. Thomas Hou

Theme 3 Applications, standards, and implementations of cognitive radio 365

CHAPTER 13 Defining cognitive radio 367

Przemysław Pawełczak and Rangarao Venkatesha Prasad

CHAPTER 15 Cognitive radio network security   **431**
Jung-Min "Jerry" Park, Kaigui Bian, and Ruiliang Chen

CHAPTER 16 Public safety and cognitive radio   **467**
Marntx Heskamp, Roel Schiphorst, and Kees Slump

CHAPTER 17 Auction-based spectrum markets in cognitive radio networks   489

Xia Zhou, Heather Zheng, Maziar Nekovee,
and Milind M. Buddhikot

CHAPTER 20 Cognitive radio evolution   **587**

Joseph Mitola III

Preface

Cognitive radio is the next disruptive radio communication and networking technology. It is currently experiencing rapid growth due to its potential to solve many of the problems affecting present-day systems. For instance, interest in cognitive radio by the industrial sector has been rapidly growing over the past couple of years, which has manifested into several forms, including:

- Regulatory agencies moving toward allowing the operation of cognitive radios in licensed television spectrum bands.
- Creation and eventual ratification of international wireless standards supporting secondary access of licensed spectrum.
- Incorporation of cognitive radio technology into existing standards.
- Active lobbying by various entities and coalitions for cognitive access to wireless spectral white spaces, including nonprofit organizations and various corporations.

Wireless device manufacturers (e.g., Motorola, Eriksson, and Nokia), telecommunication operators (e.g., BT, France Telecom), and chip makers (e.g., Intel) are all beginning to invest in this new technology, especially with respect to research and development. Outside the area of wireless communication, software-defined radio and cognitive radio technology are expected to have important applications in consumer electronics and the automotive industry. However, although there have been numerous publications, conferences, tutorials, short courses, and books in this area, there does not exist a single comprehensive textbook introducing this material to the communication networks generalist in a structured manner.

The foremost objective of this book is to educate wireless communication generalists about cognitive radio communication networks. The first two parts of this book introduce the reader to the wireless communication and networking theory involved in designing/implementing cognitive radio systems and networks. End-of-chapter questions give the reader the opportunity to apply what she or he has just learned to address problems arising in that chapter. Finally, the third part of this book, which contains numerous implementations, applications, and case studies, helps the reader synthesize the previous two parts by highlighting how these concepts fit in real-world problems.

The intended readership for this book is both wireless communication industry and public sector practitioners and researchers in electrical engineering, computer engineering, and computer science (including graduate students). Both groups are assumed to have a basic background in wireless communications and networks, although they lack any expertise in cognitive radio. This book is designed specifically to introduce communication generalists to the area of cognitive radio communications and networks via a structured approach.

For the industry practitioner, our book provides a self-contained guide that caters to their immediate needs of learning new concepts and techniques in this emerging area. While the introductory sections bring the reader up to speed with the fundamentals of cognitive radio, the third part provides an up-to-date "handbook of cognitive radio technology," as it will be in the marketplace within a few years. Sample code and implementations allow researchers to gain a hands-on knowledge of cognitive radio technology, which is much appreciated in an industrial setting. Furthermore, by having specifically two chapters on the two emerging industry standards based on cognitive radio (IEEE 802.22 and IEEE SCC41), we believe that this book will be of definite benefit to the industry practitioners in this sector. The relevance of the book to this group is further increased by one of the editors and several of the contributors being from industrial research laboratories worldwide (e.g., BT, Motorola, Cisco, Philips, Intel, Broadcom).

For academic researchers, cognitive radio communications has been the subject of much recent research and is starting to become well-established. Nevertheless, cognitive radio networks are recognized as a highly challenging area, with many open research problems remaining to be explored. The cognitive radio networks part of our book is written by internationally leading experts in the field and caters to the needs of researchers in the field who require a basis in principles and challenges of cognitive radio networks (such as architecture, security, cooperation) from where they can explore new research directions in cognitive radio. Moreover, many academic institutions are beginning to offer courses on cognitive radio, software-defined radio, and advanced wireless systems engineering. At the moment, only a half dozen universities offer entire courses on cognitive radio or software-defined radio. However, numerous courses on advance wireless communications and networks exist worldwide, possessing at least a module on cognitive radio, which will eventually evolve into full courses in the near future as this area continues to grow.

Since this book covers a wide range of topics in cognitive radio communications and networks, we arranged the presentation of the topics covered by this book such that each chapter gradually builds upon the knowledge and information of the previous chapters. As a result, someone with a basic understanding of communication systems and networks can learn about the topics covered in this book in a step-by-step approach, either in a classroom environment or via self-education. Moreover, communication engineers who already have an understanding of some of the material covered in the book can easily skip ahead to topics of interest. Finally, the chapters have been grouped together into three thematically related parts in order to provide better structure for the reader with respect to the topics covered: Radio Communications, Networks, and Implementation; Applications; and Case Studies. Note that the final chapter of this book provides an all-encompassing vision on the evolution and future directions for cognitive radio by Joseph Mitola III, the father of both software-defined radio and cognitive radio.

This book is made possible by the extensive support of numerous individuals throughout the duration of this project. First, we are deeply indebted to our contributors, who all share in our vision of educating wireless communication generalists about cognitive radio communication networks by producing well-written, informative, and high-quality chapters. Second, we would like to thank our publishers at Elsevier, especially Tim Pitts and Melanie Benson, for their hard work, guidance, and encouragement during the creation of this book. Third, we owe our special thanks to the following colleagues for taking the time to review the manuscript of this book at various stages of this project (in alphabetical order): Alireza Attar, David Cavalcanti, Si Chen, Natasha Devroye, Oliver Holland, Santosh Kawade, Devin Kelly, Michael Leferman, Shiwen Mao, Sai Shankar Nandagopala, Timothy Newman, Srikanth Pagadarai, Przemysław Pawełczak, Rangarao Venkatesha Prasad, Di Pu, Yi Shi, Jingkai Su, David Taubenheim, William Webb, Kai-Kit Wong, Liguang Xie, and Qing Zhao. Fourth, we would like to thank Honggang Zhang, Frederick Martin, and Rajarathnam Chandramouli, with whom we worked with much pleasure in organizing the Second International Conference on Cognitive Radio Oriented Wireless Communication and Networks (CrownCom 2007), for bringing us together in Orlando, Florida, USA, where the idea for this book was born. Finally, we would like to thank our families for their support and encouragement.

Alexander M. Wyglinski
Worcester Polytechnic Institute, United States

Maziar Nekovee
BT Research and University College London, United Kingdom

Y. Thomas Hou
Virginia Polytechnic Institute and State University, United States

About the Editors

A. M. Wyglinski   M. Nekovee   Y. T. Hou

Alexander M. Wyglinski is an Assistant Professor of Electrical and Computer Engineering at Worcester Polytechnic Institute (WPI), Director of the WPI Limerick Project Center, and Director of the Wireless Innovation Laboratory (WI Lab). He received his Ph.D. from McGill University in 2005, his M.S.(Eng.) from Queens University at Kingston in 2000, and his B.Eng. from McGill University in 1999, all in electrical engineering. Professor Wyglinski's current research interests include wireless communications, wireless networks, cognitive radios, software-defined radios, transceiver optimization algorithms, dynamic spectrum access networks, spectrum sensing techniques, machine learning techniques for communication systems, and signal processing techniques for digital communications. He is a member of the IEEE, IEEE Communications Society, IEEE Signal Processing Society, IEEE Vehicular Technology Society, IEEE Women in Engineering, Eta Kappa Nu, and Sigma Xi.

Maziar Nekovee leads cognitive radio research at BT (British Telecom) and is also involved in leading a number of large EU and International collaborative R&D projects on cognitive radio networks and secondary/dynamic spectrum access. These projects involve altogether over 30 industrial and academic partners from Europe, China, India, and the United States. They intend to bring very close to market the cognitive radio technology as well as researching long-term directions of the technology. Dr. Nekovee received his BSc. and MSc. (cum laude) in electrical engineering from Delft University of Technology in The Netherlands in 1990 and his Ph.D. in theoretical and computational physics from the University of Nijmegen in The Netherlands in 1995. His research cuts across several disciplines and currently focuses on theory, modeling, simulations, and development of complex networked systems, including cognitive and cooperative radio networks, wireless vehicular communication networks, and complex social and biological networks.

Dr. Nekovee is the recipient of a prestigious Industry Fellowship from the U.K.'s Academy of Science, the Royal Society, and an Honorary Senior Fellow at University College London.

Y. Thomas Hou is an Associate Professor of Electrical and Computer Engineering at Virginia Polytechnic Institute and State University (Virginia Tech), Blacksburg, VA, USA. He received his Ph.D. from Polytechnic Institute of New York University in 1998. Prof. Hou's current research interests include cross-layer design and optimization for cognitive radio wireless networks, cooperative communications, MIMO-based ad hoc networks, video communications over dynamic ad hoc networks, and algorithm design for sensor networks. He was a recipient of an Office of Naval Research (ONR) Young Investigator Award (2003) and a National Science Foundation (NSF) CAREER Award (2004) for his research on optimizations and algorithm design for wireless ad hoc and sensor networks. He has published extensively in leading IEEE/ACM journals and conferences and received five best paper awards from IEEE (including IEEE INFOCOM 2008 Best Paper Award and IEEE ICNP 2002 Best Paper Award). Prof. Hou is on the editorial boards of a number of IEEE and international journals. He is a senior member of IEEE and ACM and holds five U.S. patents.

When radio meets software

1

Alexander M. Wyglinski[1], Maziar Nekovee[2], and Y. Thomas Hou[3]
[1]Worcester Polytechnic Institute, United States
[2]BT Research and University College London, United Kingdom
[3]Virginia Polytechnic Institute and State University, United States

1.1 INTRODUCTION

Data communication networks are a vital component of any modern society. They are used extensively in numerous applications, including financial transactions, social interactions, education, national security, and commerce. In particular, both wired and wireless devices are capable of performing a plethora of advanced functions that support a range of services, such as voice telephony, web browsing, streaming multimedia, and data transfer. With the rapid evolution of microelectronics, wireless transceivers are becoming more versatile, powerful, and portable. This has enabled the development of *software-defined radio* (SDR) technology, where the radio transceivers perform the baseband processing entirely in software: modulation/demodulation, error correction coding, and compression.

Since its introduction in 1991, SDR has been defined as a radio platform of which the functionality is at least partially controlled or implemented in software. Consequently, any waveform defined in the memory of the SDR platform can be employed on any frequency [1]. Although initially constrained by the conversion process between the analog and digital signaling domains, the emergence of cheap high-speed digital-to-analog converters (DACs) and analog-to-digital converters (ADCs) has brought the ideal SDR concept of an entirely software communication system implementation (including radio frequency functionality) closer to a reality.

Wireless devices that can be described as SDR have in fact been around for several decades. They were initially employed in military applications before finding applications in the commercial sector. Military programs such as SPEAKeasy sought to enable communication and interoperability between several military standards [2]. Although ambitious, the SPEAKeasy project did produce a functional prototype, even though the design choices involved in programming waveforms using low-level assembly language meant that the software was not compatible

Doi: 10.1016/B978-0-12-374715-0.00001-0

with newer processors. Furthermore, in terms of portability, the Phase I prototype of SPEAKeasy was large enough to fit into the back of a truck [3].

One of the first significant commercial introductions of SDR platforms was the Vanu Anywave™ software radio base station, which incorporated multiple cellular access standards into a simple SDR implementation. Since the cellular standards are based in software, they can be changed "on the fly" to adapt to different user needs of each cell, rather than replacing the radio frequency (RF) hardware, which can be a prohibitively expensive upgrade. Furthermore, new standards can be uploaded to the SDR platform for immediate deployment in a cellular region [4]. To increase the effectiveness and improve the aging process of an SDR platform, most developers seek to use portable code for their software, reusable components that can work under different waveform configurations, and generic hardware that can be easily upgraded [5].

Given the ease and speed of programming baseband operations in an SDR platform, this technology is considered to be a prime candidate for numerous advanced networking applications and architectures that were unrealizable only several years ago. An SDR platform that can rapidly reconfigure operating parameters based on changing requirements and conditions and through a process of cognition is known as *cognitive radio* [6]. The term *cognitive radio* (CR) was first defined by Joseph Mitola III [7]. According to Mitola, CR technology is the "intersection of personal wireless technology and computational intelligence," where CR is defined as "a really smart radio that would be self-aware, RF-aware, user-aware, and that would include language technology and machine vision along with a lot of high-fidelity knowledge of the radio environment" [7]. Cognitive radio clearly goes hand in hand with SDR; together, they can achieve functionality considered impossible only a decade ago. Consequently, before continuing any further with respect to CR, we first provide an overview of SDR technology.

1.2 SOFTWARE-DEFINED RADIO

1.2.1 What Is Software-Defined Radio?

Before describing what SDR does, it is useful to review the design of a conventional digital radio. Figure 1.1 shows a block diagram of a generic digital radio [8], which consists of five sections:

- The antenna section, which receives (or transmits) information encoded in radio waves.
- The RF front-end section, which is responsible for transmitting/receiving radio frequency signals from the antenna and converting them to an intermediate frequency (IF).
- The ADC/DAC section, which performs analog-to-digital/digital-to-analog conversion.

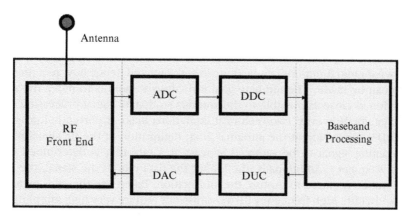

FIGURE 1.1

Schematic block diagram of a digital radio [8].

- The digital up-conversion (DUC) and digital down-conversion (DDC) blocks, which essentially perform modulations of the signal on the transmitting path and demodulation of the signal on the receiving path.
- The baseband section, which performs operations such as connection setup, equalization, frequency hopping, coding/decoding, and correlation, while also implementing the link layer protocol.

The DDC/DUC and baseband processing operations require large computing power, and in a conventional digital radio are implemented in dedicated hardware. In programmable digital radio (PDR) systems baseband operations and link layer protocols are implemented in software while the DDC/DUC functionality is performed using application-specific integrated circuits (ASICs).

Software-defined radio refers to technologies wherein these functionalities are performed by software modules running on field programmable gate arrays (FPGAs), digital signal processors (DSP), general-purpose processors (GPP), or a combination thereof. This enables programmability of both DDC/DUC and baseband processing blocks. Hence, operation characteristics of the radio, such as coding, modulation type, and frequency band, can be changed at will, simply by loading a new software. Also multiple radio devices using different modulations can be replaced by a single radio device that can perform the same task.

If the AD/DA conversion can be pushed further into the RF block, the programmability can be extended to the RF front end and an ideal *software* radio can be implemented. However, there are a number of challenges in the transition from hardware radio to software (-defined) radio. First, transition from hardware to software processing results in a substantial increase in computation, which in turn results in increased power consumption. This reduces battery life and is one of the key reasons why software-defined radios have not been deployed yet in end-user devices, but rather in base stations and access points, which can take advantage of external power resources.

Second, the question where the AD/DA conversion can be performed determines what radio functions can be done in software and hence how reconfigurable a radio can be made. The ultimate goal for software radio is to move the AD/DA conversion as close as possible to the antenna so that all signal processing can be done digitally. However, two technical limitations make it currently infeasible to the AD/DA conversion at the antenna. First, digitization of the RF signal requires the incoming signal to be sampled at least at a rate that is determined by the Nyquist frequency. Additionally, the higher the data rate of the signal, the higher the resolution required to capture the information. Taken together, this means that high-bandwidth, high-frequency RF transmissions require very high sampling rates.

The ability to support very high sampling rates, which is especially critical with the use of high-frequency signals in the gigahertz range, limits the range of what can be digitized. To give an example, the typical channels used by an 802.11 WiFi device are 20 MHz wide. To assure that the full 20 MHz is presented to the modem without distortion, it is not unusual for ADC to digitize 40 MHz or so of signal bandwidth. To capture 40 MHz of analog signal bandwidth set by the IF filters without aliasing artifacts, the ADC will probably sample the signal at a rate above 80 million samples per second (Msps). Indeed, it is only recently that sufficiently fast DSPs and wideband AD/DA chipsets have become available at affordable cost to make it feasible to contemplate AD conversions of the IF rather than the baseband signal.

SDR is currently used to build radios that support multiple interface technologies (e.g., CDMA, GSM, and WiFi) with a single modem by reconfiguring it in software. However, SDR modems are expensive, since they typically entail programmable devices like FPGAs, as opposed to the mass-produced, single-purpose ASICs used in most consumer devices today (and are key enablers for low-cost handsets). Even today's multimode devices tend to just have multiple ASICs (or multiple cores on a single ASIC). SDR is currently used mostly in military applications, where cost is less of a constraint. SDR is also a modem technology and it ignores RF design issues. In particular, the RF design of a wireless device is typically closely coupled with the underlying access technology and modem design. For example, different air interface technologies have different spectral mask requirements and different degrees of vulnerability to cochannel interference and strong adjacent channel power. A device that must work over a wide bandwidth or over a wide range of RF signal scenarios (i.e., what other devices are operating in the nearby spectrum neighborhood) will be more complex and expensive than a single-purpose device.

1.2.2 Evolution of Software-Defined Radio

Two decades ago most radios had no software at all, and those that had it didn't do much with it. In a remarkably visionary article published in 1993 [2], Joseph Mitola III envisioned a very different kind of radio: A mostly digital radio that could be reconfigured in fundamental ways just by changing the software code running on it. He dubbed this *software-defined radio*.

A few years later Mitola's vision started to become reality. In the mid-1990s military radio systems were invented in which software controlled most of the signal processing digitally, enabling one set of hardware to work on many different frequencies and communication protocols. The first (known) example of this type of radio was the U.S. military's SPEAKeasy I and SPEAKeasy II radios, which allowed units from different branches of armed forces to communicate for the first time. However, the technology was costly and the first design took up racks that had to be carried around in a large vehicle. SPEAKeasy II was a much more compact radio, the size of two stacked pizza boxes, and was the first SDR with sufficient DSP resources to handle many different kinds of waveforms [9]. SPEAKeasy II subsequently made its way into the U.S. Navy's digital modulator radio (DMR) with many waveforms and modes, able to be remotely controlled with an Ethernet interface. These SPEAKeasy II and DMR products evolved not only to define these radio waveform features in software, but also to develop an appropriate software architecture to enable porting the software to an arbitrary hardware platform, thus achieving independence of the waveform software specification and design from the underlying hardware [9].

In the late 1990s SDR started to spread from the military domain to the commercial sector, with the pace of penetration into this market considerably accelerating in the new millennium. Cellular networks were considered as the most obvious and potentially most lucrative market that SDR could penetrate. The benefits it could bring to this industry included a general-purpose and therefore more economic hardware platform, future-proofing and easier bug fixes through software upgrades, and increased functionality and interoperability through the ability to support multiple standards [10].

Companies such as Vanu, AirSpan, and Etherstack currently offer SDR products for cellular base stations. Vanu Inc., a U.S.-based company, has been focusing on the commercial development of SDR business since 1998. It received a lot of attention in 2005 with its Anywave™ GSM base station, which became the first SDR product to receive approval under the newly established software radio regulation. The Anywave base station runs on a general-purpose processing platform and provides a software implementation of the BTS (base transceiver station), BSC (base station controller), and TRAU (transcoder and rate adaptation unit) modules of the BSS (base sation subsystem). It supports GSM and can be upgraded to GPRS and Edge. The product was first deployed in rural Texas by Mid Tex Cellular in a trial, where Vanu base station showed successfully how it could concurrently run a time division multiple access (TDMA) and a GSM network, as well as remotely upgrade and fix bugs on the base station via an Internet link.

Following this successful trial other operators, such as AT&T and Nextel, expressed interest in the Anywave base station. In 2001 the 3GNewsroom was reporting SDR base stations as the key solution to the 3G rollout problem. The ability of SDR base stations to reconfigure on the fly and support multiple protocols was thought to be the safest option for rolling out 3G. In reality, SDR didn't play the key role that was anticipated. However, a closer look at the operator's infrastructure shows that programmable devices have become a key component of current 3G base stations. In March 2005 Airspan released the first commercially available SDR-based IEEE 802.16 base station. The AS.MAX base station uses picoarrays™ and a reference software implementation of the IEEE 802.16d standard. The picoarray is a reconfigurable platform that is 10 times faster in processing power than today's DSPs. The AS.MAX base station promises to be upgradeable to the next generation mobile 802.16e standard and so has the potential to offer a future-proof route to operators looking to rolling out WiMAX services.

In addition to the preceding proprietary SDR platforms developed for the military and commercial sectors, there has also been significant progress in the SDR development in the open-source research and university communities. GNU Radio is an open-source architecture designed to run on general-purpose computers. It is essentially a collection of DSP components and supports RF interface to the universal software radio peripheral (USRP), an up- and down-convertor board coupled with ADC and DAC capabilities, which can be coupled to a daughter RF board. GNU Radio has been extensively used as an entry-level SDR within the research community. Some major SDR platforms developed in the university and research communities will be described in detail in this book.

As mentioned, due to its high demand on computation and processing, SDR technology has worked only in devices that have less constraint in size and power consumption, such as base stations and moving vehicles. But there is an increasing demand for SDR to enter portable and handheld devices in the future. The main issue with introducing SDR into portable devices has been that it requires the use of programmable platforms, which are generally power hungry and hence lead to reduced battery life and large devices. However, SDR provides the ability to support multiple waveforms on a single device, and so ultimately could give an end user increased choice of services if incorporated into a portable device, such as a handset. SDR could also assist seamless roaming at the national and international levels. However, as new processing platforms emerge that overcome power and size constraints, it is very likely that SDR will make its way into portable devices. Indeed, some industry insiders are predicting that by 2015 there will be a transition from the current generation of handsets to SDR handsets.

1.3 COGNITIVE RADIO

1.3.1 What Is Cognitive Radio?

The reconfigurability offered by SDR technology enables radios to switch functions and operations. However, an SDR can do this only on demand; it is not capable of reconfiguring itself into the most effective form without its user even knowing it. In Mitola's dissertation and a number of publications, he envisioned such a self-reconfiguring radio and dubbed the term *cognitive radio* for it. According to Mitola's early vision, a CR would be realized through the integration of model-based reasoning with software radio and would be trainable in a broad sense, instead of just programmable. In analogy with the mental process of cognition, Mitola also outlined a cognitive cycle through which such radio can reconfigure itself through an ongoing process of awareness (both of itself and the outside world), perception, reasoning, and decision making. The concept of CR emphasizes enhanced quality of information and experience for the user, with cognition and reconfiguration capabilities as a means to this end. Today, however, CR has become an all-encompassing term for a wide variety of technologies that enable radios to achieve various levels of self-configuration, and with an emphasis on different functionalities, ranging from ubiquitous wireless access, to automated radio resource optimization, to dynamic spectrum access for a future device-centric interference management, to the vision of an ideal CR. Haykin, for example, defines CR as a radio capable of being aware of its surroundings, learning, and adaptively changing its operating parameters in real time with the objective of providing reliable anytime, anywhere, and spectrally efficient communication [11]. The U.S. Federal Communications Commission (FCC) uses a narrower definition for this concept: "A Cognitive Radio (CR) is a radio that can change its transmitter parameters based on interaction with the environment in which it operates. The majority of cognitive radios will probably be SDR (Software Defined Radio) but neither having software nor being field programmable are requirements of a cognitive radio."

Despite these differences in both the scope and the application focus of the CR concept, two main characteristics appear to be in common in most definitions. They are *reconfigurability* and *intelligent adaptive behavior.* Here by *intelligent adaptive behavior* we mean the ability to adapt without being a priori programmed to do this; that is, via some form of learning. For example, a handset that learns a radio frequency map in its surrounding could create a location-indexed RSSI vector (latitude, longitude, time, RF, RSSI) and uses a machine-learning algorithm to switch its frequency band as the user moves.

From this it follows that cognitive radio functionality requires at least the following capabilities:

- **Flexibility and agility**, the ability to change the waveform and other radio operational parameters on the fly. In contrast, there is a very limited extent that the current multichannel multiradio (MC-MR) can do this. Full flexibility becomes possible when CRs are built on top of SDRs. Another important requirement to achieve flexibility, which is less discussed, is reconfigurable or wideband antenna technology.
- **Sensing**, the ability to observe and measure the state of the environment, including spectral occupancy. Sensing is necessary if the device is to change its operation based on its current knowledge of RF environment.
- **Learning and adaptability**, the ability to analyze sensory input, to recognize patterns, and modify internal operational behavior based on the analysis of a

new situation, not only based on precoded algorithms but also as a result of a learning mechanism. In contrast, the IEEE 802.11 MAC layer allows a device to adapt its transmission activity to channel availability that it senses. But this is achieved by using a predefined listen-before-talk and exponential backoff algorithm instead of a cognitive cycle.

1.3.2 Evolution of Cognitive Radio

The main precursors for CR research was the seminal work by Mitola and Maguire in 1999 and early spectrum measurement studies conducted as early as in 1995 to quantify the spectrum use, both in the licensed and unlicensed band. In the United States, CR research focused quickly on dynamic spectrum access (DSA) and secondary use of spectrum as the main objectives of the initial research. This was due to the fact that it was attracting a number of early research projects (e.g., URA, SPECTRUM, and MILTON). The most notable project in the spectrum management and policy research was the XG-project funded by DARPA. The main goal of the XG-project was to study the so-called policy servers and secondary-use technologies, particularly for military purposes. However, the early success of XG was pushing the community to study more broadly the possibilities of CR. Another boost for the research was given by several vociferous researchers (such as Lessig, Reed, and Peha), who pointed out that there are possible flaws in the current regulatory domain.

In the standardization domain, three major groups have emerged to work on relevant technologies and architectures: IEEE 802.22 and SCC41 (formally P1900) working groups and more recently ETSI's Reconfigurable Radio Systsems Technical Committee on CRs and SDRs. Also, the SDR Forum as an industry group has studied some CR-related issues. Commercially, the most advanced standardization activity is IEEE 802.22 and related research that aims to provide dynamic access to vacant TV spectrum. However, IEEE 802.22 requires a rather limited level of cognition.

At the time of this writing, CR is being intensively investigated and debated by regulatory bodies as the enabling technology for opportunistic access to the so-called TV white spaces (TVWS): large portions of the VHF/UHF TV bands that become available on a geographical basis after the digital switchover. In the United States, the FCC already proposed to allow opportunistic access to TV bands in 2004 [12]. Prototype CRs operating in this mode were put forward to the FCC by Adaptrum, I^{2}R, Microsoft, Motorola, and Philips in 2008 [13]. After extensive tests, the FCC adopted in November 2008 a Second Report and Order that establishes rules to allow the operation of cognitive devices in TVWS on a secondary basis [14]. Furthermore, in what is potentially a radical shift in policy, in its recently released Digital Dividend Review Statement [15], the U.K. regulator, Ofcom, is proposing to "allow licence exempt use of interleaved spectrum for cognitive devices" [15]. Furthermore, Ofcom states that, "We see significant scope for cognitive equipments using interleaved spectrum to emerge and to benefit from international economics of scale" [15]. More recently, on February 16, 2009, Ofcom published a new consultation providing further details of its proposed cognitive access to TVWS [16]. With both the United States and United Kingdom adapting the cognitive access model, and the emerging 802.22 standard for cognitive access to TV bands [17, 18] being at the final stage, we can expect that CR may become mainstream technology worldwide in the near future.

1.4 KEY APPLICATIONS

As discussed, CRs are highly agile wireless platforms capable of autonomously choosing operating parameters based on both prevailing radio and network conditions [11, 19]. Consequently, CRs have the potential to revolutionize how devices perform wireless networking. For instance, CRs allow radios operating on different protocols and standards to communicate with each other. This is known as *interoperability* [20, 21]. Furthermore, CRs are capable of transmitting in unoccupied wireless spectrum while minimizing interference with other signals in the spectral vicinity; that is, DSA [22–24].

1.4.1 Interoperability

Today, a plethora of wireless standards, applications, and services are being employed across numerous sectors of modern society, as well as within the same sector, such as military, public safety, and emergency responders [25]. Consequently, the use of multiple (potentially incompatible) communication standards within a specific sector could seriously impact the effectiveness of coordinated operations, yielding a situation analogous to the biblical account describing the Tower of Babel. For instance, the effectiveness of the emergency responders to cope with the aftermath of Hurricane Katrina in New Orleans during August 2005 was greatly affected by the inability of their diverse range of deployed communication equipment to operate with each other, especially within a decentralized operating environment.[1] This is shown in Figure 1.2, where members of Team A employ a communications standard operating on a carrier frequency that is different from the communication equipment employed by both Teams B and C. Thus, unless these teams are coordinated with respect to operating parameters and communication standards, effective communications between them would be nearly impossible.

Nevertheless, there are several reasons why sectors such as the military and public safety still maintain a range of communication solutions, such as significant financial investment and specific performance requirements. Consequently, CR possesses the potential to undo the Tower of Babel syndrome with respect to communications between teams employing different equipment [20, 21, 25, 26].

---

[1]Without reliable electrical power, cellular base stations and other centralized communication nodes were nonfunctional after their emergency power supplies were depleted.

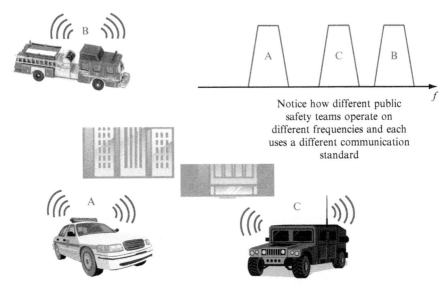

FIGURE 1.2

Example of public safety and emergency responder teams within the same geographical area operating on different center frequencies and potentially using different communication standards.

Due to its ability to rapidly assume any available radio configuration, CR platforms can reconfigure themselves to a legacy communications standard in order to communicate with any communication system deployed in the field or facilitate communications between two non-CR platforms employing different standards. Furthermore, with its onboard artificial intelligence, CR can automatically distinguish between different communication standards in the absence of any centralized control.

1.4.2 Dynamic Spectrum Access

With the increasing demand for additional bandwidth to support existing and new services, both spectrum policy makers and communication technologists are seeking solutions for this apparent spectrum scarcity. Meanwhile, measurement studies have shown that much of the licensed spectrum is relatively unused across time and frequency [1, 27–32]. Nevertheless, current regulatory requirements prohibit unlicensed transmissions in these bands, constraining them instead to several heavily populated, interference-prone frequency bands. To provide the necessary bandwidth required by current and future wireless services and applications, the FCC has commenced work on the concept of unlicensed users "borrowing" spectrum from spectrum licensees [33, 34]. This approach to spectral usage is known

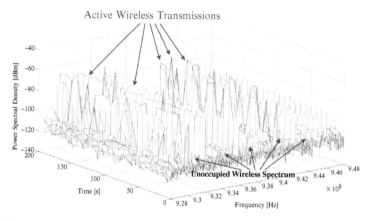

FIGURE 1.3

Wireless spectrum of 928–948 MHz in Rochester, New York, on June 19, 2008. Notice the time variations in the spectrum occupancy at several frequencies in this band.

as *dynamic spectrum access*. With recent developments in CR technology, it is now possible for these systems to simultaneously respect the rights of incumbent license holders while providing additional flexibility and access to spectrum.

An example of how DSA would work can be illustrated in Figure 1.3, where parts of the spectrum between 928 MHz and 948 MHz are occupied over both frequency and time. However, it is readily observable that there also exists portions of the spectrum that are unoccupied for a significant period of time, making them suitable candidates for secondary access by unlicensed wireless devices in a DSA framework. Nevertheless, when accessing these unoccupied frequency ranges within licensed spectrum, the secondary wireless device must ensure that it does not interfere with the operations of the primary user transmissions. Interference may occur when the out-of-band (OOB) radiation of the secondary transmission exceeds the tolerable levels, contaminating the primary user transmission if located relatively close in the frequency domain. Simultaneously, given the time-varying nature of wireless transmissions, a spectrum that might be unoccupied at one time instant could potentially be occupied at a subsequent time instant. Consequently, the CR platform must be environmentally aware and rapidly reconfigurable to prevent secondary user interference of primary user transmissions.

To achieve higher spectral efficiency, multiple access techniques can be employed such that multiple secondary users can transmit data within the same frequency range. Several techniques have been proposed to achieve multiple secondary user access, including those based on *code division multiple access* (CDMA) [35, 36], spatial multiplexing [37], and *orthogonal frequency division multiplexing* (OFDM) [38, 39]. With respect to OFDM-based techniques, the *spectrum pooling* concept can be effectively employed, where data are transmitted across unoccupied portions of frequency using a subset of active subcarriers [24].

1.5 BOOK ORGANIZATION

The chapters of this book have been grouped together into three thematically related parts to provide better structure for the reader with respect to the topics covered: Radio Communications, Networks, and Implementation; Applications; and Case Studies. These parts can be loosely divided into the following six subparts: The first subpart includes Chapters 1 and 2. It gives the reader some essential background of SDR and CR technologies and their impact on radio spectrum regulatory policies. The second subpart of this book includes Chapters 3 to 7 and focuses on the underlying physical layer technologies, spectrum sensing, reconfiguration, adaptation, and spectrum access, all of which are basic building blocks for a CR. In particular, Chapter 3 offers a review of the digital communications that are most relevant to the design of CRs. Readers who are familiar with this background may proceed directly to the other chapters.

The third subpart of this book consists of Chapters 8 to 12 and is centered around networking aspects of CRs. In particular, Chapter 8 offers a review of fundamentals of communication networks and may be skipped if readers already have this background. The fourth subpart of this book includes Chapters 13 to 17, which cover CR terminology, applications, and security issues. In particular, Chapter 14 offers an extensive discussion on how CR can share the spectrum with TV bands, while Chapter 16 discusses how CR can alleviate the interoperability problem for public safety communications. With respect to spectrum trading, Chapter 17 examines the challenges and solutions in this area with an emphasis on dynamic spectrum auctions.

The fifth subpart of the book includes Chapters 18 and 19 and focuses on CR testbed platforms. In particular, Chapter 18 gives an in-depth coverage of GNU radio and how to build a CR. Chapter 19 reviews some state-of-the-art testbed platforms for CRs. The last subpart of this book contains Chapter 20, which offers a perspective of CR architectural evolution and its future roadmap.

Cognitive radio communication techniques and algorithms

Radio frequency spectrum and regulation

Dennis Roberson[1] and William Webb[2]
[1]Illinois Institute of Technology, United States
[2]Office of Communications, United Kingdom

2.1 INTRODUCTION

The electromagnetic spectrum, and in particular the so-called radio frequency portion of this spectrum, is rapidly becoming one of our planet's most valuable natural resources. It has no mass, no inherent functionality, provides no nourishment, and is not a source of energy, yet it is viewed as a sufficiently scarce and valuable resource that relatively small portions of this commodity, as measured in spectral and geographic dimensions, command prices measured in billions of U.S. dollars [40]. This chapter provides an overview of the background behind this unusual state of affairs through a discussion of the fundamental nature of this basic resource, a brief historic context for its utilization, and the regulatory regime that has developed to govern its use. The concluding sections of the chapter describe the emerging regulatory approaches to cognitive radio technology, the spectrum occupancy studies being conducted in support of these efforts, and the benefits these approaches will likely provide.

2.2 SPECTRUM: NATURE'S COMMUNICATION HIGHWAY

The radio frequency spectrum is an abundant natural resource that uniformly covers the planet and is available for a wide variety of useful purposes. Beyond the historic voice communications and increasingly dominant multimedia and data networking focus of this text, this spectrum is regularly used for a diverse array of applications, including radar for finding large and small objects (from airplanes in the sky, to obstacles in the vicinity of your automobile, to studs in the walls of your home), excitation for illuminating spaces (sulfur lamps), monitoring and sensing applications, and even cooking food in the microwave oven in your home. The radio frequency spectrum is a component of the overall electromagnetic spectrum that stretches from roughly zero to nearly 3×10^{27} Hz (cycles per second). In this broad range, the domain from roughly 10 kHz to 300 GHz is usually described as the radio frequency spectrum [41] (though interestingly this region includes the spectral regions described as sonic and ultrasonic on the low end as well as the microwave spectral region on the high end of the range).

2.2.1 Physical Characteristics of Spectrum

The radio frequency spectrum is formed by a virtually infinite set of discreet frequencies characterized as waves with wavelengths corresponding to the frequencies according to the simple formula, frequency equals the speed of light divided by the wavelength, or

$$f = \frac{c}{\lambda}. \tag{2.1}$$

These radio waves operate as periodic waves and therefore the time-varying value of the signal at a point in space may be represented in the general form

$$s(t) = A(t)\cos(2\pi ft), \tag{2.2}$$

where s is the signal strength, t is the time, A is the amplitude of the signal, and f is its frequency. Note that A can either be a constant or vary with time.

To gain a feeling for the physical nature of invisible radio waves, it is often convenient to consider their visible cousins in the electromagnetic spectrum, namely, light waves. Specifically, by using a prism or observing a rainbow we are able to see that sunlight or even the light from fluorescent or tungsten filament light bulbs (i.e., various forms of "white light") is composed of a wide range of different colors (see Figure 2.1). These colors are actually different frequencies (or wavelengths depending on your perspective) of electromagnetic waves that have been fused together to form the single beam of white light. In similar fashion, other sections or bands of electromagnetic spectrum in the radio frequency region are brought together for a variety of communications and, as noted in the last section, other useful purposes.

Like all elements of the electromagnetic spectrum, the radio frequency component of this spectrum has the wavelike characteristics of reflection, refraction, diffusion, absorption, and scattering. These characteristics and their effect on incident light are illustrated in Figures 2.2 and 2.3. In general, these characteristics usually inhibit the use of the spectrum by redirecting the focus of the energy, creating interference or absorbing the energy all together. With appropriate care and at times considerable ingenuity, these characteristics can be used to enhance the capabilities of the radio frequency source (i.e., the transmitter) either by enabling

FIGURE 2.1

The output of a prism on a table. Each of the three defined areas in this black and white figure represent a different color (public photo courtesy of Flickr).

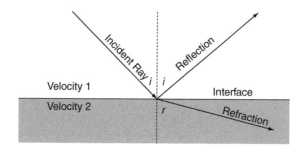

FIGURE 2.2

The reflection and refraction of an incident radio wave at the interface to a medium with a different refractive index.

its transmissions to reach receivers that are otherwise hidden from the transmitter (i.e., non–line of sight or NLOS), allowing for the reuse of spectrum, or enabling enhancements to the signal-carrying capacity of the signal.

The exciting feature of radio frequency (and other electromagnetic) waves is that all these wave-modifying interactions can be in operation at the same time. This means that if we could see these discrete RF waves around us we would observe an amazingly complicated and chaotic swirl of different frequency waves operating at different power levels being affected by all of the properties illustrated in Figures 2.2 and 2.3, and, in the process, also interfering with one another as they collide. This interference property would in turn cause waves to be locally

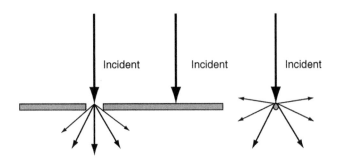

FIGURE 2.3

The impact of diffraction, absorption, and scattering on an incident radio wave.

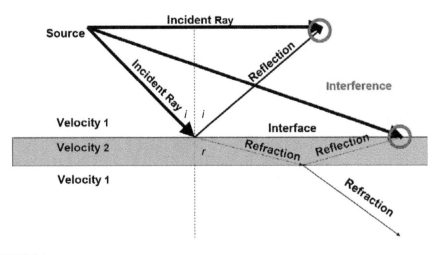

FIGURE 2.4

Self-interference in a simple reflection/refraction environment.

amplified, annihilated, or simply modulated much as the confluence of ocean waves creates a wide variety of interesting wave patterns on a rugged beach. A very simple example of self-interference is illustrated in Figure 2.4. This is shown to begin to suggest some of the challenges that exist in the multitransmitter-multireceiver world that we live in with the numerous overlapping frequencies of the various radio wave sources. The authors encourage the reader to ponder this topic and think through a few scenarios to gain a deeper understanding of the characteristics of the physical world that underlies our modern wireless network and communications environment.

2.2.2 **Implications for Communication Applications**

The utility of the spectrum is derived from its ability to be modulated in a variety of ways to transport useful information. This specifically includes applying or removing power from a specific frequency or spectral range, such as pulse modulation or amplitude shift keying (ASK); increasing or decreasing the power level applied to a frequency, such as amplitude modulation (AM); switching the transmitted power from one frequency to another, such as frequency modulation or frequency shift keying (FSK); altering the phase of the signal, such as phase shift keying (PSK); or combining these techniques to create a variety of evermore complex signal encoding structures. More generally, by combining power, frequency (shift and phase), signal timing, and spatial considerations, very powerful signal encoding, sharing, and spatial reuse capabilities can be deployed.

The critical feature here is that the transmitter and receiver share a common knowledge of how this encoding is to take place and spectrally where the receiver should look to be able to find the transmitter. There is also a need to make sure that others are not consistently trying to use the same spectrum at the same time in the same place. All this leads to the need for government regulation to ensure that the spectrum is optimally shared and the communications is efficiently and effectively transmitted while optimizing the use of the spectrum and thereby maximizing its data-carrying capacity.

2.3 **REGULATORY HISTORY AND SUCCESSES**

The history of spectrum regulation closely followed the early development and deployment of the first wireless communications systems, the wireless telegraph pioneered by Marconi and others in the late 1800s [41]. The first international wireless standards meeting, the International Radiotelegraph Conference, was organized by the International Telegraph Union (ITU), the governing body for wired telegraph operation, and held in 1906 in Berlin. At the meeting 29 nations gathered to sign the International Radiotelegraph Convention, which established the basic standards for wireless telecommunications, particularly between ships and land-based stations. This included the establishment of the famous international distress signal, namely "SOS" or ".—." in Morse code. These regulations have been expanded and revised over the years and continue today as the Radio Regulations, now updated based on the outcomes of the quadrennial World Radiocommunications Conference (WRC). These meetings are administered by and the regulations are maintained under the International Telecommunications Union (the renamed predecessor to the original ITU), which is itself an executive agency of the United Nations.

2.3.1 **Objectives and Philosophy**

Each nation administers the telecommunications system and specifically the radio spectrum for its geographic dominion through a regulatory agency. Though the form, function, and goals of these organizations vary widely, they are generally along the lines of the ITU's Mission Statement, which is summarized as

> *The ITU mission: bringing the benefits of ICT to all the world's inhabitants. ITU's mission is to enable the growth and sustained development of telecommunications and information networks, and to facilitate universal access so that people everywhere can participate in, and benefit from, the emerging information society and global economy. ITU remains dedicated to helping the world communicate.* [42]

2.3.2 Early History and Success

The initial radio standards were focused on very pragmatic protocols like the idea that all shore-based radio receivers should listen to the communications from all ships at sea independent of whether they were from the country where the receiver was located [41]. This was especially important for monitoring distress signals. Though frequencies were formally allocated as early as 1912 to initially assist in these ship-to-shore communications, in light of the increasing number of radio frequency services and the related growth in interference issues, the 1927 International Radiotelegraph Conference provided a relatively comprehensive allocation of frequency bands to the various radio services in existence at the time (fixed, maritime and aeronautical mobile, broadcasting, amateur, and experimental).

On October 15, 1947, ITU became a United Nations specialized agency, an organizational relationship that continues today. At the same time, the International Frequency Registration Board (IFRB) was established within the ITU to coordinate the increasingly complicated task of managing the radio frequency spectrum. Also in the same year, the Table of Frequency Allocations, first introduced in 1912 (based on concerns that arose after the *Titanic* disaster), became mandatory to assist, guide, and at times tabulate spectrum use in the various member countries. Over the years this table has been expanded (especially to higher and higher frequency ranges) and continuously refined by the ITU.

While the ITU provides a framework and guidance for global spectral use, each nation has its own independent regulatory body that operates under the national government of the specific country. In the United States the primary regulatory body is the Federal Communications Commission (FCC). In its founding legislation in 1934, Congress granted the Commission broad powers to regulate spectrum use in the United States in the "public interest." This body in turn replaced and expanded the charter of the Federal Radio Commission, which had been established in 1927. The FCC develops and enforces regulations in support of the laws governing the commercial use of the spectrum created by the U.S. Congress and signed into law by the president. As an important note, unlike most countries that have a single body to administer spectral use, in the United States, government use of the spectrum is administered by a separate entity, the Office of Spectrum Management in the National Telecommunications and Information

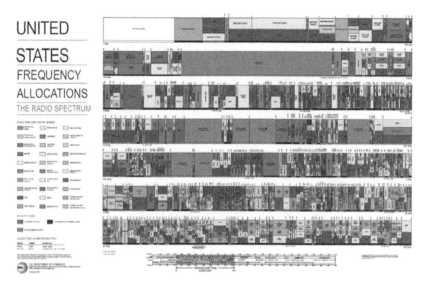

FIGURE 2.5

Frequency allocation charts for the United States (courtesy of U.S. Dept. of Commerce, National Telecommunications and Information Administration, Office of Spectrum Management, Oct. 2003) [43].

Administration (NTIA), which reports to the Department of Commerce in the executive branch of the government. Together these bodies determine, maintain, and regulate the comprehensive allocation of U.S. spectrum tables as illustrated in Figure 2.5. A similar spectrum allocation table for the United Kingdom is shown in Figure 2.6.

2.4 EMERGING REGULATORY CHALLENGES AND ACTIONS

For many years this national regulatory activity, while complex, was generally managed as a standard government bureaucracy with studied opinions modulated by the success or failure of various lobbying efforts from interested parties determining the optimal "pubic interest"–based use for the spectrum. This included decisions on the size, allocation, and ultimate assignment of the various spectral bands and channels to the critical technology-enabled purposes of the day. It also included value judgments on applications like broadcast radio and two-way radio and making decisions on the relative value of the needs in the emergency services, military, commercial, and personal use arenas. This also included the explicit reward of monopoly or oligopoly powers over portions of the spectrum to specific government, institutional, or commercial entities. Given its impact on

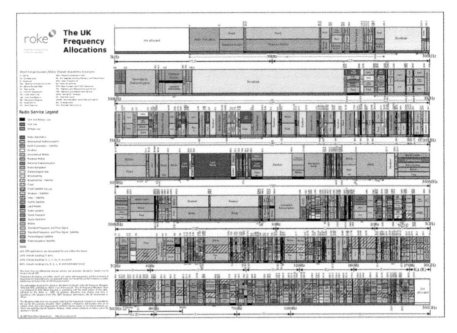

FIGURE 2.6

Frequency allocation charts for the United Kingdom (courtesy of Roke Manor Research Ltd.).

business, various institutions, and government organizations, spectrum regulation has historically been more about economics, politics, lobbying efforts, and current perceived public opinion than it is about the details of a specific promising new technology.

The ever-increasing pace of technology advancement and the related increases in the demands for the various applications the technology has enabled have rendered this "studied opinion"–based allocation scheme virtually impossible to manage in the developed countries of the world. For this reason (amplified later), since roughly the mid-1980s, most regulatory bodies have moved from the strict assessment of "public interest" value, to the use of a market-based approach for "new spectrum" assignment (i.e., various forms of auction), especially for commercial assignments such as cell phone bands. As described later, this has been significantly augmented by unlicensed regions of spectrum to quickly and inexpensively cover a broad range of spectral needs with the proviso that the user of this spectrum may experience significantly greater interference challenges than it would have in a dedicated, licensed spectral band.

2.4.1 **Era of Increasing Regulatory Challenges**

To better understand the dynamics behind this market-based direction, it is important to note that the supply of spectrum is clearly finite, but the demand for the spectrum is fundamentally unbounded. This ever-increasing demand is sometimes based on what the authors whimsically describe as the "quadruple whammy" of spectrum use. This "quadruple whammy" is composed of four elements, all focused on satisfying a critical personal or societal need. These needs include the insatiable human cravings to both communicate with others in all available forms (e.g., voice, text, music, video) and obtain information about a seemingly endless variety of topics. Businesses demand a continuing enhancement in the efficiency in conducting the transactions on which their business is based and information about their customers, partners, and suppliers to improve their understanding on how their products and services can be better optimized to meet their customer's needs and increase their profits. The military and emergency services need real-time information to better plan, coordinate, direct, execute, and assess their various activities. Finally, government and nongovernment social service organizations and educational organizations are seeking to improve their effectiveness and efficiency in providing their services.

The four elements of the quadruple whammy are listed and elaborated as:

Applications. The number and variety of different radio applications is virtually unbounded and rapidly evolving. For example, these include: broadcast communications (television, radio), commercial communications (emergency services radios), industrial communications and fleet management, aeronautical communications, military communications, personal communications (cell phones, two-way radios), wireless networks (personal, local area, metropolitan), satellite communications.

Coverage. The need to offer these applications to an ever broader audience and eliminate any spatial constraints in the use of the applications.

Duty Cycle. The most popular of these applications will be used for an ever-increasing percentage of the time, ultimately following the popular personal dictum of "always on, always connected."

Performance. Not only are these applications being deployed broadly and used all the time, the demands for ever-increasing levels of performance for the popular applications require increasing allocation of spectral bandwidth, since there is a direct correlation between allocated bandwidth and the sustainable data rate that a channel can support. In its simplest form this is described by the Nyquist bandwidth formula:

$$C = 2B \quad \text{(for binary signals)} \tag{2.3}$$

or

$$C = 2B \log_2(M) \quad \text{(for multilevel signals)}, \tag{2.4}$$

where C is the capacity, B is the bandwidth, and M is the number of signal levels.

2.4.2 Allocation, Reallocation, and Optimization

Based on the fundamental economic law of supply and demand, with a finite supply of spectrum and an ever-increasing demand for that spectrum, the value of the spectrum has risen and will continue to rise; and that rise will be directly related to the increasing demand. As such, it is not surprising that all the radio frequency spectrum has been fully allocated for a considerable period of time (since roughly 1980). Therefore, as new applications have been created the difficulty in finding a place for these applications in the spectral domain has greatly increased. For a time, there was enough available spectrum that each application could be uniquely positioned, if not in an ideal spectral band, at least in a workable band. Later, regulatory value judgments were required to determine that an existing application should be retired, replaced, relocated, or have its spectral allocation reduced to make room for a new high-priority application that could optimally operate in the spectrum used by the existing application. This is obviously an extremely difficult task. In the United States with both the NTIA and the FCC providing the allocations, this negotiations-based bureaucratic decision-making task was rendered even more challenging and more time consuming.

This increasing difficulty is a major driver of the two dominant trends that have transformed the spectrum management landscape since the early 1980s. The first of these has already been mentioned, namely, the move to government auctions as a means of using raw dollars as a dominant means (obviously coupled with public comment) of determining the most appropriate allocation of spectrum, or at least the block band assignments. This market-based approach is currently being extended to allow entities that have obtained spectral resources to resell or lease these resources to others on a long-term or temporal basis. By using these methods the market should be able to "move" spectrum ownership to those users and applications that value it the most rather than requiring the regulator to make a judgment on the optimal use of the spectrum.

Regulators have also been setting aside increasing amounts of spectrum for unlicensed access, sometimes also known as licensed-exempt spectrum or "spectrum commons." These bands are not licensed to any entity, but are available to all to use subject to "rules of entry," which typically restrict power levels in order to minimize the probability of interference. Such bands have proven very valuable for uses such as WiFi, Bluetooth, radio tagging (RFID), and a wide range of consumer products from garage door openers to baby alarm monitors. The most widely used band for unlicensed activities is at 2.4 GHz, but large new bands at 5-6 GHz are now available around the world for new applications.

The trend toward market-based spectrum management is also leading to the establishment of quasi-autonomous entities set up at arm's length from the government to manage the national communications resource with such organizations as the Office of Communications in the United Kingdom (or more commonly Ofcom) serving as a solid example. This organization was created via the Office of Communications Act of 2002 and subsequently empowered via the Communications Act of 2003, with the following general duties:

3(1) It shall be the principal duty of Ofcom, in carrying out their functions; (a) to further the interests of citizens in relation to communications matters; and (b) to further the interests of consumers in relevant markets, where appropriate by promoting competition. [44]

This organization explicitly is responsible for regulating television, radio, telecommunications, and various wireless services.

The other approach, called *dynamic spectrum access networks*, and its tightly related cousin cognitive access networks (with the primary embodiment being cognitive radios), has created a push for regulatory approval enabling temporary (usually licensed-exempt) sharing of spectrum along both spatial and, importantly, temporal dimensions between heterogeneous users. This approach is based on the important observation that most of the spectrum, in most of the places, most of the time is completely unused [45]. Given the previous quadruple whammy discussion, this is a rather amazing circumstance, but based on the measurements performed in various spectrum occupancy studies (see Section 2.6), less than 20% of the spectral capacity is actually being used, even in urban environments like Chicago and New York [1, 28–30, 46–57] (see Figure 2.7). To help reconcile these two seemingly disparate thoughts, though the quadruple whammy circumstance of crowded airways is a growing fact of life, there are many well-established wireless applications with long-term spectral band assignments that are rarely used or are used only in specific spatial locations. For example, ship-to-ship and ship-to-shore radios are rarely used in places like Beijing, Denver, Madrid, or Mexico City since they are not located on major waterways. Similarly, many military radios are used primarily for training on military bases except in time of war or during war games, and their bands could be used for other purposes outside these relevant geographies and times. The increasing recognition of this unused spectral capacity has helped drive the various initiatives directed toward identifying and supporting more efficient dynamic sharing mechanisms for the world's scarce spectrum resources.

2.4.3 Regulatory Actions

It can be argued that this trend to enable temporal as well as spatial spectrum sharing took root in the United States in 1985, when in recognition of the growing spectral scarcity challenge, the U.S. FCC modified its rules for the industrial, scientific, and medical band to enable its use for wireless communication (FCC Rules, Part 15-247) [58]. This was the first of three major initiatives to begin to address this critical issue. The other two have been the allowance of ultra-wideband (UWB) underlays, based on an FCC Report and Order [59] filed February 14, 2002 (and released April 22, 2002), and even more recently, cognitive radio overlays supported by another FCC Report and Order [60] released March 11, 2005. Other nations are following this trend, including the direction outlined in the highly regarded Spectrum Framework Review produced by Ofcom in June 2005 [61].

2.4.4 **Spectrum Task Forces and Commissions**

Through the years a variety of task forces and studies have focused on spectrum usage. In the United States these have been commissioned by such organizations as the Federal Communications Commission, the National Telecommunication and Information Administration, the National Research Council for the National Academies, the academies themselves, the National Science Foundation, and Congress. Two of the more prominent efforts conducted over the past decade were the FCC Spectrum Task Force [27] commissioned by then FCC Chairman Michael Powell and chaired by Dr. Paul Kolodzy, and the CSIS Spectrum Commission chaired by Bob Galvin, the legendary former chairman and chief executive officer of Motorola, and James Schlesinger, the former chair of the Atomic Energy Commission, director of the CIA, secretary of Defense, and secretary of Energy [62].

Both of these extensive studies came to the conclusion that the existing regulatory structure that attempts to allocate a discrete band for each application is no longer functional. Further, they pointed to the need to take advantage of the technology now becoming available to enable improved time-based utilization of the spectrum; that is, dynamic spectrum access networks. While these studies have clearly pointed the way forward and encouraged the kind of regulatory action described in the previous subsection, much more is to be done in this domain to fully unleash the emerging power of cognitive radio technology.

2.5 REGULATORY ISSUES OF COGNITIVE ACCESS

Based on the numerous initiatives described in the preceding section, it is important to look specifically at the regulatory issues and available options for addressing cognitive spectrum access. At the time of this writing, this is an emerging regulatory field; therefore this section presents an outline of the potential actions and the current set of options being explored. Over time, this section of the text will likely see significant updates as specific options become regulations in various countries and their strengths and weaknesses become more apparent. At the same time, it is a particularly good time to write this section, since a great deal of research has been done on this topic and the U.S. FCC recently took the first concrete regulatory action in this important area [63].

2.5.1 **Should a Regulator Allow Cognitive Access?**

As discussed in earlier sections, it's up to the national regulator in each country to decide who has access to the radio spectrum. The regulators enable access by either issuing licenses or exempting particular devices from the need to have a license. Therefore, cognitive access in commercial spectrum would generally be illegal until it is enabled by a regulator through a set of orders specifying the conditions under which this kind of technology might be deployed.

The regulator could adopt a variety of approaches to cognitive access, including

- Deciding not to allow it (and hence do nothing).
- Enabling existing license holders to allow cognitive access into their own bands if they chose to.
- Licensing cognitive access to particular bands.
- Exempting cognitive equipment from the need for licensing with appropriate restrictions on when, where, and how they might operate.

The approach that each regulator selects will, to some degree, depend on that person's spectrum ownership philosophy and in some cases the legal rights of the existing owners. The first decision that needs to be taken is whether regulatory intervention is appropriate. Some might argue that spectrum is the property of the state and the regulator has every right to allow additional users into a band already licensed to a particular operator. Others might consider that in purchasing a regional or national license an operator has complete rights to the exploitation of that spectrum, including subleasing it to cognitive users if they wish.

An analogy with land may be helpful here. Imagine a homeowner who decides to leave the country for six months, perhaps on business. What should happen to the home? One regime (e.g., the United States or the United Kingdom) might allow the owner to rent it out and keep any proceeds or simply leave it vacant. Another regime might require that the home be made available to the state as soon as it was empty for the state to make use of it, perhaps to house the homeless or a foreign visitor, until the owner returned. The parallels with cognitive access are obvious. A regulator who believes in the "full property rights" model would allow license holders to sublease their spectrum for cognitive access at their discretion. One who prefers the "state rights" or "reclamation" model would be inclined to enable licensed or license-exempt access to the spectrum.

There may be different interpretations for different frequency bands. For example, in many countries television transmitters are licensed on a geographical basis, with a license issued per tower. The result of this is that there are geographic areas between towers where the spectrum has not been licensed to anyone—often called *white space*. This "unowned" spectrum can be treated differently because there is no owner who can bargain with cognitive users other than, perhaps, the regulator. Hence, a different regulatory decision might be expected for such a spectrum compared to a spectrum that has been regionally or nationally licensed.

In practice, few at present imagine that cognitive users could successfully negotiate access to spectrum from licensed users. The licensed users would surely require some form of compensation for enabling this right, which would increase the difficulty in making a business case for cognitive access. The licensed user might also decide that the business and technical risks are too high or the effort involved in defining the terms of entry too great. So, while negotiated access remains possible—indeed in the United Kingdom trading regulations effectively permit it—it seems unlikely that it will be the first form of cognitive access to emerge.

The final choice for a regulator who has decided to allow cognitive access is whether to license that access or not. Most proponents of cognitive access envisage a license-exempt situation, where consumers can buy a device and immediately access the spectrum. An alternative would be to issue a license to an organization, perhaps via an auction, that had the rights to use cognitive access in a particular band. Which model is most appropriate is still far from clear. The licensed approach provides some possibility that infrastructure, such as a cellular network, will be constructed, while the license-exempt approach will tend to result in short-range communication systems. Ideally, the regulator should select the approach that maximizes the value for the country. But, at this early stage in the development of cognitive access, it is far from clear which this would be. Since most proponents currently favor the license-exempt model, then regulators who have considered cognitive access have tended to also favor the license-exempt option. However, this could change in the future and different approaches could be adopted in different frequency bands.

2.5.2 How to Determine the Rules of Entry

If the regulator decides to allow cognitive access to a particular band, perhaps a TV broadcasting band, the next step is to set the rules of entry. The regulator will typically seek rules that ensure a very low probability of interference to the incumbent users of the band while at the same time placing the minimum possible restrictions on the cognitive device. The regulator also needs to focus on the potential for interference between cognitive devices as their usage grows. Avoiding interference is generally a very difficult technical problem as well as a challenging political issue.

For example, if cognitive devices are to be based on sensing technology that measures the existing signal strength to determine whether a particular channel is currently in use, then the regulator must determine the signal level down to which the device must be able to sense to ensure that there is a very low likelihood of interference. As experience has shown, determining such a level is very complex. The "hidden node" problem, which is further discussed in Chapters 4 and 14, means that the cognitive device needs to sense to a lower signal level according to the difference in signal strength that it receives compared to the hidden node. But this difference depends on local conditions—on the buildings or other sometimes dynamic obstructions (e.g., a large truck or a train) that temporarily blocks the signal, on the exact geometry of the situation, and so on. Each situation (space, time, and frequency) is different, making any deterministic analysis impossible. Instead, widespread measurement and modeling campaigns are often used to obtain a "best estimate"-based understanding of the possible range of the signal strength values in a geographic area.

All of this is generally quite controversial. The incumbents will seek the maximum level of protection and hence seek out situations where the hidden node problem or other signal modification issues are most extreme. Even setting an appropriate level of probability of interference occurring can be fraught with difficulties. Many other rules are also needed, including the maximum in-band power, the out-of-band power limits, the bandwidth, whether transmit power control is required, the periodicity of sensing and so on. These tend to be simpler to set because they are mostly parameters that regulators address for every band or technology they consider.

2.5.3 Regulatory Implications of Different Methods of Cognition

To date, three broad techniques for identifying whether bands are free from use have been suggested. These are sensing, beacons, and geolocation, which importantly can be used discretely or in combination to effect the desired level of confidence in the attainment of a low-interference environment. Here the text will not seek to compare their efficiency, only their regulatory implications. The implications of sensing has already been mentioned: that the regulator needs to determine the appropriate device sensitivity.

Geographical Databases

An alternative to sensing is for a cognitive device to precisely know its location and have access to a database listing the frequencies it is allowed to use at each location. This overcomes the regulatory issues associated, with sensing, but leads to other regulatory issues, such as

- To what accuracy should the device know its location?
- Who will maintain the database? Will there be one provider for all bands or a separate database per band? What will the commercial arrangements be? Will there be monopoly concerns?
- What availability is needed for the database? Is it acceptable for it to be off-air for substantial periods?
- How will devices download updated versions of the database? How frequently should they do so? What will the loading on the spectrum be as a result?
- What about the dynamic use of the spectrum for frequencies that are only occasionally in use?

These issues are only now starting to be considered by regulatory authorities in countries such as the United States and the United Kingdom.

The use of databases does provide one interesting advantage for the regulators: the ability to control license-exempt devices in a manner that has never been possible before. So, for example, a regulator that wished to remove all cognitive devices from a band in a specific geography could do so by setting all frequencies in the database to "busy." The cognitive devices then would be unable to transmit in this geography. Alternatively, the regulator could set aside cognitive device-free spectrum or control certain types of cognitive devices or implement different rules for devices with different cognitive capabilities.

Beacon Reception

This approach requires the transmission of a signal from some appropriate infrastructure providing information on which frequencies are available for cognitive use in the vicinity. Cognitive devices tune to this channel and use the information provided to select their preferred frequency. This can be thought of as centralizing the database described in the previous section and setting the location to that reachable from the specific transmission source. If the cognitive device is unable to find a beacon, it is safest for it not to transmit, since it may be within the area covered by the beacon but shielded from receiving it. Hence, if it were to transmit on a randomly selected channel it could cause interference.

While this resolves the sensing problem at the cognitive device it raises many other problems:

- Who provides the beacon signal? What are the commercial arrangements, and if there is only one provider, are there competition concerns?
- How is the information the beacon transmits kept up to date, especially where the licensed services are changing rapidly?
- What spectrum is used for the beacon?
- What technical parameters and protocols are used by the beacon transmitter?
- How to prevent the beacon signal from being received outside its intended coverage area and as a result being applied incorrectly, and conversely how to make sure that it is available to all cognitive devices in the target area?
- Should there be separate beacons for separate frequency bands or one beacon for all the bands into which cognitive access is allowed?
- Is it acceptable for the use of cognitive devices to be denied access if the beacon fails or is taken off-air for any reason?

In part based on these questions, the beacon reception approach has not generated much interest in the regulatory community to date and hence few, if any, of these questions have been explored at a regulatory level.

2.5.4 Regulatory Developments to Date

At the time of writing, in July 2009, on a global basis a number of regulators have given serious consideration to cognitive access in general. Most notably in November 2008 the FCC published its Report and Order enabling cognitive access in the white space in the TV broadcast spectrum [63]. Furthermore, in July 2009, Ofcom published a statement on "Licence-Exempting Cognitive Devices Using Interleaved Spectrum" [64], following a public consultation on this topic, which was released in February 2009.

The FCC concluded that sensing alone was insufficiently proven for cognitive access. They noted that devices provided on trial often failed to identify wireless microphones and were very poor at detecting TV signals when there was a strong signal in an adjacent band. Some devices were unable to detect that a band was empty even when it was. While this is a "fail-safe" mode of failure, it did little to inspire confidence in sensing as the sole technique for determining interference-free cognitive access. Therefore, the FCC concluded that at present sensing alone would result in an unacceptable risk of interference, although it noted that, as technology improved, sensing might become sufficiently accurate to be used. As a result, it further concluded that geographical databases (termed *geolocation*) were also required. Specified details such as the "locational" accuracy (50 m) and the frequency of consulting the database (at least daily or whenever movement is detected) were also stipulated in the Report and Order document.

The broad idea of the U.S. approach is that geolocation will provide an effective means of identifying vacant television channels, but that sensing will be a valuable backup, especially for wireless microphones that may not be listed in the database. Using both techniques might also provide valuable insight into the sensing levels needed, enabling potential changes in future regulation. Regulators and users around the world now await with interest the emergence of devices that meet these rules to see whether they can work effectively and economically.

2.6 SPECTRUM MEASUREMENTS AND USAGE

Through the years innumerable efforts have been made by private and public entities around the globe to measure the utilization of specific spectral ranges in various places. These studies have most often been performed by companies seeking to understand the usage characteristics in a specific frequency band into which they are attempting to provide a new wireless service. Others have performed more significant studies over broader spectral ranges, but most of these have been short-term studies looking at specific spectral regions in specific geographies for relatively short periods of time [1, 29, 30, 46–55]. A few of these studies have been conducted over longer periods or more extensive geographies, further enhancing our collective understanding of spectrum utilization [28, 56, 57].

2.6.1 Early Spectrum Occupancy Studies

Spectrum occupancy studies of various kinds have taken place for many years in many places across the planet. These include such initiatives as the radio car procured by the Radio Division of the Department of Commerce in the United States in 1927 to support its radio inspection and operator licensing function (see Figure 2.7) [65]. Unfortunately, from a long-term spectral usage understanding perspective, most of these studies have been performed by individuals focused on very specific spectral domains and geographies to obtain the very specific information needed to advance their own commercial or regulatory purposes. Recently,

FIGURE 2.7

Radio car, circa 1927 (photo courtesy of the Institute for Telecommunications Science, NTIA, U.S. Dept. of Commerce).

however, an increasing number of studies have aimed at obtaining a broader understanding of the spectral usage with an eye toward cognitive access radios. These resolve into one of the following three categories: short-term "snapshot studies," long-term "spectrum observatory" studies, or sensor array studies. Each of these approaches is treated in its own section below.

2.6.2 Snapshot Studies

Numerous short-term spectrum occupancy studies have been conducted by a variety of organizations in different regions of the world [1, 28–30, 46–55]. Perhaps the most prominent set of measurements at this writing have been provided by the Shared Spectrum Company (SSC) often in concert with a university partner (Stevens Institute of Technology and Illinois Institute of Technology, IIT) and often funded by the National Science Foundation. This includes measurements in a very diverse set of environments including the urban environments of Chicago, Dublin, Ireland, and New York City; several suburban measurements in Virginia near Washington, D.C.; and even the radio quiet zone at the National Radio Astronomy Observatory in Green Bank, West Virginia. These studies have generally demonstrated that there is an abundance of unused or lightly used spectrum, which could potentially be exploited through use of dynamic spectrum access networks, or even static networks with carefully defined geographic boundaries (see Figure 2.8). It is anticipated that this kind of study will likely be continued to explore an ever-larger geographic domain and return to previously studied sites to explore the changes that may have occurred since the initial measurement.

2.6.3 Spectrum Observatory

The concept of a spectrum observatory is a relatively new idea in the spectrum world. Despite numerous spectrum occupancy studies, as described previously,

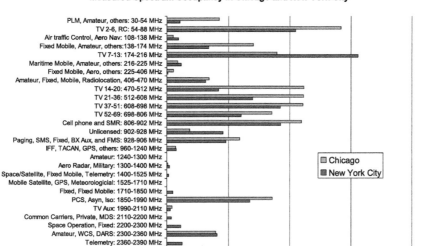

Measured Spectrum Occupancy in Chicago and New York City

FIGURE 2.8

Snapshot view of spectrum usage by spectral range in Chicago and New York City [53].

most of them have been of short duration (two days or less in any single location). "Snapshot" studies often focused on a relatively narrow spectral band. While these have been very effective in gaining general spectral information about the location or in some cases a narrow spectral band across a reasonable geographic area (mobile studies), much of the interesting information can be obtained only by looking at the spectrum over weeks or even months of time. To the authors' knowledge, the first true spectrum observatory is the Wireless Network and Communications Research Center's (WiNCom) observatory funded by NSF and located at IIT in Chicago, Illinois. This observatory is a logical extension of the work performed by Shared Spectrum Company and described already.

This spectrum observatory is dedicated to the multiyear study of the spectral usage patterns in Chicago in the spectral range from 30 MHz to 6 GHz using fixed, nomadic, and potentially mobile observatory systems. One of its fundamental purposes is to detect and characterize spectral holes in time and space that can be exploited by cognitive radio systems in the future. This information might be transmitted to cognitive radios through the beacon approach or used to populate the on-board databases described previously. The 22-story-high antenna array for the observatory is pictured in Figure 2.9.

FIGURE 2.9

IIT's spectrum observatory antenna array.

In addition to providing global spectrum occupancy information of the form shown in Figure 2.8, the spectrum observatory provides a variety of interesting views of the spectrum utilization. Figures 2.10 and 2.11 show two common examples of spectrograms showing, respectively, the television channel occupancy, and the cellular infrastructure transmission pattern. In both cases the charts show seven days worth of data with the signal strength displayed using dark shades to represent noise floor level signals and light shades representing significant power spectral densities. Turning to the figures themselves, in Figure 2.10 we can clearly distinguish the channels that had already transitioned to a digital broadcasting format (ATSC) with their solid uniform signature from those still transmitting using the old analog format (NTSC), which have higher-intensity pilot signals present.

2.6.4 Spectral Sensor Arrays

For localized observation of the dynamic use of the spectrum, sensor arrays should become a very valuable measurement tool. To the authors' knowledge, the only major program focused in this area as of the writing of this text is the European Union's Seventh Framework collaborative project called Sendora [66,67]. The information that such a system should provide would be invaluable in providing detailed

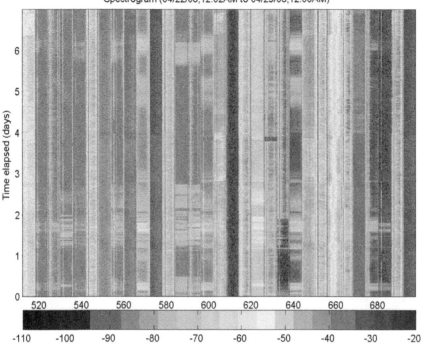

FIGURE 2.10

Spectrogram of TV channels 21–51 taken in Chicago, April 22–29, 2008 [57].

guidance to a beacon-based cognitive radio system on the spectrum that is in use or forecasted to be in use at a specific time and place. This sensor array system would be particularly valuable in high network traffic areas and of even greater value in areas that also have a highly dynamic usage pattern (e.g., mobile, automobile-based wireless systems at a busy intersection in a city).

2.7 APPLICATIONS FOR SPECTRUM OCCUPANCY DATA

The task of gathering spectrum occupancy data in a consistent manner over an extended period of time is very challenging, but when it is successfully accomplished the key question is the ultimate utility of the captured data. More to the point, what is the real value of studying spectrum occupancy? As noted in Section 2.4.2, based on basic supply/demand economics with a limited and fixed supply of spectrum and an ever-increasing demand for this commodity, it is clear that the value of the spectrum will continue to increase for the foreseeable future. This fact by itself suggests the theoretical importance of understanding the

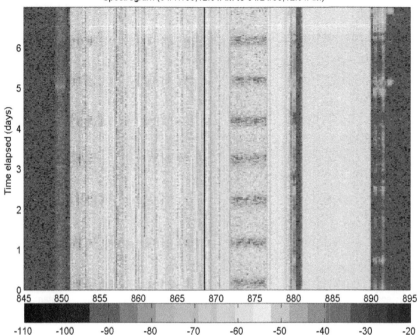

FIGURE 2.11

Spectrogram of 800 MHz cellular band in Chicago taken April 17–24, 2008 [57]. In this figure we see the operation of the infrastructure for both the Nextel (push to talk) service (851–869 MHz) and the low-band cellular service (869–894 MHz) over a seven-day period. Through this spectrogram we clearly see the power reduction approach deployed by one of the cellular service providers and to a slightly lesser degree by Sprint-Nextel during the early morning hours (midnight to 4 AM).

spectrum usage patterns. Beyond the research community, spectrum studies are very important to government regulators and the spectrum policy people in the various government agencies involved in wireless communications–related issues in most of the governments of the world. This information is also of extreme value to those engaged in the design of wireless systems, those who operate various wireless data and voice services, and increasingly those in the various domains of government who provide emergency services, care for our infrastructure, protect the country, and support the information needs of the citizens.

2.7.1 Regulatory Guidance

Spectrum occupancy data are particularly important to the various policy makers and regulatory bodies responsible for writing the spectrum allocation and usage rules for this valuable natural resource (see Section 2.5). Given the significant legal and economic consequences of the actions taken by these government (or pseudo-government) bodies and the normal challenges of budgetary constraints, these organizations usually lack the engineering resources to do detailed spectral studies to support their various spectrum decisions and therefore must depend largely on the work of academics, other government agencies, or often the various companies that have a vested interest in the particular spectral range. Ideally the advent of, and even more so the broad-based deployment of spectrum observatory systems, will provide the data needed by the regulatory bodies to dramatically enhance the value of their decisions. As a slight aside, this information should also be of considerable value in identifying sources of spectrum usage that are inconsistent with existing regulations.

2.7.2 Wireless Systems and Device Design Opportunities

Spectrum occupancy data should also enhance the quality of the designs provided by the various wireless system and device suppliers as they are provided with a more in-depth understanding of their targeted spectral environment. This should support improved designs and provide the ability to simulate and test their systems using representative data sets covering the use of the spectrum in the bands they target. This will be particularly valuable for the emerging cognitive radio designs with the particularly critical needs that they must be able to operate across a broad spectral range ultimately in a dynamic usage environment.

2.7.3 Wireless Communications and Data Service Providers

Wireless network and communications service providers should also benefit from this wealth of new information on the utilization of the spectrum. This will enable them to more optimally utilize the spectrum they control and give them a much improved understanding of the current state of the spectrum that may become available through an auction process. Finally, these providers will now have the information they need to help them assess the potential impact of dynamic spectrum access networks and the way they may wish to influence the direction for the deployment of these networks (i.e., sensor, database, beacon based) and potentially their own involvement in the operation of these future networks.

2.7.4 Societal Value

Overall, an improved spectrum occupancy measurement regime and the associated dramatic enhancement in our understanding of the spectrum use should help ensure that this important natural resource is truly being deployed in a way that optimizes the public interest. This should enable exciting new technologies to be more rapidly deployed to meet the public's ever-expanding needs and desires to communicate, to be informed, to be educated, and to provide information to others. Finally, based on a more optimal use of the spectrum, the actual cost for these improved services should be minimized, making the services available to a broader cross-section of the population.

2.8 CHAPTER SUMMARY AND FURTHER READINGS

Through this chapter the nature of the electromagnetic spectrum with specific focus on that portion of the spectrum most useful for radio networks and communications systems has been described and illustrated. A cursory review of the early use of this spectrum has been presented along with the rise in national and international regulatory bodies focused on both the allocation of the spectrum and defining the acceptable parameters for its use. The chapter also considered, in some depth, the evolution in the sophistication of the approaches needed to ensure that adequate spectral resources are available to meet the ever-growing global demands on this finite natural resource.

Specifically, the emerging unlicensed approaches to the use of the spectrum have been examined, including the emerging cognitive access approach to spectrum utilization. Here we discovered the need for an improved understanding of the spectral environment and some of the techniques for obtaining this understanding at the device and system levels. Some of the emerging techniques used to measure the spectral usage were explored as well as the approaches to communicate this information to assist cognitive devices in appropriately utilizing available spectrum. Finally, we looked at how this information might be utilized more broadly to enhance wireless network and communications systems obviously with a specific focus on cognitive systems.

The interested reader is encouraged to check out the following references with respect to additional information on wireless spectrum regulation: [62, 68–70].

2.9 PROBLEMS

1 In what five ways can wireless spectral waves be altered? What impact does each of these physical phenomenon have on the ability of a receiver to successfully capture the transmitted signal?

2 What is the ITU, where does it organizationally report, how does it operate, and why is this organization so important to the world of spectrum management?

3 What is the relationship between a national regulatory organization (e.g., the FCC in the United States or Ofcom in the United Kingdom) and the ITU? Which would you go to if you were trying to secure licensed spectrum for an application in your country? If you were trying to create a global standard, which organization might be more helpful?

4   Given the effects of the "quadruple whammy" described in the chapter, why is "most of the spectrum, most of the time, in most of the places" still unused?

5   Is spectrum scarce or plentiful? Explain your answer.

6   What is TV white space? Why has it become so important in the context of cognitive radio systems?

7   What are the three ways of cognitively determining the availability of spectrum for use by a cognitive radio system? What are the advantages and disadvantages of each approach?

8   What is a spectrum observatory and how does it work? What is its potential value to a cognitive radio system?

Digital communication fundamentals for cognitive radio

3

Si Chen and Alexander M. Wyglinski

Worcester Polytechnic Institute, United States

3.1 INTRODUCTION

Data communications is an integral part of modern society, ranging across financial transactions (e.g., ATMs, credit cards), entertainment (e.g., music and movie downloads), and other areas. Data communication systems usually need to satisfy the following requirements: being robust to error, having sufficient data rates, and meeting network capacity constraints. However, it is often the case that attempting to simultaneously achieve two or more of these requirements may result in conflicting system operating parameters. Consequently, understanding the mathematics behind these requirements in terms of physical layer operating parameters can help us better design systems that achieve a balance between these requirements.

A digital communication system can be modeled as the chain of functionally independent blocks, such as those depicted in Figure 3.1. Although this block diagram may not depict a complete transceiver chain, it does include the blocks that are relevant to the cognitive radio community. Compared to the block diagram for a traditional communication system, the boundary between the digital and analog signaling domains in this block diagram is pushed as close as possible to the radio frequency front end, resulting in many of the data transmission operations being conducted entirely in the digital signaling domain. This is a distinct characteristic of a cognitive radio system, which often employs software-defined radio technology at the core of its implementation.

Consequently, cognitive radio researchers need to understand the fundamental limits and concepts of a communication system to fully exploit its capabilities and advantages. In this chapter, we investigate the fundamental constraints and

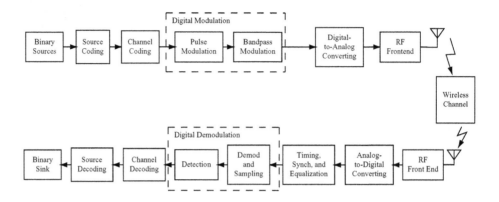

FIGURE 3.1

Anatomy of a wireless digital communication system.

properties of a digital communication system of a cognitive radio related to the physical layer of the OSI reference model.

3.2 DATA TRANSMISSION

A digital communication system is designed to transport a message from an information *source* through a transmission medium (i.e., channel) to an information *sink*. The goal is to accomplish this task such that the information is efficiently transmitted with a certain degree of reliability. In digital communication systems, the metric of reliability for a given transmission is commonly referred to as the *bit error rate* (BER) or *probability of bit error*, which is measured at the receiver output. Note that several data transmission applications require a minimum *data rate*, where the amount of information transferred from information source to information sink must be achieved within a specific time duration. Consequently, bandwidth efficiency is an important characteristic of any digital communication system.

3.2.1 Fundamental Limits

When designing a digital communication system, it is important to understand the achievable limits when transmitting data under specific operating conditions such as the *signal-to-noise ratio* (SNR) or the available bandwidth. Consequently, many digital communication system designers use the concept of *channel capacity* to mathematically determine these achievable limits. The channel capacity was first derived by Claude Shannon and Ralph Hartley and is given by

$$C = B\log_2(1 + \gamma), \tag{3.1}$$

where B is the transmission bandwidth, and γ is the SNR. The channel capacity C, measured in *bits per second* (b/s), is defined as the maximum data rate a system can achieve without error, even when the channel is noisy. Hence, Equation (3.1) is very useful for the following reasons [71]:

1. It provides us with a *bound* on the achievable data rates given bandwidth B and signal-to-noise ratio (received SNR). This can be employed in the ratio

$$\eta = \frac{R}{C}$$

where R is the signaling rate and C is the channel capacity. Note that, as $\eta \to 1$, the system becomes more efficient.
2. It provides us with a basis for trade-off analysis between B and γ.
3. It can be used for comparing the noise performance of one modulation scheme versus another.

Note that Equation (3.1) provides us with only the achievable data rate limit but does not tell us how to construct a transceiver to achieve this limit.

3.2.2 Sources of Transmission Error

A data transmission error may occur in any part of a communication system. Two common sources of error are the introduction of *noise* into the data transmission and the effects of a *band-limited* transmission medium, which are two characteristics of a communications *channel*.

In designing a digital communication system, we often represent the physical transmission channels as mathematical models. One most commonly used model is the *linear filter channel with additive noise* as illustrated in Figure 3.2. In this model, $s(t)$ is the channel input, $c(t)$ is the impulse response of the linear filter,

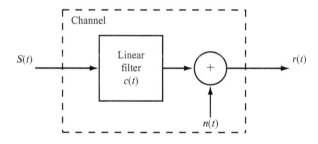

FIGURE 3.2

The linear filter channel with additive noise.

$n(t)$ is an additive random noise process, and $r(t)$ is the channel output, which can be computed as

$$r(t) = s(t) * c(t) + n(t),$$

where $*$ denotes convolution. Such channel models can be further categorized according to whether the linear filter is time variant or time invariant.

Additive White Gaussian Noise (AWGN)

By definition, *noise* is an undesirable disturbance accompanying the received signal that may distort the information carried by the signal. Noise can originate from both human-made and natural sources, such as thermal noise due to the thermal agitation of electrons in transmission lines, other wireless transmitters, or even other conductors. The combination of such sources of noise is known to possess an approximately Gaussian distribution, as shown in Figure 3.3(a). A histogram of zero-mean Gaussian noise with a variance of $\sigma_n^2 = 0.25$ is shown, with the corresponding continuous probability density function (pdf) superimposed on it. The continuous Gaussian pdf is defined as [72]

$$f_X(x) = \frac{1}{\sqrt{2\pi\sigma_n^2}} e^{-\frac{(x-\mu_n)^2}{2\sigma_n^2}}, \tag{3.2}$$

where μ_n and σ_n^2 are the mean and variance.

In most situations, the assumption that the noise introduced by the channel is *white*, which means that the noise received at any frequency is approximately the same, can justifiably be made. An example of zero-mean white Gaussian noise with a variance of $\sigma_n^2 = 0.25$ is shown in Figure 3.3(b). Even if this assumption may

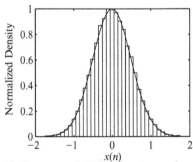

(a) Histogram of AWGN superimposed on the probability density function for a Gaussian random variable of $N(0,0.25)$.

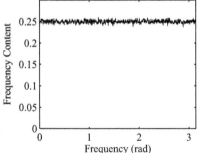

(b) Power spectral density of AWGN with $N(0,0.25)$.

FIGURE 3.3

Time and frequency domain properties of AWGN.

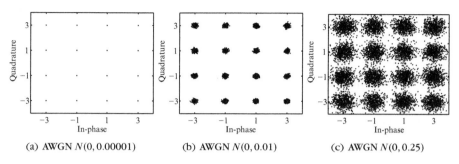

FIGURE 3.4

16-QAM signal constellations with varying amounts of noise: (a).

not hold in certain cases, it does help make the mathematical analysis of digital communication systems tractable.

The receiver has to make a decision on what has been transmitted based on the received signal, which is a mixture of noise and originally transmitted signal. Usually this is accomplished via a "nearest-neighbor" rule with a known set of symbols. However, if a large amount of noise is added to the signal, the received symbol might be shifted closer to a symbol other than the correct one, resulting in a decision error, as seen in Figure 3.4, where additive white Gaussian noise is included with the signal. It is readily observable that, as the noise power increases, the constellation points become fuzzy and begin to overlap each other. It is at this point that the system begins to experience errors.

Band-Limited Channel

In general, usable radio frequency (RF) spectrum is limited and increasingly becoming overpopulated. In addition, digital communication transceivers become increasingly expensive when handling wide transmission bandwidths. Consequently, most transmission are band limited. Narrowband transmissions use narrowband filters at the transmitting and receiving ends to permit only the modulated signal through the system. One problem with this is the noise appearing at the output of the filter.

3.3 DIGITAL MODULATION TECHNIQUES

In communication theory, *modulation* is the process of varying a periodic waveform—that is, a tone—to use that signals to convey a message. In a typical digital communication system, the modulation process usually occurs between the encoding process and the RF front end. After the encoding process, we need a unique signal waveform to represent each transmitted symbol. Since we are using binary sources and sinks, we can have up to 2^b possible symbols if each symbol can represent b bits.

3.3.1 Representation of Signals

Sinusoidal waves are often used to transmit signal waveforms. Consequently, we need a convenient mathematical framework that can represent both the signal waveforms employed by the digital communication system and noise. Two mathematical frameworks that can be employed are an *envelope/phase framework* and an *in-phase/quadrature framework*. These two frameworks are mathematically related as shown in Table 3.1.

Mathematically representing the transmitted signal and noise, we can either use a *time waveform representation* or a *vector representation*. With respect to the latter, we exploit the fact that any sinusoidal signal can be uniquely represented using three variables: ω (frequency), A (amplitude), and ϕ (phase). Furthermore,

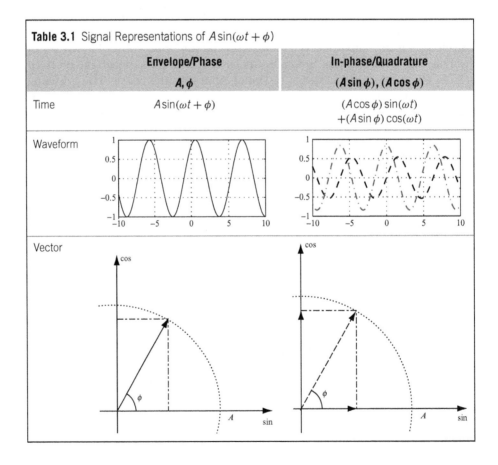

Table 3.1 Signal Representations of $A\sin(\omega t + \phi)$

	Envelope/Phase A, ϕ	In-phase/Quadrature $(A\sin\phi), (A\cos\phi)$
Time	$A\sin(\omega t + \phi)$	$(A\cos\phi)\sin(\omega t)$ $+(A\sin\phi)\cos(\omega t)$
Waveform		
Vector		

a sinusoidal signal can be decomposed into both sine and cosine functions with zero phase:

$$s(t) = A\sin(\omega t + \phi) = A\cos\phi\sin(\omega t) + A\sin\phi\cos(\omega t). \tag{3.3}$$

Since $\sin(\omega t)$ and $\cos(\omega t)$ are orthogonal, we can represent the two sinusoidal waves using one complex number of which the real part represents the sin() function and the complex part represents the cos() function:

$$A\cos\phi\sin(\omega t) + A\sin\phi\cos(\omega t) \Rightarrow [A\cos\phi, A\sin\phi]. \tag{3.4}$$

Notice how a two-dimensional vector with its head located at $[A\cos\phi, A\sin\phi]$ and its tail located at $[0, 0]$ in the Cartesian plane can be generated using amplitude, frequency, and phase information from the signal waveform.

3.3.2 Euclidean Distance between Signals

One of the key characteristics that helps define the difference between two signal waveforms is their *Euclidean distance*. This distance is useful when conducting the error analysis for a specific modulation scheme or comparing two modulation techniques. From a signal constellation plot, such as those shown in Figure 3.4, we can see how an error may occur in the transmission whenever the actually received signal vector $\vec{r} = \vec{s_1} + \vec{n}$ is incorrectly shifted to another signal constellation point by the noise contribution $\vec{n}$. Moreover, the closer we have these signal constellation points located to each other, the greater the probability that errors will be introduced into the transmission. Furthermore, as the level of noise increases, the probability of having transmission errors also increases, since a larger number of signal constellation points will be incorrectly decoded.

The definition of the Euclidean distance between any two signal waveforms $s_i(t)$ and $s_j(t)$ is given by

$$d_{ij}^2 = \int_{-\infty}^{\infty} \left[s_i(t) - s_j(t) \right]^2 dt. \tag{3.5}$$

In particular, in analyzing the error performance of a modulation scheme, we are often more interested in the minimum Euclidean distance between signals, since they have the greatest likelihood of resulting in an error. Consequently, we define the minimum Euclidean distance as

$$d_{\min}^2 = \min_{s_i(t), s_j(t), i \neq j} \int_{-\infty}^{\infty} \left[s_i(t) - s_j(t) \right]^2 dt. \tag{3.6}$$

As for the vector representation of the Euclidean distance between two signals, we can define this as

$$d_{ij} = |\vec{s_i} - \vec{s_j}|$$

and

$$d_{\min} = \min_{\vec{s}_i, \vec{s}_j, i \neq j} |\vec{s}_i - \vec{s}_j|.$$

(3.7)

3.3.3 Decision Rule

The Euclidean distance is an important metric when determining whether $\vec{r}$ should be recognized as the signal $\vec{s}_i$ based on the distance between $\vec{r}$ and $\vec{s}_i$. Consequently, if the distance is smaller than the distances between $\vec{r}$ and any other signals, we say $\vec{r}$ is $\vec{s}_i$. As a result, we can define the *decision rule* for $\vec{s}_i$ as

$$|\vec{r} - \vec{s}_i| < |\vec{r} - \vec{s}_j|, \forall j \neq i,$$

(3.8)

which can be expanded to

$$|\vec{r}|^2 + |\vec{s}_i|^2 - 2|\vec{r}||\vec{s}_i| \cos \theta_i < |\vec{r}|^2 + |\vec{s}_j|^2 - 2|\vec{r}||\vec{s}_j| \cos \theta_j,$$

(3.9)

where θ is the angle between $\vec{r}$ and $\vec{s}$. Assuming that s_i and s_j possess the same signal energy, we can rewrite the decision rule as

$$|\vec{r}||\vec{s}_i| \cos \theta_i > |\vec{r}||\vec{s}_j| \cos \theta_j$$

(3.10)

$$\vec{r} \cdot \vec{s}_i > \vec{r} \cdot \vec{s}_j$$

(3.11)

when the signal $\vec{s}_i$ was transmitted. Similarly, the time waveform representation of this decision rule can be defined as

$$\int_0^T r(t)s_i(t)dt > \int_0^T r(t)s_j(t)dt$$

(3.12)

when the signal $s_i(t)$ was transmitted.

3.3.4 Power Efficiency

Power efficiency is another important metric for a modulation technique. Although it is possible to increase the signal power level of a transmission to increase the Euclidean distance between signals, thus increasing noise immunity, most digital communication systems strive for some level of power efficiency for several reasons, including minimizing power consumption due to a limited power supply or minimizing interference with spectrally adjacent transmissions. Consequently, the power efficiency of a modulation scheme measures the largest minimum signal distance achievable by a modulation technique given the lowest transmit power available.

First, let us define the *energy per symbol*. For M different symbols, the average energy per symbol is defined as

$$\bar{E}_s = \sum_{i=0}^{M-1} P(i) \int_{-\infty}^{\infty} s_i^2(t)dt,$$

(3.13)

where $P(i)$ is the probability of transmitting symbol i. Now suppose that each symbol is composed of $\log_2 M$ bits; we can then define the average *energy per bit* as

$$\bar{E}_b = \frac{\bar{E}_s}{\log_2 M}. \tag{3.14}$$

Finally, we can mathematically define the *power efficiency* of a signal constellation set used for modulation as

$$\epsilon_P = \frac{d_{min}^2}{\bar{E}_b}. \tag{3.15}$$

To illustrate the calculation of the power efficiency for a specific modulation scheme, suppose we would like to compute the power efficiency for a 2-PAM modulation scheme, where $s_1(t) = s(t)$, $s_0(t) = -s(t)$, and T is the length of $s(t)$. Then, using the aforementioned equations yields

$$\bar{E}_b = \bar{E}_s = \int_0^T s^2(t)dt, \tag{3.16}$$

and

$$\epsilon_P = \frac{d_{min}^2}{\bar{E}_b} = \frac{\int_0^T [2 \times s(t)]^2 dt}{\int_0^T s^2(t) dt} = 4. \tag{3.17}$$

Consequently, given the mathematical expression for the signal waveform of a given modulation scheme, it is possible to compute its power efficiency and the average energy per bit. The following two subsections mathematically define a couple of common modulation schemes often used in practice.

3.3.5 *M*-ary Phase Shift Keying

Under certain operating conditions, it is possible that the amplitude values of a transmission is more readily susceptible to the channel errors than the phase values. Furthermore, the efficient design of an RF amplifier with a sufficient linear range has always been problematic. Consequently, employing a constant-amplitude modulation technique designed to modulate the entire transmission using phase information can potentially alleviate these issues. One such modulation scheme is *M-ary phase shift keying* (M-PSK), where the M signal constellation points are equally spaced around a circle of constant distance from the origin. Figure 3.5 shows an example of an M-PSK signal constellation for $M = 16$ points, where each point uniquely represents $b = \log_2(16) = 4$ bits.

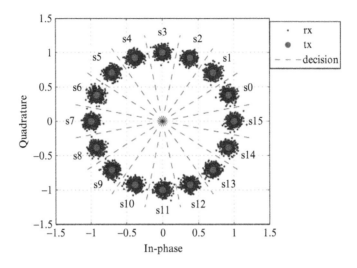

FIGURE 3.5

Constellation map of 16-PSK.

As for the general mathematical expression for the ith signal waveform for an M-PSK signal, $s_i(t)$, it is defined as

$$s_i(t) = A \cos\left(\omega_c t + \frac{2\pi \times i}{m}\right), \quad m = 1, 2, \ldots, M. \tag{3.18}$$

Note that, since all of the transmission data are represented solely by the phase information, M-PSK is well-suited for nonlinear power amplifiers and is robust to amplitude distortion channels due to the constant envelope. However, M-PSK modulation schemes do possess a smaller power efficiency than M-QAM.

3.3.6 *M*-ary Quadrature Amplitude Modulation

M-ary quadrature amplitude modulation (M-QAM) is a modulation scheme that conveys data by modulating the data transmission onto the amplitude via two carrier signals. These two signals, usually sinusoidal functions such as sine and cosine, are 90 degrees out of phase with each other and are therefore called *quadrature carriers*. Hence, the data transmission can be represented by simultaneous amplitude variations on both carrier signals. One method for creating an M-QAM waveform consists of combining two $\sqrt{M}$ pulse amplitude modulation ($\sqrt{M}$-PAM), with each modulation scheme acting on a different orthogonal carrier signal. This process of combining two $\sqrt{M}$-PAM waveforms to generate an M-QAM signal is shown in Figure 3.6. An example of an M-QAM signal constellation when $M = 16$ is shown in Figure 3.7.

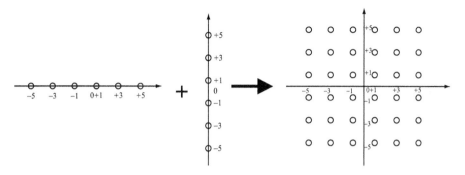

FIGURE 3.6

Derivation of QAM from two PAMs.

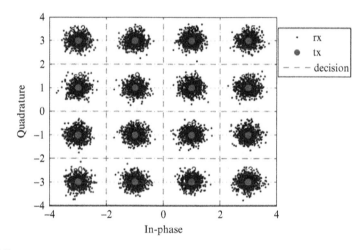

FIGURE 3.7

Constellation map of 16-QAM.

One reason that M-QAM is often used in digital communication systems is that it has a simple receiver structure as shown in Figure 3.8. The received signal $r(t)$ is passed through two $\sqrt{M}$-PAM receivers separately. Since the waves of the two $\sqrt{M}$-PAM receivers are orthogonal to each other, extraction of the two orthogonal $\sqrt{M}$-PAM signals embedded in an M-QAM signal can be achieved.

3.4 PROBABILITY OF BIT ERROR

One of the primary metrics for measuring the error robustness of a digital communication system is the *probability of bit error*, which is otherwise known as

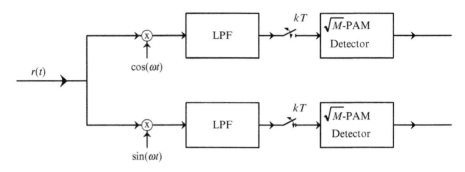

FIGURE 3.8

QAM receiver.

the *bit error rate*. In this section, we introduce the mathematical framework for evaluating the probability of bit error of a specified modulation scheme.

3.4.1 Derivation of Probability of Bit Error

To keep the mathematics tractable, let us assume that the modulation scheme of interest possesses only two signal waveform representations, namely $s_1(t)$ and $s_2(t)$, where it is assumed that the signal $s_1(t)$ was transmitted. Then, using the decision rule derived in Section 3.3.3, an error event occurs whenever the following inequality is satisfied:

$$\int_0^T r(t)s_1(t)dt < \int_0^T r(t)s_2(t)dt, \tag{3.19}$$

which possesses the equivalent vector representation:

$$\vec{r} \cdot \vec{s_1} < \vec{r} \cdot \vec{s_2}. \tag{3.20}$$

Since the received signal is equal to $r(t) = s_1(t) + n(t)$, we can expand this equation to become

$$\int_0^T s_1^2(t)dt + \int_0^T n(t)s_1(t)dt < \int_0^T s_1(t)s_2(t)dt + \int_0^T n(t)s_2(t)dt, \tag{3.21}$$

which can then be rewritten as

$$\int_0^T s_1^2(t)dt - \int_0^T s_1(t)s_2(t)dt < - \int_0^T n(t)s_1(t)dt + \int_0^T n(t)s_2(t)dt$$

$$= \int_0^T n(t)(s_2(t) - s_1(t))dt. \tag{3.22}$$

Knowing that the energy of the signal $s_1(t)$ and the correlation between signals $s_1(t)$ and $s_2(t)$ are, respectively, given by

$$E_1 = \int_0^T s_1^2(t)dt \tag{3.23}$$

and

$$\rho_{12} = \int_0^T s_1(t)s_2(t)dt, \tag{3.24}$$

and assuming that the noise term has a Gaussian distribution $n(t) \sim N(0;\sigma)$, we can rewrite Equation (3.22) as

$$E_1 - \rho_{12} < z, \tag{3.25}$$

where $z \sim N(0;\sigma)$, and

$$\sigma^2 = E\left(z^2\right) = \frac{N_0}{2}\int_0^T \left[s_1(t) - s_2(t)\right]^2 dt$$
$$= \frac{N_0}{2}(E_1 + E_2 - 2\rho_{12}). \tag{3.26}$$

The resulting expression is nothing more than the squared difference of the waveforms $s_1(t)$ and $s_2(t)$ integrated over one symbol period. This quality is also multiplied by the single-side noise density $N_0/2$. Assuming equal energy levels for $s_1(t)$ and $s_2(t)$, namely, $E_1 = E_2 = E$, we can then get the following:

$$\sigma^2 = N_0(E - \rho_{12}). \tag{3.27}$$

Since $s_1(t)$ is assumed to be transmitted, an error occurs whenever the received signal is recognized as $s_2(t)$. Consequently, the probability of bit error is equal to

$$P(e|1) = P(\text{error}|\text{"1" transmitted})$$
$$= P(z \geq E - \rho_{12}). \tag{3.28}$$

Therefore, with $z \sim N(0;\sigma)$ and $\sigma^2 = N_0(E - \rho_{12})$, the probability distribution function of z can be defined as

$$p(z) = \frac{1}{\sqrt{\pi}\sigma}e^{-z^2/2\sigma^2}$$
$$= \frac{1}{\sqrt{\pi N_0(E - \rho_{12})}}e^{-z^2/2N_0(E-\rho_{12})}. \tag{3.29}$$

Therefore, the probability of bit error can be derived to be equal to

$$P(z \geq E - \rho_{12}) = \frac{1}{\sqrt{\pi N_0 (E - \rho_{12})}} \int_{E-\rho_{12}}^{\infty} \exp\left(-\frac{z^2}{2N_0(E - \rho_{12})}\right) dz$$

$$= \frac{1}{\sqrt{2\pi}} \int_{\sqrt{\frac{E-\rho_{12}}{N_0}}}^{\infty} e^{-x^2/2} dx \qquad (3.30)$$

$$= Q\left(\sqrt{\frac{E - \rho_{12}}{N_0}}\right),$$

where $Q(x)$ is the *Q function*, which defines the area under the tail of a Gaussian pdf. The Q function is defined as

$$Q(x) = \frac{1}{\sqrt{2\pi}} \int_x^{\infty} e^{-y^2/2} dy. \qquad (3.31)$$

Note that, when the signal energy levels for $s_1(t)$ and $s_2(t)$ are not equivalent— that is, $E_{s_i} \neq E_{s_j}$—we get the minimum Euclidean distance to be equal to $d_{min}^2 = E_{s_1} + E_{s_2} - 2\rho_{12} = \int_0^T [s_1(t) - s_2(t)]^2 dt = ||\vec{s_1} - \vec{s_2}||^2$, which then yields the following probability of bit error:

$$P_e = Q\left(\sqrt{\frac{d_{min}^2}{2N_0}}\right), \qquad (3.32)$$

which is shown in Figure 3.9.

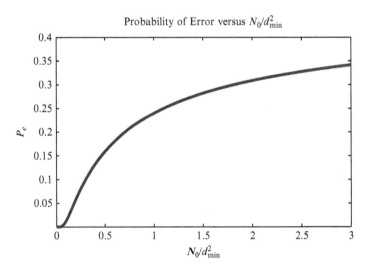

FIGURE 3.9

Theoretical values of P_e for different N_0/d_{min}^2.

Upper Bound on Probability of Bit Error

Although obtaining the mathematical expression for the probability of bit error for a modulation scheme consisting of just two signal waveforms is straightforward, there exist numerous modulation schemes that consist of more than two signal waveforms. Consequently, it is often the case that deriving the exact expression for such modulation schemes is difficult or even mathematically intractable. As a result, obtaining the theoretical *bound* of the probability of bit error for a specific modulation scheme is usually sufficient. There exists two types of bounds: an *upper bound* and a *lower bound*. In this section and the next, we focus on the derivation of both types of bounds.

Using ξ_{1j} to denote the error scenario of incorrectly deciding that s_j was intercepted at the receiver even though s_1 was actually transmitted, and using the vector representation of the modulation scheme $\vec{r} = \vec{s}_1 + \vec{n}$, where $\vec{s}_1$ is the reference vector, an error event ξ occurs when *at least one* ξ_{1j} occurs; that is,

$$\xi = \{\xi_{12} \text{ or } \xi_{13} \text{ or } \xi_{14} \text{ or } \ldots \text{ or } \xi_{1m}\}$$

$$= \bigcup_{j=2}^{m} \xi_{1j}, \tag{3.33}$$

which means ξ is the union of all *pairwise error probabilities*. Using the *law of total probability*, we can obtain the following inequality for the probability of bit error:

$$P_e = P(\xi) = P\left(\bigcup_{j=2}^{m} \xi_{ij}\right)$$

$$\leq \sum_{j=2}^{m} P(\xi_{ij}). \tag{3.34}$$

Note that, if all pairwise error events are mutually exclusive, Equation (3.34) would become an equality, since the intersection terms associated with the law of total probability would reduce to zero. Consequently, since $P(\xi_{ij}) = Q(\sqrt{d_{ij}^2/2N_0})$, we can then derive our upper bound to P_e to be equal to

$$P_e \leq \sum_{j=2}^{m} Q\left(\sqrt{\frac{d_{ij}^2}{2N_0}}\right). \tag{3.35}$$

As an example, suppose we would like to compute the upper bound on the probability of bit error for 4-QAM. Since $d_{12} = d_{14}$, we can compute this upper bound on P_e to be equal to

$$P_e \leq 2Q\left(\sqrt{\frac{d_{12}^2}{2N_0}}\right) + Q\left(\sqrt{\frac{d_{13}^2}{2N_0}}\right). \tag{3.36}$$

Lower Bound on Probability of Bit Error

Suppose now that we are interested in obtaining the lower bound on the probability of bit error for a specific modulation scheme. Defining the received vector as $\vec{r} = \vec{s_j} + \vec{n}$, allow only $\vec{s_i}$ and $\vec{s_j}$ to be transmitted, given that i and j are fixed. In other words, the receiver needs only to decide between $\vec{s_i}$ and $\vec{s_j}$. Therefore, $P(\xi)$ for the receiver is given by $Q(\sqrt{d_{ij}^2/2N_0})$.

However, in practice a receiver must consider M signals instead of just two signals. Consequently, we define the lower bound on the probability of bit error as

$$P_e \geq Q\left(\sqrt{\frac{d_{ij}^2}{2N_0}}\right), \text{where } i \neq j \text{ and } i,j \text{ are fixed.} \tag{3.37}$$

Studying this expression, we observe that if we decrease the size of d_{ij}^2, the resulting probability of error increases. If a receiver has M signals to detect instead of just two, then the lower bound on the probability of bit error is based on a pair of signals possessing the minimum Euclidean distance across all possible signal pairs in the modulation scheme; namely,

$$P_e \geq Q\left(\sqrt{\frac{d_{min}^2}{2N_0}}\right). \tag{3.38}$$

Thus, the lower and upper bounds on the probability of bit error, P_e, can be defined by

$$Q\left(\sqrt{\frac{d_{min}^2}{2N_0}}\right) \leq P_e \leq \sum_{j=2}^{M} Q\left(\sqrt{\frac{d_{ij}^2}{2N_0}}\right) \leq (M-1)Q\left(\sqrt{\frac{d_{min}^2}{2N_0}}\right) \tag{3.39}$$

where $d_{1j}^2 = ||\vec{s_1} - \vec{s_j}||^2$ and $d_{min} = \min\{d_{ij}|j \geq 2\}$. Note that the upper bound requires a uniform distance distribution, d_{ij}, and $i, i \neq j$, while the lower bound holds in general.

A Dynamic Spectrum Access Networking Example

Question: Suppose a primary user operating in a dynamic spectrum access network employs a modulation scheme with $d_{min}^2 = 0.2J$ and a center frequency at f. The noise spectral density N_0 is measured to be 1 mW/Hz. If a secondary user starts transmitting in the vicinity of the primary user, the combined spectral density of noise and the secondary user at f increases to 3 mW/Hz. What is the increase in lower bound of the probability of error of the primary user in terms of decibels?

Answer: To obtain the Q function $Q(x)$ in MATLAB based on Equation (3.31), we need to employ the mathematical relationship

$$\text{erfc}(x) = \frac{2}{\sqrt{\pi}} \int_x^{\infty} e^{-y^2} dy, \tag{3.40}$$

which can be coded in MATLAB as `Q(x)=1/2*erfc(x/sqrt(2))`. Thus, the probability of bit error of the primary user transmission prior to the introduction of the secondary user is equal to

$$Pe \geq Q\left(\sqrt{\frac{d_{\min}^2}{2N_0}}\right) = Q\left(\sqrt{\frac{0.2}{2 \times 0.001}}\right) = -54.12\, dB; \quad (3.41)$$

while after the introduction of the secondary user, the probability of bit error for the primary user now becomes

$$Pe \geq Q\left(\sqrt{\frac{d_{\min}^2}{2N_0'}}\right) = Q\left(\sqrt{\frac{0.2}{2 \times 0.003}}\right) = -23.09\, dB \quad (3.42)$$

which is an increase in probability of bit error of more than 30 dB.

3.4.2 Probability of Bit Error of *M*-ary Phase Shift Keying

Notice that the pairwise distances are equivalent. With any choice of s_j we get the same distribution of distances, d_{ij}, for $i \neq j$. Thus, our bounds on P_e depend only on d_{ij}, for $i \neq j$. We take 4-QAM as an example, we observe that $d_{\min} = d_{12} = d_{14} < d_{13}$, which yields the following bounds on the probability of bit error P_e:

$$Q\left[\left(\sqrt{\frac{d_{12}^2}{2N_0}}\right) \leq Pe \leq 2Q\left(\sqrt{\frac{d_{12}^2}{2N_0}}\right) + Q\left(\sqrt{\frac{d_{13}^2}{2N_0}}\right) \cong 2Q\left(\sqrt{\frac{d_{12}^2}{2N_0}}\right)\right]. \quad (3.43)$$

3.5 MULTICARRIER MODULATION

In this chapter we cover the mathematical analysis of modulation schemes applied to single-carrier frequency data transmission. However, over the past several decades an increasing number of data communication systems have started employing another form of transmission framework based on sending parallel streams of information in the frequency domain on different center frequencies. Employed in a wide range of applications, including digital subscriber line (DSL) modems, wireless local area networks (WLANs), digital audio broadcasting (DAB), and digital video broadcasting (DVB), *multicarrier modulation* has exhibited its potential to transmit large amounts of data across a channel while possessing reasonable error robustness. Furthermore, the mathematical analysis studied thus far can be easily extended from the single-carrier case to the multicarrier case.

One of the primary advantages of multicarrier modulation, as will be shown in Chapter 6, is its ability to tailor its *subcarrier* operating parameters on an individual or block-by-block basis. This additional flexibility over high data rate single-carrier transmission techniques makes it an excellent candidate for dynamic spectrum access, since it can deactivate those subcarriers within the vicinity of primary user transmissions, thus respecting the exclusive rights of the incumbent license holders to their allocated wireless spectrum.

As we see in this section, a variety of multicarrier modulation techniques could be employed in a dynamic spectrum access network enabled by cognitive radio communication systems.

3.5.1 Basic Theory

Multicarrier modulation (MCM) is a form of frequency division multiplexing (FDM), where data are transmitted in several narrowband streams at different carrier frequencies. However, unlike conventional FDM systems, where the narrowband subcarrier signals are separated by guard bands in the frequency domain [73], MCM systems allow for overlapping adjacent subcarriers when a certain set of conditions are satisfied [74–77] (see Section 3.5.3 for more information about these conditions). As a result, MCM systems are spectrally efficient.

A generic single-input/single-output MCM transmitter/receiver system is shown in Figure 3.10. A high-speed input data stream, $x(n)$, is parsed into N relatively slower streams and modulated using a prescribed signal constellation. The modulated streams, $d^{(k)}(n)$, $k = 0, \ldots, N-1$, are then up-sampled by a factor N, yielding the signals $y^{(k)}(n)$, $k = 0, \ldots, N-1$. They are then filtered by a bank of synthesis filters, $g^{(k)}(n)$, $k = 0, \ldots, N-1$, and the filtered signals are summed to form the composite transmit signal, $s(n)$:

$$s(n) = \sum_{k=0}^{N-1} \sum_{l=-\infty}^{\infty} g^{(k)}(l) y^{(k)}(n-l). \qquad (3.44)$$

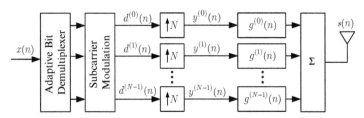

(a) Single-input/single-output MCM transmitter.

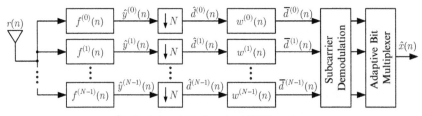

(b) Single-input/single-output MCM receiver.

FIGURE 3.10

Schematic of a generic single-input/single-output MCM system.

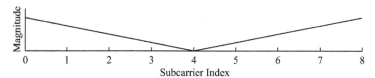

(a) Spectrum of the second subcarrier prior to up-sampling.

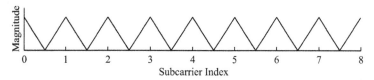

(b) Spectrum of the second subcarrier after up-sampling by a factor of 8.

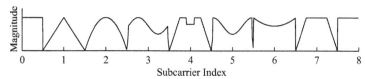

(c) Spectrum of the composite transmit signal.

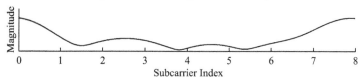

(d) Example of a frequency-selective fading channel spectrum.

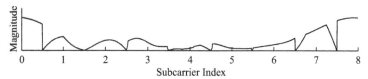

(e) Impact of frequency/selective fading channel on the composite transmit signal spectrum.

FIGURE 3.11

The effects of an eight-subcarrier MCM transmitter and frequency-selective fading channel on the individual subcarrier spectra.

Equivalently, in the frequency domain, the manipulation of the subcarrier signals at the transmitter are outlined in Figure 3.11 for an $N = 8$ subcarrier MCM system.[1] For example, given $d^{(1)}(n)$ in Figure 3.11(a), its spectrum is compressed by a factor

[1]The diagram of Figure 3.11, for simplicity, does not show any overlap between the adjacent subcarriers. However, MCM systems are capable of allowing overlap of the subcarriers. The spectral subcarrier shapes are intentionally exaggerated to identify the different subcarriers.

of 8 and repeated in frequency due to up-sampling to yield $y^{(1)}(n)$, as depicted in Figure 3.11(b). Then a copy of the compressed spectrum from each subcarrier is bandpass-filtered and placed at the corresponding center frequencies, as shown in Figure 3.11(c), forming the composite transmit signal $s(n)$. Thus, the MCM transmitter is converting a parallel set of signals in the time domain into a parallel set of signals in the frequency domain using a combination of up-samplers and synthesis filters.

Between the transmitter and receiver lies a channel, which introduces both noise and distortion (mainly due to multipath propagation) to the composite transmit signal. It should be mentioned that the channel can be modeled as a finite impulse response (FIR) filter that possesses a frequency-selective fading characteristic (see Figure 3.11(d)). As a result, when $s(n)$ passes through the channel, assuming for now that no noise is present, the channel attenuates the spectrum of $s(n)$ nonuniformly in frequency, as shown in Figure 3.11(e).

At the receiver, a bank of analysis filters, $f^{(k)}(n)$, $k = 0, \ldots, N-1$, are employed to separate the subcarriers out of the received composite signal, $r(n)$, into N individual signals, $\hat{y}^{(k)}(n)$, $k = 0, \ldots, N-1$. These signals are then down-sampled by a factor of N, yielding $\hat{a}^{(k)}(n)$, $k = 0, \ldots, N-1$. To remove the distortion introduced by the channel, equalizers $w^{(k)}(n)$, $k = 0, \ldots, N-1$ are employed on a per-subcarrier basis.[2] Several equalizer design approaches for MCM systems are discussed in Section 3.6. The outputs of the subcarrier equalizers, $\bar{a}^{(k)}(n)$, $k = 0, \ldots, N-1$, are then demodulated and the resulting binary sequences combined using a multiplexer, yielding the reconstructed high-speed output $\hat{x}(n)$ [73, 82, 83].

Although the modulation and demodulation stages of an MCM system are usually more complex than in a single-carrier system, MCM systems possess a number of advantages due to their "divide-and-conquer" nature in the frequency domain. Since the channel usually has no flat-frequency response, it is easier to compensate for the channel distortion on a per-subcarrier basis rather than on the entire received signal, as will be shown in Section 3.6. Moreover, since the channel distortion may not be equivalent for all subcarriers, adapting the transmission parameters per subcarrier (i.e., signal constellation and transmit power levels) would allow for increased throughput while guaranteeing a prescribed error performance.

A thorough comparison between single-carrier and multicarrier systems was performed by Saltzberg using a number of criteria, as summarized in Table 3.2 [84].[3] There is little difference in performance between single-carrier and multicarrier systems, since the latter can be interpreted as a linear reversible transformation of

---

[2]Although linear per-subcarrier equalizers have been employed in Figure 3.10(b), decision-feedback equalizers on each subcarrier [78] or per-subcarrier Tomlinson-Harashima precoding schemes [79–81] can also be used.

[3]The single and multicarrier implementations studied are based on ADSL systems, with the single-carrier system employing decision-feedback equalizers with Tomlinson filtering, while the multicarrier system used frequency-domain, single-tap subcarriers equalizers and adaptive bit loading.

Table 3.2 Comparison between Single-Carrier and Multicarrier Systems

Issue	Single Carrier	Multitone	Equivalent
Performance in Gaussian noise			✓
Sensitivity to impulse noise (uncoded)		✓	
Sensitivity to narrowband noise (uncoded)	✓		
Sensitivity to clipping	✓		
Sensitivity to timing jitter	✓		
Latency (delay)	✓		
Need for echo cancellation	✓		
Computations per unit time		✓	
Complexity of algorithm	✓		
Cost and power consumption in analog sections	✓		
Adaptability of bit rate		✓	

the former. However, there are a number of practical differences. For instance, multicarrier systems can perform adaptive bit loading in a straightforward fashion, which can enhance system performance with respect to maximizing throughput or increasing error robustness. However, multicarrier systems are more sensitive to the effects of narrowband noise, amplitude clipping, timing jitter, and delay. With respect to the computational complexity, FFT-based multicarrier systems employing frequency-domain, single-tap subcarrier equalizers usually use fewer multiplications and additions per unit time than single-carrier systems, which require lengthy equalizers to eliminate the distortion introduced by the channel. As a result, multicarrier systems have fewer computations per unit time. On the other hand, when the multicarrier system performs adaptive bit loading, the complexity of the algorithm increases relative to a single-carrier system due to the iterative search performed by the system for the appropriate subcarrier signal constellations.

Implementation

There exist a number of MCM implementations, as shown in Table 3.3. The implementations have been divided into two categories, depending on the choice of filters employed by the analysis and synthesis filter banks. The first category contains implementations that use the discrete Fourier transform in the filter bank

Table 3.3 Several Implementations of Multicarrier Systems

Name	Description
Discrete Fourier Transform–Based Multicarrier	Employs the discrete Fourier transform–basis functions to modulate subcarriers to different center frequencies. This can be efficiently implemented using FFTs. Several examples are • *Orthogonal frequency division multiplexing* is the name given to this technique when used in wireless applications. • In wireline applications, this technique is called *discrete multitone*.
Filter Bank Multicarrier	Employs bandpass filters at both the transmitter and receiver to filter the subcarriers prior to combining them and separating them, respectively. Several examples are • *Complex exponential-modulated filter banks* modulate a prototype low-pass filter to different center frequencies using complex exponentials. • *Cosine-modulated filter banks* modulate a prototype low-pass filter to different center frequencies using cosines. • *Transmultiplexers* are filter banks used in multirate signal processing. They are the duals of subband coders. • *Perfect reconstruction filter banks* are designed to eliminate cross-talk under ideal channel conditions. • *Oversampled filter banks* employ a sampling factor higher than the total number of subcarriers. • *Modified discrete Fourier transform filter banks* delay either the real or imaginary components of each subcarrier signal with respect to each other to minimize cross-talk.

implementation. This type of MCM implementation is widely employed in a number of wireless and wireline applications due to its efficient implementation involving the fast Fourier transform (FFT). The other category is based on employing band-pass filters at the synthesis and analysis filter banks. The techniques by which the filter banks are created define each of the implementations in this category. For instance, most implementations modulate a single prototype low-pass filter to different center frequencies in order to keep the cost of designing the filters low.

Each implementation possesses a number of advantages and disadvantages. In the following sections, details for several of these implementations will be presented.

3.5.2 Orthogonal Frequency Division Multiplexing

The first implementation is an extremely popular one due to its efficient hardware implementation using the FFT and the inverse FFT (IFFT). Known in wireless applications as *orthogonal frequency division multiplexing* (OFDM) [85, 86] or in wireline applications as *discrete multitone* (DMT) [87, 88], these systems use discrete Fourier transform (DFT) basis functions to create the synthesis and analysis filter banks. The filters in the filter banks are *odd stacked*, which means they are uniformly distributed throughout the frequency domain, with one filter centered at $\omega_0 = 0$ rad/s. Although OFDM systems could be implemented using a bank of sinusoid generators [89], practical implementations employ the FFT and IFFT, which results in a significant complexity reduction [85]. As a result, OFDM/DMT has become a popular choice in many multicarrier applications, including digital audio broadcast, digital subscriber line, digital video broadcast, and wireless local area networks such as the IEEE 802.11a/g, ETSI HiperLAN/2, and MMAC HiSWAN.

A schematic of an OFDM transceiver is shown in Figure 3.12. A high-speed input stream $x(n)$ is first demultiplexed into N data streams, $x^{(k)}(n)$, $k = 0, \ldots, N - 1$, using a serial-to-parallel converter, where $x^{(k)}(n)$ is the subcarrier data for

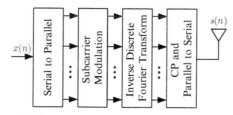

(a) OFDM transmitter with cyclic prefix adder.

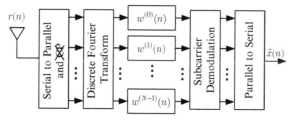

(b) OFDM receiver with cyclic prefix remover.

FIGURE 3.12

Schematic of an OFDM system employing a cyclic prefix (CP).

subcarrier k. These streams are then individually modulated using M-QAM constellations, to yield $y^{(k)}(n)$, $k = 0, \ldots, N - 1$, where $y^{(k)}(n)$ is the M-QAM-modulated subcarrier data for subcarrier k. The inverse DFT (IDFT) is then applied to the subcarriers, defined as [90]

$$s^{(l)}(n) = \frac{1}{N} \sum_{k=0}^{N-1} y^{(k)}(n) e^{j2\pi kl/N} ,$$ (3.45)

where $l = 0, \ldots, N - 1$, resulting in the subcarriers being modulated to one of N evenly spaced center frequencies in the range $[0, 2\pi)$.

Before the subcarriers are converted to form the composite signal, $s(n)$, it is necessary to add some redundancy to compensate for one of the main disadvantages of OFDM: low spectral selectivity. Since OFDM employs the DFT and its inverse, the filters applied to the subcarriers have a low stopband attenuation as the frequency responses of the filters are of the form sinc(x). Therefore, the performance of the OFDM system would significantly decrease if it were operating in a time-dispersive environment. To counteract the time-dispersiveness of the channel, a cyclic extension is employed either before the symbol (i.e., cyclic prefix) or after it (i.e., cyclic suffix) to capture this effect (the details of how the cyclic extension works is discussed in the following subsection). Without loss of generality, the system adds a cyclic prefix to the OFDM symbol.

At the receiver, the cyclic prefix is removed from the received composite signal, $r(n)$, and converted from a serial stream into a collection of parallel streams using a serial-to-parallel converter. The DFT is applied as [90]

$$\hat{y}^{(k)}(n) = \sum_{l=0}^{N-1} r^{(l)}(n) e^{-j2\pi kl/N}$$ (3.46)

for $k = 0, \ldots, N - 1$, where $r^{(k)}(n)$, $k = 0, \ldots, N - 1$ are the parallel input streams to the DFT. The subcarriers are then equalized with $w^{(k)}(n)$, $k = 0, \ldots, N - 1$ to compensate for the distortion introduced by the channel. The equalized subcarriers are then demodulated before being multiplexed together using the parallel-to-serial converter, forming the output $\hat{x}(n)$.

OFDM with Cyclic Extension

As mentioned previously, the synthesis and analysis filters of OFDM have relatively poor spectral selectivity. Therefore, a cyclic extension is appended to the OFDM symbol to compensate. Otherwise, the intersymbol interference of adjacent OFDM symbols due to the time-dispersive channel degrades the error performance of the system. Although a buffer of zeros would suffice in preventing the intersymbol interference [91], the use of a cyclic extension has the added benefit of simplifying the design of an optimal subcarrier frequency-domain equalizer (see Section 3.6.3 for details).

A graphical representation of how a cyclic extension functions is shown in Figure 3.13. In this case, the cyclic prefix is created by copying the end of the

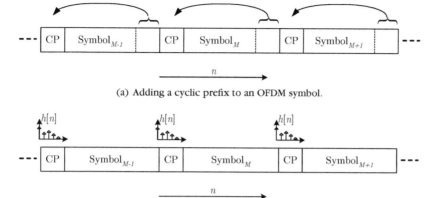

(a) Adding a cyclic prefix to an OFDM symbol.

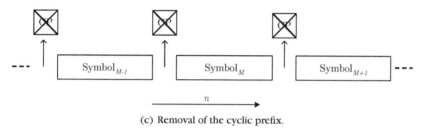

(b) Spreading by channel impulse response $h(n)$ from previous symbol into the cyclic prefix.

(c) Removal of the cyclic prefix.

FIGURE 3.13

The process of adding, capturing the intersymbol interference, and removing a cyclic prefix.

OFDM symbol and placing that copy at the beginning of the symbol, for every symbol. This process is shown in Figure 3.13(a).

With the channel impulse response (CIR) modeled as an FIR filter $h(n)$, the CIR is convolved with a sampled version of the transmitted signal with the included CP. As a result, the CIR spreads the samples of symbol $M - 1$ onto the samples of symbol M, while the samples of symbol M are spread onto the samples of symbol $M + 1$. However, observing Figure 3.13(b), if the CP is of sufficient length to capture all the interference due to the CIR, the symbols experience only the spreading of samples from within their own symbol. At the receiver, the CP is removed, as shown in Figure 3.13(c), and the OFDM symbols proceed with demodulation and equalization.

Despite the usefulness of the cyclic prefix, there are several disadvantages. First, the length of the cyclic prefix must be sufficient to capture the effects of the CIR. If not, the cyclic prefix fails to prevent distortion introduced from other symbols. The second disadvantage is the amount of overhead introduced by the cyclic prefix. By adding more samples to buffer the symbols, we must send more information across the channel to the receiver. This means that, to get the same throughput as a system without the cyclic prefix, we must transmit at a higher data rate.

3.5.3 Filter Bank Multicarrier Systems

Although an OFDM system can possess a computationally efficient implementation, its also has several drawbacks, of which poor spectral selectivity is the worst. The use of a cyclic extension to prevent the intersymbol interference, although effective, requires a significant amount of overhead in terms of the addition of redundant information to the transmitted signal. One solution to this problem is to employ an MCM system that uses synthesis and analysis filters with relatively high spectral selectivity. For instance, Rizos, Proakis, and Nguyen showed in 1994 that the interference due to overlapping frequency responses of the subcarrier filters in a DMT system was worse than a cosine-modulated filter bank system when no prereceiver processing (e.g., cyclic extension with a time-domain equalizer) was employed. Only when the DMT system employs prereceiver processing while operating in a linearly distorted channel (additive white Gaussian noise and near-end cross-talk, NEXT) does its performance exceed that of the cosine-modulated filter bank system at the cost of reduced throughput. This is the motivation of the second implementation called *filter bank multicarrier* (FB-MC) [74, 80, 92–102].

In FB-MC, a set of synthesis and analysis filters are designed such that they have both adequate spectral selectivity and bandwidth efficiency. Although each filter could be designed on an individual basis, a more efficient approach is to design a single prototype low-pass filter and modulate it to several specified center frequencies to generate the synthesis and analysis filters $g^{(k)}(n)$ and $f^{(k)}(n)$, $k = 0, \ldots, N - 1$. Usually the filters are uniformly spaced, designed to be highly spectrally selective to minimize cross-talk with adjacent subcarriers, and can either be *odd stacked* or *even stacked*; that is, no center frequency at $\omega_0 = 0$ rad/s. For example, in Figure 3.14 the subcarrier frequency responses of the synthesis filters for an OFDM system and an FB-MC system employing square-root raised cosine filters is shown for $N = 8$ subcarriers. Notice how in Figure 3.14(a) the sidelobe levels are significantly higher than the sidelobe levels found in Figure 3.14(b). With respect to cyclic extensions, FB-MC systems usually need to introduce equalization strategies to mitigate intersymbol and intercarrier interference, especially when they operate in a critically sampled mode, since a cyclic extension is not applicable.

Referring to the schematic of a generic single-input/single-output MCM system in Figure 3.10, a high-speed input stream $x(n)$ is first demultiplexed into N data streams and individually modulated using a specified signal constellation, resulting in the modulated data streams, $d^{(k)}(n)$, $k = 0, \ldots, N - 1$. The subcarriers are then up-sampled by N before being filtered by the synthesis filters $g^{(k)}(n)$, $k = 0, \ldots, N - 1$ and summed together, forming $s(n)$. At the receiver, the received composite signal, $r(n)$, is separated into the N subcarriers using the analysis filter bank, then the subcarriers are down-sampled by a factor N and equalized before being demodulated and multiplexed together, forming the reconstructed output signal, $\hat{b}(n)$.

Observing the subcarrier filter spectra in Figure 3.14, all of the filters overlap with the adjacent filters, which may give rise to distortion. However, if these filters

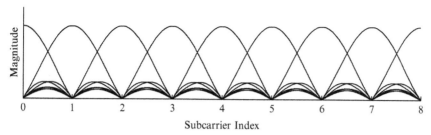

(a) OFDM subcarrier spectrum.

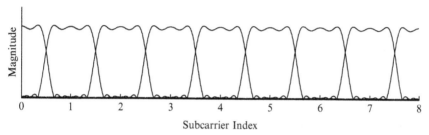

(b) FB-MC subcarrier spectrum (employing a square-root raised cosine prototype low-pass filter with a roll-off of 0.25).

FIGURE 3.14

Subcarrier spectra of $N = 8$ OFDM and FB-MC systems.

satisfy certain conditions, it is possible to have distortionless transmission between the transmitter and receiver even if the subcarriers overlap.

3.6 MULTICARRIER EQUALIZATION TECHNIQUES

From the previous sections, it is obvious that there exist a number of performance advantages when employing adaptive allocation. To improve the quality of the received signal, equalizers are employed to remove most of the distortion introduced by the channel. The design decisions involved in the implementation of an equalizer depend on a number of factors, including the type of distortion in the received signal and the implementation complexity. In this section, we describe the distortion normally found in multicarrier transmission systems and techniques to remove the distortion. Moreover, two frequency-domain equalization (FEQ) techniques are covered.

3.6.1 Interference in Multicarrier Systems

Several types of distortion are found in multicarrier signals. Pollet et al. define the three types of interference prevalent in multicarrier systems based on the

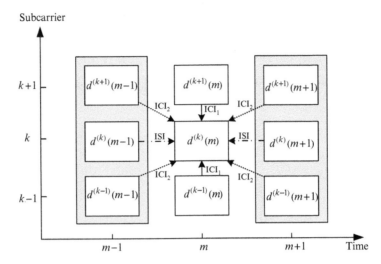

FIGURE 3.15

Different types of distortion present in multicarrier systems (from [87]).

origin of the interference [87]. A graphical relationship of these types of distortion with respect to multicarrier transmission is shown in Figure 3.15. Considering the interference on symbol m in subcarrier k, these three types are

ICI_1 This form of *intercarrier interference* (ICI) occurs when the interference contains the symbols transmitted over the other subcarriers of the mth symbol period.

ICI_2 This ICI occurs when the interference contains the symbols transmitted during periods other than the mth symbol period from subcarriers other than subcarrier k.

ISI This *intersymbol interference* (ISI) occurs when the interference contains the symbols $d^{(k)}(n)$, where $n \neq m$.

As discussed in Section 3.5.2, the cyclic prefix in OFDM systems is used to capture the effects of ISI and ICI_2. As for ICI_1, the analysis filter bank of the OFDM system transforms it into a set of complex gains affecting the subcarriers. A general discussion about intersymbol interference is provided in Section 3.7.

3.6.2 Distortion Reduction

To mitigate the effects of the distortion introduced by the channel, two techniques may be employed by the system. One technique is *channel coding*, where the data are encoded with some redundancy so as to increase the probability of correctly recovering the original data from the intercepted transmission at the receiver. Several classes of channel codes can be employed to correct for impulse errors in the transmission. Moreover, when an interleaver is employed by the encoder, channel coding can also compensate for predictable channel behavior by randomizing the errors to make them appear as burst errors.

The other technique is *channel equalization*, where the effects of the channel are inverted at the receiver. This technique is specifically designed for channels with predictable distortion behavior. In most cases, FIR filters are used at the receiver to equalize the incoming signal. One approach is to use multicarrier systems with FIR equalizers employed in each subcarrier. Due to the "divide-and-conquer" nature of MCM, where the data are transmitted in several subcarriers simultaneously, each subcarrier is affected by only a portion of the channel in terms of bandwidth, and so requires fewer taps to compensate for the distortion. For instance, if the channel is frequency selective and a single-carrier system is employed, the equalizer at the receiver requires a large number of taps to invert the channel. If a multicarrier system is employed, where N is sufficiently large, the frequency-selective fading channel is transformed into N approximately flat subchannels. Therefore, the N subcarrier equalizers may consist of nothing more than a complex gain. This complex gain compensates for the gain and phase of the channel affecting the subcarrier. This is the procedure employed in ADSL modems [82, 83, 87, 92]. The details of this type of equalizer are discussed in Section 3.6.3. Even when N is not large enough to transform the channel into flat subchannels, equalizers with lengths greater than 1 could be employed per subcarrier.

Although equalization may compensate for the channel distortion, caution must be used when dealing with the noise that is added to the received signal. Suppose that the channel frequency response at the center of subcarrier i is defined by H_i. If a single-tap equalizer is used, it is equal to $C_i = 1/H_i$. However, since there is noise in the received signal, it also gets multiplied by the equalizer. If the noise frequency response in subcarrier i is given by $V_i(\omega)$, the postequalized noise frequency response would be

$$V_i'(\omega) = \frac{V_i(\omega)}{H_i(\omega)}, \tag{3.47}$$

where $H_i(\omega)$ is the channel frequency response over subcarrier i. Since $|H_i(\omega)| < 1$ due to path loss, this means that $|V_i'(\omega)| > |V_i(\omega)|$. An illustrative example of the effect of equalization on the noise spectrum is shown in Figure 3.16 for an input signal with an initially flat spectrum. One solution is to precompensate the information for the distortion before transmission. This process is known as *preequalization* [103]. However, preequalization works only if accurate channel knowledge is available at the transmitter, thus requiring a feedback path back from the receiver.

One of the advantages that multicarrier modulation has over single-carrier modulation is that equalization can be performed on each subcarrier, as opposed to a long time-domain equalizer employed by single-carrier systems. However, several researchers have devised a number of equalization schemes for single-carrier

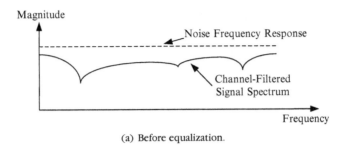

(a) Before equalization.

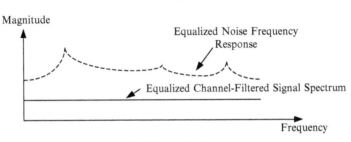

(b) After equalization.

FIGURE 3.16

The effect of equalization on the noise spectrum.

systems that possess the same complexity as a multicarrier system implementing per-subcarrier equalization. Moreover, these single-carrier systems do not suffer from the large peak-to-average ratio (PAR)[4] problem experienced by multicarrier systems.

Although single-carrier FEQ techniques may be superior to multicarrier systems with respect to PAR, single-carrier systems with FEQ cannot perform adaptive bit allocation. Moreover, although adaptive power allocation can be performed in a single-carrier system by changing the spectral pulse shape applied to the transmitted data stream, it is more difficult relative to a multicarrier system. As a result, they cannot fully exploit their advantages. For instance, Czylwik performed a comparison between an OFDM employing bit allocation and subcarrier FEQs and a single-carrier system with an FEQ. The results showed that the adaptive OFDM system outperformed the single-carrier system.

3.6.3 Optimal Single-Tap Per-Tone Equalization for OFDM Systems

Once the cyclic prefix of appropriate length has been removed, the received signal is decomposed into separate subcarriers using the DFT. Then, to equalize the gain

---

[4]*Peak-to-average ratio* is the ratio between the peak transmitted power of a signal and its average power.

of the desired signal, the subcarriers are multiplied with the inverse of the channel frequency response across each of the subcarriers.

The importance of the cyclic prefix resides in the fact that it transforms the linear convolution between the transmitted signal $s(n)$ and the channel impulse response $h(n)$ into a symbol-by-symbol circular convolution. Suppose the OFDM symbol starts at time $n = 0$. Denoting $s(0), \ldots, s(N-1)$ as the N output samples of the transmitter IDFT for the first OFDM symbol, the addition of the cyclic prefix of length K gives rise to a new signal:

$$\tilde{s}(n) = \begin{cases} s(n+N-K) & 0 \le n \le K-1 \\ s(n-K) & K \le n \le N+K-1 \end{cases}.$$

Defining $\tilde{r}(n)$ as the result of the convolution of the signal $\tilde{s}(n)$ with a channel impulse response $h(n)$ of length S, assuming $S \le K+1$, this yields

$$\tilde{r}(n) = \sum_{k=0}^{S-1} h(k)\tilde{s}(n-k)$$

$$= \begin{cases} \displaystyle\sum_{k=0}^{n-K} h(k)s(n-K-k) + \sum_{k=n-K+1}^{S-1} h(k)s(n-k+N-K), \\ \hspace{6cm} K \le n \le K+S-1 \\ \displaystyle\sum_{k=0}^{S-1} h(k)s(n-K-k), \hspace{2cm} K+S \le n \le N+K-1. \end{cases}$$

From this equation, it is observed that, after the removal of the cyclic prefix, the received sequence $r(n) = \tilde{r}(n+K)$ is

$$r(n) = \sum_{k=0}^{N-1} h(k)s\{[(n-k)]_N\} = h(n) * N\, s(n). \tag{3.48}$$

Thus, the received samples, after removal of the cyclic prefix, are made up of just the circular convolution of the sent signal (i.e., N samples per symbol) with the channel impulse response $h(n)$. If now one looks at Equation (3.48) in the frequency domain, it looks like

$$R(k) = H(k) \cdot S(k),$$

where capital letters represent N-point DFTs of the corresponding sequences. With the multiplication of the corresponding frequency samples, each of the subcarriers experiences a different complex channel "gain" $H(k)$. Therefore, what must be done is to multiply each subcarrier with a gain that is an inverse to the channel frequency response acting on that subcarrier. This is the principle behind *per-tone equalization*. Knowing what the channel frequency gains are at the different subcarriers, one can use them to reverse the distortion caused by the channel by dividing the subcarriers with them. For instance, if the system has 64 subcarriers centered at frequencies $\omega_k = 2\pi k/64$, $k = 0, \ldots, 63$, then one would take the

CIR $h(n)$ and take its 64-point FFT, resulting with the frequency response $H(k)$, $k = 0, \ldots, 63$. Then, to reverse the effect of the channel on each subcarrier, one would simply take the inverse of the channel frequency response point corresponding to that subcarrier,

$$W(k) = \frac{1}{H(k)}, \qquad (3.49)$$

and multiply the subcarrier with it.

3.6.4 Frequency-Domain Equalizers for Multicarrier Systems

Although the single-tap per-tone equalizer for OFDM has the advantage of being simple and is optimal with respect to compensating for channel distortion, it has a few drawbacks:

- The length of the cyclic prefix must be sufficiently long to capture the effects of ISI and ICI$_2$. If the channel impulse response is long, the cyclic prefix constitutes a greater percentage of the OFDM symbol.
- This type of equalization works well only with OFDM.

Therefore, other implementations should be considered when conditions for the single-tap equalizer are unfavorable. One solution is to employ multitap equalizers on each subcarrier. Since each subcarrier operates across a smaller portion of the channel in the frequency domain, the equalizer design may be less complex.

3.7 INTERSYMBOL INTERFERENCE

When transmissions are sent across a dispersive channel, it is possible for the output of that channel intercepted by the receiver to be distorted via the temporal spreading and resulting overlap of the individual symbol pulses. One consequence of such temporal spreading is the resulting inability of the receiver to accurately distinguish between different received pulse shapes. We refer to this phenomenon as *intersymbol interference*.

Suppose we represent the data transmission channel as the block diagram shown in Figure 3.17, where we have a binary information source followed by both

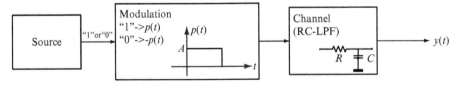

FIGURE 3.17

Simplified block diagram for a wireless channel.

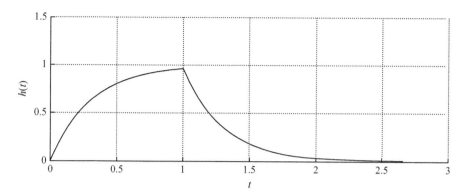

FIGURE 3.18

RC-LPF filter response of the modulation pulse shape when $T = 1, A = 1$.

modulation and low-pass filtering blocks. Depending on the binary information, a train of different pulse shapes are generated by the modulation block, which is then used as the input to the low-pass filtering block. The low-pass filtering block possesses the following impulse response:

$$ h(t) = p(t) * h(t)_{\text{LPF}} = \begin{cases} A(1 - e^{-t/RC}) & 0 \leq t \leq T \\ h(T)e^{-(t-T)} & t \geq T, \end{cases} \tag{3.50} $$

which smoothes out the incoming train of pulse shapes, causing some of the pulse shapes to spread into the subsequent pulse shapes and yielding ISI. For example, the combined impulse response of the modulation pulse shape is shown in Figure 3.18.

Thus, referring to Figure 3.17, to accurately model the effects of ISI in a data transmission across a dispersive channel, one must do the following:

1. Send pulses one after another.
2. Apply the principle of superposition such that $y(t) = \sum_{n=0}^{\infty} I_n h[t - nT)]$ in order to get $y(\tau)$.

For example, after sending a 1 followed by a 0, we get the response shown in Figure 3.19. Notice that we have a destructive interference at $t = 2T$. Similarly, the superposition of pulse shapes can also yield constructive interference as well. Therefore, we can define a generic expression for the ISI at the lth time period as

$$ y_l = \sum_{n=-\infty}^{\infty} I_n h_{l-n} = I_n * h_n, \tag{3.51} $$

where $h_{l-n} = h[(l-n)T]$, $I_l h_0$ is the term for detection, while the rest of the terms are considered to be interference.

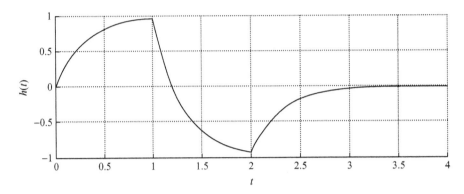

FIGURE 3.19

RC-LPF filter response of combined modulation pulses when $T = 1, A = 1$.

3.7.1 Peak Interference/Peak Distortion

Assuming zero ISI prior to the N_1 and after the N_2 time periods, the ISI level between N_1 and N_2 is equal to

$$\text{ISI term} = \text{Sum of distortion terms}$$

$$= \sum_{k=-N_2}^{N_1} h_{-k}I_k. \tag{3.52}$$

Consequently, we can define the *peak distortion D_p*, which is considered to be the worst-case ISI, as the following:

$$D_p = \sum_{k=-N_2, k\neq 0}^{N_1} |h_{-k}|, \tag{3.53}$$

which is the summation of all the absolute values of impulse response of other pulses.

3.7.2 Chernoff Bound

When $I_0 = 1$, the received signal is $z = h_0 + D + n$, where $D = \sum_{k=-N_1}^{N_2} I_k h_{-k}$. Therefore, $P_e = P(z < 0) = P(D + n > h_0)$ can be rewritten into an exponential form:

$$P(e^{s(D+n)} > e^{sh_0}) \tag{3.54}$$

since e^x is a monotonic increasing function. Using the Chebyshev inequality, namely,

$$P(x \geq t) \leq \frac{E\{x\}}{t} \quad \text{for } x \geq 0, t \geq 0, \tag{3.55}$$

and setting $x = e^{s(D+n)}$ and $t = e^{sb_0}$ while assuming D and n are independent, we get the following bound:

$$P_e \leq \frac{E\{e^{sD}e^{sn}\}}{e^{sb_0}} = e^{-sb_0}E_D\{e^{sD}\}E_n\{e^{sn}\}. \tag{3.56}$$

We know that $E_D\{e^{sD}\}$ is the characteristic function of the moment-generating function of the random variable D, and $E_n\{e^{sn}\}$ is the characteristic function of the moment-generating function of the random variable n. As a result, we can represent them as, respectively,

$$E_D\{e^{sD}\} = \Phi_D(s)$$

$$E_n\{e^{sn}\} = \Phi_n(s) = e^{\frac{s^2\sigma^2}{2}}. \tag{3.57}$$

Setting $D = I_1b_1$, where $I_1 = \pm 1$, we obtain the following expressions:

$$E\{e^{sD}\} = \frac{e^{sb_1} + e^{-sb_1}}{2} = I_D(s) = \cosh(sb_1) \tag{3.58}$$

and

$$P_e \leq e^{-sb_0}e^{s\sigma^2/2}\cosh(sb_1), \tag{3.59}$$

where the best s can be found via trial and error. Usually, a good starting point is $s = b_0 - D_{\text{peak}}/\sigma$, where

$$D_{\text{peak}} = \sum_{k=-N_1}^{N_2} |b_{-k}|. \tag{3.60}$$

In general, the Chernoff bound is equal to

$$D = \sum_{k=-N_1}^{N_2} I_k b_{-k}, \Phi_D(s) = \prod_{k=-N_1}^{N_2} \cosh(sb_{-k}). \tag{3.61}$$

3.8 PULSE SHAPING

Pulse shaping can be employed to help counteract the effects of ISI, where the transmit waveform can be bandlimited while simultaneously avoiding the introduction of ISI. This is shown in Figure 3.20, where the binary information source generates a stream of 1 and 0 values that are used as inputs to an impulse modulation block. The output of this block yields an impulse train of which the values are defined by the information source. This modulated impulse train is then used as

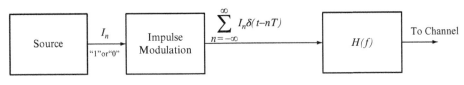

FIGURE 3.20

Basic system model.

an input to a transmit pulse shaping filter $H(f)$. By definition, given an impulse as the input to the system, the output is the impulse response of the system. It then follows that, for a train of impulses used as inputs to the system $H(f)$, the output is the time-shifted copies of the impulse response of that system superimposed upon each other. This output is then transmitted across the channel. Consequently, if the choice of pulse shaping filter $H(f)$ is done correctly, it is possible to minimize the effect of the ISI resulting from the dispersive channel.

3.8.1 Nyquist Pulse Shaping Theory

We know that the inverse Fourier transform of a rectangular signal is a sinc pulse, where

$$\text{sinc}(x) = \frac{\sin(\pi x)}{\pi x}. \tag{3.62}$$

This is shown in Figure 3.21. Suppose we let $w = 1/2T$, then $h(t) = A/T\text{sinc}(t/T)$. Since by choosing $w = 1/T$, our sampling instants are located at the equally spaced zero crossings, there will be no ISI. This gives us the *Nyquist time-domain condition for zero ISI*, which is mathematically defined as

$$b_n = b(nT) = C\delta_{n_0} = \begin{cases} C, & n = 0 \\ 0, & n \neq 0. \end{cases} \tag{3.63}$$

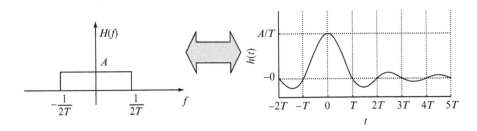

FIGURE 3.21

Inverse Fourier transform of a rectangular signal is a sinc wave.

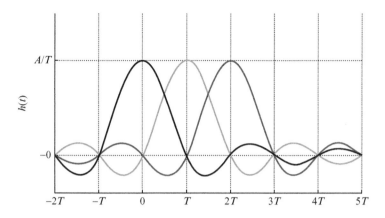

FIGURE 3.22

Superposition of waveforms and the absence of ISI.

As a result, we have the following for $n(t) = 0$ and $C(f) = H(f)$:

$$r(t) = \sum_{n=-\infty}^{\infty} I_n h(t - nT) = \frac{A}{T} \sum_{n=-\infty}^{\infty} I_n \, \text{sinc}(\frac{t}{T} - n) \qquad (3.64)$$

where $h(t) = A/T\text{sinc}(t/T)$ and $h(t - nT) = A/T\text{sinc}[(t - nT)/T] = A/T\text{sinc}[(t/T) - n]$. Consequently, when periodically sampling at $t = kT$, we get the expression to be equal to

$$r(kT) = \frac{A}{T} \sum_{n=-\infty}^{\infty} I_n \, \text{sinc}(k - n) = \frac{A}{T} I_k, \qquad (3.65)$$

which possesses no ISI at the sampling instants, as shown in Figure 3.22. This choice of $h(t)$, which is called a *Nyquist-I pulse*, gives $W = W_{\min} = 1/2T$. However, in practice, it is very difficult to realize a Nyquist-I pulse, since it is highly sensitive to timing errors and $h(t) \sim 1/t$ for large t.

From the derivation of the peak distortion, D_p, we can obtain an expression for the ISI when Nyquist-I pulses are employed:

$$D_p = \frac{A}{T} \sum_{n=-\infty, n \neq 0}^{\infty} |\text{sinc}(\epsilon' + k - n)|, \qquad (3.66)$$

where $\epsilon' = t/T$. Thus, for any $\epsilon' > 0$ and $|n| > N$, we can approximate $|\text{sinc}(\epsilon' + k - n)|$ by $1/|k - n|$. However, as $\sum 1/n \to \infty$, for any $\epsilon' > 0$, $D_p \to \infty$ since $\text{sinc} \approx 1/t$. Hence, Nyquist-I pulse is not practical. On the other hand, as we see next, there also exists a *Nyquist-II pulse*, where $W > W_{\min} = 1/2T$ possesses no ISI sensitivity issues.

Raised Cosine Pulse

The *raised cosine pulse* is one type of Nyquist-II pulse. It possesses a transfer function given by

$$H_{RC}(f) = \begin{cases} T, & 0 \leq |f| \leq \frac{1-\beta}{2T} \\ \frac{T}{2}\left\{1 + \cos\left[\frac{\pi T}{\beta}\left(|f| - \frac{1-\beta}{2T}\right)\right]\right\}, & \frac{1-\beta}{2T} \leq |f| \leq \frac{1+\beta}{2T} \\ 0, & |f| \geq \frac{1+\beta}{2T} \end{cases} \quad (3.67)$$

where β is called the *roll-off factor*, which takes values between 0 to 1, and $\beta/2T$ is called the *excess bandwidth*. Figure 3.23 illustrates the raised cosine spectral characteristics.

Note that, with the raised cosine pulse, $W > 1/(2T)$. Taking the inverse Fourier transform of $H_{RC}(f)$, the impulse response of a raised cosine pulse is defined by

$$h_{RC}(t) = \frac{\cos(\pi\beta t/T)\,\text{sinc}(\pi t/T)}{1 - (2\beta t/T)^2}. \quad (3.68)$$

As opposed to Nyquist-I pulses, Nyquist-II pulses such as the raised cosine pulse have no ISI sensitivity issue since their peak distortion, the tail of $h(t)$, converges relatively quickly. The peak distortion of a digital communication system employing raised cosine pulses is defined by

$$D_p = \sum_{n=-\infty}^{\infty} |h[\epsilon' + (n - k)]| \sim \frac{1}{n^3}. \quad (3.69)$$

Therefore, whenever a timing error occurs, the distortion will not accumulate to infinity as with Nyquist-I pulses.

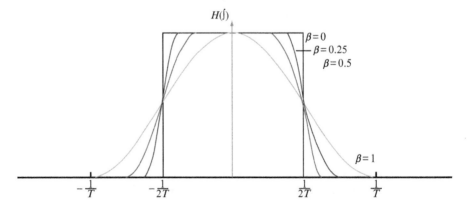

FIGURE 3.23

Transfer function of raised cosine pulse.

3.8.2 Nyquist Frequency-Domain No ISI Criterion

Recall that the *Nyquist time-domain no ISI criterion* is given by

$$b_n = b(nT) = C\delta_{n_0} = \begin{cases} C, & n = 0 \\ 0, & n \neq 0 \end{cases}. \tag{3.70}$$

However, for a bandwidth-limited pulse, $H(f) \approx 0$ for $|f| > W$. As a result, the impulse response is given by

$$b(t) = \int_{-W}^{W} H(f)e^{j2\pi ft}\,df. \tag{3.71}$$

Thus, at sample times $t = nT$, we can write this expression as

$$b_n(t) = b(nT) = \int_{-W}^{W} H(f)e^{j2\pi f_n t}\,df. \tag{3.72}$$

If we let $f = \lambda + k/T$, then we get

$$
\begin{aligned}
b_n &= \sum_{k=-N}^{N} \int_{-\frac{1}{2T}}^{\frac{1}{2T}} H\left(\lambda + \frac{k}{T}\right) e^{j2\pi\lambda_n t}\,d\lambda \\
&= \int_{-\frac{1}{2T}}^{\frac{1}{2T}} H_{\text{eq}}(\lambda)e^{j2\pi\lambda_n T}\,d\lambda,
\end{aligned}
\tag{3.73}
$$

where $H_{\text{eq}}(\lambda) = \sum_{k=-N}^{N} H(\lambda + \frac{n}{T}), |\lambda| \leq 1/2T$, is the folded spectrum of $H(\lambda)$. As a result, this forms our *Nyquist frequency-domain no ISI criterion*:

$$
\begin{aligned}
H_{\text{eq}}(f) &= C, \; |f| \leq \frac{1}{2T} \\
H_{\text{eq}}(f) &= \sum_{k=-N}^{N} H(f + k/T), \text{ for } |f| \leq \frac{1}{2T}.
\end{aligned}
\tag{3.74}
$$

Notice how, by applying the definition for the Nyquist frequency-domain no ISI criterion, we obtain no ISI for

$$
\begin{aligned}
b_n &= \int_{-\frac{1}{2T}}^{\frac{1}{2T}} C\,e^{j2\pi f_n T}\,df = \frac{C}{T}\,\text{sinc}\left(\frac{t}{T}\right) \\
&= \frac{C}{T}\,\text{sinc}(n),
\end{aligned}
\tag{3.75}
$$

which satisfies zero ISI at the sampling instants $t = nT$.

3.9 CHAPTER SUMMARY AND FURTHER READINGS

In this chapter, we describe several fundamental concepts in digital communication systems that are pertinent to cognitive radio and dynamic spectrum access. In particular, we discuss signal representations and introduce digital modulations such as M-QAM and M-PSK. The probability of bit error is introduced, and bounds for this quantity are obtained for some basic digital modulations. Subsequently, the basic mathematical theory of multicarrier modulation is described and a number of existing implementations of these schemes reviewed. Finally, we discuss inter-symbol interference, which is a critical problem in multicarrier modulations, and describe a number of approaches to mitigate this problem.

Although several fundamentals of digital communication systems are presented in this chapter, this presentation of the material is to serve as a "quick-start guide." The interested reader is strongly encouraged to study several of the excellent digital communication textbooks currently available [71, 104–108].

3.10 PROBLEMS

1. The Gaussian random variable is used extensively in digital communications with respect to the analysis of various transceiver systems. In this problem, we examine how we can generate a random signal that possesses a Gaussian distribution.

    (a) Using the `randn` function in MATLAB, generate 100,000 samples of Gaussian random variable that possesses a mean of 0.5 and a standard deviation of 2. Plot the distribution of the resulting samples.

    (b) Using the `rand` function in MATLAB, generate 100,000 samples that possess a Gaussian distribution by using the central limit theorem; that is, generate 10 sets of 100,000 samples from 10 independently and identically distributed uniform random variables. Choose the mean and variance of these uniform random variables such that the result Gaussian distribution possesses the same characteristics as in part (a). Plot the resulting distribution.

2. Find and compare the power efficiency for the three binary signal sets that follow.

    (a) $s_1(t) = B\cos(\omega_0 t + \phi)$ and $s_2(t) = B\cos(\omega_0 t - \phi)$ for $0 \leq t \leq T$ and where $\bar{E}_b \leq A_0^2 T/2$, find the best (B, ϕ).

    (b) $s_1(t) = A\cos(\omega_0 t + \theta)$ and $s_2(t) = B\cos(\omega_0 t)$ for $0 \leq t \leq T$ and where $\bar{E}_b \leq A_0^2 T/2$ and A_0 is known, find the best (A, B, θ).

    (c) $s_1(t) = A\cos(2\pi(f_0 + \Delta)t + \theta)$ and $s_2(t) = A\cos(2\pi(f_0 - \Delta)t - \theta)$ for $0 \leq t \leq T$, find the best peak frequency deviation, Δ, in terms of T.

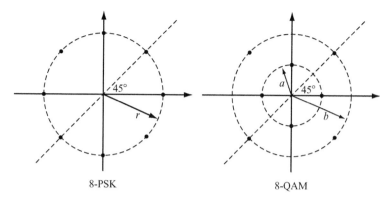

FIGURE 3.24

Two octal signal constellations: 8-PSK and 8-QAM.

3   Suppose a primary user is using 4-QAM with $d_{min}^2 = 0.1$ J and a center frequency at f. The noise spectral density N_0 is measured to be 1 mW/Hz. If a secondary user comes in and the combined spectral density of noise and the secondary user at f increases to 3 mW/Hz, what is the increase in upper bound of the probability of error of the primary user in terms of decibels?

4   (a)   Plot the capacity of an additive white Gaussian noise channel with a bandwidth of $W = 3000$ Hz as a function of P/N_0 for values of P/N_0 between -20 dB and 30 dB.
    (b)   Plot the capacity of an additive white Gaussian noise channel with $P/N_0 = 25$ dB as a function of W. In particular, what is the channel capacity when W increases indefinitely?

5   Consider the octal signal point constellations in Figure 3.24.
    (a)   The nearest-neighbor signal points in the 8-QAM signal constellation are separated in distance by A units. Determine the radii a and b of the inner and outer circles.
    (b)   The adjacent signal points in the 8-PSK are separated by a distance of A units. Determine the radius r of the circle.
    (c)   Determine the average transmitter powers for the two signal constellations and compare the two powers. What is the relative power advantage of our constellation over the other? (Assume that all signal points are equally probable.)

6   The 16-point constellation adopted in the V.29 standard for 9600 b/s voice-band modems is shown in Figure 3.25.
    (a)   Draw decision zones for the AWGN channel. Hint: They are not as simple as in the rectangular 16-QAM design.
    (b)   Calculate the average energy of the signal set in terms of a, noting that each signal point does not have the same set of nearest neighbors.

Determine whether this constellation is more energy efficient than the standard square 16-QAM design.

(c) One advantage of this constellation is a smaller degradation under carrier phase error in the demodulation, which causes a rotation of signal space relative to the decision boundaries. Discuss qualitatively the effect for the V.29 and standard 16-point designs.

7 Recall from Problem 6 the 16-point constellation shown in Figure 3.25, which was adopted in the V.29 standard for 9600 b/s voiceband modems. Determine the upper and lower bounds on the probability of symbol error, P_s. Note that each signal point does not have the same set of nearest neighbors.

8 A computer transmits binary data at a rate of 56 kb/s using a baseband binary PAM system employing a raised cosine spectrum.

(a) Determine the theoretical transmission bandwidth required for each of the following roll-off factors: $\alpha = 0.25, 0.3, 0.5, 0.75, 1.0$.

(b) Using MATLAB, plot the baseband waveform of the binary data stream 101100 for each of the roll-off factors $\alpha = 0.25, 0.3, 0.5, 0.75, 1.0$. Hint: The baseband binary PAM system maps 1 to a signal amplitude of 1 and 0 to a signal amplitude of -1.

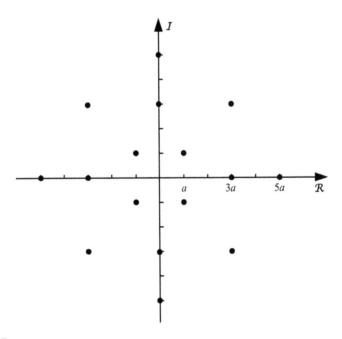

FIGURE 3.25

16-point constellation of a V.29 standard 9600 b/s voiceband modem.

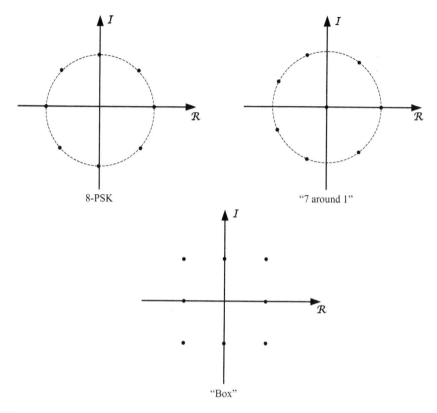

FIGURE 3.26

Three eight-point signal constellations (8-PSK, "7 around 1," and "Box").

9   An analog signal is sampled, quantized, and encoded into a binary PCM wave. The number of representation levels used is 128. A synchronizing pulse is added at the end of each code word representing a sample of the analog signal. The resulting PCM wave is transmitted over a channel of bandwidth 12 kHz using a quaternary PAM system with raised cosine spectrum. The roll-off factor is unity.
   (a)   Find the rate (b/s) at which information is transmitted through the channel.
   (b)   Find the rate at which the analog signal is sampled. What is the maximum possible value for the highest frequency component of the analog signal?

10   Three eight-point constellations are proposed as shown in Figure 3.26. Draw the appropriate decision boundaries for each technique in two dimensions, and express the probability of symbol error, P_s, in terms of the peak energy-to-noise density ratio. Repeat for an average energy normalization. Which of the demodulators would be easiest to implement?

Spectrum sensing and identification

4

Qing Zhao[1] and Ananthram Swami[2]
[1] University of California, Davis, United States
[2] Army Research Laboratory, United States

4.1 INTRODUCTION

Multiple measurement campaigns reveal that much of the licensed spectrum remains unused—both in time and in frequency: traffic in wireless networks tends to be bursty. Hence, efficient exploitation of the spectrum requires the ability to exploit instantaneous opportunities at a rather fine time scale [109]. For cognitive networks to operate efficiently, secondary users should be able to exploit radio spectrum that is unused by the primary network. A critical component of cognitive networking is thus *spectrum sensing*. The secondary user (SU) should sense the spectrum efficiently, quickly seize opportunities to transmit, and vacate the spectrum should a primary user (PU) reoccupy the spectrum. As noted in [110], a critical component of opportunistic spectrum allocation is the design of the spectrum sensor for opportunity detection.

In this chapter, we discuss how to detect spectrum opportunities by detecting primary signals and highlight the difference between these two. Fundamental trade-offs in spectrum sensing—throughput performance versus interference constraint and sensing accuracy versus sensing overhead—are discussed in detail.

This chapter is organized as follows. In Section 4.2, we present different types of detectors for primary signal detection. In Section 4.3, we consider the detection of spectrum opportunities based on the detection of primary signals. We highlight their differences by establishing that perfect detection of primary signals does not necessarily lead to perfect detection of spectrum opportunities. In Section 4.4, we discuss the trade-off between performance and interference constraint. We show that the optimal operating characteristic (probability of false alarm versus probability of miss detection) of the spectrum opportunity detector should be designed based on MAC layer performance in terms of throughput and probability of colliding with primary users. In Section 4.5, we discuss the trade-off between sensing accuracy and sensing overhead. A brief summary and further readings are given in Section 4.6, followed by several problems.

4.2 PRIMARY SIGNAL DETECTION

In this section, we discuss the detection of primary signals, which, while not equivalent to the detection of spectrum opportunities, constitutes a basic step in spectrum opportunity detection as shown in Section 4.3. In later sections, we discuss how primary signal detection can be translated to the problem of spectrum opportunity detection.

The spectrum sensor essentially performs a binary hypothesis test on whether or not there are primary signals in a particular channel.[1] The channel is idle under the null hypothesis and busy under the alternate:

$$\mathcal{H}_0 \text{ (idle)} \quad \text{vs.} \quad \mathcal{H}_1 \text{ (busy)} \tag{4.1}$$

Under the idle scenario, the received signal is essentially the ambient noise in the radio frequency (RF) environment, and under the busy scenario, the received signal would consist of the PU's signal and the ambient noise; thus,

$$\mathcal{H}_0: \quad y(k) = w(k)$$

$$\mathcal{H}_1: \quad y(k) = s(k) + w(k)$$

for $k = 1, \ldots, n$, where n is the number of received samples, $w(k)$ represents ambient noise, and $s(k)$ represents the PU signal. It seems natural that the received signal will have more energy when the channel is busy than when it is idle; this is the underlying concept in the energy detector, which we discuss in detail in Section 4.2.1. When aspects of the signal structure are known, one can exploit the structure; a special case leads to the cyclostationary detector discussed in Section 4.2.2. When the PU's signal $s(n)$ is fully known, one can use the matched filter; see Section 4.2.3.

Regardless of the precise signal model or detector used, sensing errors are inevitable due to additive noise, limited observations, and the inherent randomness of the observed data. False alarms (or Type I errors) occur if an idle channel is detected as busy, and missed detections occur when a busy channel is detected as idle. A false alarm may lead to a potentially wasted opportunity for the SU to transmit. A missed detection (or a Type II error), on the other hand, could potentially lead to a collision with the PU, leading to wasted transmissions for both PU and SU. A clear tension lies between the need to protect the primary licensed

---

[1]Here we use channel in a general sense: It represents a signal dimension (time, frequency, code, etc.) that can be allocated to a particular user.

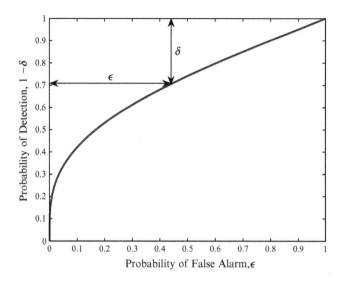

FIGURE 4.1

Typical ROC.

user and to provide service to the secondary user. In this chapter, we represent the protection afforded to a PU by placing a cap on the allowed probability of missed detection. The optimization problem then is to minimize the probability of false alarm, subject to a constraint on the probability of missed detection. This is a slight twist in the classical formulation of the binary Neyman-Pearson test. The performance of a detector is characterized by two parameters, the probability of missed detection (P_{MD}) and the probability of false alarm (P_{FA}), which are defined as

$$\epsilon = P_{FA} = \text{Prob} \{\text{Decide } \mathcal{H}_1 | \mathcal{H}_0\};$$

$$\delta = P_{MD} = \text{Prob} \{\text{Decide } \mathcal{H}_0 | \mathcal{H}_1\} .$$

A typical receiver operating characteristic (ROC), which is a plot of $1 - \delta$, the probability of detection (P_D), versus P_{FA}, is shown in Figure 4.1.

Choosing different sensors, detection algorithms, or sensing parameters (such as an observation window) leads to different ROCs. The choice of operating point, however, should be dictated by the MAC layer performance in terms of the throughput of the SU and the interference constraint for protecting the PU (see Section 4.4).

To motivate the difficulty of the problem, the sensing requirements in the 802.22 Draft are summarized in Table 4.1 [111, 112]. Note that the signal-to-noise ratio (SNR) is very low in the operating regime. We next consider several detectors for detecting primary signals.

Table 4.1 Sensing Requirements in the 802.22 Draft Standard

Parameter	Digital TV	Wireless Microphone (Part 74)
Channel detection time	≤2 sec	≤2 sec
Channel move time	2 sec	2 sec
Detection threshold (required sensitivity)	−116 dBM (over 6 MHz)	−107 dBm (over 200 KHz)
Probability of detection	0.9	0.9
Probability of false alarm	0.1	0.1
SNR	−21 db	−12 dB

4.2.1 Energy Detector

In many cases, the signaling scheme of the PU may be unknown to the SU; this may correspond to the case where an agile PU has considerable flexibility and agility in choosing its modulation and pulse shaping. In such a case the signal can be modeled as a zero-mean stationary white Gaussian process, independent of the observation noise, which is also modeled as white Gaussian. In this setting, the energy detector is optimal in the Neyman-Pearson sense [113]. The spectrum sensing problem is now one of distinguishing between two mutually independent and identical Gaussian sequences:

$$\mathcal{H}_0: \quad y(k) = w(k)$$
$$\mathcal{H}_1: \quad y(k) = s(k) + w(k)$$

for $k = 1, ..., n$. Here $w(k)$ and $s(k)$ are zero-mean complex Gaussian random variables with variances σ_w^2 and σ_s^2 per dimension. Let $\mathbf{y} = [y(1), ..., y(n)]'$ denote the vector of the n observed samples. We find it convenient to denote $\sigma_0^2 = \sigma_w^2$, and $\sigma_1^2 = \sigma_s^2 + \sigma_w^2$. The Neyman-Pearson detector is a threshold detector on the likelihood ratio or equivalently the log-likelihood ratio (LLR):

$$\text{LLR} = \log\left(\frac{p(\mathbf{y}|\mathcal{H}_1)}{p(\mathbf{y}|\mathcal{H}_0)}\right) \gtrless \tau',$$

where τ' is a suitably chosen threshold. Given the independent and identical assumption, the detector is easily seen to be equivalent to deciding $\mathcal{H}_1$ if

$$z = \frac{1}{2n\sigma_0^2} \sum_{k=1}^{n} |y(k)|^2 > \tau. \tag{4.2}$$

This is simply an energy detector (ED). The statistic z is a scaled version of a standard χ^2 random variable with $2n$ degrees of freedom. We next evaluate the ROC for the ED.

If x_i are independent real Gaussian variables with zero means and unit variances, then

$$x = \sum_{i=1}^{2n} x_i^2$$

is χ^2 distributed with $2n$ degrees of freedom and probability density function (pdf),

$$p(x) = \frac{1}{2^n (n-1)!} x^{n-1} \exp(-\frac{x}{2}).$$

Thus, the tail probability $\Pr(x > \tau)$ can be computed via integration by parts as

$$\Pr(x > \tau) = \int_\tau^\infty \frac{1}{2^n (n-1)!} x^{n-1} \exp(-\frac{x}{2}) \, dx$$

$$= e^{-\frac{\tau}{2}} \sum_{k=0}^{n-1} \frac{1}{k!} \left(\frac{\tau}{2}\right)^k$$

$$= \Gamma_u(\tau/2, n)$$

where $\Gamma_u(\cdot, \cdot)$ is the upper incomplete gamma function defined by

$$\Gamma_u(a, n) = \frac{1}{\Gamma(n)} \int_a^\infty x^{n-1} e^{-x} dx \qquad (4.3)$$

Now the test statistic $z \times 2n$ has the same pdf as a χ^2 variable with $2n$ degrees of freedom. Hence, P_{FA} and P_{MD} for the energy detector can be obtained as

$$\epsilon = P_{FA} = \Gamma_u(n\tau, n) \qquad (4.4)$$

$$\delta = P_{MD} = 1 - \Gamma_u\left(\frac{n\tau}{1 + \text{snr}}, n\right) \qquad (4.5)$$

where $\text{snr} = \sigma_s^2 / \sigma_w^2$.

Figure 4.2 shows the performance of the energy detector at $\text{snr} = -21$ dB, for $N = 5 \times 10^4$, 10^5, and 2×10^5 samples. As expected, performance improves monotonically with increasing N.

See Problems 1 and 2 for a discussion on computing the threshold τ in Equation (4.4). See Problems 3 and 4 for a discussion of how the energy detector just discussed should be modified if the signal $s(n)$ is correlated and the covariance matrix is known. One disadvantage of the energy detector is that at low SNR the number of samples required to achieve specified performance metrics (P_D, P_{FA}) is proportional to $1/\text{snr}^2$; see Problem 2.

The Nonzero-Mean Case

In the 802.22 standard, the primary signal contains known synchronization sequences. Often these sequences are repeated to facilitate detection. When the

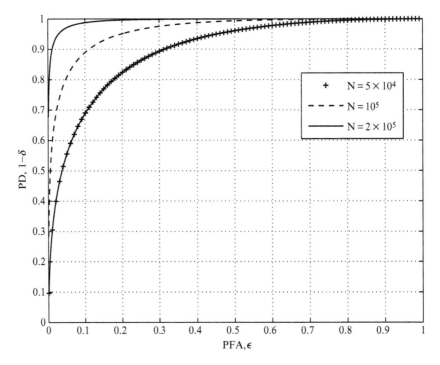

FIGURE 4.2

ROC of energy detector for various values of N, SNR $= -21$ dB.

SU is frame synchronized, the known sync sequences can be used in a matched filter as discussed later in Section 4.2.3. How does the energy detector perform in the case where frame synchronization has not been achieved? To this end, we now consider the performance of the ED, treating the signal component $s(k)$ as the mean of the received signal under $\mathcal{H}_1$. Under $\mathcal{H}_1$, $y(k) = s(k) + w(k)$ is Gaussian with mean $s(k)$ and variance σ_w^2. The P_{FA} for the ED is given by Equation (4.4) as before. Under $\mathcal{H}_1$, z is noncentral χ^2 distributed with $2n$ degrees of freedom. Define the noncentrality parameter:

$$\mu^2 = \frac{1}{2n\sigma_w^2} \sum_{i=1}^{n} |s(i)|^2. \tag{4.6}$$

Then, $P_D = $ Prob $(z > \tau | \mathcal{H}_1)$ is given by the tail probability of this distribution, which has been expressed as [114; 108, Chapter 2.1.4; 113, Section 2.2.3]

$$P_D = Q_n(\mu, \sqrt{2n\tau}), \tag{4.7}$$

where $Q_n(a, b)$ is the generalized Marcum Q function,

$$Q_n(a, b) = \frac{1}{a^{n-1}} \int_b^\infty x^n e^{-\frac{x^2+a^2}{2}} I_{n-1}(ax) \, dx,$$

where I_n is the modified Bessel function of the first kind, of order n. This function can be evaluated in MATLAB via `marcumq.m`.

The mean and variance of z are given by $1 + \mu^2$ and $(1 + 2\mu^2)/n$. For a large n, the central limit theorem can be invoked and the ROC can be found via the Gaussian approximation. See Problem 1. It is important to note that the signal $s(n)$ is not assumed known in implementing the ED; further, evaluation of the performance of the ED requires knowledge only of the SNR parameter μ. This corresponds to a special case of the Bayesian linear model discussed in [113, Section 5.4].

Energy Detection under Fading

Let us next consider a block fading environment. Under $\mathcal{H}_1$, the received signal is expressed as $y(k) = hs(k) + w(k)$, where the random variable h, assumed fixed over the block of n samples, represents fading. In such an environment, the SNR μ is a random variable. In the typical case of Rayleigh fading, μ is exponentially distributed with parameter $\bar{\mu}$, the average SNR. The P_{FA} expression remains unchanged, since there is no fading under $\mathcal{H}_0$. Under $\mathcal{H}_1$, the average probability of detection can be computed from

$$E\{P_D\} = \int f_\mu(\mu) Q_n(\mu, \sqrt{2n\tau}) \, d\mu, \tag{4.8}$$

where f_μ is the pdf of the SNR. For the case of Rayleigh fading, a closed-form expression is given in [114].

Although we discussed a time-domain implementation, the ED can also be conveniently implemented in the frequency domain. An important advantage of the ED is the ease of implementation. The SNR must be known to set the ED threshold and characterize its performance.[2] Significant weaknesses are its lack of robustness when the variances are unknown or time varying. When the noise variance is known only within a range, say (a, b), the threshold must be conservatively set to protect the PU. This means that P_{MD} is dictated by the higher value b and P_{FA} by the lower value a. As we explore in Problem 5, at low SNR, the uncertainty lowers the effective SNR from snr to snr $- (b - a)$. At low SNR, this means that performance degradation is dramatic when the uncertainty range is comparable to the SNR. This phenomenon has been called the *SNR wall* [115]. The robustness issue is explored in Problem 5. In the context of weak signal detection, the ED can be outperformed by higher-order detectors, particularly when the uncertainty range $(b - a)$ is large [116]. Another weakness is that, at low SNR, the number of samples

---

[2]In the typical detection scenario, a constraint on P_{FA} is imposed and hence only the noise variance needs to be known to set the threshold. In the PU/SU scenario, P_{MD} is specified; hence the SNR must be known to set the threshold.

required increases as $1/\text{snr}^2$, as we have seen in Problem 2. Yet another issue is that the ED cannot differentiate between signals from primary and secondary users.

These motivate us to consider detectors that can exploit additional signal features.

4.2.2 Cyclostationary Feature Detector

Often quite a bit about the PU's signal structure is known. For example, the data rates, the modulation type, the carrier frequency, and location of guard bands may be known. Digitally modulated signals have periodic features that may be implicit or explicit. The carrier frequency and symbol rate can easily be estimated via square-law devices. In some standards, the PU network uses a pilot tone frequency that can be exploited by the SU. The use of a cyclic prefix also leads to periodic signal structures. The means and correlation sequences of such signals exhibit periodicity and are, hence, called *cyclostationary signals* [117, 118]. The test statistic in a cyclic detector is

$$S(f; \tau) = \frac{1}{N} \sum_{n=1}^{N} y(n) y(n + \tau) e^{-j2\pi fn}.$$

If the received signal $y(n)$ can be written as $y(n) = \sum_k s_k(n) \exp(j2\pi f_k n) + w(n)$, where $s_k(n)$ are mutually independent zero-mean wide-stationary processes, independent of the circularly symmetric white noise sequence $w(n)$, then for a large N

$$S(f; \tau) \approx \sum_k R_k(\tau) e^{j2\pi f_k \tau} \delta(f - 2f_k),$$

where $R_k(\tau) = E\{s_k(t)s_k(t + \tau)\}$. This detector is easily implemented via fast Fourier transforms (FFTs). Knowledge of the noise variance is not required to set the detection threshold; hence the detector does not suffer from the "SNR wall" problem of the energy detector. However, the performance of the detector degrades in the presence of timing and frequency jitters (which smear out the spectral lines) and RF nonlinearities (which induce spurious peaks). Representative papers that consider the approach are [119–121].

4.2.3 Matched Filter

Often the pilot or sync sequences used in the primary network are known to the SU. For example, the WRAN 802.22 standard specifies these sequences. Let $s(n)$, $n = 1, ..., N$, denote the known pilot sequence. Assuming perfect synchronization, the received signal at the SU can be written as $y(n) = hs(n) + w(n)$, where $w(n)$ is additive white Gaussian noise (AWGN) and h represents an unknown channel gain. For this AWGN setting, the optimal detector is the matched filter [113]. The test statistic is

$$z = \frac{1}{N} \sum_{n=1}^{N} y(n)s^*(n).$$

Using the results of Problem 1, it can be shown that the performance of the detector is given by

$$Q^{-1}(P_{FA}) - Q^{-1}(P_D) = \sqrt{N} \text{ snr}, \tag{4.9}$$

where SNR is defined as snr $= |h|^2 \sum_{n=1}^{N} |s(n)|^2/N$.

At low SNR the number of required samples is of order $1/$snr in contrast with the $1/$snr^2 samples required by the energy detector. This is a significant advantage. However, performance is degraded in the presence of frequency and timing offsets, as well as fading (the gain $|h|^2$ is now random) and delay spread (which requires equalization). The matched filter approach has been explored in the context of cognitive radios in [122–124].

4.2.4 Cooperative Sensing

In earlier sections we noted that the performances of a single detector can be severely degraded due to fading, shadowing, or a faulty sensor. This is one motivation for cooperative sensing, where observations from multiple SUs are combined to improve detector performance. Let the received signal at the kth SU be given by

$$y_k(n) = \theta h_k s(n) + w_k(n), \; n = 1, ..., N, \;\; k = 1, ..., K,$$

where $\theta = \ell$ under $\mathcal{H}_\ell$ and K is the number of cooperating users. The noise sequences are assumed to be independent and identically distributed in time n and mutually independent across the sensors. The channel gain coefficient h_k is assumed to be independent across the sensors; in other words, this represents a diversity system with K degrees of freedom. The performance of such a system has been well studied in the context of bit detection (e.g., $s(n) = 1$ versus $s(n) = -1$). Also a significant body of previous work exists on this topic under the area of distributed detection, particularly distributed radar systems. Our detection problem is to test whether $\theta = 0$ or 1.

Because of the rich history of distributed detection/estimation in classical distributed radar systems and more recently in wireless sensor networks, there has been lot of activity in this area in the context of cognitive radios. Representative papers include [125–131]. Related work is on the use of multiple antennas; here again the idea is to exploit diversity in the fading [132].

Considerable overhead would be required to transmit all of the $\{y_k(n)\}$. As such, there is interest in performing local detection and passing only binary decision variables to a fusion center (FC). Alternatively, the LLR could be quantized and the quantized LLR sent to the FC. Censoring ideas could also be applied: a user transmits its LLR only if the absolute value of the LLR is above a threshold. Depending on the level of complexity that can be supported, the FC has many choices for its fusion rule. If the SUs transmit binary decisions, it could implement a simple OR rule (the channel is declared to be busy if at least one sensor says it is busy) or a weighted majority rule. If raw data or quantized LLRs are sent, the fusion rule becomes more complex. The cooperative scheme requires a control channel (which could be in-band and frequency hopped) and a trusted spectrum broker. Latency is an important issue: The time required to sense the channel, report the measurements to the FC, and for the FC to detect white space and allocate spectrum to a user must be considerably less than the channel free time. Distributed implementations are also possible in which SUs exchange information with all their neighbors and each separately makes a decision.

Let ϵ_k and δ_k denote the local performance indices of the kth sensor. If sensor observations, and thus local decisions, are conditionally independent, then the performance of the OR detector is readily obtained:

$$P_{\text{FA}} = 1 - \prod_{k=1}^{K}(1 - \epsilon_k) \approx \sum_{k=1}^{K} \epsilon_k,$$

where the approximation holds if the ϵ_k's are small. Similarly,

$$P_{\text{MD}} = \prod_{k=1}^{K} \delta_k.$$

The individual δ_k, ϵ_k depend on the specific detector as has been discussed in earlier sections.

If the FC has access to the individual sensor statistics, it can combine them in many classical ways: equal gain combining (EGC), selection combining, etc. For simplicity, consider the case where the K sensor statistics are independent. In EGC, the effective SNR is the sum of the individual SNRs, and adding K noncentral χ^2 random variables each with $2N$ degrees of freedom and noncentrality parameter μ_k^2 results in a noncentral χ^2 variable with $2NK$ degrees of freedom and a noncentrality parameter $\mu^2 = \sum_{k=1}^{K} \mu_k^2$. Then P_D can be obtained via Equation (4.7), and the P_D averaged over a fading channel can be obtained from Equation (4.8). P_{FA} is given by Equation (4.4) as before.

In addition to its ability to cope with the hidden node problem, cooperative sensing can be used to *localize* the active transmitters.

4.2.5 Other Approaches

The implicit frequency nonselective flat fading assumption made so far in this chapter essentially assumes a narrowband channel model. When the spectrum to be sensed is wideband, there are multiple challenges. First, one may want to consider (partially overlapping) subchannels for each of which the flat fading assumption would be reasonable. How many channels should be monitored? This is dictated partly by the affordable complexity of the receiver, the traffic usage in the primary network, and the desired rates for the secondary user. If the primary traffic is heavy, SU would seek to monitor multiple bands. But this entails increased sampling rates, receiver complexity, and energy consumption. A related issue is whether the SU can sense (and transmit on) multiple (possibly well-) separated channels or whether the channels should be contiguous.

Multiresolution- and wavelet-based methods have been proposed to deal with the wideband problem. The power spectral density is smooth within each subchannel but possibly discontinuous across subchannel boundaries. By using the wavelet transform, the discontinuities can be identified and thus spectrum activity detected [133, 134]. Compressed sensing ideas are exploited in [135] by exploiting the sparseness of the signal spectrum; sub-Nyquist sampling schemes, in conjunction with wavelet-based edge detection, are used to provide coarse estimates of spectrum occupancy and transmitter location.

4.3 FROM DETECTING PRIMARY SIGNALS TO DETECTING SPECTRUM OPPORTUNITIES

In this section, we highlight the differences between detecting primary signals and detecting spectrum opportunities. We show that, besides noise and fading, the geographic distribution and traffic pattern of primary users have significant impact on the performance of spectrum opportunity detection.

4.3.1 Definition and Implications of Spectrum Opportunity

A rigorous study of cognitive radio systems must start from a clear definition of *spectrum opportunity* and *interference constraint*. An initial attempt in defining these two central concepts can be found in [136]. To protect primary users, an interference constraint should specify at least two parameters $\{\rho, \zeta\}$. The first parameter ρ is the maximum allowable interference power perceived by an active primary receiver; it specifies the noise floor and is inherent to the definition of *spectrum opportunity* as shown later. The second parameter ζ is the maximum outage probability that the interference at an active primary receiver exceeds the noise floor. Allowing a positive outage probability ζ is necessary due to opportunity detection errors. This parameter is crucial to secondary users in making transmission decisions based on imperfect spectrum sensing as shown in [137] and discussed in Section 4.4.

Spectrum opportunity is a local concept defined with respect to a particular secondary transmitter and its receiver. Intuitively, *a channel is an opportunity to a pair of secondary users if they can communicate successfully without violating the interference constraint imposed by the primary network*. In other words, the existence of a spectrum opportunity is determined by two logical (binary

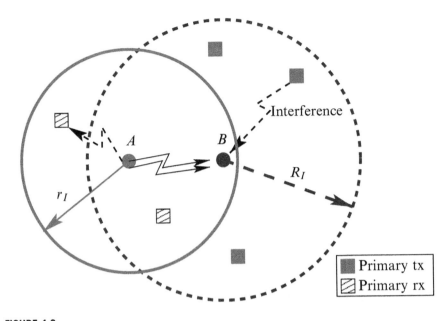

FIGURE 4.3

The *spectrum opportunity* definition.

value) conditions: the reception at the secondary receiver being successful and the transmission from the secondary transmitter being "harmless." Deceptively simple, this definition has significant complications in cognitive radio networks where primary and secondary users are geographically distributed and wireless transmissions are subject to path loss and fading.

For a simple illustration, consider a pair of secondary users (*A* and *B*) seeking to communicate in the presence of primary users as shown in Figure 4.3. A channel is an opportunity to *A* and *B* if the transmission from *A* does not interfere with nearby *primary receivers* in the solid circle, and the reception at *B* is not affected by nearby *primary transmitters* in the dashed circle. The radius r_I of the solid circle at *A* depends on the transmission power of *A* and the parameter ρ of the interference constraint, whereas the radius R_I of the dashed circle depends on the transmission power of primary users and the interference tolerance of *B*.

The use of a circle to illustrate the interference region is simplistic and immaterial. This definition applies to a general signal propagation and interference model by replacing the solid and dashed circles with, respectively, the subset of primary receivers that are potential victims of *A*'s transmission and the subset of primary transmitters that can cause interference with the reception at *B*. The key message is that spectrum opportunities must be defined jointly at the transmitter and the receiver. It is a function of (1) the transmission powers of both primary and secondary nodes, (2) the geographical locations of these nodes, and (3) the interference constraint. From this definition, we arrive at the following properties of spectrum opportunity.

Property 1 Spectrum Opportunity

P1.1 Spectrum opportunity depends on both transmitting and receiving activities of primary users.

P1.2 Spectrum opportunity is, in general, asymmetric: A channel that is an opportunity when A is the transmitter and B the receiver may not be an opportunity when B is the transmitter and A the receiver.

P1.1 shows clearly the difference between detecting primary signals (i.e., the presence of primary transmitters) and detecting spectrum opportunities. P1.2 leads to a complex relationship between opportunity detection performance at the physical layer and the link throughput and interference constraint at the MAC layer. As shown in Section 4.4, this relationship varies with the application type (for example, whether acknowledgment is needed to complete a successful data transmission) and the use of handshaking such as RTS/CTS (request to send/clear to send). In other words, it depends on whether the roles of the transmitter and the receiver need to be reversed during the process of communicating a data packet.

4.3.2 Spectrum Opportunity Detection

Spectrum opportunity detection can be considered a binary hypothesis test. We adopt here the disk signal propagation and interference model as illustrated in Figure 4.3. The basic concepts presented here, however, apply to a general model.

Let $\mathbb{I}(A, d, \mathrm{rx})$ denote the logical condition that there exist primary receivers within distance d to the secondary user A. Let $\overline{\mathbb{I}(A, d, \mathrm{rx})}$ denote the complement of $\mathbb{I}(A, d, \mathrm{rx})$. The two hypotheses for opportunity detection are then given by

$$\mathcal{H}_0: \text{opportunity exists; that is, } \overline{\mathbb{I}(A, r_I, \mathrm{rx})} \cap \overline{\mathbb{I}(B, R_I, \mathrm{tx})},$$

$$\mathcal{H}_1: \text{no opportunity; that is, } \mathbb{I}(A, r_I, \mathrm{rx}) \cup \mathbb{I}(B, R_I, \mathrm{tx}),$$

where $\mathbb{I}(B, R_I, \mathrm{tx})$ and $\overline{\mathbb{I}(B, R_I, \mathrm{tx})}$ are similarly defined, and R_I and r_I are, respectively, the interference range of primary and secondary users under the disk model. Notice that $\overline{\mathbb{I}(A, r_I, \mathrm{rx})}$ corresponds to the logical condition on the transmission from A being "harmless" and $\overline{\mathbb{I}(B, R_I, \mathrm{tx})}$ the logical condition on the reception at B being successful. Also notice the difference in the definition of the hypotheses for spectrum opportunity detection as compared to those for primary signal detection given in Equation (4.1).

Detection performance at the physical layer is measured by the probabilities of false alarm P_{FA} and miss detection P_{MD}: $P_{\mathrm{FA}} = \Pr\{\text{decides } \mathcal{H}_1 \mid \mathcal{H}_0\}$, $P_{\mathrm{MD}} = \Pr\{\text{decides } \mathcal{H}_0 \mid \mathcal{H}_1\}$. The trade-off between false alarm and miss detection is captured by the receiver operating characteristic, which gives $P_D = 1 - P_{\mathrm{MD}}$ (probability of detection or detection power) as a function of P_{FA} (see Figure 4.5,

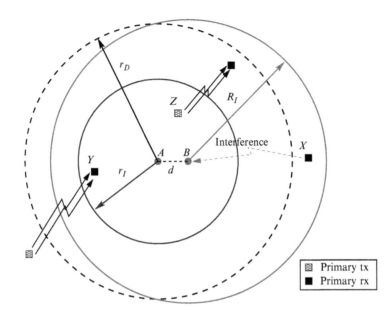

FIGURE 4.4

Spectrum opportunity detection.

later, for an illustration). In general, reducing P_{FA} comes at the price of increasing P_{MD} and vice versa. Since false alarms lead to overlooked spectrum opportunities and miss detections are likely to result in collisions with primary users, the trade-off between false alarm and miss detection is crucial in the design of cognitive radio systems [137].

Without assuming cooperation from primary users, the observations available to the secondary user for opportunity detection are the signals emitted from primary *transmitters*. This basic approach to opportunity detection is commonly referred to as *listen before talk* (LBT); that is, detecting spectrum opportunities by detecting primary signals as discussed in Section 4.2. As shown in Figure 4.4, A infers the existence of spectrum opportunity from the absence of primary transmitters within its detection range r_D, where r_D can be adjusted by changing, for example, the threshold of an energy detector. The probabilities of false alarm P_{FA} and miss detection P_{MD} for LBT are thus given by

$$P_{FA} = \Pr\{\mathbb{I}(A, r_D, \text{tx}) \mid \mathcal{H}_0\}, \quad P_{MD} = \Pr\{\overline{\mathbb{I}(A, r_D, \text{tx})} \mid \mathcal{H}_1\}. \tag{4.10}$$

Uncertainties, however, are inherent to such a scheme, even if A listens to primary signals with perfect ears (i.e., perfect detection of primary transmitters within its detection range r_D). Even in the absence of noise and fading, the geographic distribution and traffic pattern of primary users have significant impact on the performance of LBT. Specifically, there are three possible sources of detection

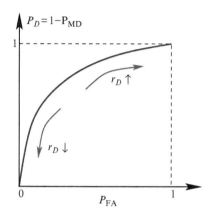

FIGURE 4.5

ROC of spectrum opportunity detection (the ROC is obtained by varying the detection range r_D).

errors: hidden transmitters, hidden receivers, and exposed transmitters. A *hidden transmitter* is a primary transmitter located within distance R_I to B but outside the detection range of A (node X in Figure 4.4). A *hidden receiver* is a primary receiver located within the interference range r_I of A but its corresponding primary transmitter is outside the detection range of A (node Y in Figure 4.4). An *exposed transmitter* is a primary transmitter located within the detection range of A but transmits to a primary receiver outside the interference range of A (node Z in Figure 4.4). For the scenarios shown in Figure 4.4, even if A can perfectly detect the presence of signals from any primary transmitter located within its detection range r_D, the transmission from the exposed transmitter Y is a source of false alarms, whereas the transmission from the hidden transmitter X and the reception at the hidden receiver Z are sources of miss detections. As illustrated in Figure 4.5, adjusting the detection range r_D leads to different points on the ROC. It is obvious from Equation (4.10) that P_{FA} increases but P_{MD} decreases as r_D increases.

Example 1
Spectrum Opportunity Detection in Poisson Primary Networks

As an example, we analyze the performance of LBT for spectrum opportunity detection in a Poisson primary network with uniform traffic. To highlight the difference between spectrum opportunity detection and primary signal detection, we study the performance of LBT with perfect ears. This allows us to illuminate the impact of primary users' geolocation and traffic pattern on the performance of detecting spectrum opportunities and separate it from noise and fading that affect the performance of detecting primary signals.

Consider that primary users are distributed according to a two-dimensional homogeneous Poisson process with density λ. Transmissions are slotted. In each slot, a primary user X has a probability p to become a transmitter. Its receiver is chosen with equal probability from primary users located within a distance R_p (the transmission range) to X. Based on the thinning theorem and the displacement theorem for marked Poisson processes [138], both primary transmitters and receivers form a two-dimensional homogeneous Poisson process with density $p\lambda$. Note that these two Poisson processes are not independent.

Consider a pair of secondary users A and B that are distance d apart and opportunity detection is performed by the transmitter A via LBT. A disk interference model is used, where the interference ranges of primary and secondary users are R_I and r_I, respectively (see Figure 4.3). We assume that A can detect perfectly the presence of primary transmitters within a distance r_D. We can then obtain closed-form expressions for probabilities of false alarms and miss detections. The derivations are lengthy and omitted due to space limits. Details can be found in [139].

Specifically, let d be the distance between A and B. Let $S_I(d, r_1, r_2)$ denote the common area of two circles centered at A and B with radii r_1 and r_2, respectively, and $S_C(d, r_1, r_2)$ denote the area within a circle with radius r_1 centered at A but outside the circle with radius r_2 centered at B. The probabilities of false alarms and miss detections of LBT with a detection range of r_D are given in Equations (4.11)–(4.13), where $S_o = S_C(d, r_D, R_I) \cap S_C(d, r_I + R_p, R_I)$. As shown in [139], the double integrals in Equations (4.11)–(4.13) can be reduced to a single integral.

$$P_{FA} = 1 - \exp\left\{-p\lambda\left[\pi r_D^2 - S_I(d, r_D, R_I) - \iint_{S_o} \frac{S_I(r, R_p, r_I)}{\pi R_p^2} r\, dr\, d\theta\right]\right\} \quad (4.11)$$

$$P_{MD} = \frac{1}{1 - \Pr[\mathcal{H}_0]}\left\{\exp(-p\lambda\pi r_D^2) - \exp\left[-p\lambda\left(\pi(r_D^2 + R_I^2) - S_I(d, r_D, R_I)\right.\right.\right.$$

$$\left.\left.\left. + \iint_{S_c(d, r_I + R_p, R_I) - S_o} \frac{S_I(r, R_p, r_I)}{\pi R_p^2} r\, dr\, d\theta\right)\right]\right\} \quad (4.12)$$

$$\Pr[\mathcal{H}_0] = \exp\left\{-p\lambda\left[\iint_{S_c(d, r_I + R_p, R_I)} \frac{S_I(r, R_p, r_I)}{\pi R_p^2} r\, dr\, d\theta + \pi R_I^2\right]\right\} \quad (4.13)$$

By varying the detection range $r_D \in (0, r_I + R_p]$, we obtain the ROC curve for LBT in Poisson primary networks. An example is given later in Figure 4.8.

4.4 FUNDAMENTAL TRADE-OFFS: PERFORMANCE VERSUS CONSTRAINT

A fundamental question in designing the spectrum opportunity detector is how to choose the detector operating point (P_{FA}^*, P_{MD}^*) to achieve the optimal trade-off between false alarms and miss detection. Such a trade-off, however, should be addressed in terms of MAC layer performance: the throughput of the secondary user and the probability of colliding with primary users. A translation from the physical layer performance in terms of $\{P_{FA}, P_{MD}\}$ to the MAC layer performance in terms of throughput and interference constraint is therefore crucial in choosing the optimal operating point of the spectrum opportunity detector. The issue is addressed in this section, where we consider separately the so-called global and local interference models introduced in [140].

4.4.1 MAC Layer Performance Measures

The MAC layer performance is measured by the throughput of the secondary user and the interference to the primary users. The design objective is to maximize the throughput under a constraint on the maximum outage probability ζ that the interference at an active primary receiver exceeds the noise floor ρ. We refer to such events as *collisions with primary users*.

The figures of merit at the MAC layer are given by the probability P_S of successful data transmission and the probability P_C of colliding with primary users. The objective and constraint at the MAC layer is thus given by

$$\max P_S \quad \text{subject to } P_C \leq \zeta.$$

We first consider P_S, which is application dependent. For applications requiring guaranteed delivery, an acknowledgment (ACK) signal from B to the secondary transmitter A is required to complete a data transmission. Specifically, in a successful data transmission, the following three events should occur in sequence: A detects the opportunity $[\overline{\mathbb{I}(A, r_D, \text{tx})}]$ and transmits data to B; B receives data successfully $[\overline{\mathbb{I}(B, R_I, \text{tx})}]$ and replies to A with an ACK; A receives the ACK $[\overline{\mathbb{I}(A, R_I, \text{tx})}]$, which completes the transmission. We thus have

$$P_S = \Pr\{\overline{\mathbb{I}(A, r_D, \text{tx})} \cap \overline{\mathbb{I}(B, R_I, \text{tx})} \cap \overline{\mathbb{I}(A, R_I, \text{tx})}\} \tag{4.14}$$

$$= \Pr\{\overline{\mathbb{I}(A, r_E, \text{tx})} \cap \overline{\mathbb{I}(B, R_I, \text{tx})}\}, \tag{4.15}$$

where $r_E = \max\{r_D, R_I\}$.

For best-effort delivery applications [141], acknowledgments are not required to confirm the completion of data transmissions. In this case, we have

$$P_S = \Pr\{\overline{\mathbb{I}(A, r_D, \text{tx})} \cap \overline{\mathbb{I}(B, R_I, \text{tx})}\}. \tag{4.16}$$

The probability of collision is defined as[3]

$$P_C = \Pr\{A \text{ transmits data} \mid \mathbb{I}(A, r_I, \text{rx})\}. \tag{4.17}$$

Note that P_C is conditioned on $\mathbb{I}(A, r_I, \text{rx})$ instead of $\mathcal{H}_1$. Clearly, $\Pr[\mathbb{I}(A, r_I, \text{rx})] \leq \Pr[\mathcal{H}_1]$.

Since the secondary transmitter A transmits data if and only if A detects no nearby primary transmitters, we have

$$P_C = \Pr\{\overline{\mathbb{I}(A, r_D, \text{tx})} \mid \mathbb{I}(A, r_I, \text{rx})\}. \tag{4.18}$$

4.4.2 Global Interference Model

Consider first a global interference model where the transmission from every primary user of interest affects the reception at B and the transmission from A affects the reception at every primary user. Under this condition, an opportunity occurs if and only if no primary users are transmitting. Spectrum opportunities are thus symmetric, and detecting primary signals is equivalent to detecting spectrum opportunities. Furthermore, we have the following properties, assuming that A transmits in a slot if and only if the channel is detected as an opportunity at the beginning of this slot.

Property 2 PHY-MAC Translation under the Global Interference Model

P2.1 Successful transmissions from A to B can result only from opportunities
   (i.e., $\mathcal{H}_0$).
P2.2 Every correctly identified opportunity leads to a successful transmission.
P2.3 Every miss detection results in a collision with primary users.

These properties lead to the following simple relationship between $\{P_{\text{FA}}, P_{\text{MD}}\}$ and $\{P_S, P_C\}$.

$$P_S = (1 - P_F) \Pr[\mathcal{H}_0], \quad P_C = P_{\text{MD}}. \tag{4.19}$$

With this relationship, to maximize P_S under the constraint of $P_C \leq \zeta$, we can obtain the optimal operating point $(P_{\text{FA}}^*, P_{\text{MD}}^*)$ for the spectrum sensor. The joint design of the spectrum sensor at the physical layer and the tracking and access decisions at the MAC layer are addressed in [137], which shows that the optimal detector operating characteristic is given by $P_{\text{MD}}^* = \zeta$ and the optimal access policy at the MAC layer is to simply trust the spectrum detector: Transmit if and only if the channel is detected as an opportunity.

---

[3]In obtaining the definition of P_C, we have assumed that the interference caused by the ACK signal is negligible due to its short duration.

4.4.3 **Local Interference Model**

When the transmissions from primary and secondary users have local effect, the statements and the relationship between $\{P_{FA}, P_{MD}\}$ and $\{P_S, P_C\}$ given in Equation (4.19) no longer hold. The relationship between PHY and MAC has complex dependency on the applications and the use of MAC handshaking.

Impact of Application

We illustrate here the impact of applications on the relationship between PHY and MAC. Specifically, we compare applications requiring guaranteed delivery with those relying on best effort (for example, media streaming and network gaming). For the former, we assume immediate acknowledgment is required at the end of each slot to complete a successful data transmission. For the latter, acknowledgments are not necessary. Due to the asymmetry of spectrum opportunities and the local effect of transmissions, we have the following relationship between $\{P_{FA}, P_{MD}\}$ and $\{P_S, P_C\}$.

Property 3 PHY-MAC Translation under the Local Interference Model

P3.1 For both types of applications, $P_C \neq P_{MD}$.

P3.2 For applications with guaranteed delivery, correctly detected opportunities may lead to failed data transmission, and miss detections may lead to successful data transmission; that is,

$$\Pr[\text{success} \mid \mathcal{H}_0] \leq 1 - P_{FA}, \ \ 0 < \Pr[\text{success} \mid \mathcal{H}_1] \leq P_{MD}.$$

P3.3 For best-effort delivery, correctly detected opportunities always result in successful data transmission, and miss detections may also lead to successful data transmission; that is,

$$\Pr[\text{success} \mid \mathcal{H}_0] = 1 - P_{FA}, \ \ 0 < \Pr[\text{success} \mid \mathcal{H}_1] \leq P_{MD}.$$

Example 2
PHY-MAC Translation in Poisson Primary Networks

Consider the same Poisson primary network given in Example 4.1. We obtain the closed-form expressions for P_C and P_S for LBT with perfect ears given here. Detailed derivations can be found in [139].

$$P_C = \frac{\exp(-\rho\lambda\pi \, r_D^2)\{1 - \exp[-\rho\lambda\pi(r_I^2 - I(r_D, r_I, R_p))]\}}{1 - \exp(-\rho\lambda\pi \, r_I^2)}, \tag{4.20}$$

$$P_S = \begin{cases} \exp\{-\rho\lambda[\pi(r_E^2 + R_I^2) - S_I(d, r_E, R_I)]\}, & \text{guaranteed delivery,} \\[2mm] \exp\{-\rho\lambda[\pi(r_D^2 + R_I^2) - S_I(d, r_D, R_I)]\}, & \text{best-effort delivery,} \end{cases} \tag{4.21}$$

where $I(r_D, r_I, R_p) = \int_0^{r_D} 2r[S_I(r, r_I, R_p)/\pi R_p^2]dr$.

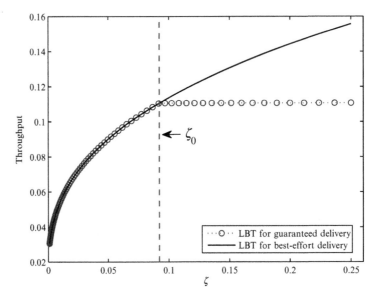

FIGURE 4.6

Success probability versus interference constraint.

Based on this result, we can study the impact of applications on the MAC layer performance; that is, P_S (representing link throughput) under a collision constraint $P_C \le \zeta$. Shown in Figure 4.6 is P_S as a function of the collision constraint ζ. We observe that, even though the detection performance at the physical layer is the same, the MAC layer performance can be different depending on the applications.

As shown in Figure 4.6, when the collision constraint is tight, the throughput is the same for these two types of applications. The collision constraint ζ has a critical value ζ_0 above which the throughput for best-effort delivery is higher than that for guaranteed delivery. Figure 4.7 shows ζ_0 as a function of the primary traffic load $\rho\lambda$ (or the density of active primary transmitters). We can see that ζ_0 is a decreasing function of $\rho\lambda$. This suggests that primary systems with heavy traffic are more suitable for spectrum overlay with best-effort delivery applications.   ∎

Impact of MAC Handshaking

The fundamental deficiency of detecting spectrum opportunities from detecting primary signals resembles the hidden and exposed terminal problem in the conventional ad hoc networks of peer users. It is therefore natural to consider the use of a RTS/CTS handshaking to enhance the detection performance of LBT. We show here that, although RTS/CTS signaling can improve the performance of opportunity detection at the physical layer, it may lead to decreased throughput at the MAC layer for best-effort delivery applications.

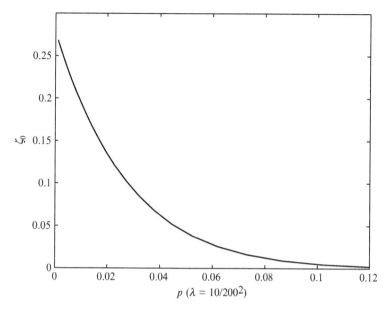

FIGURE 4.7

Critical value of ζ versus primary traffic load.

For RTS/CTS-enhanced LBT, spectrum opportunity detection is done jointly by A and B through the exchange of RTS/CTS signals. Specifically, the transmitter A first detects a chosen set of primary transmitters. If there are no signals from this set, it transmits an RTS to B. Upon receiving the RTS (which automatically indicates the absence of interfering primary transmitters), B replies with a CTS. A successful exchange of RTS/CTS indicates an opportunity, and A starts to transmit data to B. For this RTS/CTS-enhanced LBT, we have the following relationship between $\{P_{FA}, P_{MD}\}$ and $\{P_S, P_C\}$.

Property 4 PHY-MAC Translation with RTS/CTS Signaling

P4.1 $P_C = \frac{Pr[\mathcal{H}_1]}{Pr[\mathbb{I}(A,\mathrm{rx})]} P_{MD} \geq P_{MD}$.

P4.2 *Correctly detected opportunities always result in successful data transmission, as well as miss detections; that is,*

$$P_S = (1 - P_{FA}) \Pr[\mathcal{H}_0] + P_{MD} \Pr[\mathcal{H}_1].$$

The PHY and MAC performance of RTS/CTA-enhanced LBT in a Poisson primary network with uniform traffic can be similarly analyzed [139]. An example ROC curve is shown in Figure 4.8. Note that (0, 0) does not belong to the ROC curve of RTS/CTS-enhanced LBT. This is because the effective detection range is bounded above R_I, since to receive the CTS signal successfully, no primary transmitters can

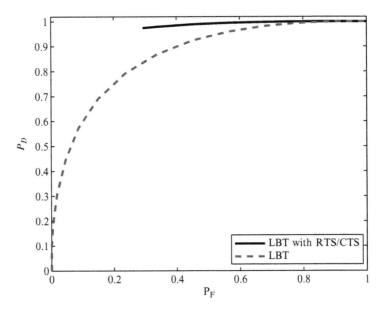

FIGURE 4.8

ROC performance comparison.

be within R_I of A. In other words, a detection range $r_D \leq R_I$ leads to the same (P_{FA}, P_{MD}) as $r_D = R_I$.

It can be shown that the ROC performance of RTS/CTS-enhanced LBT is always better than or equal to that of LBT when $r_D \geq R_I$. However, at the MAC layer, RTS/CTS-enhanced LBT may lead to lower throughput when the collision constraint is less restrictive and the application requires only best-effort delivery, as shown in Figure 4.9. Note that, using RTS/CTS-enhanced LBT, the throughput is the same for guaranteed delivery and best-effort delivery. This suggests that whether to adopt handshaking at the MAC layer depends on the applications and the interference constraint ζ.

4.5 FUNDAMENTAL TRADE-OFFS: SENSING ACCURACY VERSUS SENSING OVERHEAD

In this section, we discuss the trade-off between sensing accuracy and sensing overhead. Increasing the sensing time improves the fidelity of the sensing outcomes, thus reducing overlooked spectrum opportunities. On the other hand, increasing the sensing time results in less transmission time. The trade-off between sensing accuracy and sensing overhead depends on the SNR level, the duration of spectrum opportunities, and the interference constraint ζ as illustrated in [137].

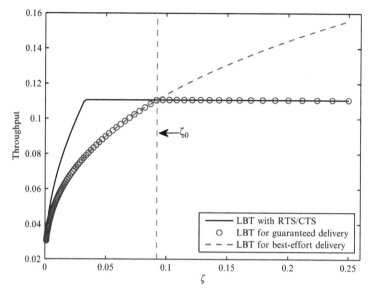

FIGURE 4.9

Throughput comparison.

Consider a slotted primary network. Let N represent the slot length in some arbitrary units, and n the duration of the sensing window. Let $P_{FA}(n)$ and $P_{MD}(n)$ denote the performance metrics based on a sensing window of length n. Assuming that the channel is free, the fractional time that the channel will be accessed is $1 - P_{FA}(n)$, and the fractional slot available for such transmission is $(N - n)/N$. In [126, 142], a spectral efficiency metric defined as

$$\eta(n) = \frac{N - n}{N}[1 - P_{FA}(n)]$$

was introduced. For a specified P_D (interference constraint), P_{FA} can be written as

$$P_{FA}(n) = Q\left[(1 + \text{snr})Q^{-1}(P_D) + \text{snr}\sqrt{n}\right],$$

which decreases monotonically in n. The metric thus consists of two terms: the first one decreases as n increases and the second one increases with n. One seeks the n that maximizes $\eta(n)$. Figure 4.10 shows throughput η versus sensing window duration n for various values of the interference constraint, δ, and various values of SNR: $-15, -10, -5, 0, 5$ dB depicted by markers x, o, +, $\square$, and $\Diamond$, respectively. As the interference constraint becomes tighter (i.e., as the specified δ decreases), the optimal n^* increases and η^* decreases. As SNR increases, the required sensing window length decreases and efficiency increases. As the slot length N increases, η^* increases, since the cost of sensing is amortized over a longer slot duration; see Figure 4.11.

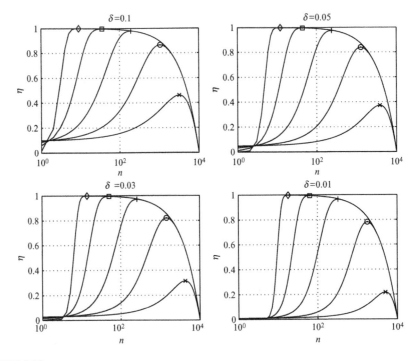

FIGURE 4.10

Optimal sensing time: efficiency η versus sensing window length n for various SNRs and P_{MD}.

4.6 CHAPTER SUMMARY AND FURTHER READINGS

In this chapter, we discuss primary signal detection and spectrum opportunity detection and highlight their relationship. Fundamental trade-offs in spectrum sensing, including performance versus constraint and sensing accuracy versus sensing overhead, are studied in detail. The translation from the physical layer detection performance to the MAC layer performance is investigated, demonstrating the complex dependency of the relationship between PHY and MAC on the applications and the use of MAC handshaking such as RTS/CTS.

A related and important issue in spectrum sensing is tracking time-varying spectrum opportunities. Due to hardware limitations and the energy cost of spectrum monitoring, a secondary user may not be able to sense all the channels in the spectrum simultaneously. A sensing strategy is therefore needed for intelligent channel selection to track the rapidly varying spectrum opportunities. The purpose of a sensing strategy is twofold: to find idle channels for immediate access and to gain statistical information on the spectrum occupancy for better opportunity tracking in the future. The optimal sensing strategy should therefore strike a balance

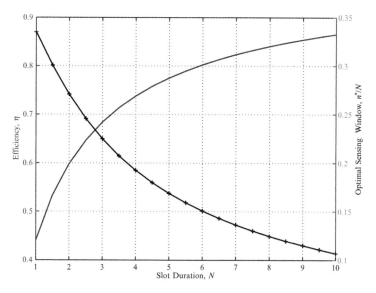

FIGURE 4.11

Optimal sensing time: efficiency η (solid curve) and optimal window length n^*/N (marked +) versus slot length N.

between these two often conflicting objectives. This problem has been addressed within a decision-theoretic framework in [137, 143, 144].

Another related problem is the impact of the transmission power of secondary users on the reliability of opportunity detection as investigated in [139]. It is shown in [139] that reliable opportunity detection is achieved in the two extreme regimes of the ratio between the transmission power p_{tx} of secondary users and the transmission power P_{tx} of primary users: $P_{tx}/P_{tx} \to 0$ and $P_{tx}/P_{tx} \to \infty$.

4.7 PROBLEMS

1 Consider the problem of detecting between two real Gaussian random variables with means μ_i and variances σ_i^2, $i = 0, 1$. Show that the P_{FA} is given by $\Pr(Z > \tau | \mathcal{H}_0) = Q(\frac{\tau - \mu_0}{\sigma_0})$, where $Q(\tau) = (1/\sqrt{2\pi}) \int_\tau^\infty e^{-t^2/2} dt$ is the standard Gaussian tail function. Hence, the ROC can be expressed as

$$P_D = 1 - \delta = Q\left(\frac{\sigma_0 Q^{-1}(\epsilon) + \mu_0 - \mu_1}{\sigma_1}\right),$$

where $Q^{-1}(x)$ is the inverse Q function. The $Q(\cdot)$ and $Q^{-1}(\cdot)$ functions are available in MATLAB as `qfunc.m` and `qfuncinv.m`.

2 For a given P_D, one can compute the threshold τ in Equation (4.5) as follows. The equation for Γ_u in Equation (4.3) can be written as a summation via explicit integration by parts. Next, one can use a fixed-point iteration scheme to find τ. An easier method is as follows. When the degrees of freedom N is large enough, the χ^2 distribution can be approximated by the Gaussian. Essentially, when N is large enough and under some regularity conditions, one can invoke the central limit theorem and approximate the statistic z in Equation (4.2) as a Gaussian. Note that it is not necessary to assume that $y(.)$ is Gaussian. Using results from Problem 1, verify that the ROC of the energy detector can be approximated by

$$P_D = 1 - \delta = Q\left(\frac{Q^{-1}(\epsilon) - \text{snr}\sqrt{N}}{1 + \text{snr}}\right).$$

This expression for P_D clearly demonstrates that desired performance metrics can be achieved at any SNR provided N is large enough. To explore this further, assume that P_D and P_{FA} are fixed. Write an expression for the number of required samples N. How does this behave at low SNR?

3 Suppose that the signal vector is zero-mean Gaussian but has time-varying variance given by $\sigma_s^2(n)$. Simplify the LLR to show that the optimal detector is a weighted energy detector.

4 Suppose that the signal vector $\mathbf{s} = [s(1), ..., s(N)]^T$ has covariance matrix Σ_s. Use the eigenvalue decomposition of $\Sigma_s = V\Lambda_s V^T$ and simplify the LLF. Show that the LLF can be interpreted as a two-step process: the signal is prewhitened by projecting it onto the eigenvector matrix V^T; the resulting output is passed through a weighted energy detector, $\tilde{z} = \sum_{k=1}^{N} w(n)|\tilde{y}(n)|^2$ where $\tilde{\mathbf{y}} = V^T\mathbf{y}$.

5 The ED threshold must be set to ensure that P_{MD} does not exceed a specified threshold δ^*. This requires knowledge of the SNR; that is, the variance of the noise as well as the signal power. First consider the case where the signal variance σ_s^2 is known, but the noise variance σ_w^2 is unknown. One approach is to estimate the unknown parameters conditioned on the hypothesis then treat them as known parameters; this is known as the generalized likelihood ratio test (GLRT); see, for example, [113]. Suppose that the signal power σ_s^2 is known but not the noise power σ_w^2. Since the noise is zero-mean independent and identical Gaussian, the maximum likelihood (ML) estimate of its variance is the sample variance; this estimator is unbiased and the variance of the estimator goes down as $1/N$. For large N, one can roughly say that the estimate $\hat{\sigma}_w^2 \in [a\sigma_w^2, b\sigma_w^2]$, where $0 < a < 1 < b$. Rewrite the ED statistic as $z' = \frac{1}{2n}\sum_{k=1}^{n}|y(k)|^2$. Evaluate the mean and variance of z' under the two hypotheses. Invoking the central limit theorem, write an expression for P_D and P_{FA}. Note that the worst case P_{FA} is obtained when the noise variance is $b\sigma_w^2$, and that the worst case P_D is obtained when the noise variance is $a\sigma_w^2$. Using these, verify that the number of samples required to ensure the desired performance (P_D, P_{FA}) over the uncertainty range of the noise variance is proportional to $1/[\text{snr} - (b - a)]^2$, which amounts to a degradation in SNR by the normalized uncertainty range $(b - a)$. Further, performance breaks down when the SNR is close to $(b-a)$. This effect has been called the *SNR wall* [115]. How do these results change if there is uncertainty in both the signal and noise powers? Only in the signal power?

6   Use MATLAB simulations to obtain the P_{FA} and P_D of LBT with perfect ears in a Poisson primary network with uniform traffic and compare the simulated results with the analytical results given in Example 4.1.

7   Use MATLAB simulations to study how the ROC curve of LBT with perfect ears changes with the transmission power p_{tx} of the secondary user. Assume a Poisson primary network with uniform traffic and the disk propagation and interference model as given in Example 4.1. Note that $p_{tx}^{2/\alpha} \propto r_I^2$, where α is the path loss exponent. Compare your findings with the analytical results given in [139].

Spectrum access and sharing

5

Alireza Attar[1], Oliver Holland[2], and Hamid Aghvami[2]
[1] University of British Columbia, Canada
[2] King's College London, United Kingdom

5.1 INTRODUCTION

This chapter[1] deals with methods of accessing and sharing radio spectrum in wireless communications. Wireless communication is achieved by transmission and reception of electromagnetic waves utilizing swaths of radio spectrum ranging from (in some exceptional cases) as low as the extremely low-frequency (ELF) band (3–30 Hz) to as high as the extremely high-frequency (EHF) band (30–300 GHz). Due to the physical characteristics of different frequency bands, such as wavelength, information capacity, and propagation, the 30–3000 MHz portion of the radio spectrum is the most sought-after resource for various mobile and wireless applications.

It is the task of international standardization bodies, such as the International Telecommunications Union (ITU) or the European Conference of Postal and Telecommunications Administrations (CEPT), to *allocate* spectrum bands to different wireless technologies in order to improve transnational coordination and interoperability. National regulatory bodies, for instance, the Federal Communications Commission (FCC) in the United States or the Office of Communications (Ofcom) in the United Kingdom, *assign* allocated bands to a specific technology and stakeholder, such as mobile operators or TV broadcasters. The assigned spectrum can then be accessed by those who have the right to do so. Given the high value of radio spectrum, the *access* to this natural resource at the user level is generally *shared*; for example, a number of subscribers to a specific mobile operator

---

[1] Some of the work reported in this chapter formed part of the Delivery Efficiency Core Research Programme of the Virtual Centre of Excellence in Mobile and Personal Communications, Mobile VCE (www.mobilevce.com). This research has been funded by EPSRC and by the Industrial Companies that are members of Mobile VCE. Fully detailed technical reports on this research are available to Industrial Members of Mobile VCE.

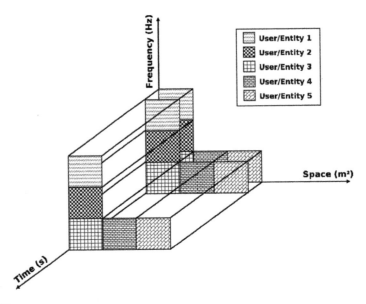

FIGURE 5.1

Dimensions of spectrum access and sharing.

or wireless Internet service provider (WISP) might share the same resource, under the control of that operator or WISP.

Given that multiple users, or ultimately multiple entities, need to access the radio spectrum for a variety of purposes, mechanisms should exist to allow the sharing of this resource, facilitating maximum utilization of every frequency band. A number of spectrum sharing scenarios can be envisaged, whereby the physical dimensions upon which spectrum sharing can be realized include time, frequency, and space, as shown in Figure 5.1. The propagation characteristics of electromagnetic waves do not, however, generally allow precise decomposition of the radio spectrum into orthogonal components as depicted in Figure 5.1; for instance, it is generally not possible to confine the propagation of radio waves to a specific location, such as an urban or rural area. The raw principles of spectrum sharing do, however, normally assume the possibility of such decomposition. Examples of real-life systems exploiting temporal spectrum sharing principle include half-duplex radios such as a push-to-talk "walkie-talkies" or IEEE 802.11-x wireless local area networks (WLANs) incorporating CSMA-style random access. Frequency-domain spectrum sharing is widely used in the form of frequency division multiple access (FDMA) and orthogonal FDMA (OFDMA)–based systems. Finally, frequency reuse in cellular systems, where a specific channel in one cell is also used in other distant cells, is a good example of spatial spectrum sharing.

A different way of looking at spectrum sharing classifications is from the spectrum access rights point of view. In this context, the first possibility is sharing the *licensed* spectrum resource, for instance, within a radio access network (RAN) among all users. To this end, depending on the radio access technology (RAT) used, multiple access control (MAC) techniques, such as time division multiple access (TDMA), frequency division multiple access (FDMA), or code division multiple access (CDMA), can be utilized. In either network-to-user communications, referred to as *downlink* (DL) communications in many cellular contexts, or user-to-network communications, referred to as *uplink* (UL) in many contexts, a unique set of spectrum resources, such as time slots, subchannels, codes, or a combination of these, are allocated to each user.

Another spectrum sharing possibility from the spectrum access rights point of view is the coexistence of several radio access technologies or service providers in the same licensed frequency band. Since all participating systems have equal rights in accessing the spectrum, such an approach is one manifestation of *horizontal spectrum sharing*. One example realization of this concept is "lite-licensing," recently introduced by the FCC [145].

Finally, a *vertical spectrum sharing* technique in a licensed band can be envisioned as a primary system, which is generally the license holder of that specific band, having the highest priority in accessing the resource, coexisting with a lower-priority secondary system that can access the spectrum only by complying with the primary's stringent requirements. Such requirements might include interference avoidance rules or maximum allowable transmission power, among other constraints.

A further possibility is the allocation of specific chunks of spectrum for *license-exempt* access, in which case all systems have the same right to access the band without implied spectrum ownership. Such an ideology is therefore also an example of horizontal spectrum sharing, whereby bands to which this paradigm currently applies include the industrial, scientific, and medical (ISM) bands (e.g., the 2.4 GHz ISM band) and the unlicensed national information infrastructure (UNII) bands (e.g., the 5 GHz UNII band). Wireless access technologies that are developed to operate in such license-exempt bands often employ random-access mechanisms such as carrier sense multiple access (CSMA) with collision avoidance (CA) or frequency-hopping spread spectrum (FHSS). However, the "license-exempt" status of such bands means that there exists no interference management mechanism to coordinate the resource accesses of the coexisting systems.

The term *spectrum sharing*, as used in the literature and implied previously, can be very broadly defined (see, e.g., [146]). In the interests of efficiency improvement through freer spectrum access and advancement in regulations, this chapter constrains the term to refer to the vertical, horizontal, and (to some extent) license-exempt spectrum sharing cases just mentioned. Hence, we use the following definition for spectrum sharing throughout the rest of this chapter:

Definition: Spectrum sharing is the simultaneous usage of a specific radio frequency band in a specific geographical area by a number of independent entities, leveraged through mechanisms other than traditional multiple- and random-access techniques.

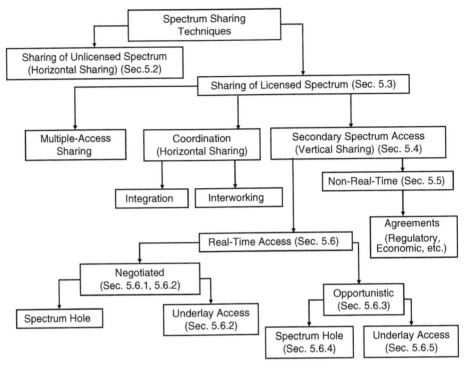

FIGURE 5.2

Classification of spectrum sharing techniques.

The key factors in this definition of *spectrum sharing* are twofold: First, the independence assumption of coexisting systems means legacy MAC mechanisms, which are used to share resources among users in a cellular system for example, are out of the scope of this chapter. Second, the emphasis is on scenarios where "mechanisms" to facilitate spectrum sharing exist. For instance, in this context license-exempt spectrum sharing in general is not of interest except where coordination techniques among coexisting systems can be envisaged.

The classification of various spectrum sharing techniques is depicted in Figure 5.2. For ease of reference, under each technique, the section(s) of this chapter relevant to that technique is (are) indicated. We elaborate on these approaches in more detail in the following sections. The rest of this chapter is structured as follows. Section 5.2 discusses unlicensed spectrum sharing and presents a simple example analyzing the collision probability for shared access in unlicensed spectrum. Section 5.3 discusses licensed spectrum sharing, and Section 5.4 expands on this in the context of secondary spectrum access. Secondary spectrum access is itself classified into the realms of non-real-time access and real-time access in

Sections 5.5 and 5.6; the latter is further split into negotiated access and opportunistic access. Throughout the discussion of licensed spectrum access and secondary spectrum access in Sections 5.3 and 5.4, numerous examples are given, primarily to highlight analytical considerations in realizing pertinent spectrum sharing techniques. A summary of this chapter can be found in Section 5.7, and complementary problems can be found in Section 5.8.

5.2 UNLICENSED SPECTRUM SHARING

Unlicensed frequency bands are chunks of spectrum set aside to be used by devices that wish to operate in a way that is not constricted by licenses and the associated complicated transmission/ownership rules and are therefore prone to interference. The most commonly used unlicensed bands are the 2.4 GHz ISM band, used by IEEE 802.11 b/g/n and Bluetooth devices, and the 5 GHz UNII band, as used by IEEE 802.11a and the European HyperLAN standards. As mentioned previously, as they are unlicensed, none of these bands is solely reserved for specific wireless communications technologies; for instance, most microwave ovens also operate in the 2.4 GHz ISM band, thereby creating additional interference.

Various spectrum measurement campaigns have shown unlicensed bands to be among the most heavily used frequency bands [147, 148]. Surprising as this might seem, there are actually a number of reasons as to why the usage of such unprotected bands is so popular. First, there is the ease of developing innovative technologies to operate in unlicensed bands, since a cumbersome approval process similar to licensed technologies is not involved; second, there is no cost to the consumer of using such bands. There are, however, certain transmission power caps, as described by ITU-R 5.138, 5.150, and 5.280. Country-specific requirements might also be stipulated; for example, Part 18 of the FCC rules governs ISM bands in the United States, where Part 15 Subpart B also contains rules for unlicensed communication devices that might be applicable to ISM bands. Such mandatory transmission power caps result in localization of interference, which in turn makes it possible to have numerous unlicensed devices operating in those bands as long as they are spatially/temporally distributed. To better understand the importance of spatial and temporal distribution of users of unlicensed bands, consider the following example.

Example 1

Let's consider a simple WLAN scenario where N nodes are trying to access the medium. Each node is using p-persistent CSMA [149], whereby at each frame time (referred to as a *time slot*), a specific node that wants to send packets first senses the channel. If the channel is clear, this node, with a probability of p, sends its packet, or else it waits until the next slot with probability $q = 1 - p$.

Now, assume each node is generating a traffic load based on a Poisson process, with a mean of G packets per time slot. The probability of generating k packets at a specific time slot t at any node is

$$P_t(k) = \frac{G^k e^{-G}}{k!}.$$

Assuming the availability of $D_n(t)$ packets from previous time slots at node n, the probability of the *silence* of node n in this time slot, $P_{s,n}(t)$, equals the probability of not having any packets to send plus the probability of having traffic but waiting for the next time slot due to the p-persistence nature of this random access procedure; that is,

$$P_{s,n}(t) = \left[\{1 - P_t(0)\} + D_n(t) \right] (1 - p) + P_t(0).$$

The probability of collision is then determined as the probability of at least two nodes trying to access the medium simultaneously. This probability can be calculated (as shown next) as 1 minus the sum of probabilities that either no node transmits at time slot t or only one node transmits. The backlogged traffic at each node is a function of capacity of the channel (itself a function of transmission power, channel gain, and noise at the receiver) and the collision probability in previous time slots. If, for simplicity, we assume all nodes have a similar backlog of data at time slot t—that is, $D_n(t) = D(t)$—we can calculate the collision probability as

$$P_{\text{collision}}(t) = 1 - \prod_{n}^{N} P_{s,n}(t) - \binom{N}{1} \prod_{n}^{N-1} P_{s,n}(t).$$

Using the probability expressions inferred previously, the collision probability can be simplified to

$$P_{\text{collision}}(t) = 1 - \left[pe^{-G} + q(1 + D(t)) \right]^{N-1} (N + pe^{-G} + q[1 + D(t)]).$$

Note that increasing the average traffic load generated G or the total number of nodes sharing the band N increases the collision probability, as expected.

This simple analysis indicates that the suitability of unlicensed shared spectrum is critically dependent on the spatial/temporal density of the systems sharing the band. Furthermore, this analysis does not consider the fact that interference in the unlicensed band might stem from other coexisting systems not controlled by the random-access technique. As more and more technologies are being developed to (at least partially) operate in unlicensed bands, there is a danger of reaching a saturation point, whereby unlicensed bands such as ISM and UNII will become useless due to the prohibitive interference level. This is a phenomenon often referred to as the *tragedy of commons* [150]. ■

The successful deployment and rapid growth of WLAN technologies, such as the IEEE 802.11 family, is one of the fruits of the creation of unlicensed bands. Based on this success story, many observers recommend the assignment of more unlicensed bands to further propel innovative solutions. Some regulators, however, such as Ofcom in the United Kingdom, oppose such views. In its 2004 "Spectrum Framework Review," Ofcom predicts that the 4.3% share of spectrum that is unlicensed in

2000 in the United Kingdom will increase to only 6.9% by 2010 [151]. In contrast, the 95.3% of the spectrum governed by command and control approaches in 2000 will significantly decrease to 21.6% of spectrum by 2010. The rest of the bands, that is, 71.5% of the spectrum, will be governed by market-based mechanisms in 2010 [151].

As the standardization of technologies for 4G communications reaches its final stages, whereby IEEE 802.16 (a.k.a., WiMAX) and the 3GPP's UMTS long-term evolution (LTE) are the main candidates, the process of allocating appropriate operating bands to these technologies is also being finalized. Although both systems require licensed bands to protect the quality of service of their users, IEEE 802.16h is also considering coexistence strategies for use in unlicensed bands.

5.3 LICENSED SPECTRUM SHARING

As discussed in Section 5.1, licensed bands are frequency bands assigned exclusively to a licensee, for instance, a specific mobile operator. Traditionally, such a license award also stipulates a specific technology to be used in the band, for example, GSM or UMTS. More recently, however, regulators such as Ofcom in the United Kingdom have shown an interest in assigning spectrum bands as "technology neutral." Depending on the technology used in the licensed band of a specific service provider, various MAC techniques are used to allow end users to share the medium. As discussed in Section 5.1, such spectrum sharing techniques are not discussed in this chapter.

Consider a pool of spectrum shared among a number of equal-priority RANs; that is, a horizontal spectrum sharing scenario. A central entity, such as a spectrum server [152, 153] or centralized spectrum coordinator (CSC) [154], allocates bandwidth to each RAN, where these allocations are updated at periodic intervals. The spectrum allocation decision is based on the requests of every RAN, which itself is based on traffic load prediction for the next period in the corresponding RANs. Traffic load predictions are calculated using information about traffic in previous periods, among other possible parameters. To achieve a reliable traffic prediction, various estimation techniques, such as autoregressive (AR), moving average (MA), or autoregressive moving average (ARMA) modeling, can be utilized [154].

Suppose the traffic load of a specific RAN is represented by the discrete-time wide sense stationary (WSS) process $x(t)$, where t denotes the sampling time of traffic load, or equivalently $t_2 - t_1$ represents the resource allocation period for all $t_1, t_2 \in t$. A WSS process is a stochastic process with a constant mean value and an autocorrelation that is a function of $\tau = t_2 - t_1$. Such simplified mathematical modeling assumptions are generally used in the literature to study the characteristics of complicated real-life systems, as used for instance in Example 5.1. A WSS process like $x(t)$ can be represented as the output of a minimum-phase system, such as $L(s)$, with a white noise process input $i(t)$. In this case, $x(t)$ can be modeled as an autoregressive process by

$$x(t) + a_1 x(t-1) + \cdots + a_N x(t-N) = b_0 i(t),$$

where a_i for $i \in \{1, 2, \ldots, N\}$ and b_0 coefficients can be calculated from Yule-Walker equations [72]. Hence, by having $N - 1$ samples of the past traffic load, it is possible to estimate the traffic at the Nth period using this model. Another approach is to use a moving average model, which needs samples of only the $i(t)$ process to estimate the next period's traffic, according to

$$x(t) = b_0 i(t) + \cdots + b_M i(t-M).$$

The weights here can be chosen using the autocorrelation function of $x(t)$ or, arbitrarily, for instance, as $b_i = 1/M$ (i.e., equal weights). Further to the preceding, it is also possible to create an autoregressive moving average model by combining these two techniques [72]. Other linear or nonlinear estimation techniques might also be employed [154].

As the total available bandwidth of the spectrum pool is limited, the optimum utilization of the spectrum is achieved when the participating RANs have highly uncorrelated traffic patterns. When systems experience fairly similar traffic loads, it is unlikely that one system will have sufficient capacity to lend spectrum to another, thereby defeating the purpose of spectrum sharing.

Example 2

Consider two different normalized traffic patterns transported by up to three RANs, sharing a pool of 5 MHz spectrum as shown in Table 5.1. Compare the resource allocation for each case, considering the correlation of traffic among participating RANs.

Assume that RAN m predicts the need for D_m units of spectrum, where one unit of spectrum is equal to the minimum bandwidth transferable to/from the spectrum pool. For instance, if a 200 kHz channel (equivalent to a GSM channel) is designated as one unit of spectrum, then a 5 MHz UMTS channel equals 25 units of spectrum. The graphs in Figures 5.3 and 5.4 illustrate the results for Scenarios 1 and 2, respectively. As can be seen in the case of Scenario 1 (Figure 5.3), the correlation of traffic among the different systems shows a similar pattern, especially toward the middle periods of resource allocation. This traffic similarity reduces the effectiveness of spectrum sharing as the resource requests will likely exceed available resources, as in Period 3 of Figure 5.3. On the other hand, when the traffic patterns of different systems show less correlation, as in Figure 5.4, there is a better utilization of resources in the pool.

Interestingly, the average traffic load of each system (computed from the estimated traffic loads over the five observation periods) in both scenarios is the same. Hence, the inefficiency of spectrum sharing among systems with a similar traffic pattern is evident from Figure 5.3 as compared with Figure 5.4.

Note that the overall resource request pattern is dictated by the system with the largest operating bandwidth, in our case System 3. This needs 5 times more resource units than System 1, and 25 times more than System 2.

Table 5.1 Normalized Predicted Traffic Load in 5 Consecutive Periods for Each System in Scenario 1 and in Parenthesis for Scenario 2.

	System 1	System 2	System 3
Operating bandwidth	1 MHz	200 KHz	5 MHz
Normalized predicted traffic load in Period 2/Scenario 1 (Scenario 2)	0.9 (0.2)	1.0 (0.5)	0.2 (0.6)
Period 2/Scenario 1 (Scenario 2)	0.75 (1.2)	1.6 (0.7)	0.45 (0.4)
Period 3/Scenario 1 (Scenario 2)	1.6 (1.7)	2.0 (2.0)	0.95 (0.7)
Period 4/Scenario 1 (Scenario 2)	1.05 (1.9)	0.7 (1.7)	0.55 (0.5)
Period 5/Scenario 1 (Scenario 2)	1.2 (0.5)	0.6 (1.0)	0.65 (0.8)
Average traffic load Scenario 1 (Scenario 2)	1.10 (1.10)	1.18 (1.18)	0.56 (0.56)

Besides the predicted level of traffic, each RAN should base its decisions about resource requests on a revenue versus cost analysis. Assume that the revenue of service provisioning for a unit of spectrum is R_s and that the cost of borrowing a unit of spectrum from the pool is C_b. This cost can be interpreted as the aggregate cost associated with the cost of the rented spectrum for the duration of resource allocation and the costs of signaling overhead, retuning the transmitter/receivers to new channels, and the delay during the transition period from the old spectrum allocation to the new channels. Let us take the specific instant of time when the choice is made as the starting time for our calculations, at which time RAN m already has N_m units of spectrum and an accumulated revenue $W_m(0)$. Example 5.3 shows the profitability pattern of this RAN as a function of traffic load.

Example 3

Considering the parameters introduced already, and assuming the traffic pattern of System 1, Scenario 1 in Table 5.1, derive the revenue for System 1.

The revenue at each resource allocation period can be calculated as the sum of balances from all previous periods (i.e., the net profit or loss) with the revenue in the current period minus costs. For period i we can therefore derive

$$\text{Revenue} = D_m R_s,$$

$$\text{Cost} = (D_m - N_m) C_b.$$

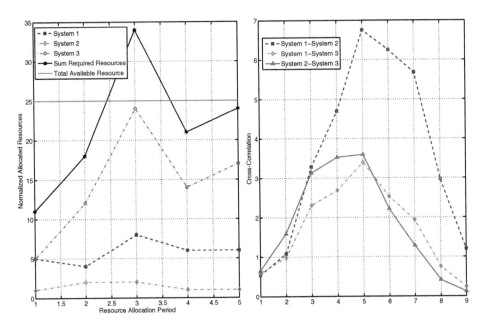

FIGURE 5.3

The required resource units for systems in Example 5.2, Scenario 1, and the correlation of the traffic pattern among these systems.

If, in a specific traffic load, more resources are available than required—that is, $N_m > D_m$—this cost reflects the revenue from renting the excess channels. To capture this effect, we write

$$W_m(i) = W_m(0) + D_m R_s - \frac{\frac{D_m}{N_m} - 1}{\left|\frac{D_m}{N_m} - 1\right|}(D_m - N_m)C_b.$$

The graph in Figure 5.5 plots results assuming $R_s = \$0.5$, $W_m(0) = \$1$, the initial availability of 10 resource units for this operator, and $C_b \in \{0.25, 0.5, 0.75, 1\}$. It is interesting to observe that, as long as the cost of borrowing a resource unit is equal to or greater than the achievable revenue per that resource unit, despite considerable discrepancies in the costs of borrowing, the final revenue at the end of the last resource allocation period will be the same. More discussion of the issue can be found in Problem 4 at the end of this chapter. ∎

As shown in Figure 5.1, spatial coordination is another important factor in sharing radio spectrum. Consider two wireless systems that want to share a specific channel: a cellular network (e.g., GSM/GPRS/EDGE) with an average cell radius of 1 km, and a broadcasting system (e.g., DVB-T/DVB-H) with an average cell radius of 15 km. Clearly, if the cellular system uses a frequency reuse plan as shown in

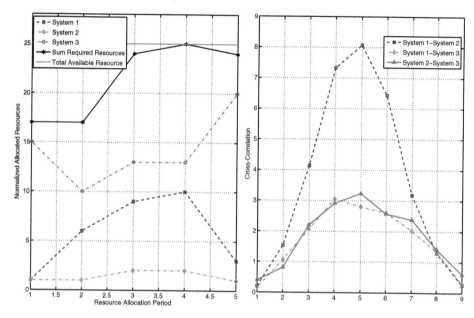

FIGURE 5.4

The required resource units for systems in Example 5.2, Scenario 2.

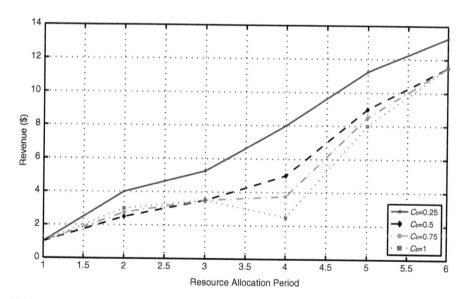

FIGURE 5.5

Accumulated wealth for System 1 with traffic pattern as in Scenario 1 of Example 5.3, considering different values for rental costs of a resource unit from the spectrum pool.

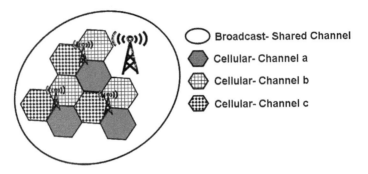

FIGURE 5.6

Spatial coordination considerations.

Figure 5.6, it is not possible to share channels a, b, or c with the broadcasting system due to interference implications. Even in the case where spectrum sharing is among systems with a similar coverage area, misplacement of the broadcast signals of the two systems can have a negative impact in terms of cochannel interference (CCI) [155]. This is because resource management of the systems will not be properly coordinated in the way that it would be for cells in a single system, issues such as cell-breathing affecting coverage, and transmission power differences among cells and among the systems.

5.4 SECONDARY SPECTRUM ACCESS

A promising approach to tackle the problem of spectrum underutilization is the secondary spectrum analysis (SSA) paradigm. The two categories of SSA solutions are real-time and non-real-time SSA, elaborated on in the following sections. A viable enabling technology to harness the potential of SSA is cognitive radio (CR). More detailed discussions on real-time and non-real-time SSA are presented in the next section.

There are two classes of SSA solutions, based on the method of coexistence of the primary and secondary systems. In the *overlay* approach, the secondary is allowed to access bands only where and when no primary communication is available. The second possibility is the *underlay* approach, where the primary accepts the possibility of interference up to a predefined (and agreed) threshold. In the former case, the main challenge is the accurate and timely sensing of the shared channel to identify the existence of the primary transmissions (for a more detailed analysis of spectrum sensing techniques, please refer to Chapter 4). After reliable detection of spectrum opportunities is achieved, adaptive transmission schemes should be employed to efficiently "aggregate" the randomly distributed spectrum holes into a wideband, high-data-rate communication channel. Multicarrier modulation schemes are invaluable enabling mechanisms to this end. In Section 5.6.4, we introduce a multicarrier CDMA transceiver capable of cognitive communication. In the latter case of primary–secondary coexistence—that is, the underlay approach—the interfering effect of the secondary's transmissions on the available primary receivers should be determined. This in turn mandates acquiring channel state information (CSI) to every primary receiver within the transmission range of the secondary transmitter. We investigate the underlay transmission principle through an OFDM-based underlay scheme in Section 5.6.5.

Secondary spectrum access, as leveraged through the deployment of cognitive radio technology, provides a vertical spectrum sharing solution in the sense that the secondary user of the spectrum is allowed to access the licensed band of the primary user only in adherence to strict access rules and the requirements of the primary system. The main requirement here is that of interference avoidance, so as not to degrade the primary system's quality of service. While real-time SSA is currently hotly researched in the literature and extensively discussed in Section 5.6, it is possible to provision for non-real-time SSA solutions too.

5.5 NON-REAL-TIME SSA

In some situations, the license owner of a specific band might be willing to allow a temporary secondary system to access that band for a specific time period and in a specific location. This spectrum sharing paradigm, which is referred to as *non-real-time SSA*, can be interpreted as the temporary suballocation of the rights to transmit in a band, at a certain time and space, to a system other than the license holder of the band.

Consider, for instance, a local council/school/university/church, which has received a license to broadcast its public service programs, such as training courses/public information, in a particular band. Such services are generally arranged for a limited duration of time, such as the opening time of the council. If the regulations concerning the award of the band's license allow, the license owner could rent the unutilized spectrum in a specific time (and location) when (and where) it is not being used to raise additional capital. This approach to sharing the medium has been used extensively in wired broadcasting services, such as cable TV, whereby a number of content providers share a specific cable channel to broadcast their content. Therefore, spectrum regulations are, perhaps, the most pertinent obstacle to this spectrum sharing approach, rather than specific technical solutions.

5.6 REAL-TIME SSA

The rest of this chapter focuses on technical solutions regarding the application of real-time SSA. In general, it is possible to classify such technical solutions into *negotiated* or *opportunistic* access approaches.

5.6.1 Negotiated Access

Whenever a medium for interaction between the primary and secondary systems exists, negotiated SSA is a possibility, depending on the negotiating terms. One such scenario is SSA among homogeneous wireless systems, say, a number of UMTS or WiMAX operators. In fact, many next-generation wireless standards are proposing architectures and tools facilitating resource sharing, such as described by the 3GPP [156]; such tools might, ultimately, be developed in contexts such as 3GPP LTE or IEEE 802.16h.

One interesting possibility assisting negotiated access is the availability of a universal signaling channel for cognitive purposes. Such a cognitive pilot channel (CPC) has been vigorously studied in, for instance, [157] and [158] and debated in collaborative colloquia and elsewhere [159]. One task of such a signaling channel could be to facilitate overlay SSA by indicating the availability of spectrum opportunities in a licensed band. To this end, the primary system might broadcast information (including access constraints, such as the permitted transmission power level and usage time interval) about an idle resource on the CPC, for potential secondary users to seize the chance to transmit on the offered resource [6]. Alternatively, under a different access paradigm, which puts the onus for protecting the spectrum on the primary system, the primary system might broadcast information about busy resources, so as to ban the secondary transmitter from interfering in those bands at specific times or locations. Such a paradigm is implicit in ideas like spectrum sharing though beacons [160]. Note that identification of an idle band, such as through broadcasts on the CPC, does not guarantee optimum prevention of interference to the primary receivers (or indeed optimum secondary use of resources), due to the "hidden node" problem, for example (see Figure 5.7).

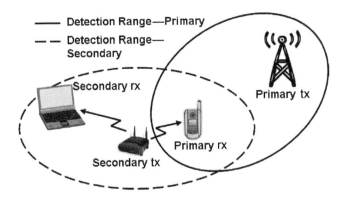

FIGURE 5.7

Hidden node problem, pertinent to primary–secondary spectrum access with the transmitter-centric interference detection approach.

Conversely, an essential consideration for secondary transmitters in an underlay SSA regime is the determination of the interference effects of their transmissions on the primary receivers. This calls for a receiver-centric approach to interference mitigation, as opposed to the current transmitter-centric approach. To this end, instead of identifying an idle spectrum band for secondary access, the CR needs to determine which transmission power should be used in a specific channel in order to mitigate interfering with the primary receivers.

Few studies in the literature have focused on developing a CPC for underlay SSA. We proposed the concept of the universal resource awareness channel (RAC) to address this issue [161]. The RAC is a common channel upon which all primary *receivers* must broadcast information about their resource usages and to which all cognitive radios must listen to assess the effects of their prospective transmission. A fixed transmission power for broadcasting information on RAC by all primary nodes is used, which facilitates the CRs estimating the channel quality to each primary receiver. Further information on the RAC concept can be obtained from [161].

Example 4

Investigate the scalability of the RAC concept with respect to the transmission power of a secondary system.

First, assume that the power of received signals at a radio can be expressed as

$$P_{rx} = P_{tx} \times kd^{-\alpha},$$

where P_{tx} and P_{rx} are the transmitted and received powers, respectively, and the constant k and path loss exponent α depend on the channel (generally, $\alpha \in [2, 4]$). This received power must be more than a threshold $P_{threshold}$ for a message transmitted on the RAC to be received. At the limit of operation, the coverage distance of RAC transmissions from a radio can therefore be formulated as

$$d = \left(\frac{kP_{tx}}{P_{threshold}} \right)^{1/\alpha}.$$

Assuming a fixed density ρ of radios per unit area, the number of radios m that should be able to be heard on the RAC by a reference radio (including the reference radio hearing its own transmissions) can therefore be ascertained as

$$m = \rho.\pi \left(\frac{k.P_{tx}}{P_{threshold}} \right)^{2/\alpha}.$$

If it is assumed that the bit length of the overhead (synchronization sequence) is D_J bits and the information part of the signaling frame is D_D bits, with frames of T second periods, the maximum number of radios n that can be served by the RAC is then given by

$$n = \left\lfloor \frac{C.T}{D_J + D_D} \right\rfloor,$$

where C is the capacity (in bits per second) of the RAC, and $\lfloor . \rfloor$ indicates the floor operation. In defining an acceptable success probability $P(\text{success})$ for a radio trying to transmit on the RAC, we therefore arrive at

$$
P(\text{success}) = \begin{cases} \dfrac{\left\lfloor \frac{C.T}{D_J+D_D} \right\rfloor}{\rho.\pi \left(\frac{k.P_{\text{tx}}}{P_{\text{threshold}}} \right)^{2/\alpha}}, & n < m, \\[2em] 1, & \text{otherwise.} \end{cases}
$$

■

5.6.2 Is Quality of Service Provisioning Possible in a Shared Band?

An important incentive to develop negotiated SSA solutions is the potential to guarantee quality of service (QoS), not only for the primary system (which should be an inherent design consideration in any SSA approach), but also for the secondary system, when possible. To demonstrate such a QoS guarantee possibility, let us discuss a negotiated SSA scenario as follows [162, 163].

Consider the interference channel set up in Figure 5.8, where it is assumed tx_1–rx_1 are the primary users and tx_2–rx_2 are the secondary users of a shared band. The total channel bandwidth B is divided into N subchannels, all with equal bandwidth w. Each subchannel is narrow enough to assume flat fading, and channel gains follow an independent and identical exponential distribution with unit mean. Furthermore, it is assumed that an underlay secondary spectrum access regime is utilized in this band, whereby the secondary link is allowed to transmit simultaneously with the primary system (we elaborate further on underlay SSA in Section 5.6.5). In such an underlay scheme, the goal of interference limit is to protect the primary receiver; however, the effect of interference at the secondary receiver should not be ignored. Hence, to generalize the problem formulation, we also consider a received interference threshold at the secondary receiver. As will be explained later, this constraint can simply be relaxed by proper choice of interference thresholds to ignore any limitation for the primary transmitter.

Consider the case where the objective of resource allocation in each link in Figure 5.8 is to minimize its transmission power so that a minimum rate (as a measure of QoS level) for that link can be guaranteed. Note that the constraint here is defined on the QoS rather than received interference, which was used in the introduction of underlay SSA. Here, the upper bound limit on interference in the underlay SSA definition is translated to a lower bound limit on the achievable rate.[2] The optimization problem (for link t) that can be used to model this scenario is defined as follows:

---

[2]The optimization problem of minimizing transmitted power subject to received interference limit results in the trivial solution of not transmitting at all.

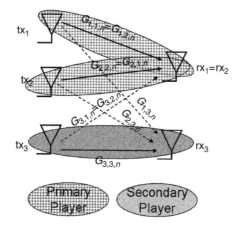

FIGURE 5.8

Two-link interference channel setup, pertinent to many spectrum sharing scenarios.

$$\underset{p_{i,n}}{\text{Minimize}} \sum_{n=1}^{N} \mathbb{E}_{g_{i,n}} \left\{ p_{i,n}\left(g_{i,n}\right) \right\}, \tag{5.1}$$

subject to

$$R_{\text{QoS},i} - \sum_{n=1}^{N} R_{i,n} \leq 0, \tag{5.2}$$

and

$$\mathbb{E}_{g_{i,n}} \left\{ \sum_{n=1}^{N} p_{i,n}\left(g_{i,n}\right) \right\} \leq P_{\text{max},i}, \tag{5.3}$$

where $\mathbb{E}_x(.)$ denotes expected value with respect to random variable x, $R_{\text{QoS},i}$ is the guaranteed rate of link i, $R_{i,n}$ is defined in Equation (5.4), and $p_{i,n}\left(g_{i,n}\right)$ is the allocated power, which is determined based on the direct channel gain of link i in subchannel n; that is, $g_{i,n}$. Constraint (5.2) guarantees an *instantaneous*[3] minimum rate for each link. Note that the power allocation in Equation (5.1) is not "explicitly" a function of the cross-channel gain $h_{i,n}$. The effect of interference of user i in this case is implicitly captured in the achievable rate of user j, where $i \neq j$. The achievable rate in each subchannel is given by

$$R_{i,n} = w \log \left[1 + \frac{p_{i,n}\left(g_{i,n}\right) g_{i,n}}{\sigma_i^2 + p_{j,n}\left(g_{j,n}\right) h_{j,n}} \right]. \tag{5.4}$$

---

[3]A more general *average* rate guarantee is an alternative goal, which provides more flexibility in allocating the resources.

To solve this resource allocation problem, we can rewrite constraint (5.2) as

$$\sum_{n=1}^{N} \left\{ \alpha_n R_{\text{QoS},i} - R_{i,n} \right\} \leq 0, \tag{5.5}$$

where $\alpha_n \in [0, 1]$ and $\sum_{n=1}^{N} \alpha_n = 1$ determines the allocation level of the guaranteed rate in subchannel n. Corresponding to this distribution of rate over subchannels, the total available power in Equation (5.3) should also be distributed as

$$\sum_{n=1}^{N} \left[\mathbb{E}_{g_{i,n}} \left\{ p_{i,n} \left(g_{i,n} \right) \right\} - \beta_n P_{\text{max},i} \right] \leq 0, \tag{5.6}$$

where $\beta_n \in [0, 1]$ and $\sum_{n=1}^{N} \beta_n = 1$. For simplicity of analysis, let us assume uniform distribution of power and rate over all subchannels; that is, $\alpha_n = \beta_n = 1/N$. This simplifying assumption suffices for our purpose of demonstrating the QoS guarantee capability of negotiated SSA. Then, the problem of minimizing the transmitted power subject to instantaneous rate guarantee for link $i \in \{1, 2\}$ is effectively decomposed into N subproblems; that is,

$$\underset{p_{i,n}}{\text{Minimize}} \, \mathbb{E}_{g_{i,n}} \left\{ p_{i,n} \left(g_{i,n} \right) \right\}, \tag{5.7}$$

subject to

$$\alpha_n R_{\text{QoS},i} - R_{i,n} \leq 0 \tag{5.8}$$

and

$$\mathbb{E}_{g_{i,n}} \left\{ p_{i,n} \left(g_{i,n} \right) \right\} \leq \beta_n P_{\text{max},i}. \tag{5.9}$$

This new subchannel-based problem can be easily solved by rewriting constraint (5.8) in terms of the power instead of the rate for both links; that is,

$$\mathbf{S}_n \times p_n \succeq \begin{bmatrix} \sigma_1^2 \left(e^{\alpha_n R_{\text{QoS},1}/w} - 1 \right) \\ \sigma_2^2 \left(e^{\alpha_n R_{\text{QoS},2}/w} - 1 \right) \end{bmatrix}, \tag{5.10}$$

where $\succeq$ denotes elementwise inequality, $\mathbf{S}_n$ is defined as

$$\mathbf{S}_n = \begin{bmatrix} g_{1,n} & -\left(e^{\alpha_n R_{\text{QoS},1}/w} - 1 \right) b_{2,n} \\ -\left(e^{\alpha_n R_{\text{QoS},2}/w} - 1 \right) b_{1,n} & g_{2,n} \end{bmatrix}, \tag{5.11}$$

and $p_n = [p_{1,n} \, p_{2,n}]^T$.

To satisfy the minimum rate requirement in Equation (5.8) and the consequent inequality (5.10) with a nonnegative minimum power allocation, the matrix $\mathbf{S}_n$ should be invertible and elementwise positive, resulting in the minimum power allocation

$$\begin{bmatrix} p_{1,n}^* \\ p_{2,n}^* \end{bmatrix} = \mathbf{S}_n^{-1} \times \begin{bmatrix} \sigma_1^2 \left(e^{\alpha_n R_{\text{QoS},1}/w} - 1 \right) \\ \sigma_2^2 \left(e^{\alpha_n R_{\text{QoS},2}/w} - 1 \right) \end{bmatrix}. \tag{5.12}$$

A closer look at the structure of $\mathbf{S}_n$ in Equation (5.11) reveals this matrix belongs to the Z-matrix class. A Z-matrix is a square matrix with nonpositive off-diagonal elements [164]. It is known that, for a Z-matrix to have an elementwise positive inverse, it should also be a P-matrix [164]. A P-matrix is a Z-matrix with all positive principal minors. In the case of matrix $\mathbf{S}_n$ in Equation (5.11), if the determinant of $\mathbf{S}_n$ is positive, it will be a P-matrix; that is, we should have

$$g_{1,n}g_{2,n} > h_{1,n}h_{2,n}\left(e^{\alpha_n R_{QoS,1}/w} - 1\right)\left(e^{\alpha_n R_{QoS,2}/w} - 1\right). \tag{5.13}$$

Inequality (5.13) provides an optimal coexistence rule in subchannel n for the primary and secondary links. In other words, each of the N subchannels of the shared band that satisfy condition (5.13) can sustain a minimum level of QoS for the primary and secondary links, for the duration of that resource allocation period. Recall that we assume, during each resource allocation period, the channel-fading states remain constant. Those subchannels not satisfying this condition will not be shared, and only the primary system will use them. It is obvious that, as the fading states of the channel vary over time, so does the relation of channel gains in Equation (5.13), making this condition also a function of time. The power allocation defined in Equation (5.12) is the optimal power allocation for problem (5.7) subject to Equations (5.8) and (5.9). This optimal power should be allocated for each link independently, as the primary and the secondary links operate independently. However, each link can decide only if channel conditions are desirable—that is, Equation (5.13) is satisfied—if both links share their CSI information as well as QoS requirement through the SSA negotiation medium.

Condition (5.13) is an interesting generalization of the similar relationship between direct and cross-channel gains, as reported in [165] and [166]. Both [165] and [166] show that in a two-link scenario with shared spectrum, if the multiplication of cross-channel gains is higher than the multiplication of direct-channel gains between two transmitter/receiver pairs, the sharing of this channel causes more interference associated with cross-channel gains than desired transmissions associated with the direct-channel gain. Hence, the optimum rate in those channel conditions can be achieved by *orthogonal* transmission of the two links (i.e., nonoverlapping transmissions of the primary and secondary signals in the time, frequency, or code domain) instead of sharing the whole band. In our case, Equation (5.13) shows that the same concept is true in situations where a specific QoS level for the primary and secondary system is guaranteed. Therefore, among other possibilities, this equation can be used as the criterion for negotiating the guaranteed level of QoS for the secondary system, as described in the application scenario example that follows.

The QoS requirements of the primary users are determined based on the service application and are fixed. If the channel condition is desirable—that is, if there are high direct channel gains for both primary and secondary links and low cross-channel gains—a level of QoS for the secondary user can also be guaranteed.

In cases where the interfering signal of the secondary user dominates the desired signal of the primary user, the secondary user is not allowed to operate in the band as the inequality (5.13) demands. Therefore, the channel gains and the level of QoS form a trade-off in deciding whether or not to share a subchannel.

Example 5

In what sorts of applications can such a negotiated SSA approach be utilized?

Consider a licensed operator as a primary system, which is keen to earn extra revenue by increasing the utilization of its licensed band, hence is willing to accommodate a nonzero QoS level for secondary users when possible. The admittance of secondary users with nonzero QoS might decrease the maximum feasible throughput of the primary users; however, under no circumstance should it result in QoS levels falling below the desired threshold of the primary users.

As Figure 5.9 shows, guaranteeing a certain level of QoS for the primary system means that the maximum achievable rate for the secondary users is restricted. Knowing its best-effort QoS status, the secondary system uses this secondary spectrum access for its noncritical traffic such as web browsing, file downloads, or peer-to-peer file sharing with other secondary users. In each resource allocation period, the primary and secondary users exchange their CSI as well as requested rate; that is, $R_{QoS,i}$. Upon checking condition (5.13), the feasibility of requested QoS of the links is ascertained. If, due to channel condition, $R_{QoS,2}$ cannot be satisfactorily sustained, the secondary user is not allowed to access the band.

In this situation the secondary system has two choices. Either it can wait for the following resource allocation periods, when channel conditions might improve to support the requested rate, or it can reduce its rate requirement and renegotiate the secondary spectrum access. Note that another trade-off for the secondary link exists in this case: The waiting time for a better channel condition can be rewarded with a shorter data transfer period due to the higher achievable data rates for the secondary user over that period. ■

Example 6

Investigate the effect of the primary user's QoS level on the allocated power and achievable rates for the simple two-user scenario shown in Figure 5.9.

Solving the optimization problem (5.7) subject to Equations (5.8) and (5.9), it is clear from Figure 5.10 that the optimum power allocation tends to converge to a point where the achievable rates are near (or equal to) the minimum rate requirement. As the trade-off between channel condition and minimum QoS level changes—for example, due to new QoS requirements as in Figure 5.10—the resource allocation might reach an infeasible channel sharing region. In such cases, the primary system does not allow the secondary system to access that channel, and all the resources are allocated to the primary users. Hence,

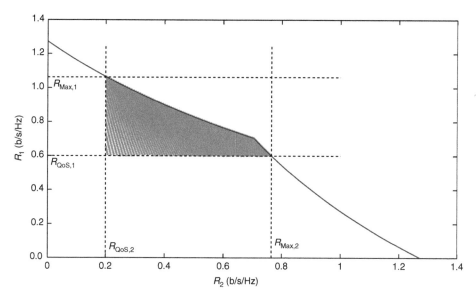

FIGURE 5.9

Feasible rate region (shaded) for the case where $g_1 = g_2 = 1$, $h_1 = h_2 = 0.25$, $P_{max}^1 = P_{max}^2 = 5$ W and the minimum rate requirements of the primary player $R_{QoS,1} = 0.6$ b/s/Hz, the requested rate of the secondary player $R_{QoS,2} = 0.2$ b/s/Hz, and $\sigma_1^2 = \sigma_2^2 = 1$.

condition (5.13) provides a useful feasibility test before attempting to solve the relevant optimization problem for sharing subchannels between primary and secondary systems. Obviously, given any specific channel condition for the primary user, a very high QoS level for the primary user might not be sustainable, even if the secondary user is not allowed to access the band and the primary user transmits with the maximum transmission power. The region of such unguaranteed high QoS levels is a subset of the infeasible channel sharing region, shown in Figure 5.10.

5.6.3 Opportunistic Access

In many real-time SSA scenarios, the possibility of interaction between primary and secondary systems does not exist. This might be due to the lack of a signaling channel such as the CPC or RAC, as discussed in the previous section, or simply might be a result of the nature of the primary system, for instance, the primary system being a TV channel or a radar system. Under such circumstances, the secondary system might instead "opportunistically" identify and utilize idle resources, as appropriate. Note that it is also possible to envision other opportunistic SSA scenarios, applicable to cases where signaling channels do exist.

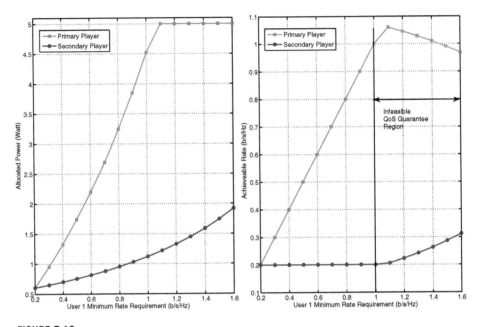

FIGURE 5.10

Optimum allocated power and achievable rate for the case of Figure 5.9 with $R_{QoS,2} = 0.2$ b/s/Hz and variable $R_{QoS,1}$.

Spectrum opportunities in opportunistic SSA contexts arise in two forms: overlay access, also known as *spectrum holes* or *white spaces*, where a portion of the band is left completely idle by the primary users, and *underlay access*, also known as *gray spaces*, where although the primary system is active on a specific band, it can be determined by the secondary system that its transmissions will not impose an inadmissible level of interference to the primary system [11]. To identify transmission opportunities within either the spectrum hole or the underlay approach, reliable spectrum sensing techniques can be exploited. The reliability of such sensing techniques is most challenging when uncertain elements exist in the channel due to shadowing and fading or in the receiver due to thermal noise [115, 167].

Spectrum sensing techniques are covered in Chapter 4 of this book. Therefore, in this chapter, we propose technical solutions to *utilize* the identified opportunities.

5.6.4 Overlay Approach

A CR should ideally sense a large swath of spectrum, to identify sufficient secondary access opportunities. There is a high probability that, at any given time and location, a CR will identify several smaller and randomly located idle bands, rather than a single large chunk of idle bandwidth that it can opportunistically access. It is therefore necessary to "aggregate" smaller noncontiguously located spectrum opportunities to create a higher data rate transmission for cognitive communication. To this end, multicarrier modulations schemes are appropriate tools for opportunistic spectrum access. We discuss some such schemes later.

Currently, many wireless communication standards are being developed based on orthogonal division frequency multiplexing (OFDM). The capability of OFDM modulation for spectrum sharing has been known for some time [24]. In [24], using an allocation vector—that is, a vector of zeros and ones for the bins of the inverse fast Fourier transform (IFFT) block at the OFDM transmitter—the use of portions of the band can be banned to mitigate interference. Furthermore, [168] proposes an OFDM-based design for cognitive operations, namely, discontiguous OFDM. The proposed technique in [168] also follows the methodology of deactivating fixed-bandwidth subbands overlapping with the primary system, to mitigate interference. However, even taking into account this subband deactivation at the secondary transmitter, there is still the possibility of interference in the nullified subbands, due to the side lobes of adjacent OFDM subbands. These side lobes do not impose any interference on the secondary OFDM-based receiver, as they are zero at the sampling instants (i.e., the orthogonality of OFDM subbands). However, given that the primary receiver is not synchronized with the secondary transmitter, they do cause interference to the primary receiver. One solution to eliminate these interfering side lobes is to use a filter after the IFFT process in the secondary transmitter; however, this causes performance degradation due to the loss of orthogonality of the OFDM subbands [168]. Alternatively, not only the overlapped subchannels, but also adjacent OFDM subchannels to the primary's signal should be deactivated in a bid to create a "guard band" between the primary and secondary systems. This approach, however, is not spectrally efficient, and spectral efficiency is the very issue that primary–secondary spectrum sharing is trying to solve.

Another multicarrier modulation approach is to use the principles of direct sequence CDMA (DS-CDMA) to create multicarrier CDMA (MC-CDMA) [169]. In single-carrier DS-CDMA, data symbols are spread using a unique, very long spreading code per user, which results in distribution of signal power over a large bandwidth. MC-CDMA uses shorter spreading codes, spreading data symbols over a series of disjoint carrier frequencies. Use of such multicarrier modulations is desirable due to the achieved frequency diversity and effective mitigation of frequency selective fading.

Let us now focus on the problem of aggregating a number of spectrum holes without creating uncontrolled interference to the primary receivers. Interference to primary receivers when transmitting using a MC-CDMA system can be avoided through using the subband deactivation technique, similarly to the OFDM case. A comparison of noncontiguous OFDM (NC-OFDM) and noncontiguous MC-CDMA (NC-MC-CDMA) has been investigated in [170]. It is shown that, as the number of deactivated subbands increases, the bit error rate (BER) performance degradation

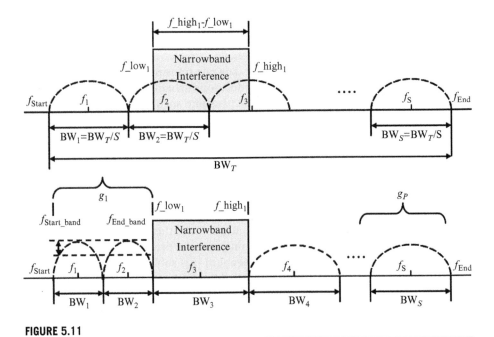

FIGURE 5.11

Comparison of spectrum partitioning for legacy and adaptive MC-CDMA in the presence of a narrowband interfering signal; that is, the primary signal in the channel.

of NC-MC-CDMA becomes more than for the NC-OFDM approach. We refer the reader to Chapter 6 of this book for a detailed discussion of agile transmission techniques.

Instead of deactivating fixed-bandwidth subbands, we proposed a cognitive MC-CDMA that can adaptively change its transmission parameters, such as bandwidth and power of subbands, according to the interference pattern in the shared channel [171, 172]. This approach is shown in Figure 5.11. Compared to the subband deactivation method, our method increases spectrum usage efficiency considerably, because it excludes only the portion of the band affected by the interference. Also, the proposed method avoids the OFDM interference leakage problem [168]. The bottleneck for cognitive MC-CDMA, however, is the availability of sharp adjustable filters and appropriate spreading codes. It is worth noting that due to the spread-spectrum nature of this approach, the imperfect implementation of filters results in only minimal interference on legacy systems.

To demonstrate the spectrum aggregation requirements in general and more specifically the capabilities of the cognitive MC-CDMA system just proposed, consider the following scenario focusing on DL operation of this cognitive MC-CDMA system. Suppose a number of legacy (primary) systems are operating in different bands of a chunk of spectrum, while at the same time a cognitive system wishes to utilize the idle bands. Note that, by assuming the primary systems to be *legacy*, we imply that no means of negotiated SSA exist. As an example, consider the primary spectrum band to be dedicated to TV broadcast, whereby each TV station occupies a different channel in that band. We assume that the primary systems' usage of spectrum varies temporally (and spatially), but at much longer intervals compared with the resource allocation period of the cognitive MC-CDMA system. Referring back to our TV broadcasting example, this assumption is equivalent to assuming different TV stations might broadcast their programs at different times covering different areas. We further assume that the bandwidth and the frequency allocation of the primary systems are available at the cognitive MC-CDMA transmitter.[4]

Upon receiving the information of the radio frequency (RF) environment, the cognitive MC-CDMA base station (BS) adaptively changes its transmission parameters and informs CR users about changes in its impulse modulator, chip wave-shaping filters, and RF modulators. The cognitive MC-CDMA BS forms the data to be sent to K active secondary users. A different spreading code—that is, a signature sequence—is assigned not only to each user (in compliance with the CDMA nature of the system), but also for each subband allocated to that user, due to the variation of the bandwidth from subband to subband. The variable bandwidth alters the spreading code length and makes it impossible to use the same spreading code in all subbands, for a given user; this is a major difference compared to conventional MC-CDMA designs [169], where the same chip rate is applied in all subbands. The chips associated with the signature sequences have the same duration, $T_{c,s}, s = 1, 2, \ldots, S$, where S is the number of subbands. Therefore, in total, $K \times S$ codes per cell are needed, as opposed to the use of K codes in the conventional case. We assume that perfect codes with the required length exist.

The spread data symbols of secondary users modulate an impulse train with energy per chip equal to E_c. The amount of energy saved by not transmitting in certain bands is used to increase E_c in the other bands, another major difference with conventional MC-CDMA. A chip wave-shaping filter, with impulse response $h_s(t)$ for the sth subband, creates the Nyquist pulse shape for the modulated impulse train. The summed pulse shape of all users is up-converted to the target RF subband with corresponding subband amplitude A_s, which is adaptively calculated according to power constraints. Finally, all modulated signals are added to create the multicarrier wideband signal.

We assume the waveform $h_s(t)$ has a unit-energy and root raised-cosine shape with a roll-off factor of α_s ($0 \leq \alpha_s \leq 1$) [108]. The choice of root raised-cosine filter is common in many communications systems due to its desirable characteristics.

---

[4]There are two main approaches to obtaining the spectral occupancy information of the primary system. As an example for accessing TV white spaces, cognitive radios (CRs) either use sensing techniques or access geolocation databases to determine in a specific locale which channels are occupied and where the TV transmission towers are located. In our scenario, such information can be obtained in an efficient centralized fashion—that is, by the base station of the cognitive system—based on the measurements of the cognitive users, similar to the IEEE 802.22 approach [18].

One such characteristic is that if, in the receiver, a matched filter is utilized, the received signal will have a raised cosine shape, ideal for intersymbol interference (ISI) mitigation. In the frequency domain we have

$$X_s\left(f\right) = H_s\left(f\right)H_s^*\left(f\right),$$

or equivalently in the time domain

$$x_s(t) = F^{-1}\{|H_s(f)|^2\}, \tag{5.14}$$

where F^{-1} denotes inverse Fourier transform and $H_s(f)$ is the Fourier transform of $h_s(t)$. The assumption of unit energy means

$$\int_{-\infty}^{+\infty} |H_s\left(f\right)|^2 df = 1.$$

The bandwidth of the subbands—that is, BW_s—might change from band to band, based on the size and the location of the narrowband primary signals. Bandwidth is related to the chip duration and the roll-off factor as

$$BW_s = (1 + \alpha_s)\frac{1}{T_{c,s}}, \quad s = 1, 2, \dots, S. \tag{5.15}$$

The overall bandwidth of the subbands should satisfy the constraint $\sum_{s=1}^{S} BW_s = BW_T$, where BW_T is the fixed total available bandwidth.

Let us index subbands in ascending order. Starting from the lowest frequency and accounting for the location of the (interfering) primary signals, we denote the set of indices of M subbands, where interfering signals occur in subbands specified by the set G. In this formulation, the transmitted signal can be written as

$$S_{MC}(t) = \sum_{k=1}^{K} \sum_{\substack{s=1 \\ s \notin G}}^{S} \sum_{n=-\infty}^{+\infty} \sum_{i=1}^{N_s} A_s d_k(n) c_{k,s}(i) \times h_s(t - iT_{c,s}) \cos(\omega_s t + \theta_s),$$

where $d_k(n)$ represents the nth data symbol of the kth user, and $c_{k,s}(i)$ is the ith chip of the sth subband spreading code (signature sequence) for $i = 1, 2, \dots, N_s$. We denote the number of chips per symbol by N_s and assume binary phase shift keying (BPSK) signaling.

Consider a slow-varying, frequency-selective Rayleigh fading channel with delay spread of T_m and coherence bandwidth $(\Delta f)_c$. It is assumed, to avoid ISI distortion, that each band's chip duration, $T_{c,s}$, is greater than the delay spread, and that each subband width (BW_s) is greater than the coherence bandwidth of the channel. Hence, we can relate the roll-off factor α_s, $T_{c,s}$ and T_m, as

$$\alpha_s \geq \left(\frac{T_{c,s}}{T_m} - 1\right), \quad s = 1, 2, \dots, S, \text{ and } s \notin G, \tag{5.16}$$

where

$$T_m \leq T_{c,s} \leq 2T_m, \quad s = 1, 2, \dots, S, \text{ and } s \notin G, \tag{5.17}$$

to satisfy $0 \leq \alpha_s \leq 1$. Therefore, given a desired subband width BW_s, we can compute the required roll-off factor α_s and chip duration $T_{c,s}$, using Equations (5.15), (5.16), and (5.17). It is desirable to keep the roll-off factor in the range $0.18 \leq \alpha_s \leq 0.25$.

Based on these assumptions, the complex baseband equivalent impulse response of the sth channel, as seen by the kth user, is $c_s(t) = \alpha_{k,s} \exp(j\beta_{k,s})\delta(t)$, where $\alpha_{k,s}$ and $\beta_{k,s}$ are independent and identical random variables. The coefficients $\alpha_{k,s}$ have Rayleigh distribution with second-order moment of unity and the $\beta_{k,s}$ follow uniform distribution over $[0, 2\pi)$. The received signal is a combination of the MC-CDMA signal, narrowband primary interference, and noise, as

$$r(t) = \sum_{k=1}^{K} \sum_{\substack{s=1 \\ s \notin G}}^{S} \sum_{n=-\infty}^{+\infty} \sum_{i=1}^{N_s} \left\{ A_s \alpha_{k,s} d_k(n) c_{k,s}(i) \times h_s(t - iT_{c,s} - \tau_k) \cos(\omega_s t + \theta'_{k,s}) \right\}$$
$$+ n_w(t) + \sum_{l \in G} n_{j,l}(t),$$

where $n_w(t)$ is additive white Gaussian noise with a double-sided power spectral density (psd) of $N_0/2$, $n_{j,l}(t)$ is the interfering narrowband signals in subband $l \in G$, and ω_s is the center frequency of the subband s. The overall received phase is $\theta'_{k,s} = \theta_s + \beta_{k,s}$, where θ_s is the transmitted phase associated with ω_s.

Following adaptations at the transmitter, the cognitive receiver varies the bandwidth and the center frequency of each subband filter. Novel solutions for practical implementation of such adaptive and adjustable filters have been vigorously studied in past work, for example, using RF-MEMS [173]. In our case, it is assumed that the receiver obtains information on the center frequency and bandwidth of subbands through a signaling channel from the transmitter. The required output of the bandpass filter for the sth subband ($s \notin G$) can be written as

$$y_s(t) = \sum_{k=1}^{K} \sum_{n=-\infty}^{+\infty} \sum_{i=1}^{N_s} \left\{ \sqrt{2E_c} A_s \alpha_{k,s} d_k(n) c_{k,s}(i) \times x_s(t - iT_{c,s} - \tau_k) \cos(\omega_s t + \theta'_{k,s}) \right\}$$
$$+ BP\{n_w(t)\},$$

where $BP\{.\}$ represents the bandpass filtering operation and $x_s(t)$ is given in Equation (5.14). As this equation shows, the narrowband interfering signals—that is, the primary system's signal in the channel—have been removed at the secondary receiver. In practice, however, since the roll-off of practical filters is not steep, it might be necessary to allocate appropriate guard bands between the narrowband signals and the subbands of the cognitive MC-CDMA system to avoid leakage.

As discussed in this section, the main responsibilities of CRs in an overlay SSA scheme are to first identify the spectrum holes reliably and aggregate the identified transmission opportunities, for instance, through adaptive multicarrier modulation techniques. In the next section, we study the problem of underlay SSA in more detail.

5.6.5 Underlay Approach

While identifying and using spectrum holes provides higher insurance in interference mitigation toward primary users, far more spectrum capacity can be leveraged through the secondary system transmitting in portions of band that are being actively used by the primary system. This, of course, must be done such that the operation of the secondary system does not cause performance degradation on the primary system through imposing an unacceptable level of interference to it.[5]

A *deterministic* solution for such an underlay scheme is ultra-wideband (UWB) transmission. By *deterministic*, we are referring to the fact that transmission power and the occupied bandwidth of UWB systems is independent of any primary system and fixed through regulations. The stringent transmission power mask defined by spectrum regulatory bodies for UWB transmission, together with the relatively high operating frequency of UWB, ensures a very localized interference pattern toward any nearby primary receiver. The specifications and performance analysis of UWB systems is out of scope of this book.

Another interesting solution for opportunistic underlay SSA can be envisioned in cases where the secondary transmitter can receive and "decode" the primary transmitter's message *before* the primary receiver. In such situations, powerful coding techniques, such as dirty paper coding [174], can be utilized to allow the secondary link to share the band with infinitesimal interaction with the primary system [175]. In this context, the knowledge of the channel as well as transmitted data are effectively utilized to undo the received interference at the primary receiver, given a known channel condition. A thorough analysis of this approach is presented in Chapter 11.

Further opportunities for underlay access can be identified through random variations of the shared channel. More specifically, if the signal power of the secondary transmissions is considerably attenuated before reaching the primary receiver, through channel propagation phenomena such as path loss, shadowing, or fading, coexistence of primary and secondary systems is indeed possible. Such desirable channel conditions, however, follow random patterns, usually characterized by their statistical characteristics. Consider Figure 5.8 depicting an interference channel where $tx_1 - rx_1$ are the primary and $tx_2 - rx_2$ are the secondary users of a band. There are N subchannels in a total channel bandwidth B, all with the same bandwidth w. Each subchannel is narrow enough for flat fading to be a realistic

---

[5]In an overlay cognitive transmission scheme, as discussed previously in this chapter, the secondary cognitive transmitter identifies the absence of the primary transmission in the band before starting its own transmission. In such scenarios there is a nonzero probability that the secondary transmitter erroneously decides a band is free to use while primary signal is still transmitting in the band. Such unintentional interference is outside the scope of this chapter but is discussed in Chapter 4.

assumption, and channel gains follow an independent and identical distribution with unit mean. We have already used this model to study a negotiated SSA approach in Section 5.6.1, which was also an underlay solution.

In this section we extend our analysis to an opportunistic underlay case. In such an underlay scheme, the goal of having an interference limit is to protect the primary receiver, but the effect of interference at the secondary receiver should not be ignored. Hence, to generalize the problem formulation we also consider a received interference threshold at the secondary receiver. As will be explained later, this choice can simply be relaxed by proper setting of the interference thresholds to remove any limitation for the primary transmitter.

As the primary and secondary links are operating independently, we seek distributed mechanisms for resource allocation as opposed to the negotiated case presented in Section 5.6.1. Considering N subchannels in this scenario, the instantaneous throughput of link i in Figure 5.8 can be derived as

$$R_i = \sum_{n=1}^{N} R_{i,n} = \sum_{n=1}^{N} w \log\left(1 + \frac{p_{i,n}\left(g_{i,n}, h_{i,n}\right) g_{i,n}}{\sigma_i^2 + p_{j,n}\left(g_{j,n}, h_{j,n}\right) h_{j,n}}\right), \tag{5.18}$$

where $g_{i,n}$ and $h_{i,n}$ are the independent and identical random values of the direct and cross-channel gains for link i in subchannel n, respectively, for all $i, j \in \{1, 2\}$ and $i \neq j$. We assume Rayleigh fading channels, which means $g_{i,n}$ and $h_{i,n}$ have independent exponential distributions. Without loss of generality, we assume the noise level at receiver i is equal in all subchannels. Note that, in contrast to Equation (5.4), the power allocation is a function of both direct and cross-channel gains, $g_{i,n}$ and $h_{i,n}$, respectively. The ergodic capacity of this link can be defined as follows:

$$\mathbb{C}_{ER,i} = \underset{g_{i,n}, h_{i,n}}{\text{Max}} \; \mathbb{E}_{g_{i,n} h_{i,n}}\left\{\sum_{n=1}^{N} R_{i,n}\right\}, \tag{5.19}$$

subject to

$$\begin{cases} \mathbb{E}_{g_{i,n}, h_{i,n}}\left[\sum_{n=1}^{N} p_{i,n}\left(g_{i,n}, h_{i,n}\right) h_{i,n}\right] \leq \Gamma_{\text{avg},i} \quad \text{(a)}, \\ \sum_{n=1}^{N} p_{i,n}\left(g_{i,n}, h_{i,n}\right) h_{i,n} \leq \Gamma_{\text{inst},i} \qquad \text{(b)}, \end{cases} \tag{5.20}$$

and

$$\mathbb{E}_{g_{i,n}, h_{i,n}}\left\{\sum_{n=1}^{N} p_{i,n}\left(g_{i,n}, h_{i,n}\right)\right\} \leq P_{\text{max},i}. \tag{5.21}$$

Here $P_{\text{max},i}$ is the average transmit power limit of transmitter $i \in \{1, 2\}$ and $\Gamma_{\text{avg},i}$ and $\Gamma_{\text{inset},i}$ are the average and instantaneous total interference limits for link i, respectively. The interference limit constraint defined in Equation (5.20a) is active when an average interference limit policy is imposed by the primary licensee, whereas Equation (5.20b) is valid if an instantaneous interference limit is desired. We assume only one of these types of interference limit constraints can be chosen by the primary link in the shared band. By defining $\Gamma_{avg,1} = +\infty$ or $\Gamma_{inset,1} = +\infty$, the effect of received interference limit on primary transmitter is relaxed. In the rest of this section, without loss of generality, we assume a bounded value for the interference threshold for the primary transmitter.

Note that our setup incorporates several important differences from available results in literature such as [176, 177, and 178]. First we investigate the "mutual" interfering effect of the primary and secondary links, whereas the problem formulations in the aforementioned references ignore the effect of the primary link on the secondary's performance. In our approach, on the other hand, this interference effect is studied, as it will affect the power allocation policy of the secondary link. Second, we assume two transmit power caps in our scenario are available simultaneously. The average transmit power of each link is limited, as in Equation (5.21), due to regulation or battery constraints. Furthermore, constraint (5.20a) or (5.20b) imposes a second transmit power cap due to the received interference limit policy of the band. The importance of constraint (5.21), not present in [176, 177, and 178], is its limiting factor when deep fades are reducing the cross-channel gain from transmitter i to receiver j—that is, $h_{i,n}$—to near zero values. Then, constraint (5.20) will result in unbounded values for transmission power $p_{i,n}$, which is obviously not realistic. Finally, while [177 and 178] studied only the secondary link's capacity, we elaborate on both the primary and secondary links' capacity. The analysis of the primary link's capacity, beside that of the secondary, enable us to better recognize the trade-offs involved in instantaneous versus average interference limit policies, resulting in adoption of more efficient resource allocation solutions.

Instantaneous Interference Limit

In this subsection we address the optimization problem defined by (5.19) subject to Equations (5.20b) and (5.21). Since we assume independent primary and secondary links, cooperative transmission or interference cancellation techniques are not exploited in this scenario. To solve this problem, one can decompose the problem into N subproblems as follows. First, constraint (5.20b) is rewritten as

$$\sum_{n=1}^{N} \left\{ p_{i,n} \left(g_{i,n}, h_{i,n}\right) h_{i,n} - \alpha_n \Gamma_{inst,i} \right\} \leq 0, \tag{5.22}$$

where $\alpha_n \in [0, 1]$ and $\sum_{n=1}^{N} \alpha_n = 1$ determines the distribution of the total interference limit over the N subchannels. Corresponding to this distribution of interference over subchannels, the total available power in Equation (5.21) should also be distributed as

$$\sum_{n=1}^{N} \left[\mathbb{E}_{g_{i,n}, h_{i,n}} \left\{ p_{i,n} \left(g_{i,n}, h_{i,n}\right) \right\} - \beta_n P_{max,i} \right] \leq 0, \tag{5.23}$$

where $\beta_n \in [0, 1]$ and $\sum_{n=1}^{N} \beta_n = 1$. We assume uniform distribution of power and rate over all subchannels; that is, $\alpha_n = \beta_n = 1/N$ in this chapter.

Now in the spectrum sharing scenario with instantaneous received interference limits over a specific subchannel n, it is straightforward to observe that the optimum power allocation for link i is obtained via the greedy algorithm as

$$p_{i,n}^{\text{inst}*} = \text{Min}\left(P_{\max,i}, \frac{\Gamma_{\text{inst},i}}{b_{i,n}}\right), \quad \text{for all } i \in \{1,2\}. \tag{5.24}$$

Note that the greedy power allocation Equation (5.24) depends on only the fading states of the cross-channel from transmitter i to receiver j, i.e., $b_{i,n}$. The ergodic capacity of each link in this case can be calculated by substituting Equation (5.24) into Equation (5.19). This yields

$$\mathbb{C}_{ER,i}^{\text{inst}} = \int_0^\infty \int_{b_{i,n} > \frac{\Gamma_{\text{avg},i}}{P_{\max,i}}}^\infty R_{i,n}\left(\frac{\Gamma_{\text{avg},i}}{b_{i,n}}\right) f_b\left(b_{i,n}\right) f_g\left(g_{i,n}\right) \, db_{i,n} dg_{i,n}$$

$$+ \int_0^\infty \int_{b_{i,n} < \frac{\Gamma_{\text{avg},i}}{P_{\max,i}}}^\infty R_{i,n}\left(P_{\max,i}\right) f_b\left(b_{i,n}\right) f_g\left(g_{i,n}\right) \, db_{i,n} dg_{i,n}, \tag{5.25}$$

where $f_b(\cdot)$ and $f_g(\cdot)$ denote the probability density functions (pdf) of the channel gains $b_{i,n}$ and $g_{i,n}$, respectively.

Average Interference Limit

Now let us investigate the optimal power allocation strategies under the average received interference limit; that is, (5.19) subject to Equations (5.20a) and (5.21). The Lagrangian function of this optimization problem can be written as

$$L\left(p_{i,n}(g_{i,n}, b_{i,n}), \lambda_0, \eta_0\right) = \mathbb{E}_{g_{i,n}b_{i,n}}\left\{\sum_{n=1}^N w\log\left(1 + \frac{p_{i,n}\left(g_{i,n}b_{i,n}\right)g_{i,n}}{\sigma_i^2 + p_{j,n}\left(g_{j,n}b_{j,n}\right)b_{j,n}}\right)\right\}$$

$$- \lambda_0\left(\mathbb{E}_{g_{i,n},b_{i,n}}\left[\sum_{n=1}^N p_{i,n}\left(g_{i,n}, b_{i,n}\right)b_{i,n}\right] - \Gamma_{\text{avg},i}\right)$$

$$- \eta_0\left(\mathbb{E}_{g_{i,n},b_{i,n}}\left\{\sum_{n=1}^N p_{i,n}\left(g_{i,n}, b_{i,n}\right)\right\} - P_{\max,i}\right). \tag{5.26}$$

Here λ_0 and η_0 denote the Lagrange multipliers for the constraints (5.20a) and (5.21), respectively. The optimal power allocation satisfies the necessary condition

$$\frac{\partial L\left(p_{i,n}(g_{i,n}, b_{i,n}), \lambda_0, \eta_0\right)}{\partial p_{i,n}} = 0,$$

which yields

$$p_{i,n}^*\left(g_{i,n}b_{i,n}\right) = \left(\frac{w}{\lambda_0 b_{i,n} + \eta_0} - \frac{\sigma_i^2 + p_{j,n}b_{j,n}}{g_{i,n}}\right)^+, \tag{5.27}$$

for all $i,j \in \{1,2\}$, $i \neq j$ and $(x)^+ = \text{Max}(x, 0)$. This is an iterative water-filling-like solution where the variables λ_0 and η_0 determine the water level, $w/(\lambda_0 b_{i,n} + \eta_0)$.

However, this water-filling solution is performed simultaneously over two time-varying channels, $b_{i,n}$ and $g_{i,n}$. If the average instantaneous received interference constraint were not present, the water level would be limited by Equation (5.21) alone. Hence, constraint (5.21) defines the water level in the dimension[6] of direct channel gain $g_{i,n}$. Similarly, constraint (5.20) determines the water level in the dimension defined by $b_{i,n}$.

As mentioned before, we assume all channels are experiencing independent and identical Rayleigh fading with unit mean. The channel gains are, then, samples of independent and identical exponential distributions with unit mean. For user i, denote the received noise plus the interference from user j by $I_{j,n} = \sigma_i^2 + p_{j,n}b_{j,n}$. Then, over the direct channel dimension of this two-dimensional water-filling solution, by replacing Equation (5.27) into Equation (5.21), we get

$$P_{\max,i} = \sum_{n=1}^{N} \left[\int_0^\infty \int_{\frac{I_{j,n}(\lambda_0 b_{i,n}+\eta_0)}{w}}^\infty \left(\frac{w}{\lambda_0 b_{i,n}+\eta_0} - \frac{I_{j,n}}{g_{i,n}} \right) e^{-g_{i,n}} e^{-b_{i,n}} \, dg_{i,n} db_{i,n} \right]$$

$$= \sum_{n=1}^{N} \left[\int_0^\infty \left[\frac{w}{\lambda_0 b_{i,n}+\eta_0} e^{\frac{-I_{j,n}(\lambda_0 b_{i,n}+\eta_0)}{w}} + \mathcal{E}_I \left(\frac{-I_{j,n}(\lambda_0 b_{i,n}+\eta_0)}{w} \right) \right] e^{-b_{i,n}} \, db_{i,n} \right]$$

$$= \sum_{n=1}^{N} \left[\frac{1}{\lambda_0} e^{\frac{\eta_0}{\lambda_0}\left(1+\frac{I_{j,n}\lambda_0}{w}\right)} \mathcal{E}_I \left[\frac{\eta_0}{\lambda_0} \left(1 + \frac{I_{j,n}\lambda_0}{w}\right) - b_{i,n}\left(1+\frac{I_{j,n}\lambda_0}{w}\right) \right] \right|_0^\infty$$

$$+ e^{\frac{\eta_0}{I_{j,n}\lambda_0}} \mathcal{E}_I \left[-\left(1+\frac{w}{I_{j,n}\lambda_0}\right)\left(b_{i,n}+\frac{\eta_0}{I_{j,n}\lambda_0}\right) \right] - e^{-b_{i,n}}\mathcal{E}_I \left(-\frac{I_{j,n}(\lambda_0 b_{i,n}+\eta_0)}{w} \right) \Big|_0^\infty \right]$$

$$= \sum_{n=1}^{N} \left[-\frac{1}{\lambda_0} e^{\frac{\eta_0}{\lambda_0}\left(1+\frac{I_{j,n}\lambda_0}{w}\right)} \mathcal{E}_I \left(\frac{\eta_0}{\lambda_0}\left(1+\frac{I_{j,n}\lambda_0}{w}\right) \right) - e^{\frac{\eta_0}{I_{j,n}\lambda_0}} \mathcal{E}_I \left(-\left(1+\frac{w}{I_{j,n}\lambda_0}\right)\frac{\eta_0}{I_{j,n}\lambda_0} \right) \right.$$

$$\left. + \mathcal{E}_I \left(-\frac{I_{j,n}\eta_0}{w} \right) \right]. \tag{5.28}$$

In Equation (5.28), $\mathcal{E}_I(x)$ is the exponential integral function, which is defined as

$$\mathcal{E}_I(x) = \int_{-x}^\infty \frac{e^{-t}dt}{t}.$$

Similar to calculating Equation (5.28), we can substitute Equation (5.27) in Equation (5.20a) to obtain the second required relation between η_0 and λ_0. Due to the similarity of steps for this calculation to Equation (5.28), we omit the details and present only the final result here:

---

[6]Some references use the term *two-dimensional water-filling* to indicate allocation of power over time-varying fading channels (cf. [179]). In that context Equation (5.27) provides a three-dimensional solution.

$$\Gamma_{\text{avg},i} = \sum_{n=1}^{N} \Bigg\{ w e^{\frac{-I_{j,n}\eta_0}{w}} \Bigg[\frac{1}{\lambda_0 \left(1 + \frac{I_{j,n}\lambda_0}{w}\right)} + \frac{\eta_0}{\lambda_0^2} e^{\frac{\eta_0}{\lambda_0}\left(1 + \frac{I_{j,n}\lambda_0}{w}\right)} \mathcal{E}_I \left(\frac{-\eta_0}{\lambda_0} \left(1 + \frac{I_{j,n}\lambda_0}{w}\right)\right) \Bigg]$$

$$+ I_{j,n} \Bigg[-\left(1 - \frac{\eta_0}{\lambda_0}\right) e^{\frac{\eta_0}{\lambda_0}} \mathcal{E}_I \left(-\eta_0 \left(\frac{I_{j,n}}{w} + \frac{1}{\lambda_0}\right)\right)$$

$$+ \frac{I_{j,n}\lambda_0}{w} \mathcal{E}_I \left(\frac{-I_{j,n}\eta_0}{w}\right) + \frac{1}{1 + \frac{I_{j,n}\lambda_0}{w}} e^{\frac{I_{j,n}\eta_0}{w}} \Bigg] \Bigg\}. \tag{5.29}$$

As can be observed from Equations (5.28) and (5.29) the interrelation between the Lagrangian multipliers η_0 and λ_0 are nonlinear and hence cannot be solved in closed form. Instead we can use numerical approaches such as Gauss-Newton, Levenberg-Marquardt, or trust-region methods to find the value of these variables [180]. The ergodic capacity in this case can be calculated by substituting the optimal power allocation Equation (5.27) into Equation (5.18) and use the resulting maximum achievable rate in Equation (5.19):

$$\mathbb{C}_{ER,i}^{\text{avg}} = \sum_{n=1}^{N} \Bigg\{ \int_0^\infty \int_{\frac{I_{j,n}(\lambda_0 b_{i,n} + \eta_0)}{w}}^{\infty} R_{i,n}\left(p_{i,n}^*\left(g_{i,n}, b_{i,n}\right)\right) \times f_g\left(g_{i,n}\right) f_b\left(b_{i,n}\right) dg_{i,n} db_{i,n} \Bigg\}. \tag{5.30}$$

The description of underlay techniques brings us to the conclusion of this chapter. The following summarizes the topics studied in this chapter. The set of problems at the end of this chapter is designed to further familiarize the interested reader with concepts and techniques used in this chapter.

5.7 CHAPTER SUMMARY

This chapter elaborates on spectrum access and sharing techniques. It first classifies various spectrum sharing approaches from the point of view of their physical dimensions, as well as from the spectrum access rights point of view. It then proceeds to a more detailed examination of such techniques.

Spectrum sharing in unlicensed bands has already set the scene for many novel solutions, such as the widespread deployment of WLAN hotspots, which helped extend Internet access to the public domain and promoted innovative enterprising applications. Unfortunately, however, the very benefits of these bands are under threat, as more and more applications and systems are being developed to operate on them. Since the only guideline currently applied to such unlicensed bands is transmission power caps, coordination of interference in such a heterogeneous environment is nearly impossible. It is therefore appropriate for the benefits of spectrum sharing to be extended to other realms, beyond unlicensed access.

In licensed bands, spectrum can be shared among a number of equal-priority RANs (radio access networks), constituting a horizontal spectrum sharing scenario.

In this context, an accurate estimation of traffic load in each participating system is necessary to determine the resource requirement in the next resource allocation period. Moreover, a cost versus revenue analysis helps reduce the final cost and achieve higher profitability through spectrum sharing, whereby temporal and spatial coordination among systems sharing a band is necessary to mitigate interference.

Another viable licensed spectrum sharing scenario studied in this chapter is secondary spectrum access. Depending on the technical capability and willingness of the primary and secondary systems to achieve such paradigms, negotiated or opportunistic SSA solutions can be envisioned. The signaling channel required in negotiated SSA ideally provides a receiver-centric interference control mechanism, given that the secondary transmitter should detect and measure its interfering effect on the primary receivers (i.e., reception is the point at which interference actually occurs). Whether through this or other approaches, interference awareness is a challenging task given problems such as hidden node awareness.

A major advantage of negotiated SSA is the capability of guaranteeing QoS, not only for the primary system, but also for the secondary system when possible. This chapter introduces a negotiated scenario with the sum power transmission minimization objective and shows that, under specific channel conditions, the QoS levels of both primary and secondary systems are sustainable.

In the opportunistic SSA case, besides the spectrum sensing problem pertinent to almost all SSA solutions, a spectrum aggregation technique should be used to help provide the secondary system with a higher bit rate through intelligent exploitation of sporadic spectrum opportunities composed of disjoint small slices of spectrum. Multicarrier modulation techniques are powerful tools toward this goal. This chapter discusses a MC-CDMA approach, based on a cognitive radio system, that can tailor its transmission in the DL dependent on the interference pattern of the channel. Through using a different spreading code for each subchannel of each user, a variable-bandwidth MC-CDMA system is realized and shown to effectively combat interference in an opportunistic SAA context.

Finally in this chapter, the case of underlay opportunistic SSA is discussed. Considering an interference channel problem with priority access, it is possible to exploit interference diversity of the channel to facilitate opportunistic coexistence with a primary system. Some other underlay approaches include powerful coding techniques and the use of UWB techniques.

5.8 PROBLEMS

1   Providing one example of each, compare the licensed and unlicensed spectrum sharing approaches. What are the main advantages and disadvantages of each paradigm?

2   A biomedical manufacturing company is developing a health monitoring system (HMS) composed of two parts. The first part is a network of sensors attached to the patient's body; communication among the sensors is done using a propitiatory wireless personal area network (WPAN) technology. The gathered information is then conveyed via a second part, a low-power WLAN installed in the patient's room, itself connected to the Internet via an asynchronous digital subscriber line (ADSL) modem.

   (a)   Given the high cost of acquiring a licensed band for such a system, the manufacturer is looking at spectrum sharing solutions for both WPAN network connectivity (Case 1), and WPAN to WLAN communications via a WLAN gateway on one of the WPAN sensors (Case 2). Which spectrum sharing approach might best serve the purpose of each case, and why?

   (b)   If a mobile operator allows the operation of such HMS devices within its licensed band, conditioned on usage of the band if and only if no subscribed mobile station of that operator is operating within 10 meters of the HMS, what kind of spectrum sharing scheme is realized?

   (c)   The mobile operator agrees that in emergency cases the HMS can transmit simultaneously with any ongoing call/session of the operator, provided that HMS transmits a (predefined and agreed) warning signal to the operator's base stations. This scenario is a manifestation of which spectrum sharing approach?

3   Use the derived equations for collision probability in Example 5.1.

   (a)   Write a MATLAB M-file to compute the collision probability for a specific node, given knowledge of the size of file to be transferred. Assume the file is packetized into packets of X byte size, with a frame period of Y msec (Use X and Y based on different standards, for example WIFi or WiMAX). Furthermore, assume that 10 other nodes are operating in the unlicensed band, all producing traffic with an average of two packets per frame time. The channel access attempt probability, p, is 0.7 for all the nodes, and there is no backlogged traffic at any node at the start time, although each collided packet is added to the packets for the next frame time.

   (b)   Building upon the M-file developed for part a, create an M-file to compare the collision probabilities for $p \in \{0.5, 0.7, 0.9\}$. All other parameters are as defined for part a.

4   To better understand the cost versus revenue trade-off in spectrum sharing, as described in Example 5.3, write an M-file to create a three-dimensional graph of the total revenue of operator m with $N_m = 10$ units of spectrum and $W_m(0) = \$1$, as a function of revenue per unit of spectrum $R_s \in [\$0, \$1]$ and spectrum demand (in terms of units of spectrum) $D_m \in [0, 20]$. Assume the cost of borrowing one unit of spectrum is $0.5.

5   Solving the optimization problem (5.1) subject to Equations (5.2) and (5.3) is generally non-polynomial time (NP)-hard. To simplify the task, this problem

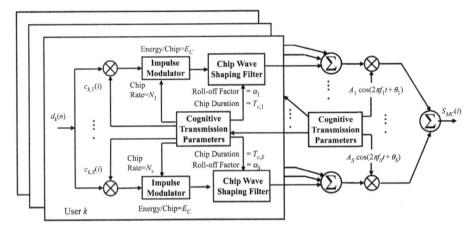

FIGURE 5.12

Downlink transmitter of cognitive MC-CDMA system.

can be decomposed either by solving the problem in each subchannel for all users (i.e., N subproblems with K users in each subproblem) or by solving the problem for all subchannels or based on each user (i.e., K subproblems for each user operating in N subchannels).

(a) Write the Lagrangian dual problem of optimization (5.1), subject to Equations (5.2) and (5.3).

(b) Decompose the Lagrangian in part a, into N subproblems based on each subchannel. What is the primal optimization problem associated with this decomposed form?

(c) Repeat part b, but for decomposition based on each user.

6 For the scenario presented in Section 5.6.1 "Negotiated Access," can you derive an admissibility criterion similar to Equation (5.13), for the situation where there are three users, two primary and one secondary? Attempt to do the same for a generalized K user case, where the subset K_1 are primary users.

7 In the opportunistic spectrum hole approach, discussed in Section 5.6.4, assume that the total transmit power for the cognitive MC-CDMA technology is fixed at P_{Total}, that E_b denotes the transmitted energy per symbol, and without the presence of any primary signal in the channel, and that S' subcarriers are used in the MC-CDMA system. If it is instead assumed that there are M primary signals in the channel, what is the amplitude of the up-conversion signal over S subchannels necessary for the transmitted cognitive MC-CDMA to successfully operate at the upper limit of its transmission power specification (see Figure 5.12)? For simplicity, assume that all subchannels have the same bandwidth and transmitted amplitude.

Agile transmission techniques

Srikanth Pagadarai[1], Rakesh Rajbanshi[2], Gary J. Minden[3], and Alexander M. Wyglinski[1]

[1]Worcester Polytechnic Institute, United States
[2]Cisco Systems, United States
[3]The University of Kansas, United States

6.1 INTRODUCTION

In Chapter 3, the physical (PHY) layer fundamentals concerning the use of orthogonal frequency division multiplexing (OFDM) are described. Several underlying principles and the resulting advantages confirm the efficiency with which OFDM enables high-speed wireless communications. However, an important assumption in Chapter 3 is that the radio frequency (RF) spectrum utilized by the transceiver system is based on the traditional command-and-control allocation policy followed by most regulatory bodies like the Federal Communications Commission (FCC). That is, contiguous spectrum is available for use. However, recent spectrum measurement studies have shown that the utilization of spectral resources over time, frequency, and space in such a scenario is quite low. For example, Figure 6.1 shows a snapshot of the power spectral density (PSD) from 88 MHz to 2686 MHz measured on July 11, 2008, in Worcester, Massachusetts. This figure clearly shows the presence of several unused portions of the spectrum. The presence of unused but licensed bands creates an artificial scarcity of available spectrum. Several methodologies have been proposed to efficiently utilize these noncontiguous portions of spectrum to improve spectral efficiency. This chapter focuses primarily on one family of these spectral efficiency techniques; however, the PHY layer design issues that need to be taken into consideration when those techniques are employed in the transceiver are explored in the following sections.

The choice of a physical layer transmission technique is a very important design decision when implementing a cognitive radio. In particular, the technique must be sufficiently agile to enable unlicensed users to transmit in a licensed band while not interfering with the incumbent users. Moreover, to support throughput-intensive

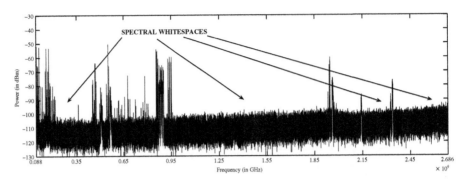

FIGURE 6.1

A snapshot of PSD from 88 MHz to 2686 MHz measured on July 11, 2008, in Worcester, Massachusetts ($N42^o16.36602$, $W71^o48.46548$).

applications, the technique should be capable of handling high data rates. One technique that meets both these requirements is a variant of orthogonal frequency division multiplexing called *noncontiguous OFDM* (NC-OFDM) [181]. Compared to other techniques, NC-OFDM is capable of deactivating subcarriers across its transmission bandwidth that could potentially interfere with the transmission of other users. Moreover, NC-OFDM can support a high aggregate data rate with the remaining subcarriers and simultaneously maintain an acceptable level of error robustness. Despite the advantages of NC-OFDM, two critical design issues are associated with this technique. First, the detection of the white spaces in the licensed bands for secondary-user transmissions. Radio parameter adaptation and hardware reconfiguration are another crucial requirement.

As mentioned earlier in this chapter, we discuss the techniques that need to be employed in a dynamic, spectrally agile, hardware-reconfigurable software-defined radio (SDR) to alleviate some of the problems arising due to secondary transmissions in an already licensed band. This chapter is organized as follows. Section 6.2 presents a classification of the spectrum sharing techniques in the existing literature. Next, in Section 6.3, we describe the transceiver system that employs these spectrum sharing techniques. In Section 6.4, we discuss some of the issues resulting from the use of noncontiguous bands, such as interference to the primary users, the need for fast Fourier transform (FFT) pruning, and the need for peak-to-average power ratio (PAPR) reduction. We then conclude the chapter with several remarks and comments in Section 6.5.

6.2 WIRELESS TRANSMISSION FOR DYNAMIC SPECTRUM ACCESS

Figure 6.2 shows a dynamic spectral access (DSA) scenario that is viewed as a solution to the problem of the artificial spectral scarcity. As shown in this figure, at any time instant, several noncontiguous spectral regions are left unused. These

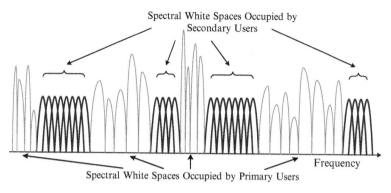

FIGURE 6.2

The utilization of noncontiguous regions of spectrum for wireless transmission.

unused portions can be used by secondary users for high-speed wireless communications while simultaneously ensuring that the primary user's rights are not violated. This idea of using multiple noncontiguous portions of spectrum is referred to as *spectrum pooling*.

6.2.1 Spectrum Pooling

The notion of "spectrum pooling," first introduced in [6], is a mechanism for pooling the spectral resources from different spectral owners and renting these spectral resources to unlicensed users during idle periods. However, such a lease of licensed spectral resources to rental users, while providing additional revenue to the licensed users, brings forth many technological, juristical, economic, and political questions concerning the regulatory aspects of spectrum pooling. The technical challenges that need to be solved to make spectrum pooling practical have been the research focus of numerous groups at universities all over the world.

Flexible pooling of the spectral resources is an important requirement for future cognitive radios to enable efficient secondary utilization of the spectrum [182]. Such a cognitive radio needs to employ agile physical layer transmission techniques to respect the rights of the incumbent licensed users and reconfigurable hardware that makes the adaptation to changing environmental conditions feasible [183]. Moreover, a formal radio etiquette needs to be formulated, which is a framework to moderate the use of the RF spectrum for guaranteeing the rights of the licensed users as well as for the flexible coordination between the unlicensed users.

6.2.2 Underlay and Overlay Transmission

Spectrum sharing techniques can be classified into *underlay* and *overlay* spectrum sharing based on the spectrum access techniques. Underlay systems use ultra-wideband (UWB) [184, 185] or spread-spectrum techniques, such as code division multiple access (CDMA) [186], to transmit the signal below the noise floor of the

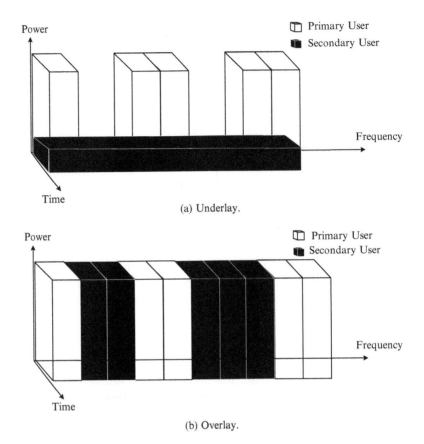

FIGURE 6.3

Overlay and underlay spectrum sharing.

spectrum [187]. An example of the time- and frequency-domain information of an underlay spectrum sharing system is shown in Figure 6.3(a). In this figure, we see that the underlay systems use wideband low-power signals for transmissions. However, this technique can increase the overall noise temperature and thereby worsen error robustness of the primary users as compared to the case without underlay systems. To avoid any interference to the primary users, the underlay system can use interference avoidance techniques, such as *notching* [188] and *waveform adaptation* [189].

The spectrum holes[1] filled in by secondary transmissions in an overlay system are shown in Figure 6.3(b). When interference among the users is high, it has been shown that frequency division multiplexing is an optimal technique [190].

---

[1]A *spectrum hole* is an unused portion of the licensed spectrum [11].

As shown in this figure, the overlay systems use the unoccupied portions of the spectrum with guard intervals for secondary transmissions, keeping the interference to the primary users to a minimum. Since the licensed system has privileged access to the spectrum, it must not be disturbed by any secondary transmissions. This results in two main design goals for an overlay system [191]:

- Minimum interference to licensed transmissions.
- Maximum exploitation of the gaps in the time–frequency domain.

To achieve these goals, the overlay system needs information about the spectrum allocation of the licensed systems, for example, by regularly performing spectrum measurements. As explained previously, spectral pooling represents the idea of storing this information by merging spectral ranges from different spectrum owners (military, trunked radio, etc.) into a common pool, where users may temporarily rent spectral resources during idle periods of licensed users, thereby enabling the secondary utilization of already licensed frequency bands [24]. In a spectrum pooling system, a centralized entity can collect measurement information gathered by the secondary-user terminals during the detection cycle and maintain the spectrum usage information. The centralized entity is responsible for making decisions on granting portions of the spectrum to the secondary users. With the use of a centralized entity, the information management of a spectrum access network would be relatively simple. However, this same entity can also easily be a bottleneck for the network due to the associated information exchange overhead. Since the overlay systems can readily exploit the unused portions of the spectrum without interfering with the incumbent users and without increasing the noise temperature of the system, we consider only overlay systems in this chapter from this point forward.

One of the most challenging problems of spectrum sharing systems is their successful coexistence in the same frequency band; that is, an overlay system should not degrade the performance of systems already working in the target frequency band. For instance, out-of-band radiation has to be reduced to enable coexistence. The transmitter spectral mask is a measure of the transmitter spectral profile to verify that the device is not transmitting excessive amounts of energy outside its assigned channel bandwidth. Several approaches have been proposed in literature for suppressing the side lobe levels, such as the deactivation of subcarriers lying at the borders of an OFDM spectrum [192], windowing [193], subcarrier weighting [194], and insertion of cancellation carriers [195].

Despite being a solution to the problem of the apparent spectrum scarcity, dynamic spectrum access puts additional design constraints on the wireless transceiver. This is because, as multiple pockets of wireless spectrum are being utilized, the noise characteristics differ substantially across the noncontiguous bands of spectrum. Hence, a spectrally agile but extremely robust modulation technique is required for use in wireless transceivers employed in a DSA scenario. As mentioned earlier, the noncontiguous OFDM proposed in [181] satisfies these requirements. While conventional multicarrier (MC) CDMA has proven to be effective compared to conventional OFDM systems because of its superior multiuser interference limiting capabilities, NC-OFDM has been shown to be better than noncontiguous MC-CDMA (NC-MC-CDMA) [196]. This is because the deactivation of subcarriers corresponding to primary-user transmissions causes a loss of orthogonality in NC-MC-CDMA, leading to a worse bit error rate (BER) performance than NC-OFDM systems. Therefore, in the following discussion, our focus for spectrally agile modulation techniques is NC-OFDM.

6.3 NONCONTIGUOUS ORTHOGONAL FREQUENCY DIVISION MULTIPLEXING

A general schematic of an NC-OFDM transceiver is shown in Figure 6.4. Without loss of generality, a high-speed data stream, $x(n)$, is modulated using M-ary phase shift keying (MPSK).[2] Then, the modulated data stream is split into N slower data streams using a serial-to-parallel (S/P) converter. Note that the subcarriers in the NC-OFDM transceiver do not need to be all active as in conventional OFDM. Moreover, active subcarriers are located in the unoccupied spectrum bands, which are determined by dynamic spectrum sensing techniques [197]. The inverse fast Fourier transform (IFFT) is then applied to these modulated subcarrier signals. Prior to transmission, a guard interval with a length greater than the channel delay spread is added to each NC-OFDM symbol using the cyclic prefix (CP) block to mitigate the effects of intersymbol interference (ISI). Following the parallel-to-serial (P/S) conversion, the baseband NC-OFDM signal, $s(n)$, is passed through the transmitter radio frequency chain, which amplifies the signal and up-converts it to the desired center frequency.

The receiver performs the reverse operation of the transmitter, mixing the RF signal to the baseband for processing, yielding the signal $r(n)$. Then, the signal is converted into parallel streams using the S/P converter, the CP is discarded, and the FFT is applied to transform the time domain data into the frequency domain. After compensating distortion introduced by the channel using equalization, the data in the active subcarriers are multiplexed using a P/S converter, and demodulated into a reconstructed version of the original high-speed input, $\hat{x}(n)$.

From this system overview, we observe that the spectrum sensing, spectrum shaping, peak-to-average power ratio, radio parameter adaption, and efficient radio implementation are critical issues associated with an OFDM-based cognitive radio. In the next section, we will describe these issues and ways to mitigate them in order to develop efficient OFDM-based cognitive radios.

---

[2]Other forms of digital modulation, including M-ary quadrature amplitude modulation, can also be employed by the transceiver.

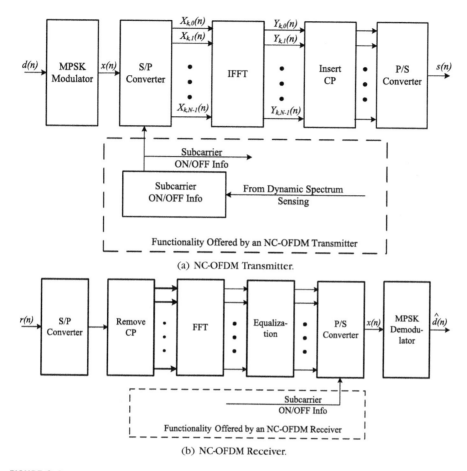

(a) NC-OFDM Transmitter.

(b) NC-OFDM Receiver.

FIGURE 6.4

An NC-OFDM transceiver.

6.4 NC-OFDM-BASED COGNITIVE RADIO: CHALLENGES AND SOLUTIONS

The digital modulation scheme based on orthogonal frequency division multiplexing is the natural approach for DSA due to inherent frequency subbanding. OFDM spectrum access is scalable while keeping users orthogonal and noninterfering provided the users are synchronized. However, the conventional OFDM scheme does not provide truly band-limited signals due to spectral leakage caused by sync-pulse-shaped transmission resulting from the IFFT operation. This may cause interference to the adjacent band primary users proportional to the power allocated to the cognitive user on the corresponding adjacent subcarrier. Therefore, an OFDM transmitter must be adapted.

6.4.1 Interference Mitigation

Figure 6.5 shows the power spectral density of an OFDM-modulated carrier. This figure shows the subcarrier spacing and the interference power due to the first side lobe in the first adjacent band. It is observed that, as the distance between the location of the subcarrier of the rental system and the considered subband increases, the interference caused by it reduces monotonically, which is a characteristic of the *sinc* pulse. However, it should also be noted that, in a practical scenario consisting of N subcarriers, the actual value of the interference caused in a particular legacy system subband is a function of the random symbols carried by the *sinc* pulses and N.

With respect to the interference caused by the unlicensed user to the licensed user, the important issue that needs to be taken into consideration when designing an OFDM-based overlay system is that its impact on the legacy system should be very small. Thus, the basic aim of any algorithm for side lobe suppression is to reduce the side lobe power levels while causing little or no effect to the other

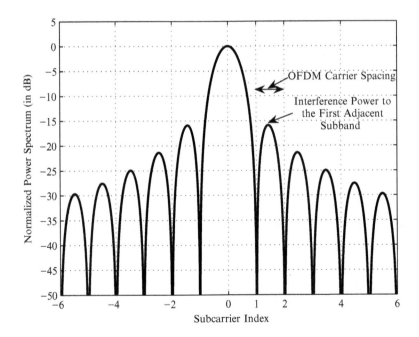

FIGURE 6.5

The interference due to one OFDM-modulated carrier.

secondary system parameters. Before moving on to a summary of the existing algorithms for side lobe suppression, a brief mathematical representation of the interference to the legacy system and two simple techniques for mitigating its effects are provided in this section.

Assuming the transmit signal $s(t)$ on each subcarrier of the OFDM transceiver system is a rectangular non-return-to-zero (NRZ) signal, the power spectral density of $s(t)$ is represented in the form [108]

$$\Phi_{ss}(f) = A^2 T \left(\frac{\sin \pi f T}{\pi f T} \right)^2, \tag{6.1}$$

where A denotes the signal amplitude and T is the symbol duration that consists of the sum of symbol duration, T_S, and guard interval, T_G. The assumption that the transmit signal $s(t)$ on each subcarrier is a rectangular NRZ signal is valid since it matches the wireless local area network (LAN) standards [198, 199]. Now, assuming that the legacy system is located in the vicinity of the rental system, the mean relative interference, $P_{\text{Interference}}(n)$, to a legacy system subband is defined as [192]

$$P_{\text{Interference}}(n) = \frac{1}{P_{\text{Total}}} \int_{n}^{n+1} \Phi_{ss}(f) df, \tag{6.2}$$

where P_{Total} is the total transmit power emitted on one subcarrier and n represents the distance between the considered subcarrier and the legacy system in multiples of Δf.

The idea of interference calculation for the case of one subcarrier can be extended to a system with N subcarriers. Let $s_n(x)$, $n = 1, 2, 3, ..., N$ be the subcarrier of index n represented in the frequency domain. Then,

$$s_n(x) = a_n \frac{\sin[\pi (x - x_n)]}{\pi (x - x_n)}, \qquad n = 1, 2, ..., N. \tag{6.3}$$

In this equation, $\mathbf{a} = [a_1 \, a_2 \, \cdots \, a_N]^T$ is a data symbol array, and x is a normalized frequency given by

$$x = (f - f_0)T,$$

where f denotes the frequency and f_0 is the center frequency. Also, x_n is the normalized center frequency of the nth subcarrier. Again, the signal in the time domain at the transmitter is assumed to be in a rectangular NRZ form. Now, the OFDM symbol in the frequency domain over the N subcarriers is

$$S(x) = \sum_{n=1}^{N} s_n(x). \tag{6.4}$$

The power spectral density of this signal is given by

$$\Phi_{ss}(f) = |S(x)|^2 = \left| \sum_{n=1}^{N} a_n \frac{\sin[\pi (x - x_n)]}{\pi (x - x_n)} \right|^2. \tag{6.5}$$

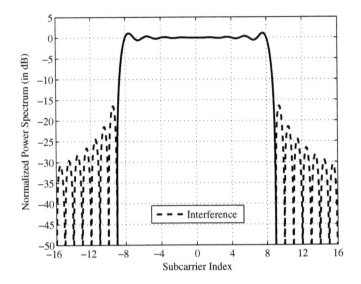

FIGURE 6.6

The interference in a BPSK-OFDM system with 16 subcarriers.

As an example, a BPSK-OFDM system with $N = 16$ subcarriers is considered. When the vector $\mathbf{a} = [1\ 1\ 1\ 1\ 1\ 1\ 1\ 1\ 1\ 1\ 1\ 1\ 1\ 1\ 1\ 1]^T$, Figure 6.6 shows the normalized OFDM power spectrum. As shown in this figure, the portion of the signal indicated in dashed lines represents the potential interference causing side lobes, resulting from summing up the sinc pulses that carry the symbols from the data vector. Also, Figure 6.6 is for the case where the data vector consists of ones and, hence, depending on the random distribution of the symbols, the side lobe power levels decay at different rates.

Windowing

One of the simplest and the earliest solutions offered to counter the effects of OOB interference is windowing the OFDM transmit signal in the time domain [192, 193]. A raised cosine window defined by

$$w(t) = \begin{cases} \frac{1}{2} + \frac{1}{2}\cos\left(\pi + \frac{\pi t}{\beta T}\right), & \text{for } 0 \leq t < \beta T \\ 1, & \text{for } \beta T \leq t < T \\ \frac{1}{2} + \frac{1}{2}\cos\left(\frac{\pi(t-T)}{\beta T}\right), & \text{for } T \leq t < (1+\beta)T \end{cases} \tag{6.6}$$

is a commonly used window type where β is defined as the roll-off factor. Applying the transmit filter, $w(t)$, the OFDM signal in time domain is as shown in Figure 6.7. It can be noted from this figure that the postfix needs to be longer than βT to maintain the orthogonality within the OFDM signal. That is, the application of windowing to reduce the out-of-band radiation of the OFDM signal has the adverse

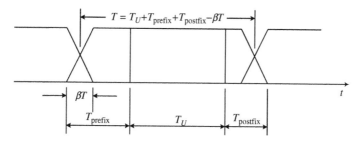

FIGURE 6.7

Structure of the temporal OFDM signal using a raised cosine window.

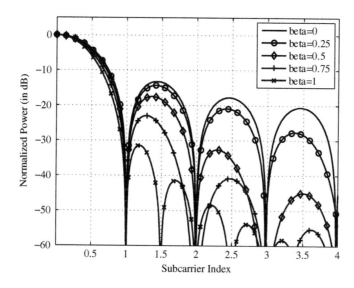

FIGURE 6.8

Impact of roll-off factor on the PSD of the rental system signal.

effect of expanding the temporal symbol duration by $(1+\beta)$, resulting in a lowered system throughput for the unlicensed user.

The impact of the roll-off factor of the raised cosine window on the side lobe power levels of the OFDM symbol is depicted in Figure 6.8. It can be observed from this figure that, for smaller values of β, the suppression achieved in the side lobe power levels of the first adjacent band is very small. As the distance between the location of the subcarrier of the rental system and the considered subband increases, the suppression achieved also increases. Also, for very large values of β, the suppression achieved is considerable even in the case of the first adjacent band. However, the symbol duration in time is also increased, which reduces the system throughput. Therefore, windowing can be applied as an additional means to suppress the high side lobes, but more powerful techniques need to be developed.

Insertion of Guard Bands

Another technique for mitigating the effects of the side lobes from the secondary user's OFDM symbols on the system performance of the legacy system is to deactivate additional subcarriers in the vicinity of the licensed user that are allotted to the unlicensed user in addition to those that are deactivated due to licensed user accesses [193]. With this technique, the already scarce spectral resources are wasted. Moreover, the reduction achieved is not significant enough, as shown in Figure 6.9.

In this figure, a BPSK-OFDM system with $N = 64$ subcarriers is considered. The simulations were performed over 20,000 symbols. From Figure 6.9(a), it can be observed that, by inserting two guard carriers (GCs) on each side of the spectrum, the achievable average reduction of the maximum interference causing side lobe is only 2.8 dB, and by inserting eight guard carriers, the reduction achieved is around 4.6 dB. A significant reduction of around 10 dB can be achieved, by giving up 25% of the allocated bandwidth; that is, by using 16 subcarriers out of 64 for inserting guard bands. The complementary cumulative distribution function (CCDF) of Figure 6.9(b) also illustrates the same point. On average, 99.9% of the side lobe power is below -3.8 dB in the original case, whereas by inserting two guard carriers on each side of the OFDM spectrum, the value is -8.7 dB, and by inserting eight guard carriers, the value is around -15 dB.

Existing Solutions

Insertion of cancellation carriers is one of the most common approaches to achieve side lobe suppression. The working principle behind this technique is described in the following paragraphs.

Suppose we define the total number of subcarriers that can be transmitted by a secondary user in a spectral white space as $L = L_A + L_{CC}$, where L_A is the number of active subcarriers used for signal transmission, and L_{CC} is the total number of subcarriers reserved for inserting cancellation subcarriers. As a result, the equations describing the individual subcarriers and the cumulative OFDM signal can be expressed as

$$s_{l_a}(y) = d_{l_a} \frac{\sin\left[\pi\left(y - y_{l_a}\right)\right]}{\pi\left(y - y_{l_a}\right)}, \quad l_a = -L_A/2, ..., L_A/2, \tag{6.7}$$

and

$$S(y) = \sum_{l_a = -L_A/2}^{L_A/2} s_{l_a}(y). \tag{6.8}$$

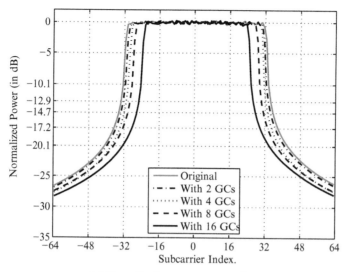

(a) Normalized power spectrum plot.

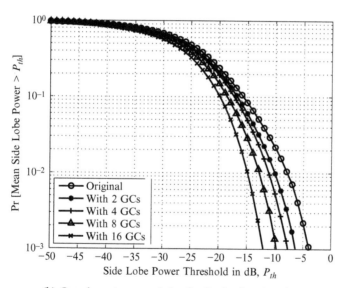

(b) Complementary cumulative distribution function plot.

FIGURE 6.9

Interference suppression in a BPSK-OFDM system with 64 subcarriers by inserting guard subcarriers.

Since the frequency response of an OFDM subcarrier is represented by the *sinc* function, the side lobe power levels of the composite signal, at any frequency location that consists of superposed and frequency-translated subcarriers, can be algebraically computed as the sum of the side lobe powers of each *sinc* function at that location given the input sequence. Therefore, if I_k represents the side lobe amplitude level at the kth frequency index (in the out-of-band, OOB, region) normalized to the subcarrier bandwidth, then we can express this amplitude level as

$$I_k = \sum_{l_a=-L_A/2}^{L_A/2} s_{l_a}(k). \tag{6.9}$$

Suppose the amplitude level of the cancellation subcarrier inserted at $j = L_A/2 + 1$ to nullify the side lobe at the $k = L/2 + 1$ frequency index is selected in such a way that it possesses a side lobe at the kth frequency index, which is equal in amplitude but opposite in sign to I_k. In other words, we select C_j such that it reduces the interference I_k. A simple solution is to select C_j such that $C_k = -I_k$. Figure 6.10 illustrates this procedure for a 16-subcarrier BPSK system, where $L_A = 14$ and $L_{CC} = 2$. Suppose the symbol sequence transmitted over the active subcarriers is [1 1 1 1 1 1 1 1 1 1 1 1 1 1]. The superscripts r and l are used to signify, respectively, the side lobe power levels on the right and left sides of the OFDM signal spectrum. While choosing the symbol values over subsequent cancellation subcarriers (CCs) in a scenario where multiple cancellation subcarriers

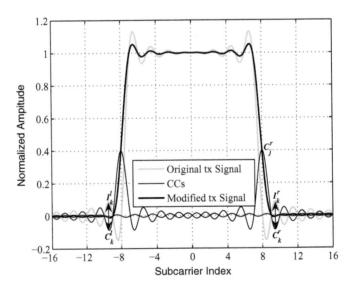

FIGURE 6.10

Side lobe power reduction with cancellation carriers.

are inserted on each side of the signal spectrum, the side lobe regrowth over the locations nulled by the previously inserted CCs is constrained to be minimal.

In the case of multiple cancellation subcarriers, the symbols over cancellation subcarriers are computed iteratively for minimizing the side lobe power. The symbol over the first cancellation subcarrier is computed such that it nullifies the first OOB side lobe amplitude. The symbol over the second cancellation subcarrier is computed for minimizing the second OOB side lobe amplitude. The procedure can be continued for a given number of cancellation subcarriers until the desired side lobe power levels are achieved. If the desired suppression is higher, a greater portion of the bandwidth has to be allocated to insert additional cancellation subcarriers. However, significant side lobe power suppression can be achieved with a small number of CCs, resulting in a reasonable trade-off between bandwidth reduction and achievable interference suppression. This technique described is illustrated in the form of a flow diagram in Figure 6.11.

Another approach proposed in [201] is the constellation expansion (CE) technique, wherein the symbols obtained by modulating the input bit sequence to a 2^k constellation space are mapped to an expanded constellation space consisting of 2^{k+1} constellation points. In other words, for every constellation point in the original symbol sequence, there are two points to choose from in the expanded constellation space. Selecting one of the points on a random basis, each symbol in a sequence of N symbols is mapped to N symbols from the expanded symbol set. An underlying assumption with the proposed CE technique is that the transmitter and the receiver have prior knowledge of the points of the expanded constellation that are associated with the points in the original constellation. Hence, after the demodulation process, the symbols can be remapped to the points of the original constellation. With this knowledge, no side information needed to be shared between the transmitter and the receiver. As an example, an approach for mapping QPSK symbols to an expanded constellation space is shown in Figure 6.12. The rationale behind this association of points is to take advantage of the randomness involved in selecting one of the two points, and hence the combination of different in-phase and quadrature-phase components from all the subcarriers results in a sequence with fewer side lobes.

Several other techniques for side lobe suppression based on complex optimization procedures have been proposed, such as (1) singular value decomposition optimization approaches for either inserting cancellation subcarriers [195] or weighting the subcarriers [194] to reduce the side lobe power levels over the neighboring RF spectrum, (2) reserved tones–based convex optimization technique [202] for suppressing side lobe power levels as well as the peak-to-average power ratio, and (3) a two-stage constellation expansion combined with a suboptimal cancellation carrier technique [200].

In (1), the weights are computed by selecting an optimization region, which is the portion of the neighboring RF spectrum over which the side lobes need to be suppressed and finding an optimal solution. The optimization is formulated as a linear least squares problem, solved using a singular value decomposition approach.

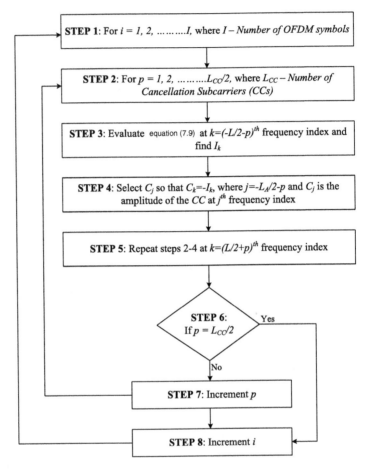

STEP 1: For $i = 1, 2, \ldots\ldots\ldots I$, where $I - Number\ of\ OFDM\ symbols$

STEP 2: For $p = 1, 2, \ldots\ldots\ldots L_{CC}/2$, where $L_{CC} - Number\ of$ Cancellation Subcarriers (CCs)

STEP 3: Evaluate equation (7.9) at $k=(-L/2-p)^{th}$ frequency index and find I_k

STEP 4: Select C_j so that $C_k=-I_k$, where $j=-L_A/2-p$ and C_j is the amplitude of the CC at j^{th} frequency index

STEP 5: Repeat steps 2-4 at $k=(L/2+p)^{th}$ frequency index

STEP 6: If $p = L_{CC}/2$ — Yes

No

STEP 7: Increment p

STEP 8: Increment i

FIGURE 6.11

The algorithm used for inserting CCs [200].

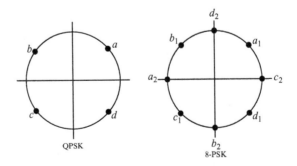

FIGURE 6.12

A mapping of symbols from QPSK constellation to an expanded constellation space.

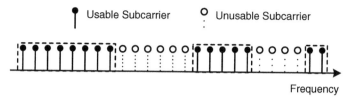

FIGURE 6.13

Subcarrier distribution over wideband spectrum.

In (2), a technique based on *reserved tones* is employed and the optimization problems for reducing PAPR and side lobe power levels are combined. A vector-valued objective function is obtained, and by employing a convex optimization algorithm, the optimal solution is achieved. Finally, in (3), symbols from a higher constellation are associated with symbols of the original constellation and the randomness in choosing points is exploited to achieve a symbols sequence with low side lobes. The results are combined with a suboptimal cancellation carrier approach to yield further reduction in the side lobe power levels. A summary of approaches that manipulate the data symbols to achieve a low side lobe power level is given in [203].

6.4.2 FFT Pruning for NC-OFDM

In a wideband communications system, a large portion of frequency channels may be occupied by transmissions from incumbent or unlicensed users. Systems that desire to operate within these occupied channels must avoid placing subcarriers in occupied or licensed spectrum. Therefore, to avoid interfering with these other transmissions, the subcarrier within the vicinity of the given transmission is turned off or nulled (Figure 6.13). In the case of systems like OFDM, these null subcarriers are represented as zero-valued inputs to the FFT and IFFT blocks. When available spectrum is sparse, the number of zero-valued inputs in the FFT may be significant relative to the total number of usable subcarriers. When the relative number of zero-valued inputs is quite large, significant hardware resources can be saved by pruning the FFT algorithm.

Channel conditions and incumbent spectrum occupancy (ISO) often vary over time, so efficient FFT pruning algorithms should be able to generate an optimized FFT implementation every time the channel conditions and ISO change. Given that the hardware resources of small form-factor cognitive radios are limited, this FFT pruning algorithm is very beneficial.

Fundamentals of FFT Pruning

In Figure 6.14 an eight-point decimation in frequency (DIF) FFT butterfly structure is shown, where a_i represents the ith input signal to the FFT block. Suppose the

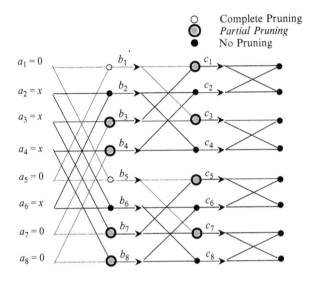

FIGURE 6.14

FFT butterfly structure. A value of 0 denotes a zero-valued subcarrier and x denotes a data-bearing subcarrier. The dotted lines represent the computations that can be pruned.

incumbent users are located at subcarriers a_1, a_5, a_7, and a_8. Therefore, input data over all these carriers must always be zero. For a conventional FFT algorithm, the total number of multiplications and additions are $N \log_2 N$. However, with an FFT pruning algorithm, the unnecessary multiplications and addition operations at stages b_1 and b_5 can be pruned as their values will always be zero. Moreover, multiplications and additions at nodes b_3, b_4, b_7, and b_8 can be replaced with a simple "copy" operation, whereas addition operations in nodes c_1, c_3, c_5, and c_7 can be pruned to save the FFT computation time. Therefore, the FFT computation time can be significantly improved with partial and complete pruning.

In wideband communication systems, the channel conditions and ISO vary over time. Therefore, the FFT pruning algorithm should be able to design an efficient FFT implementation every time the channel condition and ISO change.

Existing Solutions

Alves, Osorio, and Swamy proposed an FFT pruning algorithm that operates on any zero-valued input distribution [204]. Suppose we have a radix-2 FFT algorithm with N levels (2^N FFT points). A matrix M_I, with N columns and 2^N rows, is generated. Each element of the matrix corresponds to an addition/multiplication node of the FFT flow graph. The node needs to be computed if the corresponding element in the matrix M_I is nonzero. On the other hand, if the element of the matrix is zero, the corresponding node need not be computed. To obtain the matrix M_I, we need a subcarrier input vector with 2^N elements, where each element of this vector corresponds to each input element. If the input element is nonzero, the corresponding vector element will be unity, and if the input element is zero, the corresponding vector element will be zero. By using this input vector, we can compute the first column of the matrix M_l. In turn, by using the first column of the matrix M_l, the second column of matrix M_l can be obtained, and so forth.

To save execution time, the algorithm proposed in [181] builds upon the previous algorithm by avoiding the use of conditional statements. This algorithm is based on the Cooley-Tukey divide-and-conquer algorithm that uses in-place computation [205]. For a radix-2 FFT, the Cooley-Tukey algorithm divides the problem size into two interleaved halves with each recursive stage. This approach requires the computations to be proportional to $N \log_2 N$, whereas the equivalent discrete Fourier transform (DFT) would require the computations proportional to N^2. In [181], the proposed algorithm operates in a similar manner. Additionally, this algorithm prunes the unnecessary multiplication and addition operations at the nodes in the FFT flow graph to reduce the execution time for the FFT computations.

6.4.3 Peak-to-Average Power Ratio Problem in NC-OFDM

One critical issue of an OFDM-based implementation is the potential for high PAPR. If unchecked, high PAPR could lead to *amplitude clipping*,[3] resulting in a substantial amount of distortion being introduced to the transmitted signal. Furthermore, as the number of subcarriers increases, so does the PAPR values. Therefore, several researchers developed digital baseband PAPR reduction techniques [206–212] in an attempt to solve this problem.

Due to the null subcarriers in the NC-OFDM signal, its PAPR characteristics are slightly different from those of conventional OFDM signals. Moreover, the PAPR reduction algorithms developed for conventional OFDM signals need to be modified for its usage in NC-OFDM systems. The PAPR reduction algorithms should also be aware of the deactivated subcarriers, thereby avoiding any interference to the primary user transmissions.

Definition of Peak-to-Average Power Ratio

The complex envelope of a baseband NC-OFDM signal, consisting of all N contiguous subcarriers over a time interval $[0, T]$, is given by

$$s(t) = \frac{1}{\sqrt{N}} \sum_{k=0}^{N-1} A_k e^{j2\pi kt/T},\tag{6.10}$$

where A_k is the symbol of the kth subcarrier,[4] T is the OFDM symbol duration, and $j = \sqrt{-1}$. The symbol over the kth deactivated subcarrier is $A_k = 0$.

---

[3]This occurs when the dynamic range of the digital-to-analog converter or the power amplifier is insufficient.

[4]For example, $A_k \in \{1, -1\}$ for BPSK signaling, and $A_k \in \{\pm 1, \pm j\}$ for QPSK signaling.

The PAPR of Equation (6.10) is defined as the ratio between the maximum instantaneous power and the average power of an OFDM signal, namely [213],

$$\text{PAPR}(s(t)) = \frac{\max\limits_{0 \leq t \leq T} |s(t)|^2}{E\left\{|s(t)|^2\right\}}, \tag{6.11}$$

where $E\{.\}$ denotes the expectation operator. Without loss of generality, we can safely neglect the cyclic extension from the analysis, since it does not contribute to the PAPR problem. The continuous time PAPR of $s(t)$ can be approximated using the discrete time PAPR, which is obtained using samples of the OFDM signal, $s(n)$. It has been shown that an oversampling factor of 4 is sufficient to estimate the continuous PAPR of a BPSK system [214].

As an illustration, consider a 16 subcarrier BPSK-OFDM transceiver system. When all the input symbols are ones, the normalized power of the OFDM symbol in the time domain is as shown in Figure 6.15(a). From this figure, the mean power of the signal can be calculated to be 0.0625 and the peak power is unity. The PAPR of the signal is 16. Now, consider an input random sequence, [1 1 1 −1 1 1 −1 −1 1 1 −1 −1 1 −1 −1 −1]. The normalized power of the OFDM symbol in the time domain is shown in Figure 6.15(b). The mean power of the signal remains 0.0625 as the total power of the signal remains constant. However, the peak power is 0.3025. Thus, the PAPR is equal to 4.8407. This figure illustrates that the random sequence being transmitted has an effect on the PAPR of the signal. Moreover, it has been suggested that the sequences with the maximum correlation yield a very high PAPR value [215]. Some of the algorithms proposed in the literature aim at reducing the correlation of the sequence to reduce the PAPR. Furthermore, the PAPR of a system is directly related to the number of subcarriers in the system. The greater the number of subcarriers, the larger is the PAPR.

Existing Solutions

Subcarrier power adjustment–based techniques are one of the most popular approaches in the technical literature for combating PAPR. Power adjustment approaches can be further classified into approaches that constrain the total power of all the subcarriers [216, 217] and those that constrain the power of subcarriers in a window [218, 219].

Total power-constrained power adjustment implies that, if power of any subcarrier is reduced or turned off, the excess power allocated to it can be transferred to the remaining active subcarriers. Let π_k be the transmitted power of the kth subcarrier ($k = 0, 1, \ldots, N - 1$). If the total number of subcarriers is N, the power constraint is given by

$$\sum_{k=0}^{N-1} \pi_k = \pi_{\text{total}}. \tag{6.12}$$

The aggregate bit rate is approximately maximized if the bit error rates in all the subbands are equal, whereas BER performance is optimized when all the

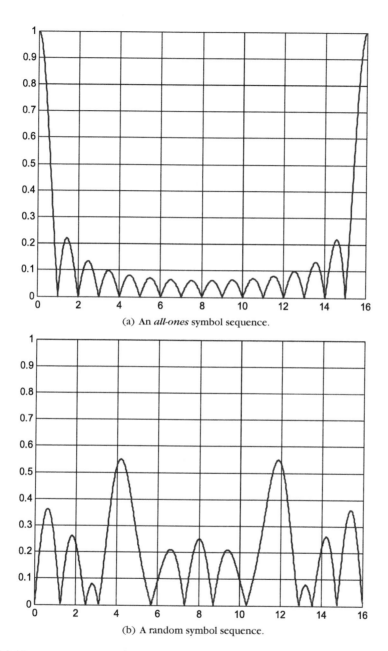

(a) An *all-ones* symbol sequence.

(b) A random symbol sequence.

FIGURE 6.15

Time-domain waveforms of a 16-subcarrier BPSK-OFDM transceiver system.

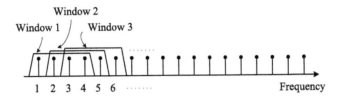

FIGURE 6.16

Subcarrier power window.

subcarriers have equal power [82]. Without power adjustment, π_k is assumed to be equal for all subcarriers. In case of total power-constrained power adjustment, it is possible that all the power could be concentrated on a single subcarrier.

In window power-constrained power adjustment, if the power level of a certain subcarrier is lowered, the excess power can be transferred to the active subcarriers within a certain predefined sliding window. The total power for every grouping of M subcarriers needs to be below the regulatory requirement, say π_{max}. Then, the subcarrier power constraints are [218]

$$\sum_{k=l}^{l+M-1} \pi_k \leq \pi_{max}, \quad \text{for all } l \tag{6.13}$$

and

$$\sum_{k=0}^{N-1} \pi_k \leq \pi_{total}. \tag{6.14}$$

For example, consecutive windows of subcarriers are shown in Figure 6.16. If the power level of a subcarrier in Window 1 is lowered, the power can be transferred to the other active subcarriers only within Window 1. When the power level of a subcarrier (say Subcarrier 3) is adjusted, the subcarrier power level must satisfy the window power constraints for all the member sliding windows (Window 1, Window 2, Window 3). A subcarrier power adjustment algorithm is shown in Figure 6.17 [220].

1. Initialize $\pi(0:N-1) = Z \sim f(x)$
2. Generate L different power adjustment factors, $\{\pi_j\}_{j=0}^{L-1}$, or interleave π using L random interleavers (or other interleavers)
3. Check the FCC power constraint for each (interleaved) subcarrier power window π_j, using the sliding window algorithm presented in Figure 6.18
4. Modified symbols, $X_{mod} = X \cdot \sqrt{\pi_j}$
5. Choose π_j, which yields lowest PAPR

FIGURE 6.17

A subcarrier power adjustment based PAPR reduction algorithm.

```
1.   for k = 0 to (N − M + 1)
2.     for L = K to (K + m − 1)
3.       if π(1) < π_min
4.          π(1) = π_min
5.       end if
6.     end for
7.     S = Σ_{l=k}^{k+M−1} π(1)
8.     if S > π_max
9.        π(k : k+M−1) = π(k : k+M−1) * π_max / S
10.    end if
11.  end for
12. return π
```

FIGURE 6.18

Sliding window power constraint.

A practical transmit power constraint is usually enforced as shown in Figure 6.18 to limit the total power across a frequency window of a specified width. For instance, the FCC has imposed requirements based on the amount of transmit power across a specified bandwidth in the UNII band [219]. These requirements are imposed, since these bands are usually unlicensed and the users are uncooperative.

Statistical Properties of PAPR

Being a variant of OFDM, NC-OFDM also suffers from a high PAPR problem. When the number of deactivated subcarriers is large compared to the number of active subcarriers, the common assumption of the input symbols being identically and independently distributed does not hold. This results in different statistical properties for the PAPR of NC-OFDM signals relative to that for OFDM signals. The conventional OFDM systems inherently assume a contiguous set of subcarriers, whereas in a NC-OFDM system, the active subcarriers are colocated with the occupied subcarriers. Therefore, the PAPR reduction techniques proposed for the OFDM systems may need to be modified to avoid interfering with existing user transmissions. Moreover, the design requirements for PAPR reduction techniques are different from that of conventional OFDM systems. In this section, we present statistical analysis of PAPR for NC-OFDM signals and elaborate the design requirements of the PAPR reduction techniques for the NC-OFDM signals.

PAPR Distribution of an NC-OFDM Signal

Assume that the total number of subcarriers, N, is large. Then, applying the central limit theorem (CLT), $s(n)$ can be modeled as a zero-mean Gaussian distributed random variable with variance $\sigma^2 = N_u \sigma_D^2 / N$, where N_u is the number of active subcarriers and σ_D^2 is the variance of the input sequence [221]. Assuming the symbols are independent and identically distributed by CLT, the real and imaginary parts of the N-point IFFT output have mutually independent Gaussian probability distribution function, $\mathcal{N}(0, \sigma^2)$. The instantaneous power of baseband signal, $s(n)$, is given by

$$\lambda = \Re\{s(n)\}^2 + \Im\{s(n)\}^2. \tag{6.15}$$

Therefore, the instantaneous power can be characterized as χ^2 distribution with two degrees of freedom [72]:

$$f(\lambda) = \frac{1}{\sigma^2} \exp\left(-\frac{\lambda}{\sigma^2}\right), \quad \lambda \geq 0. \tag{6.16}$$

So, the cumulative distribution function (CDF) is given by [72]

$$Pr[\lambda < \lambda_0] = \int_0^{\lambda_0} f(\lambda)d\lambda$$
$$= 1 - e^{-\frac{\lambda_0}{\sigma^2}}. \tag{6.17}$$

If $E\{|x(n)|^2\}$ is normalized to unity, then the CCDF of the PAPR is given by

$$Pr[\lambda > \lambda_0] = 1 - \left(1 - e^{-\frac{\lambda_0}{\sigma^2}}\right)^N. \tag{6.18}$$

However, this expression is not close to experimental results because the assumption made in deriving CCDF that the samples should be mutually uncorrelated is not true anymore when oversampling is used [222].

Several attempts have been made to more accurately determine the distribution of the PAPR for OFDM signals. In [222], it was claimed that the approximated CCDF for large N is given by

$$Pr[\lambda > \lambda_0] \approx 1 - \left(1 - e^{-\frac{\lambda_0}{\sigma^2}}\right)^{\alpha N}, \tag{6.19}$$

where $\alpha = 2.8$. This approximation is fairly close to the experimental results when the number of active subcarriers is large; that is, $N \geq 64$.

Maximum PAPR of an NC-OFDM Signal

It is known that the PAPR of an MPSK OFDM signal is always less than or equal to N, where N is the total number of subcarriers. Note that, in OFDM, all of N subcarriers are active. Now, consider the NC-OFDM signal with p active subcarriers, where the input data symbols are chosen from an MPSK constellation such that $|A_k| = 1$.

The peak power of the NC-OFDM signal is given by [213]

$$\max_{0 \leq t \leq T} |s(t)|^2 = \max_{0 \leq t \leq T} \left| \frac{1}{\sqrt{N}} \sum_{k=0}^{N-1} A_k e^{j2\pi kt/T} \right|^2$$

$$\leq \left(\frac{1}{\sqrt{N}} \sum_{k=0}^{N-1} \max |A_k| \right)^2 \tag{6.20}$$

$$\leq \frac{p^2}{N}.$$

Using the Parseval's relationship the average power of the NC-OFDM signal is given by [72]

$$E\{|s(t)|^2\} = E \left\{ \left| \frac{1}{\sqrt{N}} \sum_{k=0}^{N-1} A_k e^{j2\pi kt/T} \right|^2 \right\}$$

$$= \frac{1}{N} \sum_{k=0}^{N-1} \left\{ E|A_k|^2 \right\} \tag{6.21}$$

$$= \frac{p}{N}.$$

Then, the PAPR of the NC-OFDM signal is given from Equation (6.11) as follows:

$$PAPR(s(t)) = \frac{\max\limits_{0 \leq t \leq T} |s(t)|^2}{E\{|s(t)|^2\}}$$

$$\leq \frac{p^2/N}{p/N} \tag{6.22}$$

$$\leq p.$$

Therefore, the maximum value of PAPR for the MPSK-modulated NC-OFDM signal with p active subcarriers is equal to p, regardless of the total number of subcarriers, N.

Design Requirements of PAPR Reduction Techniques for an NC-OFDM Signal

The conventional PAPR reduction techniques for OFDM systems inherently assume a contiguous set of subcarriers. Therefore, PAPR reduction techniques proposed for OFDM systems need to be adapted to a system employing NC-OFDM. In spectrum opportunistic systems, the active subcarriers are colocated with the occupied subcarriers. As a result, both intersymbol interference and intercarrier interference may cause distortion in the primary user transmissions. Therefore, time-domain-based or distortion-based techniques, such as clipping and filtering [223], and frequency-domain-based techniques assuming contiguous subcarriers, such as coding [207], cannot be used for reducing the PAPR of NC-OFDM signals. However, frequency-domain PAPR reduction techniques are better suited, since it is easier to sort out the nulled subcarriers, avoiding any interference to existing user transmissions. The techniques, such as interleaving [224], SLM [210], and partial transmit sequences [211], need to be aware of the locations of the active subcarriers. Moreover, in a dynamic spectrum access network, the total number of active subcarriers and their locations might change continuously and the PAPR reduction techniques should be able to adapt to these changes.

6.5 CHAPTER SUMMARY AND FURTHER READINGS

In this chapter, we analyze several critical issues associated with OFDM-based cognitive radios and briefly discuss several proposed solutions. There are several significant ideas to draw from this chapter. First, spectral agility aimed at improving the utilization efficiency puts forth additional design constraints to conform to the spectral masks imposed by the regulatory agencies, to minimize the computations involved, and to avoid signal distortion during amplification due to the high PAPR nature of the multicarrier signals. In most cases, even though solutions have been proposed for the traditional contiguous subcarrier case, novel approaches need to be formulated for the noncontiguous subcarrier case.

Given that spectral agility is an important requirement for cognitive radio-oriented transceivers, it has to be noted that DFT-based OFDM is not the only modulation technique that achieves it. Several other techniques have been proposed by researchers over the past decade or so and compared with the classic DFT-based OFDM in terms of performance over widely differing channel conditions, as well as in terms of the involved implementation complexity of the transceiver. In [225], the authors introduce and show that M-band filter banks with overlapping basis functions can be used to design transmultiplexers with superior frequency-domain properties to those achievable with DFT filter banks. Consequently, the authors also show that the wavelet-transform-based discrete wavelet multitone has the DFT as a special case. A comparison of the BER performance in the presence of narrowband interference is reported in [226]. Other results on the waveform development using DWMT are reported in [227]. In [228, 229], equalization schemes for DWMT are proposed. While [228] presents simulated BER results to show that postdetection combining provides better performance, [229] presents linear and decision-feedback fractionally spaced receiver designs using modified filter banks. Reference [230] provides an overview of different orthogonal synthesis/analysis transform configurations. In [231], a comparison of the DFT-OFDM scheme with cosine-modulated filter banks (CMFBs) in terms of the prereceiver processing and performance in nonlinear phase channels is performed. Another paper that provides a comparison of different filter bank multicarrier techniques from the context of cognitive radio is [232]. In [233], a novel blind equalization algorithm for CMFB-based multicarrier transceivers is proposed and its BER performance is compared with the conventional OFDM system.

6.6 PROBLEMS

1   Using MATLAB, simulate the results shown in Figure 6.8, which is the impact of the raised cosine window with different roll-off factors on the side lobe power levels. Then, experiment with several other standard window functions and understand the roll-off nature of those windows.

2   Using MATLAB, simulate the efficiency of guard subcarriers shown in Figure 6.9 for the noncontiguous case. That is, assume several alternating sets of subcarriers occupied by licensed and unlicensed users and study the impact of the spectral spillage from the unlicensed bands to licensed portions of the spectrum. Vary the number of subcarriers in each noncontiguous set occupied by secondary users and observe the results.

3   Implement the standard Cooley-Tukey algorithm for the case of alternating licensed and unlicensed bands. Assign zeros to subcarriers used by licensed users. Store the number of additions and multiplications performed. Vary the number of zeros—that is, the number of licensed users in a given band of interest—and understand how the Cooley-Tukey algorithm is ineffective in reducing the number of computations as the unlicensed subcarriers reduce.

4   Figure 6.15 shows a plot of two discrete cases of the impact of PAPR for a BPSK-OFDM transceiver. Using Equation (6.11), plot the dependency of PAPR on the modulation scheme for a fixed number of subcarriers. Next, fix the modulation scheme and plot the PAPR as a function of the number of subcarriers.

Reconfiguration, adaptation, and optimization

7

Timothy R. Newman[1], Joseph B. Evans[2], and Alexander M. Wyglinski[3]

[1]Virginia Polytechnic Institute and State University, United States
[2]The University of Kansas, United States
[3]Worcester Polytechnic Institute, United States

7.1 INTRODUCTION

Cognitive radios take advantage of the reconfigurable attributes of a conventional software-defined radio (SDR) by using an "intelligent" control method to automatically adapt operating parameters based on learning from previous events and current inputs to the system. For cognitive radios to properly reconfigure, adapt, and optimize the system, several key parameters of the system must be identified. System parameters such as the transmission controls and environment measurements that are used to optimize the system must be identified. An optimization method, or "intelligent" control method, must be selected that can be run practically in real time to meet quality-of-service (QoS) requirements. The definition of these transmission parameters, environment parameters, and optimization techniques is at the core of the current cognitive radio research. The momentum of research efforts, due in part to the current spectrum scarcity problem, as well as a Department of Defense initiative [234] to develop a flexible software radio approach for military communications, has yielded numerous initiatives and programs by researchers in academia [235] and industry [236]. The resulting plethora of cognitive radio solutions range from cognitive radio components and radio network test beds [235] to complete radio systems [237].

In this chapter we look at the commonly used wireless communication parameters that can be used by cognitive radios to optimize the performance of the system. We explore different objectives of communication to identify common cognitive radio goals that must be related to the parameters. These relationships give the cognitive engine the information needed to adapt the radio parameters to optimize the communication objectives. We introduce several methods of optimizing wireless communication through the cognitive adaptation engine. We focus the discussion on the most commonly used techniques, including a heuristic-based method that uses genetic algorithms, an expert system technique that uses a rule-based approach, a case-based reasoning system that makes decisions based on past experiences, and hybrid techniques that use multiple combinations of these approaches.

7.2 ADAPTATION ENGINE

The previous chapters discuss the primary application of cognitive radios in much detail. Dynamic spectrum access is currently a main motivating factor in the development of cognitive radios and the market is beginning to see the fruit of this research in the radios developed by Shared Spectrum Company [238] and the cognitive engines from Wireless@Virginia Tech [239–241]. However, what many overlook is that cognitive radios have the potential to do much more than frequency adaptation and spectrum policy applications. Due to the time-varying radio channel characteristics, as well as the spectrum band availability, cognitive radios need to support time-varying QoS requirements. In addition to maximizing the efficiency of spectrum usage, other goals exist that improve wireless communication as a whole, including minimizing the bit error rate (BER), maximizing the data throughput, minimizing the power consumption, and minimizing the interference. To achieve these goals, a cognitive radio needs to optimize many related and unrelated radio transmission parameters based on the user requirements, or QoS objectives, as well as the environmental parameters.

As discussed in previous chapters, dynamic spectrum access (DSA) can be done in many ways and is not always considered a cognitive radio application. A cognitive radio implies that the wireless communication device has some sort of learning and knowledge retention capabilities. The most simple secondary-user DSA application is to sense the desired channel, transmit if no one is using it currently, and change channels if a primary user begins transmission. This application alone has no cognition and requires no cognitive adaptation engine. However, if we add the capability to remember the status of several channels and provide the capability to optimize the transmission parameters for each channel individually because each channel has its own channel characteristics, then this would require a cognitive adaptation engine. Ongoing questions that need to be answered when developing a cognitive radio include: What controllable parameters are available and which ones will I be using? Also, what characteristics of the environment can I measure? Probably the most important question is, What QoS goals will be taken into account when attempting communication? These goals guide the radio to the proper transmission parameter set. This guidance is done through a cognitive engine that can be implemented in many ways. In the following sections, we introduce common parameters a cognitive radio would take into account and describe different methods of implementing a cognitive adaptation engine.

7.3 OPERATING PARAMETERS

In developing a cognitive radio system, several inputs must be defined. The accuracy of the decisions made by an artificial intelligence (AI) method are based on the quality and quantity of inputs to the system. More inputs to the system make the radio more informed, thus allowing the decision-making process to generate decisions that are more accurate. Several categories of parameters exist. Environmental parameters represent information about the current wireless environment. For the cognitive engine to make decisions about a certain output, the current wireless environment must be modeled internally. This model is created using environmentally sensed data received by the system using an external sensor or by receiving feedback from another node in the system.

Several devices exist to detect characteristics of the wireless environment. The DARPA XG program has hardware for sensing environment characteristics, including spectrum usage [242]. This information is useful if the radio is trying to maximize spectral efficiency. Other sensors detect important characteristics such as the current noise floor or signal-to-noise ratio (SNR), or determine the BER of the current running configuration. In the following sections, we provide a list of environmentally sensed parameters used to aide in the decision-making process of the cognitive controller.

Decision variables are another important input to AI methods. In the cognitive radio case, these variables represent the transmission parameters that can be controlled by the system. Examples of these variables include transmission power and modulation type. The values of these variables are the output of the cognitive engine. The overall goal is for the cognitive engine to output transmission parameter values that optimize the communication objectives of the system.

In addition to the environmental data used to model the wireless channel and the transmission parameters, communication performance objectives must be determined to define how the system should operate. The objectives of the system are the roadmap for determining the fate of the system. They provide the means for the controller to steer the system to a specific state. For example, one basic objective is to minimize the bit error rate of the system. This can be done by manipulating the transmission parameters in such a way as to provide the lowest possible BER given the current environment. Section 7.4.1 goes into detail about five common communication performance objectives for wireless cognitive radio applications.

7.3.1 Transmission Parameters

Cognitive radios take advantage of the control parameters made available by the underlying software-defined radio system. These control parameters are used by the cognitive engine, along with the environmental parameters and communication objectives to make communication decisions. One way this is done is by using the parameters as input to a fitness function. This fitness function inputs a set of parameter values and outputs a scalar score that represents how well the control parameters achieve the given objectives. Developing the fitness functions used by a cognitive radio engine requires defining a specific list of transmission parameters that must be available to the system. These transmission parameters are equivalent to the control parameters made available by the software radio components. The term *transmission parameters* is used in this chapter to refer to the list of parameters used to control the individual radio components.

Creating a complete list of transmission parameters and generating a generic "one size fits all" fitness function usable by all radios is not possible. Radios are developed for many reasons, and depending upon the application of the radio, each will possess a unique list of parameters. This is why the selection of transmission parameters for a cognitive radio that ideally can be used for a multitude of applications can be difficult. This difficulty is due to the increased system complexity that occurs when unnecessary parameters are used in the cognitive engine decision-making process [243]. The transmission parameters presented here represent parameters commonly cited in the research literature as transmission parameters that can be used by cognitive radios to control communication characteristics [240, 244–247].

The sample set of parameters presented in this chapter represents only a fraction of possible controllable parameters within a cognitive radio. Smaller resolution parameters, such as linear amp gain or filter coefficients, provide fine-grained control. In addition, much higher-layer parameters exist that change on the order of hours, such as transmission formats (e.g., OFDM or CDMA), encryption (e.g., WEP or PGP), or error control techniques (e.g., Turbo coding or convolutional coding). This chapter covers parameters that commonly change on the order of hundreds of milliseconds, depending on the state of the wireless channel. An example list of parameters that could be used by a cognitive engine to control communications is shown in Table 7.1.

Although the focus of this chapter is on these transmission-level parameters and not system-level parameters such as using CDMA or FDMA, those higher-order system parameters may still be passed to the cognitive system to allow filtering out several possible parameter values. For example, if the cognitive component is informed that the system should be using iterative coding and OFDM modulation, this restricts the modulation type and the channel coding rate possibilities, which in turn relaxes computational constraints on the cognitive engine.

7.3.2 Environmental Measurements

Environmental measurements inform the system of the surrounding environment characteristics. These characteristics include internal information about the radio operating state and external information representing the wireless channel environment. Both types of information can be used to aide the cognitive controller in

Table 7.1 Transmission Parameter List

Parameter Name	Description
Transmit power	Raw transmission power
Modulation type	Type of modulation format
Modulation index	Number of symbols for given modulation scheme
Bandwidth	Bandwidth of transmission signal in hertz
Channel coding rate	Specific rate of coding scheme
Frame size	Size of transmission frame in bytes
Time division duplexing	Percentage of transmit time
Symbol rate	Number of symbols per second

making decisions. The environmental measurements can be classified into two categories. The first are environment variables used directly by the cognitive engine as primary input parameters. An example of this type of parameter is the noise power of the channel, which affects several communication objectives. These parameters directly affect the decision of the cognitive engine. The second class of environment parameters are trigger parameters. These parameters are monitored by the system, and decisions about the objective function are triggered based on their values. A good example of this is the battery life parameter. The system may monitor this parameter while it decreases below a specified threshold. Then, the system may alter the weighting on the objective functions to provide a higher weighting on the minimize power consumption objective. An example list of environmental measurements that could be used by a cognitive engine to model the environment and trigger decisions is shown in Table 7.2.

The path loss is the reduction in power density of the signal as it travels through space. Path loss may be due to effects such as free-space loss, refraction, diffraction, reflection, and absorption. Path loss can also be influenced by the terrain and environment. The noise power parameter informs the system of the approximate power of the noise in decibels of the measured power referenced to 1 mW. Battery life and power consumption are both internal parameters. These parameters are used to determine when the system should place more emphasis on minimizing the power. The primary external parameter is spectrum occupancy information. This parameter consists of information from cognitive radios within the local network identifying the spectral location of other signals in frequency bands of interest. This information is used to improve the spectral efficiency of the transmission and the spectral occupancy of the frequency band [181, 248].

Table 7.2 Environmentally Sensed Parameter List

Parameter Name	Description
Path loss	Amount of signal degradation lost due to the channel path characteristics
Noise power	Size in decibels of the noise power
Battery life	Estimated energy left in batteries
Power consumption	Power consumption of current configuration
Spectrum information	Spectrum occupancy information

The trigger parameters represent an important characteristic of cognitive radio systems. Much research on the cognitive engines focuses on decision making using active parameters, and the objective-steering trigger parameters are often overlooked.

7.4 PARAMETER RELATIONSHIPS

This section introduces the multiobjective problem representation and describes several common radio performance objectives. Section 7.4.1 describes several common objectives that a radio system wants to achieve. Mathematical representations of these objectives are presented and the problems of competing objectives is discussed.

In general, a multiobjective fitness function problem can be presented as trying to determine the correct mapping of a set of m parameters to a set of N objectives. This can be seen algebraically as

$$\vec{y} = \langle f_1(\vec{x}), f_2(\vec{x}), f_3(\vec{x}), \ldots, f_N(\vec{x}) \rangle \qquad (7.1)$$

subject to

$$\vec{x} = \langle x_1, x_2, x_3, \ldots x_m \rangle \in X$$
$$\vec{y} = \langle y_1, y_2, y_3, \ldots y_N \rangle \in Y,$$

where x is the set of decision variables with X as the parameter space, and y is the set of objectives with Y as the objective space. Depending on the cognition algorithm, each $f_i(x)$ is used in a different manner. For example, in the case of a multiple-objective evolutionary algorithm, each $f_i(x)$ represents the fitness function for a single objective. The goal is to combine them to get a single fitness function, $f(x)$, taking into account all parameters and objectives. First we introduce different radio objectives and discuss their mathematical representations individually.

7.4.1 Single Radio Performance Objectives

In a wireless communications environment, the radio system may want to achieve several desirable objectives. We present five common communication objectives that a cognitive engine may use to guide the system to an optimal state. The five objectives are given in Table 7.3.

Minimizing the BER is a common communications goal in the error-prone world of wireless communications. This objective represents minimizing the amount of errors in relation to the amount of bits being sent. In general, this objective represents improving the communications signal of the radio. To gather this information, along with many other QoS values, feedback from the receiver must be received. Maximizing the throughput deals with the data throughput rate of the system. Emphasis on this objective improves system throughput. Several transmission parameters affect the overall throughput of the system. From the common list of parameters shown in Table 7.1, the modulation index may have the most impact on the maximize throughput objective. Minimize power consumption is self-explanatory and used to direct the system to a state of minimal power consumption.

The last two objectives focus on the spectral domain of wireless communications. Minimizing interference encompasses avoiding areas of the spectrum with a high noise floor or high possibility of interference being present. Similarly, emphasis on the maximize spectral efficiency objective reduces the spectral space used by the transmitted signal.

For a multicarrier system with N independent subcarriers, the objective functions are defined as

$$f_{\text{min\_ber}} = 1 - \frac{\log_{10}(0.5)}{\log_{10}(\overline{P_{\text{be}}})},\tag{7.2}$$

Table 7.3 Cognitive Radio Objectives

Objective Name	Description
Minimize bit error rate	Improve the overall BER of the transmission environment
Maximize throughput	Increase the overall data throughput transmitted by the radio
Minimize power consumption	Decrease the amount of power consumed by the system
Minimize interference	Reduce the radio's interference contributions
Maximize spectral efficiency	Maximize the efficient use of the frequency spectrum

where $\overline{P_{be}}$ is the average BER over N independent subcarriers.

$$f_{\text{max\_tp}} = \frac{\sum_{i=1}^{N} \left[\frac{L_i}{L_i + O + H} \times (1 - P_{\text{ber}_i})^{(L_i + O)} \times R_{c_i} \times \text{TDD}_i \right]}{N}, \tag{7.3}$$

where L_i is the packet length, O and H are static packet overheads for the PHY and MAC layers, R_{c_i} is the coding radio, and TDD_i is the time division duplex value for channel i. For the maximum throughput multicarrier fitness function in Equation (7.3), all carrier fitness scores are summed together then divided over the total number of carriers, N, to get the average fitness score over all subcarriers. This forces the system to improve the overall system fitness. The same follows for the minimize power fitness function in Equation (7.4), the minimize interference fitness function in Equation (7.5), and the maximize spectral efficiency fitness function in Equation (7.6).

$$f_{\text{min\_power}} = \left[1 - \alpha \times \frac{\sum_{i=1}^{N} (P_{\max} + B_{\max}) - (P_i + B_i)}{N \times P_{\max} + B_{\max}} \right.$$

$$+ \beta \times \frac{\sum_{i=1}^{N} \log_2(m_{\max}) - \log_2(m_i)}{N \times \log_2(m_{\max})}$$

$$\left. + \lambda \times \frac{\sum_{i=1}^{N} R_{s_{\max}} - R_{s_i}}{N \times R_{s_{\max}}} \right] \tag{7.4}$$

$$f_{\text{min\_interference}} = 1 - \frac{\sum_{i=1}^{N} [(P_i + B_i + \text{TDD}_i) - (P_{\min} + B_{\min} + 1)]}{N \times (P_{\max} + B_{\max} + 100)} \tag{7.5}$$

$$f_{\text{max\_spectralefficiency}} = \frac{\sum_{i=1}^{N} \frac{m_i \times R_{s_i} \times B_{\min}}{B_i \times m_{\max} \times R_{s_{\max}}}}{N}, \tag{7.6}$$

where P_i is the transmit power on subcarrier i, N is the number of carriers, and $P_{\max}$ is the maximum possible transmit power for a single subcarrier. Similarly, m_i is the modulation index used on subcarrier i and $m_{\max}$ is the maximum modulation index available, B_i is the bandwidth allocated to channel i, and $B_{\max}$ and $B_{\min}$ are, respectively, the maximum and minimum bandwidths the radio can transmit over instantaneously. R_{s_i} is the symbol rate on channel i and $R_{s_{\max}}$ and $R_{s_{\min}}$ are the maximum and minimum symbol rates available. Notice that the objectives that minimize the goal have a $1-$ term, which is needed to provide a positive fitness score as the raw value decreases.

7.4.2 Multiple Objective Goals

In real-world problems, the objectives under consideration might conflict with each other. For example, minimizing power and minimizing BER simultaneously creates a conflict due to the single parameter, transmit power, affecting each objective in a different way. Determining the optimal set of decision variables for a single objective, such as minimize power, often results in a nonoptimal set with respect to other objectives, such as minimize BER and maximize throughput. The optimal set for multiple-objective functions lies on what is known as the *Pareto optimal front* [249–251]. This front represents the set of solutions that cannot be improved upon in any dimension. The solutions on the Pareto front are optimal and coexist due to the trade-offs among the multiple objectives. A graphical example of a Pareto front using a simple cognitive radio parameter scenario is shown in Figure 7.1.

The *x*-axis in the figure represents the score of the single objective fitness function for minimizing BER in the case of several modulation types, while the *y*-axis is the score for the single objective fitness function to minimize power. The parameter *x* represents the decision variable vectors used as inputs to the fitness functions. In this case, transmit power and modulation are used as decision variables. For each curve, as the fitness score to minimize power decreases, the score to minimize BER objective increases. This trade-off represents the core of the

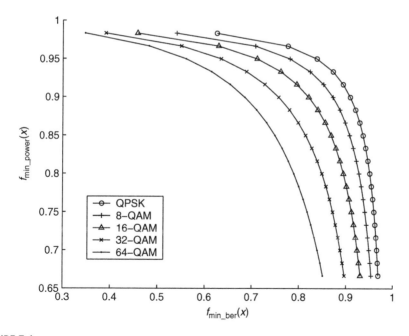

FIGURE 7.1

Pareto front trade-off.

multiple-objective optimization problem. The QPSK curve represents the Pareto front, because no parameter set on that curve can be improved upon to gain a better objective score in respect to both objectives. The other modulation curves represent the dominated solutions to the biobjective optimization problem.

To direct the system to a specific solution, we must attach preference information to each objective. Otherwise, simply minimizing both BER and power results in a set of solutions instead of a single solution. This is because minimizing BER and minimizing power have different solutions. Therefore, the objectives must also contain a quantifiable rank representing the importance of each. This allows the fitness function to characterize the trade-offs among the objectives by ranking the objectives in order of importance. Several approaches exist for determining the preference information of a set of objectives [249].

In addition to needing preference information for each objective, the scalarization of the objective vector is also necessary. Evolutionary algorithms need scalar fitness functions that provide a single scalar value for the given parameter set. In many optimization problems, when no global criteria, such as goals, for the parameters exist, objectives are often combined, or aggregated, into a scalar function. This aggregation optimization method has the advantage of providing a single scalar solution for the fitness function. As a result, this requires no extra interaction with the evolutionary algorithm to determine the optimality of a given parameter set.

There have been several approaches to the optimization of aggregated functions. Weighted sum approaches are presented in [252, 253]. The weighted sum approach attempts to minimize the sum of the positively normalized, weighted, single-objective scores. In [254], target vector optimization is developed. Target vector optimization requires a vector of goal values. The optimization is driven toward the shortest distance between any candidate solution and the goal vector. Goal programming was also studied by several authors [255, 256]. In goal programming, one objective is minimized while constraining the remaining objectives to be less than the target values. However, choosing appropriate goals for the constraints can be difficult.

As a reference example we show how the weighted sum approach can be used. The weighted sum approach allows us to combine the single-objective functions into one aggregate multiple-objective function. Equation (7.7) shows that each objective is multiplied by a weight w_i and summed together to give a single scalar value for approximating the value of a parameter set. For the single-objective equations, we form the multiple-objective function for multiple carriers given in Equation (7.7):

$$f_{\text{multicarrier}} = w_1 \times \left(f_{\text{min\_ber}} \right) + w_2 \times \left(f_{\text{max\_tp}} \right) + w_3 \times \left(f_{\text{min\_power}} \right)$$
$$+ w_4 \times \left(f_{\text{min\_interference}} \right) + w_5 \times \left(f_{\text{max\_spectralefficiency}} \right). \quad (7.7)$$

The weighting values, w_1, w_2, w_3, w_4, and w_5, determine the search direction for the algorithm. Note that, because the individual objective functions are normalized, they are unitless. To help aid in the creation of example simulations

Table 7.4 Example Weighting Scenarios

Scenario	Weight Vector [w_1, w_2, w_3, w_4, w_5]
Low-power mode (minimize power)	[0.10, 0.20, 0.45, 0.15, 0.10]
Emergency mode (minimize BER)	[0.50, 0.10, 0.10, 0.10, 0.20]
Dynamic spectrum access mode (minimize interference)	[0.10, 0.20, 0.10, 0.50, 0.10]
Multimedia mode (maximize throughput)	[0.15, 0.50, 0.10, 0.15, 0.10]

we define four example weight vectors representing common scenarios in which a cognitive radio may be placed. Each weight vector shown in Table 7.4 emphasizes different objectives, causing an algorithm using this fitness function to evolve toward solutions pertaining to the specific objective.

7.5 COGNITIVE ADAPTATION ENGINES

The cognitive adaptation engine is at the core of the cognitive radio. It is the intelligence that drives the decision-making process. Many options exist for the implementation of the engine. In the following sections we cover several options for implementation and highlight the advantages and disadvantages of each. The primary engines we look at are based on genetic algorithms, rule-based systems, and case-based reasoners. Genetic algorithms are a class of artificial reasoning whereby the search is performed in a manner similar to genetic evolution. In general, solutions to a problem set are represented by binary strings. These strings then are allowed to act in a manner similar to genetic growth; strings considered "good" split and recombine with other good strings to form new solutions, while "poorer" strings are allowed to "die" out of the solution set. This decision is made by the fitness function, which inputs the parameters and outputs a score based on the specific goals of the radio. Strings undergo a process called *mutation*—that is, a random flipping of bits—to help prevent local minimization. Genetic algorithms are typically used as a method of problem optimization [257, 258]. However, given its random nature, fast computation time, and ability to spontaneously generate unique solutions, genetic algorithms are an appealing candidate for cognitive radios. Input and output parameters can easily be mapped to a binary form and the size of the genetic population is customizable to the space available within any given configuration [258]. Genetic algorithms are used mainly when the search space is too large to be simply brute force searched to determine the parameter set.

Another approach to implementing a cognitive engine is by using a rule-based system. Genetic algorithms rely on the ability to use a small amount of memory and large amounts of processing to evolve to a solution. Rule-based systems, in contrast, use more memory and a small amount of processing to make decisions, as do case-based reasoners that hold past experiences in a database for future decision-making purposes.

A rule-based system (RBS) uses a simplistic model based on a set of IF/THEN statements to implement an expert system. These systems are widely used in many fields with the primary concept being that the knowledge of an expert is coded into the rule set. When the expert system comes across a data set, it should behave the same way as the expert that populated the database. We want to explore the feasibility of using such an RBS in the context of cognitive radio decision making. Rule-based systems are at a disadvantage when the rule base becomes large, because it becomes hard to manage and the rule base itself may consume too much memory. Rule-based systems are typically also constrained by discrete values. This can be overcome by using fuzzy sets and defined ranges for the rules to match on. However, this then requires a specific range to be determined for each rule. In the following sections, we explore the implementation of both of these techniques in the context of cognitive radios.

7.5.1 Expert Systems

An expert system uses nonalgorithmic expertise to solve certain problems [259, 260]. Expert systems have a number of primary system components that must interface with people in various ways. Figure 7.2 shows major components and the

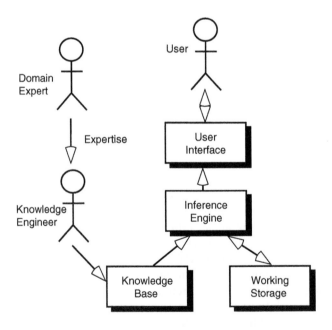

FIGURE 7.2

Expert system diagram.

individuals who interact with the system. The *knowledge base* is the representation of the expertise. Each piece of expertise is typically termed a *rule* and represented using an IF/THEN format. For example, one rule in our system may be, IF frequency band of interest is currently in use, THEN alter frequency. The expertise or the rules are created by the *domain expert*. The domain expert would say specifically what frequency is the optimal to use. Typically, as shown in Figure 7.2, a knowledge engineer is used to encode the expert's knowledge into a form that can be used by the expert system. For some projects the domain expert and the knowledge engineer may be the same person, or in our case the information produced by the domain expert actually comes from an equation.

The second major component in the system is the *working storage*. This component holds the problem-specific information for the problem currently being solved. This information is typically termed the *facts*. For example, a fact may exist stating that frequency band 5.325 GHz is currently in use. This information can be added anytime the system is running. In our case, the facts are brought in from sensors on the wireless system. Information such as the battery life, channel noise figure, and received signal-to-noise ratio can all be represented as facts.

The inference engine is also a primary system component in an expert system. This component includes the code that combines the information from the working storage and the knowledge base to find the solution. This component is accessed by the user through the *user interface*. The user interface is simply the code that controls the dialog between the user and the system.

Many expert systems are implemented as products called expert system *shells*. The shell is the software that contains the user interface, a format for the rules and facts, and an inference engine. A commonly used shell is the CLIPS expert system tool [261]. CLIPS provides a complete environment for the construction of rule-based expert systems. The primary reasons CLIPS is used in this research include knowledge representation, the rule-based environment is already implemented, portability, CLIPS is written in C and runs on several different operating platforms, interactive environment, and most important, full documentation.

A major advantage of the RBS is the ability to generate the rules offline when time is not a key factor. This is a good thing because finding the optimal transmission parameters for each of the environment scenarios requires a complete search of all combinations of transmission parameters to find the combination with the largest fitness score. To do this full search we use the fitness function as the expert and perform a full search space run over all possible parameters and find the optimal transmission parameters for each possible environment.

7.5.2 Genetic Algorithms

The methodology of a genetic algorithm (GA) can be broken up into four stages. The first stage is the initialization of the population. Initially the population is randomly generated to form the first generation of possible solutions. The choice of population size is based loosely on the specific problem we are dealing with, however, a common set of genetic algorithm settings has been defined and used in several GA implementations with slight variations [262, 263]. Traditionally, the initial population for the first generation is selected at random. However, it has been shown that, by taking advantage of previous runs and seeding the initial generations, better performance can be achieved when compared to randomly generated initial populations [243]. This technique decreases the number of generations needed for the algorithm to converge. The rate of improvement depends on the environmental variation factor (EVF) and the amount of seeding. Figure 7.3 shows how the fitness convergence can be improved by seeding the GA initial population with varying amounts of the population from previous runs with a 10% change in the environment.

The second stage is the selection stage. During each generation a proportion of the population is selected to breed a new generation. Individual solutions are run through a fitness function that assigns a fitness score to the solution representing its value. Several selection methods exist, such as tournament selection,

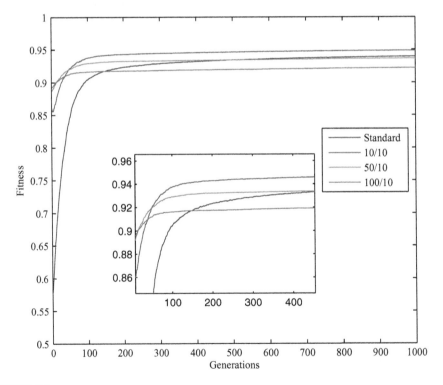

FIGURE 7.3

Fitness convergence where X/Y represent the ratio of seeding percentage and amount of variation in the environment between seedings.

stochastic remainder selection, and roulette wheel selection, that use the fitness scores to select the solutions to be used to form the next generation of solutions. In the following examples, we choose to use the stochastic remainder selection method [258,264,265]. This method uses the ratio between the fitness of an individual solution and the average fitness of the population to determine the probability of the solution moving onto the reproduction stage.

The third stage is the reproduction stage. In this stage, the next generation of solutions is generated from the previously selected group of solutions. This process is completed through genetic operators such as crossover and mutation. *Crossover* is a method of combining two possible solutions to create a new solution. *Mutation* is the process of randomly mutating a solution, typically randomly flipping a bit in the genetic sequence. Mutation provides the means for the GA to avoid local minima by preventing the solutions from becoming too similar to each other, which can slow or even stop evolution. For each new solution in the next generation a pair of solutions is selected to be the "parents." One-point and two-point crossover are two possible approaches of combining the "parents." One-point crossover selects a random point in the genetic sequence in which the "parents" swap all data beyond the selected points. Two-point crossover is similar except that two points are selected and all data between the two points are swapped. Dejong [262] shows that two-point crossover provides a better mechanism for combining and mixing the chromosomes and produces better results than the single-point crossover technique. Along with the crossover function, each new solution has a typically small chance to have a bit mutated. These processes result in the next generation of solutions. Generally, the average fitness has increased since mostly the higher-scoring solutions are selected to breed, along with a small proportion of less-fit solutions to provide for a more diverse search. Figure 7.4 shows the general flow, summarizing the previously described steps, for the genetic algorithm process.

The final stage is the termination stage. The genetic algorithm process detailed previously continues until a termination condition has been reached. Common termination conditions include:

- A solution that satisfies a minimum criteria is found.
- A fixed number of generations is reached.
- A specified computation time is reached.
- The fitness scores have flattened such that successive generations show no improvement.
- Combinations of these conditions.

7.5.3 Case-Based Reasoning Systems

Case-based reasoning (CBR) refers to the reasoning process based on previous recorded experiences (cases). A case-based reasoner is an entity that performs case-based reasoning. In the discussion of this chapter, CBR can refer to either

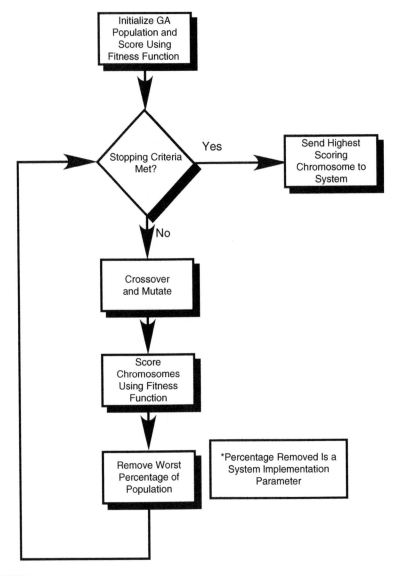

FIGURE 7.4

Genetic algorithm system process flow.

case-based reasoning or case-based reasoner according to the context. In general, CBR consists of case representation and indexing, case selection and retrieval, case evaluation and adaptation, case learning, and case database maintenance [266,267]. In the following, we will discuss the details of these modules in our CBR-CE development.

Concept of Case

A case, the fundamental element in a CBR system, represents an experienced situation. All other components in a CBR system take a case either as input or as output. The definition of a case is given in [266] as follows: "A case is a contextualized piece of knowledge representing an experience that teaches a lesson fundamental to achieving the goals of the reasoner." It is important to point out that, although a case represents an experience, not every experience is a useful case. The essence of a case lies in its capability of "teaching a useful lesson" [266]. In other words, an experience that is worth remembering by the reasoner is what helps the reasoner achieve a goal, warns the reasoner against a potential failure, or specifies an unexpected situation.

A case usually consists of two parts: the content and the indexes [266]. The content of a case records the experience or the lesson it teaches, and the index describes the context where this experience is gained and where this experience might be useful in the future. The content of a case usually consists of the problem or the situation description, the solution, and the outcome.

Case Representation and Indexing

Case representation is responsible for formatting the input information such that this information can be understood by other modules in CBR. As discussed already, a case consists of two parts: the content and the index. The content of a case contains the following information [266]:

- **Problem description.** The problem description describes the relevant experience. It specifies the detailed information about the problem, including the radio environment and the service request with its QoS requirement.
- **Solution.** The solution explains how the problem was solved in the past. It specifies a possible radio configuration for the problem specified in the problem description.
- **Outcome.** The outcome records the result of applying the solution. It specifies the feedback from the real environment (e.g., success or failure) after the solution was applied to the CPE.

Figure 7.5 gives a visual representation of a case structured for a cognitive engine.

The "problem" field specifies the general event (e.g., new CPE service request and PE detected), channel condition, and QoS requirement. The "solution" field includes information on the modulation and coding scheme, interleaving, transmission power, and bandwidth use. The "outcome" field includes feedback information.

The other important aspect of a case is its index, which specifies the context where the content of a case is useful [266]. It describes the distinguishing features of a case. In a cognitive engine development, the index is selected based on the scenario and radio environment. This choice of index facilitates the case retrieval

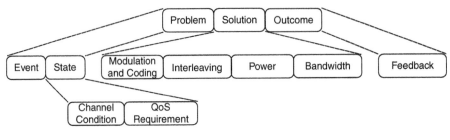

FIGURE 7.5

Case structure.

process. It is important to point out that an appropriate index can ease the case retrieval process and improve the performance of the CBR greatly. In present research, the primary choice of indexes is mainly done manually by an expert for a specific application of the CBR [266].

Case Selection and Retrieval

The case selection and retrieval module is the fundamental building block of the CBR. It searches the case database for cases that satisfy the request to a certain extent. The definition of the retrieval criterion is the most important aspect in effective case retrieval [267]. General weighted sum–based retrieval criteria do not work well in our particular environment. For example, a shorter Euclidean distance between two situations does not always guarantee a closer match of the two situations, because there may be a nonlinear relationship between the input parameters, environment parameters, and the objectives. This means that we cannot guarantee that changing a single parameter moves us closer to our goal in a linear fashion. A small parameter change may actually move us much further away from our operation goal.

The case selection and retrieval module selects the best-matched case or cases among the valid cases. The selection in the cognitive engine is based on the evaluation of the performance metrics discussed in Section 7.4. The matched case with highest utility is selected. In addition, multiple cases can be selected based on the utility metric if necessary. For example, all top three cases in utility rank can be selected and returned for further processing. Other criteria or metrics can be adopted easily in the case selection process in the modular design of the CBR. For example, it is easy to implement a case selection functionality that employs a criterion on transmission power, such as selecting a case with the minimum transmission power. The definition of the selection criterion is still an open question in many applications of CBR. The utility metric we use is just one choice with acceptable performance, as we discuss in the following section. Other metrics and utility functions such as those discussed in [268] can also be used.

Case Evaluation and Adaptation

The case evaluation module evaluates the performance of the retrieved case either by using a performance model or by applying the solution and observing the outcome. If the performance of the retrieved case is not satisfactory, the case will be modified by the case adaptation module. The case retrieved by the case selection and retrieval module is not necessarily very suitable for the new problem, because this solution is recorded as a solution to a previous problem and retrieved as a possible solution to the new problem due to the similarity between the old problem and the new one. The hope is that the retrieved case also is applicable to the new problem. However, the applicability or the performance is not guaranteed. Note that the case evaluation and case adaptation can be recursive to obtain an appropriate solution.

To tailor the solution to the new problem, the case adaptation module is necessary. In a cognitive engine implementation, the case adaptation module can fine-tune the parameters in the solution to deliver a better performance. The fine-tuning of the parameters can be implemented in several ways and is largely a research issue. The simplest way is a linear adjustment of the parameters using some a priori information about how the parameters affect the environment. This may not be the best way if the relationships are nonlinear. Complex functions can be defined offline to guide this adjustment process. However, this analysis again assumes that the relationships are well known. The major motivation of using cognitive radios is that the wireless environment is unknown and unpredictable, so any use of static functions to adapt the parameters may not be optimal. A more reasonable alternative is using a heuristic algorithm such as a genetic algorithm. No matter which algorithm is used, the selected solution from the case selection and retrieval module provides a good starting point for any algorithm. Note that this case adaptation process can be bypassed if the selected solution provides good performance.

Case Learning and Case Library/Database Maintenance

A case library is an ensemble of similar cases. Several case libraries can be organized in a hierarchical structure into a case database for efficient retrieval. In other words, the case database can contain more than one case library.

Experiences are remembered by the CBR system as cases in the case library. The CBR gains additional information, or learns, by solving new problems or receiving feedback. With the increase of the experience, more cases are accumulated in the case library/database. Also, new experience may be incorporated in the case database by updating the existing cases. In this way, the size of the case database does not increase linearly with the number of new problems and finally converges to some value. Although a larger case database does not always guarantee better performance for a specific problem, a CBR system can generally return a better solution with a larger case database than with a smaller one. This performance improvement reflects the learning capability of a CBR system. As the experience increases, the decision becomes better.

The case database maintenance module is critical to the learning process and reasoning process of the cognitive engine. When a solution to a new problem is formulated, this solution needs to be "remembered" by the CBR for future use. Also, the outcome of applying this solution to the new problem should also be "remembered." As more and more cases are recorded in the case database, the CBR gets more and more efficient [266] in the sense that it can utilize cases in the case database to solve the new problem instead of finding a solution from scratch using computationally intensive optimization. Another important consideration in case database maintenance is the balance between the number of cases in the case database and the amount of time used in case retrieval. A larger case database covers a larger problem space. However, this usually also means a longer case retrieval time and larger memory space. Although memory size is becoming less a concern in modern system design due to advancements in memory size, processing time is always a big concern in many wireless applications. For example, the cognitive radio–based IEEE 802.22 applications have a stringent time requirement specified in the standard to protect the primary user. Therefore, an appropriate case granularity needs to be defined and redundant cases removed from the case database to improve execution time.

7.6 CHAPTER SUMMARY

The cognitive adaptation engine is the core of the cognitive radio. This is the module that makes the transmission parameter decisions. This complex process involves relating the controllable parameters, the available environmental meters, and the communication objectives to determine the optimum parameter set. This optimization may take many cycles, such as in a heuristic algorithm, or it may take a single step, such as in the case of a rule-based system. We introduce only a small sample of possible transmission and environmental parameters along with some common communication objectives. The selection of the parameters included in the cognitive engine plays an important role in the quality and efficiency of the engine. Selecting inappropriate parameters may result in inaccurate solutions, while using too many parameters may result in overly complex and unnecessary computations. When designing a cognitive engine it is important to keep in mind the potential applications so as to restrict the complexity to only what is needed.

We describe several cognitive engines, from the heuristic approaches such as genetic algorithms, which rely on finding suboptimal solutions in a short amount of time, to the expert systems such as rule-based systems, which have the potential of requiring large amounts of storage yet can output a solution in a single stage. Also covered are case-based systems that build solutions from previous experiences by locating the closest previous experience and adapting it to the current situation. The difficulty in this method is how to do the adaptation. This research question requires much more work and is still wide open for debate.

The current state of several existing implementation systems is provided so the reader can understand the state of technology for cognitive engines. As shown by the limited number of implementations, this work is still in its infancy with much more time needed to advance into a state where cognitive engines are a trusted choice for use in wireless communications. The goal of this chapter is to introduce readers to the cognitive engine and help them understand the complexity involved in developing a fully functional cognitive engine. Many important research questions are pointed out that ideally will encourage readers to follow up.

7.7 PROBLEMS

1 One major application for cognitive radios is in an emergency disaster scenario, similar to Hurricane Katrina. The primary goals of the cognitive radios are to provide interoperability among different agencies and guarantee communication. List several transmission parameters and environment parameters that a cognitive radio in this scenario would need to meet these goals.

2 Given the parameters in Table 7.5 and their ranges:
   (a) How many combinations of transmission parameters are possible?
   (b) What is the optimal transmission parameter combination for the following fitness function, where $w_1 = 0.80$, $w_2 = 0.10$, and $w_3 = 0.10$:

$$f_{ex} = w_1 \times \left[1 - \frac{(P_i + B_i)}{P_{max} + B_{max}} \right]$$
$$+ w_2 \times \left(\frac{P_i \times R_{s_{max}} \times B_i \times m_i}{B_{max} \times P_{max} \times m_{max}} \right) + w_3 \times \left(\frac{0.1 \times P_i \times m_i \times R_{s_i} \times B_{min}}{B_i \times m_{max} \times R_{s_{max}}} \right)$$

   (c) Repeat (b) with $w_1 = 0.10$, $w_2 = 0.80$, and $w_3 = 0.10$.
   (d) Repeat (b) with $w_1 = 0.33$, $w_2 = 0.33$, and $w_3 = 0.34$.

3 If using an expert system such as a rule-based system for a cognitive engine implementation, who is the expert who defines the rules?

Table 7.5 Transmission Parameter List

Parameter Name	Min	Max	Step Size
Transmit power	0 dBm	20 dBm	1 dBm
Modulation index	1	5	1
Bandwidth	1 Mhz	20 Mhz	1 Mhz
Symbol rate	125 Ksps	1 Msps	125 Ksps

4 In Section 7.5.2 a list of the stopping criteria for a genetic algorithm was given. For each of these stopping criteria describe a situation in which is it applicable.

5 Explain the step-by-step process of how a case-based reasoning system determines the proper operating parameters.

6 Discuss the advantages and disadvantages of using a heuristic-based approach, such as genetic algorithms, to design a cognitive engine as opposed to using a knowledge-based approach such as a rule-based system or a case-based reasoning system.

7 What are the advantages of using hybrid design approaches for cognitive engine implementations?

Cognitive radio network theory

Fundamentals of communication networks

8

Shiwen Mao

Auburn University, United States

8.1 INTRODUCTION

A communication network consists of a set of interconnected nodes that exchange data with each other. In this chapter, we discuss the fundamentals of communication networks with a focus on wireless networks. We first review the general protocol architecture and common building blocks of communication networks. We next discuss new challenges brought about by wireless communications, namely, wireless transmissions, mobility, and energy efficiency, which make wireless networks fundamentally different from wireline networks. Through sampling the rich literature, we examine the design and optimization of various layers in wireless networks while considering their unique characteristics, including mobility modeling, scheduling, multiple access, routing, and congestion control. We conclude this chapter with a discussion of cross-layer design and optimization in the context of wireless networks.

8.2 ARCHITECTURE AND BUILDING BLOCKS

8.2.1 Protocol Architecture

As illustrated in Figure 8.1, a generic communication network consists of communication *devices* (or, *hosts*) and network *nodes*, which are interconnected with communication *links*. Applications are executed at the communication devices, which generate application data that need to be delivered to one or more remote devices. Network nodes exchange control information with each other and collaboratively carry application data from a source device to a destination device.

Doi: 10.1016/B978-0-12-374715-0.00008-3

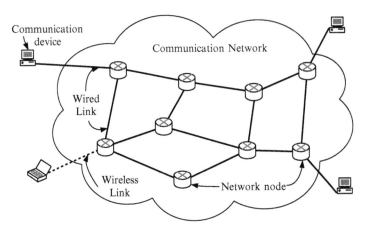

FIGURE 8.1

A generic communication network.

Communications among network nodes and communication devices are through the links that interconnect them.

Metcalfe's law states that the value of a telecommunications network is proportional to the square of the number of connected users of the system.[1] However, for a communication network to scale to large size, the task of information exchange among communication devices involves various functions and has tremendous complexity. It would be impractical, if not impossible, to implement all these functions as a single module. Rather, a *divide-and-conquer* approach would be more appealing and has actually been adopted. The communication task is broken down into subtasks; these subtasks are organized in a hierarchical way according to their dependencies to each other. The subtasks, each of which is responsible for a facet of communication, are organized into a number of stacked *layers*. Each higher layer uses the service provided by its lower layers and in turn provides service to the layer above it. Such services are provided transparently, while heterogeneity and details are hidden from the higher layer. A protocol is used for communication between entities in different systems, which typically defines the operation of a subtask within a layer. With this approach, the design of a communication network is greatly simplified, since each subtask can now be designed and optimized independently, and various competing technologies and implementations are allowed, as long as the common interfaces are strictly followed.

TCP/IP protocols, also known more formally as the *Internet protocol suite*, facilitates communications across interconnected, heterogeneous computer networks. It is a combination of a large number of protocols, which normally are

---

[1]Robert Metcalfe is the inventor of Ethernet.

organized into four layers, including the application layer, transport layer, network layer, and link layer.

8.2.2 Switching Technologies

Existing communication networks can be classified into three categories based on the switching technology employed: circuit switching networks, packet switching networks, and virtual circuit switching networks. In a *circuit switching* network, a *dedicated* communication path, called a *circuit*, is established between two stations through nodes within the network. A circuit occupies a fixed capacity of each link for the entire lifetime of the communication session, while the capacity unused by the session cannot be used by other circuits. A circuit switching communication session involves three phases. (1) Circuit establishment: Network nodes allocate resources (e.g., bandwidth and buffer) to set up a circuit for the session. (2) Data transfer: Data are transmitted as a flow of bits along the circuit. Since the circuit has its dedicated bandwidth, no delay is incurred at the network nodes. (3) Circuit termination: The network nodes tear down the circuit and release the resources for future use.

In a *packet switching* network, data are transmitted as *packets* or *datagrams*, which are blocks of data bits appended with a header or trailer carrying control information. Each packet carries routing information and is forwarded through the network from node to node along a certain path. At each node the entire packet is received, stored briefly, then forwarded to the next node (i.e., *store and forward*). Network resources are allocated on demand for each packet, and there is no need for dedicated resources. In a packet switching network, each packet is processed independently at network nodes. Therefore, the processing at the network nodes is simplified, but a sequence of packets may be received in a different order, since they may traverse different routes.

Virtual circuit switching is a hybrid technology combining features of both circuit switching and packet switching. As in circuit switching, a virtual circuit is established before data transmission begins and torn down when data transmission is over. As in packet switching, data are transmitted as packets. As in circuit switching, all packets of the same session follow the same path—that is, the virtual circuit—and are delivered in sequence. As in packet switching, packets from different virtual circuits may be interleaved. The link bandwidth is thus shared more efficiently.

Each of the switching technologies has its advantages and limitations. As a simple example, consider two computers exchanging some data. If the data volume is large, circuit switching is more efficient, since the overhead of establishing and tearing down the circuit can be amortized and no addition control overhead or delay is incurred during data transmission. However, if only a small amount of data is to be transmitted, packet switching may be a better choice, since it would be costly to establish a circuit in this case. The traditional public switched telephone network (PSTN) and most cellular networks are circuit switching networks. The

Internet, wireless local area network (LAN), wireless mesh networks, and wireless sensor networks are based on the packet switching technology. The asynchronous transfer mode (ATM) network is a virtual circuit switching network.

8.2.3 Encapsulation and Multiplexing

At a source host, application data are sent down through the layers in the protocol stack, while each layer adds a header (and may be a trailer) to the data received from its higher layer (called *protocol data unit*, PDU). The *encapsulation* process is shown in Figure 8.2. When the packet arrives at the destination, it is sent up through the protocol stack. At each layer, the corresponding header and trailer are stripped and processed for control functions in that layer. The recovered higher-layer data are delivered to the upper layer.

In TCP/IP, different higher-layer protocols can use the service provided by the same lower-layer protocol, and the same higher-layer protocol can use the service provided by different lower-layer protocols. In the first case, each packet sent down to the lower layer should have an identifier indicating to which higher-layer module it belongs. Multiplexing and demultiplexing are performed at different layers using information carried in packet headers. For example, a communication process running in a host is assigned a unique *port number*, which is carried by all the packets generated by or destined to this process. Transport-layer protocols, such as TCP or user datagram protocol (UDP), determine whether a packet is destined for this process by checking the port number field in the transport layer header. In the IP case, each protocol using IP is assigned a unique *protocol number*, which is carried in the *protocol* header field in every packet generated by the protocol. By examining the value of this field of an incoming IP datagram, the type of payload can be determined. A field called *frame type* in the Ethernet header is used for multiplexing and demultiplexing at this level.

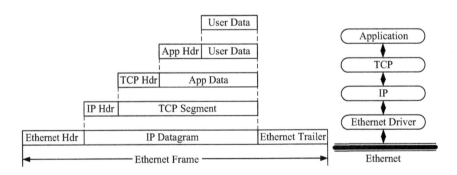

FIGURE 8.2

Encapsulation of user data through the layers.

8.2.4 **Naming and Addressing**

To enable processes in different computers to communicate with each other, naming and addressing are used to uniquely identify them. Specifically, a process running in a host can be identified by its port number. A host is identified by a *domain name*, while each network interface is assigned a unique IP address and a *physical* address.

In the application layer, an alphanumeric *domain name* is used to identify a host, such as *www.google.com* identifies a Google server. Since this layer directly interacts with users, a domain name is more user-friendly than a numeric address. Domain names are hierarchically organized into a tree structure. The root node has a null label, while each nonroot node has a label of up to 63 characters. The domain name system (DNS) is used to resolve—that is, translate—a domain name to the corresponding IP address. Then the resolved IP address, rather than the domain name, is used in the TCP/IP kernel.

Port numbers are used as addresses for user processes running in the application layer. The value of the *port number* field in the transport layer protocol header, TCP or UDP, is used to decide to which application process the data belong. Most network applications are implemented following the client-server architecture. A server is always running and uses a *well-known* port number (smaller than 1024). A client exists for only the period when service is requested and received, therefore it uses *ephemeral* port numbers that are randomly chosen and are larger than 1023.

Each host interface in the Internet has a unique IP address. A host with multiple interfaces and hence multiple IP addresses is called a *multihomed* host. An IPv4 address is a 32-bit number written in the *dotted-decimal* notation; that is, as four decimal numbers, one for each byte, separated by three periods.[2] For flexibility in network administration and operation, the *subnetting* technique was introduced, where an IP address is divided into three levels: a network ID, a *subnet ID*, and a host ID. The network ID is often referred to as *network prefix*. When subnetting is used, the combination of the network ID and subnet ID is called *extended-network prefix*. With the combination of an IP address and a port number, a process running in a host is uniquely identified in the global Internet, since the IP address is unique in the Internet and the port number is unique within the host. The combination of an IP address and a port number is called a *socket*. IP addresses are carried in every packet and are used by routers to forward the packets toward their destinations.

The physical address, also called *hardware address*, or *medium access control* (MAC) address, is used in the link layer to uniquely identify a network interface. Unlike IP addresses, a MAC address contains no location information. Different link layer protocols use different MAC addresses. The Ethernet MAC address is 48 bits long and globally unique. The first 24 bits of an Ethernet address are called *vendor component*, while the remaining 24 bits are called *group identifier*. An Ethernet

---

[2]IPv6 uses 128-bit addresses to provide sufficient IP addresses for all devices that need to connect to the Internet. IPv6 addresses are represented as eight groups of four hexadecimal digits, where each group is separated by a colon, such as 2001:0db8:85a3:0000:0000:8a2e:0370:7334.

interface card vendor is assigned with a block of Ethernet addresses, starting with a unique vendor component. Each card made by the vendor has the common vendor component, followed by a different group identifier. The MAC address uses the hexadecimal notation, such as 0x8:0:20:87:dd:88.

8.2.5 Multiple Access

The simplest way of interconnecting two computer hosts is using a point-to-point link with a host on each end. As the number of hosts increases, this approach may be inadequate, since a large number of links (i.e., $N(N-1)/2$) are needed to fully connect N hosts. In this case, a *broadcast* network, where all the hosts share a common transmission media, is more efficient.

To share the common media (e.g., a cable or a wireless channel) efficiently, all hosts must follow a set of rules to access the media. Hosts should be able to check the availability of the media and resolve collisions. In addition, since the bandwidth of the media is limited, it is desirable to share it efficiently in terms of the aggregate throughput of all the hosts. Furthermore, each host should have a fair chance to access the media and should not be allowed to take it forever. The sharing rules are defined as *media access control* protocols, which are implemented in the link layer.

8.2.6 Routing and Forwarding

Routing and forwarding are the main functions of the network layer. The IP modules in the hosts and the Internet routers are responsible for delivering packets from their sources to their destinations. Routing and forwarding consists of two closely related parts: maintaining network topology information and forwarding packets. Hosts and routers must learn the network topology to know where the destinations are, by exchanging information on connectivity and the quality of network links. Routing information is derived from network topology information and stored in a data structure called *routing tables* in hosts and routers. Routing tables are created and maintained either manually or by dynamic routing protocols. When there is a packet to deliver, a host or a router consults the routing table to find out where to forward the packet. An end-to-end path consists of multiple routers. Each router relays a packet to the next hop that brings it closer to its destination.

8.2.7 Congestion Control and Flow Control

Internet routers forward packets using the *store-and-forward* technique. Usually a router buffer is shared by many communication sessions with different source/destination pairs. If, in a short period, a large number of packets arrive, the output port may be busy for a while and the buffer may be fully occupied by packets waiting their turn to be forwarded. The router is *congested* in this case.

A similar situation may occur at a destination host, which is receiving packets from multiple sources. Received packets are first stored in a buffer, then sent to

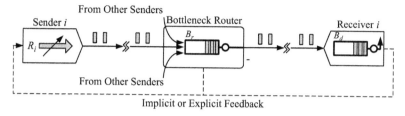

FIGURE 8.3

Flow control and congestion control in the Internet.

application processes. If the packet arriving rate is higher than the rate at which the packets are removed from the buffer, the receiving buffer will be fully occupied by packets waiting to be processed. These scenarios are likely to happen, since many hosts and routers are heterogeneous in terms of their processing capability and network bandwidth. In the case of a fast transmitter and a slow receiver, the receiver's buffer may get full. When the buffer, at either the receiver or an intermediate router, is full, arriving packets have to be dropped, since no space is left to store them.

Congestion control and flow control are used to cope with these problems. The basic idea is to let the source be adaptive to the buffer occupancies in the routers and the receiver (see Figure 8.3, where the router has a finite buffer size B_r and the receiver has a finite buffer size B_d.). For example, the receiver may notify the sender how much data it can receive without causing buffer overflow. Then the sender will not send more data than the amount allowed by the receiver. In the router case, the sender may be explicitly notified about the congestion in the router, or infer congestion from feedback, which could be implicit or explicit. Then the source will reduce its sending rate until congestion is dissolved. TCP uses *slow start* and *congestion avoidance* to react to congestion in the network, and *windows-based flow control* to avoid receiver buffer overflow.

8.2.8 Error Control

Many TCP/IP protocols use the *checksum* algorithm to detect bit errors in the header of a received packet. Suppose the *checksum* header field is K bits (e.g., $K = 16$ in IP, UDP, and TCP). At the source node, the field is first set to 0. Then, the K-bit *one's complement sum* of the header is computed, by considering the header as a sequence of K-bit words. The K-bit one's complement of the sum is stored in the checksum field and sent to the receiver. The receiver, after receiving the packet, computes the checksum of the header (including the checksum field) using the same algorithm. The result will be all ones if the header is error free. Otherwise, the header is damaged and the received packet should be discarded.

Ethernet, on the other hand, uses the *cyclic redundancy check* (CRC) technique to detect errors in the entire frame. With CRC, the entire frame is treated as a single number and divided by a predefined constant, the CRC *generator*. The remainder of the division operation is appended to the frame (as the trailer) and sent to the receiver. The receiver performs the same division operation using the entire received frame. If the remainder is zero, there is no error in the frame. Otherwise, the frame is damaged and should discarded.

Sequence numbers are used to detect lost packets. With this technique, the sender and receiver first negotiate an *initial sequence number*. Then the sender assigns a unique sequence number to each packet it sends, starting with the initial sequence number and increasing it by 1 for each packet sent. The receiver can detect which packets are lost by ordering the received sequence numbers and searching for gaps. When a packet loss is detected, the receiver may notify the sender to request a retransmission of the lost packet. Alternatively, the sender can use other error control schemes, such as forward error correction (FEC), in the application layer for better protection of the application data.

8.3 NEW CHALLENGES IN WIRELESS NETWORKS

Wireless networks refer to the class of communication networks that adopts wireless communications. In such networks, nodes exchange information using unguided transmission of light waves, microwave, radio waves, or acoustic waves. Wireless networks have become one of the most important forms of access networks that provide connection to the Internet for mobile users. Such networks bring about a whole new level of technical challenges, which are discussed in this section.

8.3.1 Wireless Transmissions

In wireless networks, the concept of communication link is quite different from that in wireline networks. A transmission through the open space medium results in a "footprint" of radio energy. In addition to the target receiver, all other nodes within the footprint overhear the transmission, resulting in interference and possible collision at these nodes. That is, wireless transmissions consume radio frequency spectrum in the network area. To avoid collision, the maximum number of concurrent transmissions in a network is upper bounded, which translates to bounded transport capacity of wireless networks [269].

It would be nontrivial to fit wireless transmissions to the traditional *link* definition. The received signal at a receiver i, denoted $y_i(t)$, is the combination of received signals from all transmissions and noise; that is,

$$y_i(t) = \sum_j b_{ji}(t)x_j(t) + n_i(t), \tag{8.1}$$

where $x_j(t)$ is the transmitted signal from node j, $b_{ji}(t)$ is the channel gain for the (overheard) transmission from node j to node i, and $n_i(t)$ is the noise at node i. Usually the gains $h_{ji}(t)$ are stochastic processes, due to path attenuation, Doppler effect, multipath fading, and shadowing. Another implication is that the channels of all the users are closely coupled. An $N \times N$ channel matrix is required to characterize a wireless network with N nodes, with $h_{ij}(t)$ as its elements. The capacity of the wireless link between a pair of nodes depends on the *signal-to-interference-plus-noise ratio* (SINR) at the receiver; that is,

$$\text{SINR}_i = \frac{h_{ii}(t)x_i(t)}{\sum_{j \in \{N \setminus i\}} h_{ji}(t)x_j(t) + n_i(t)}. \tag{8.2}$$

The specific data rate of the link also depends on the modulation and channel coding schemes adopted. Clearly, both the capacity and data rate are affected by the transmission behavior of neighboring nodes. Such a complex link model is drastically different from that of wireline links, which can be conveniently approximated as a pipe with a fixed bandwidth, and poses great challenges to networking protocol design for wireless networks.

8.3.2 Mobility

The adoption of wireless communications frees users from a socket and cable. Unguided wireless transmission implies that wireless links or connectivity are closely coupled with node location and distance, which may be constantly changing when nodes move.

In infrastructure-based wireless networks, each mobile user is associated with a fixed base station for transmitting and receiving data. The area covered by the base station's transmission is called a *service area* or a *cell*. When a user moves within a cell, the quality of the wireless link between itself and the base station may vary over time (e.g., due to Doppler shift, varying obstacles, reflecting and diffracting environment, and varying distance). When a user moves across cells, *mobility management* is needed to tear down the association with the original base station and establish a new association with the base station of the cell it enters [270]. Such *handoff* should be performed with minimum interruption to the ongoing communication session. Mobility has a significant impact on the dimensioning and resource allocation of infrastructure-based wireless networks.

In infrastructureless wireless networks, there is no fixed base station anymore. An existing link between two nodes is broken when they move away from each other, while a new link is established when two nodes move close to each other. The changing topology causes constant route failures and the routing engine has to be continuously executed to find new routes. Such constant rerouting not only incurs significant control overhead but also seriously degrades the performance of higher-layer protocols. In addition, mobility also changes the traffic distribution within the network (i.e., the *traffic matrix*), since each node is a combination of a source, destination, and router. Temporary hot spots are formed as a result, and the interference pattern within the network constantly changes.

On the other hand, mobility can be exploited to improve the performance of wireless networks. For example, the asymptotic transport capacity of wireless networks found by Gupta and Kumar is shown to be improved by a simple two-phase protocol that exploits mobility [271]. By predicting mobility, routing and clustering in mobile ad hoc networks can be made more efficient [272–274]. It is important to consider mobility in the design of wireless network protocols.

8.3.3 Energy Efficiency

Mobile nodes are usually powered by the batteries they carry. The battery power is used for processing and wireless communications. Since the batteries have limited power, a node dies when its battery dies. Therefore, the battery power and energy consumption determine how long the node operates. In infrastructure-based wireless networks, when a node dies, there is no significant impact on the network topology, except that its own service is interrupted. In infrastructureless wireless networks, the problem is more serious, since each node is a combination of source, destination, and router. When nodes die, the network topology changes and the network may even be partitioned.

It is ultimately important to use power efficiently in wireless networks, while energy efficiency is imbued in the protocol design in every layer. In the physical layer, transmit power increases with distance, following power law behavior. Suitable *power control* not only conserves energy but also reduces interference. In the MAC layer, an efficient multiple-access scheme can reduce collisions and save energy. Since each node is also a router, an energy-efficient routing protocol can spread traffic evenly throughout the network and thus avoid depleting the batteries of some nodes located at certain conjunctions. An energy-aware congestion control protocol also prevents energy waste caused by loss packets due to congestion.

An effective means of conserving energy is to schedule nodes to sleep (i.e., turning off their radios) when they are not needed. This idea has been incorporated in IEEE 802.11 as the power-saving mode [275] and in wireless sensor network MAC protocols [276]. Such sleep scheduling incurs additional delay. A new challenge brought about by sleep scheduling is how to synchronize the sleep cycles of the nodes such that connectivity is not sacrificed and the associated delay is minimized.

Although interesting work has been reported that enables wireless devices to harvest energy from their operating environment, this is still an open problem area and the technologies are not ready for practical use yet [277].

8.4 MOBILITY MODELING
8.4.1 Mobility Models

Mobility is one of the new issues brought about by wireless communications. For evaluating the performance of various wireless network systems, such as network planning, it is important to understand user mobility and develop suitable mobility models. Over the years, considerable efforts have focused on recording user mobility traces and extracting meaningful models from the traces [278–280]. In parallel to these efforts, various empirical mobility models have been proposed for studying wireless networking algorithms (see [281] for a survey). For example, mobility models can be classified as *single-node mobility models*, where the movement of a single node is specified, and *group mobility models*, which prescribe how users move in loosely defined groups. The network area can be a square, a unit disk, a rectangle (to force formation of long paths), or a torus (to eliminate the boundary effect). In the following, we discuss one of the most popular mobility models, the *random waypoint* model [282], and its application in simulation of wireless networks.

8.4.2 The Random Waypoint Model

The random waypoint mobility model was first proposed in [282] and has been used by many researchers for performance evaluation of wireless network systems [281, 283]. Following the random waypoint model, a node moves as follows:

1. Step 1. The node assumes an initial location (x_0, y_0), and $k = 1$.
2. Step 2. A destination location (x_k, y_k) is chosen within the (convex) network area, and a speed s_k is chosen uniformly from an interval $[v_{min}, v_{max}]$. Then the node moves along the line segment between the two locations toward (x_k, y_k) at a constant speed s_k.
3. Step 3. If pausing is enabled, the node pauses for t_k^p after reaching (x_k, y_k), where pausing time t_k^p is drawn from a general distribution with density function $f_p(t)$.
4. Step 4. $k = k + 1$ and go to Step 2.

As an example, in Figure 8.4 we plot the first three moves of a node following the random waypoint model. A special strength of this model is its simplicity, such that it can be implemented with a few lines of code and does not significantly increase the simulation time. It is also a parsimonious model, with only a few parameters, but is sufficiently flexible for a wide range of wireless networks [281].

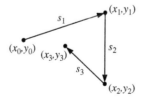

FIGURE 8.4

Trajectory of a node following the random waypoint model.

With additional constraints on the feasible values of (x_k, y_k), the random waypoint model can be used for city scenarios (i.e., the *Manhattan model*) or highway scenarios.

When a node moves around following a mobility model, its speed s_k and location (x_k, y_k) vary over time following certain distributions. Supposing time is sufficiently long (i.e., letting $k \rightarrow \infty$), both speed and location should converge to their *steady-state* or *stationary* distributions. An interesting problem is how to derive the steady-state distributions for the mobility model, which is important for properly configuring simulations and understanding simulation results. For example, if s_k is uniformly chosen from $[v_{min}, v_{max}]$, will it converge to $(v_{min} + v_{max})/2$ when k is sufficiently large? Counterintuitively, the answer is usually "no." In the following, we derive the stationary distributions of a random waypoint model following the analysis in [284], then discuss its implication on the simulation of wireless networks. Interested readers can also see [285–287] for more details.

Consider the case *without pausing*, where each node immediately chooses a new destination and starts moving after it arrives at a destination. We first derive the stationary distribution of speed. Let $F_T(s)$ denote the expected portion of time in the interval $(0, T)$ that the speed of the nodes is less than or equal to s. The cumulative distribution function of speed is $F(s) = \lim_{T \rightarrow \infty} F_T(s)$, while the corresponding probability density function is denoted as $f(s)$.

Consider a snapshot of the node movement, where the node is moving at speed s along a line segment of length l. The conditional density function $f(s|l)$ is proportional to the time spent on this path, l/s. To obtain the unconditional density function $f(s)$, we need to average $f(s|l)$ over the distribution of l. Since speed and location (i.e., path length l) are independently chosen,[3] $f(s)$ is proportional to $E[l](1/s)$. Normalizing $f(s)$, we have that

$$f(s) = \begin{cases} \frac{1}{s \log(v_{max}/v_{min})}, & v_{min} < s < v_{max} \\ 0, & \text{otherwise,} \end{cases} \quad (8.3)$$

where $v_{min} > 0$. With Equation (8.3), we can easily compute the average speed as

$$E[s] = \frac{v_{max} - v_{min}}{\log(v_{max}) - \log(v_{min})}, \quad (8.4)$$

which, interestingly, is not equal to the average of v_{max} and v_{min} in general.

We next consider the stationary distribution of location. Due to symmetry, we need to consider only the x-coordinate x. At any time, the node is moving along a path from, say, (x_1, y_1) to (x_2, y_2). Since the speed is constant, the x-coordinate x is uniformly distributed in $[x_1, x_2]$, conditioned on the endpoints of the path being (x_1, y_1) and (x_2, y_2). That is,

$$g(x|x_1, x_2) = \begin{cases} \frac{1}{|x_2 - x_1|}, & \min\{x_1, x_2\} < x < \max\{x_1, x_2\} \\ 0, & \text{otherwise.} \end{cases} \quad (8.5)$$

---

[3]It is shown in [288] that speed and location are independent in the stationary domain.

Now consider the joint probability density function of x_1, x_2, y_1, y_2. Again, this is proportional to the time spent on the path from (x_1, y_1) to (x_2, y_2); that is, l/s. Since speed and locations are independently chosen, we have $E[l/s] = l \cdot E[1/s]$. It follows that the joint density is given by

$$b(x_1, x_2, y_1, y_2) = \lambda \left[(x_2 - x_1)^2 + (y_2 - y_1)^2 \right]^{1/2}, \tag{8.6}$$

where λ is computed by normalizing $b(x_1, x_2, y_1, y_2)$ over the unit square; that is,
$1/\lambda = \int_0^1 \int_0^1 \int_0^1 \int_0^1 \left[(x_2 - x_1)^2 + (y_2 - y_1)^2 \right]^{1/2} dx_1 dx_2 dy_1 dy_2$.
Finally, the density function of the x-coordinate is found to be

$$
\begin{aligned}
g(x) &= \int_0^x \int_x^1 \int_0^1 \int_0^1 g(x|x_1, x_2) b(x_1, x_2, y_1, y_2) dy_1 dy_2 dx_1 dx_2 \\
&\quad + \int_0^x \int_x^1 \int_0^1 \int_0^1 g(x|x_1, x_2) b(x_1, x_2, y_1, y_2) dy_1 dy_2 dx_2 dx_1 \\
&= 2 \int_0^x \int_x^1 \int_0^1 \int_0^1 \frac{\lambda \left[(x_2 - x_1)^2 + (y_2 - y_1)^2 \right]^{1/2}}{x_2 - x_1} dy_1 dy_2 dx_1 dx_2. \tag{8.7}
\end{aligned}
$$

Due to symmetry, the y-coordinate y has the same stationary distribution as the x-coordinate. The location distribution (8.7) is not uniform in the unit square. Actually, the location of the node is more likely to concentrate around the center of the region [285]. To avoid such boundary effect, a torus may be used that provides a closed surface.

The stationary distributions for the case with pausing is also derived in the literature [284, 286, 287], which are omitted for brevity. However, we leave the derivation of the stationary pausing probability as a homework problem.

8.4.3 Perfect Simulation

The random waypoint model has received wide adoption in the wireless research community. However, until recently, it was found that the stationary distribution of the speed differs from the uniform distribution [285]. In addition, the authors show that, when $v_{min} = 0$, the average node speed approaches zero over time. Generally, it takes a long time (i.e., the transient phase) for the speed to converge, while simulation data should be discarded during this period to avoid biased simulations. For example, it is shown in [284] that this period could be as long as 1000 s of simulation time when the minimum speed is low, while many simulations reported in the literature run for less than 900 s of simulation time. If the minimum speed is zero, ironically, all the nodes finally stop moving, although following the "mobility" model.

Therefore, it is important to remove the transient phase of a mobility model for efficient simulation and precise interpretation of simulation results. In so-called perfect simulation, the standard method is to (1) guarantee the existence of stationary distributions and (2) remove the transient phase and ensure convergence to the stationary distributions [287].

In the previous section we derive the stationary distributions for the random waypoint model. It is shown that, if the initial speed and location are samples from the corresponding stationary distributions, convergence to the stationary regime is immediate [284, 286, 287, 289]. In the following, we show how to initialize the random waypoint model for a perfect simulation.

We first consider initializing speed s. From probability theory, if the closed-form expression for a cumulative distribution function $F(s)$ exists, denoted as $F^{-1}(u)$, we can generate random values following distribution $F(s)$ by first generating U uniformly from $(0, 1)$ and letting $S = F^{-1}(U)$. From (8.3), the cumulative distribution function of speed is $F(s) = \int_{v_{min}}^{s} f(s)ds = \frac{\log(s) - \log(v_{min})}{\log(v_{max}) - \log(v_{min})}$. The inverse of $F(s)$ is $F^{-1}(u) = v_{max}^{u}/v_{min}^{u-1}$. Therefore, the initial speed for the node is $s_1 = F^{-1}(U)$, while U is uniformly chosen from $(0, 1)$.

The location can be initialized as follows. Observing the conditional distribution (8.5), we first choose an initial path then uniformly choose a point on that path as the initial location. It is more involved to choose the initial path (i.e., the two endpoints (x_1, y_1) and (x_2, y_2)), since the distribution (8.7) is not in closed-form. In [284], the authors suggest using *rejection sampling*. The idea is to generate endpoints uniformly from the unit square in such a way that the joint probability density of the points is proportional to the distance of the resulting path l. The procedure is as follows [284]:

1. Step 1. Generate (x_1, y_1) and (x_2, y_2) uniformly in the unit square.
2. Step 2. Compute $r = [(x_2 - x_1)^2 + (y_2 - y_1)^2]^{1/2}/\sqrt{2}$.
3. Step 3. Generate a random value U_1 uniformly in $(0, 1)$.
4. Step 4. If $U_1 < r$, accept (x_1, y_1) and (x_2, y_2). Otherwise, go to Step 1.
5. Step 5. Generate a random value U_2 uniformly in $(0,1)$.
6. Step 6. The initial location is $[U_2 x_1 + (1 - U_2)x_2, U_2 y_1 + (1 - U_2)y_2]$.
7. Step 7. The node then travels to (x_2, y_2) at the initial speed. Upon reaching (x_2, y_2), subsequent speeds and destinations are chosen from the uniform distribution.

In this procedure, r is the ratio of l and the maximum distance in the unit square, $\sqrt{2}$. Steps 1-4 generate the initial path l, while Steps 5 and 6 choose the initial location uniformly along the path.

Interested readers can see [286] for more treatment of this topic and [287] for an interesting framework based on *Palm calculus* [290].

8.5 POWER CONTROL AND MULTIUSER DIVERSITY

Wireless network users are freed from a socket and cable. A direct consequence of this freedom is time-varying wireless channels; that is, fast fading due to constructive and destructive interference between multiple paths and slow fading due to shadowing effects and varying distances [291]. An effective means of combating fading is exploiting *diversity*. The basic idea is to create and use multiple *independent* propagation paths between the transmitter and receiver for improved performance. Diversity can be achieved over time by interleaving coded bits, over frequency by combining multiple paths in spread-spectrum systems, and over space by adopting multiple antennas [292].

In addition to these forms of diversity, a dense multiuser wireless network also offers the so-called multiuser diversity, by dynamically identifying and allocating network resources (e.g., power) to the user with the best channel. The resulting improvement in spectrum efficiency or system capacity is termed *multiuser diversity gain*. Multiuser diversity was first demonstrated through an information-theoretic analysis by Knopp and Humblet [293] for uplink transmissions, and further developed by other researchers for downlink transmissions [292, 294, 295] and downlink multiple-input, multiple-output (MIMO) channels [296]. This approach relies on the best, rather than the average, channel among all users. When channels are independent (since users are at different locations), it is more likely to find a strong channel when the number of users is large. It is shown that the capacity of the system, defined as the sum of data rates of all users, scales as a double logarithmic function of the number of users in the system.

In this section, we present the analysis presented in [293] to demonstrate multiuser diversity and discuss its implications on scheduling in wireless networks. Consider the uplink of an infrastructure-based wireless network with one base station and K mobile users. The received signal at the base station is

$$y = \sum_{i=0}^{K-1} h_i x_i + n, \tag{8.8}$$

where h_i and x_i are the channel gain and information for the ith user, and n is a zero-mean Gaussian random variable with variance σ^2 representing noise. If we consider frequency-flat Rayleigh fading, the h_i values are assumed to have Rayleigh distributions and the signal-to-noise ratio (SNR), denoted γ_i, for user i follows the exponential distribution [291]

$$f(\gamma_i) = \left(1/\gamma_i^0\right) e^{-\gamma_i/\gamma_i^0}, \quad \gamma_i > 0, \tag{8.9}$$

where γ_i^0 is the average received SNR for user i.

We assume some feedback between the base station and users such that users can adjust their transmit power. Let the transmit power of user i be $p_i(\vec{\gamma})$, where $\vec{\gamma} = [\gamma_0, \gamma_1, \dots, \gamma_{K-1}]$. We define the system capacity as the sum of the rates of all users, as

$$C = \sum_{i=0}^{K-1} R_i. \tag{8.10}$$

First consider the case with no power control. The sum-of-rate capacity is given by [293, 297]:

$$C_{\text{npc}} = \frac{1}{2}\text{E}\left[\log\left(1 + \sum_{i=0}^{K-1}\gamma_i\right)\right]_{\vec{\gamma}}; \tag{8.11}$$

that is, an average of the rate sum over each feasible realization of $\vec{\gamma}$. Under the so-called perfect power control law, the received powers at the base station are equalized. The transmit power of User i is $p_i(\vec{\gamma}) = \gamma_i^0/\gamma_i$. The system capacity under the perfect power control can be derived from the known capacity region of a Gaussian multiuser channel, as [293, 297]

$$C_{\text{ppc}} = \frac{1}{2}\log\left(1 + \sum_{i=0}^{K-1}\gamma_i^0\right). \tag{8.12}$$

Intuitively, we have $C_{\text{ppc}} \geq C_{\text{npc}}$. We leave the proof of this result as a homework problem.

Finally, we consider the system capacity under optimal power control, denoted as C_{opc}. The problem can be formulated as maximizing the rate sum. For practical systems, we adopt a constraint that the average transmit power for each user is finite; that is, unity.

$$\text{Maximize}\quad C_{\text{opc}} = \frac{1}{2}\int\int\cdots\int\log\left[1 + \sum_{i=0}^{K-1}p_i(\vec{\gamma})\gamma_i\right]f(\vec{\gamma})d\vec{\gamma} \tag{8.13}$$

$$\text{Subject to}\quad \int\int\cdots\int p_i(\vec{\gamma})f(\vec{\gamma})d\vec{\gamma} = 1, \quad i = [0, \cdots, K-1] \tag{8.14}$$

$$p_i(\vec{\gamma}) \geq 0, \quad i = [0, \cdots, K-1]. \tag{8.15}$$

Introducing Lagrange multipliers λ_i, corresponding to each constraint in Equation (8.14), we obtain the following system of inequalities:

$$1 + \sum_{j=0}^{K-1}p_j(\vec{\gamma})\gamma_j \geq \frac{\gamma_i}{\lambda_i}, \quad \text{with equality if and only if } p_i(\vec{\gamma}) > 0, \quad i = [0, \cdots, K-1]. \tag{8.16}$$

Assuming that the values of γ_i are distinct, we have that $p_i(\vec{\gamma}) \neq 0 \rightarrow p_j(\vec{\gamma}) = 0$, for all $j \neq i$, and $\gamma_i \geq (\lambda_i/\lambda_j) \cdot \gamma_j$. The power control law for User i is

$$p_i(\vec{\gamma}) = \begin{cases} \frac{1}{\lambda_i} - \frac{1}{\gamma_i}, & \gamma_i > \lambda_i, \ \gamma_i > \frac{\lambda_i}{\lambda_j}\gamma_j, \ j \neq i \\ 0 & \text{otherwise,} \end{cases} \tag{8.17}$$

where λ_i can be computed by solving the constraint equations (8.14) numerically.

Equation (8.17) indicates that more (less) power should be allocated to a user when its channel is stronger (weaker). Assuming all users have the same average received power, we have $\lambda_i = \lambda_j$ for all $i \neq j$ by symmetry. The power control law implies that, at any time instance, only the user with the largest instantaneous

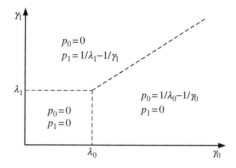

FIGURE 8.5

Scheduling in a two-user system: transmit powers for the users.

power is allowed to transmit, while all other users should not transmit. It is also worth noting that this result is quite general: It does not depend on the channel model. In Figure 8.5, we plot the transmit powers for a two-user system for various $\vec{\gamma}$ values. From the power control law (8.17) and the nonnegativeness of $p_i(\vec{\gamma})$, we can see that λ_i serves both as a cutoff SNR (when SNR is below λ_i, no transmission is allowed) and as a scaling factor in determining the "strongest" user [293].

The preceding analysis of multiuser diversity considers the overall system capacity, the rate sum. To achieve the global optimal, some users with weak channels are sacrificed, resulting in unfair resource allocation and even starvation of some users. For practical system design, the challenge is how to fully exploit multiuser diversity gain while achieving fairness among users with asymmetric channel statistics. Interested readers can see [292, 298, 299] for more details.

8.6 MULTIPLE ACCESS SCHEMES

In a wireless network, mobile users share the wireless channel. An important network control function is medium access control; that is, how to share the channel among the users efficiently (e.g., with high spectrum efficiency or overall throughput) and fairly (i.e., users should have equal chance of accessing the channel). A classification of multiple access schemes is given in Figure 8.6. Channelization is widely adopted in traditional wireless networks. In channelization, the wireless channel is partitioned into a number of subchannels, each being, for example, a frequency band, a time slot, or a spreading code. Then the subchannels are assigned to users. In this section, we discuss polling and several important random access schemes, including ALOHA/slotted ALOHA, carrier sense multiple access, and carrier sense multiple access with collision avoidance, which are important schemes especially for wireless data networks.

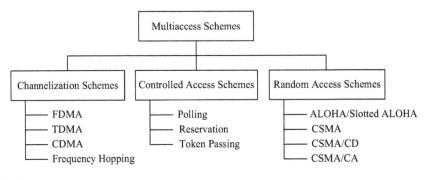

FIGURE 8.6

Classification of multiaccess schemes.

8.6.1 **Polling**

A polling system is a special type of queueing system with one server and m stations (users), where one station is served at a time. A polling system is illustrated in Figure 8.7. In the figure, each station is represented by a queue temporarily storing arriving customers, while customers arrive at each station following a random process. Each customer requests service from the server and departs the system when its service is completed. The service times also follow a certain distribution. Stations in a polling system may have one-customer buffers, finite buffers, or infinite buffers. In a one-customer buffer system, at most one outstanding customer can be buffered at each station. Any customer that arrives when the buffer is full is dropped. In an infinite buffer system, all customers that arrive at the system can be buffered without loss and will be finally served by the server at a later time.

When the server finishes serving a station, it may decide which station to serve next by following a fixed order (e.g., cyclic) or a random selection. The time between the end of a service and the beginning of the next service is the *switchover time*. There are three kinds of service policies in a polling system. In the *limited service* polling system, the server continuously serves a station until

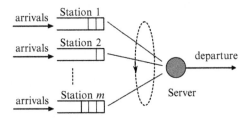

FIGURE 8.7

A polling system with one server and m stations.

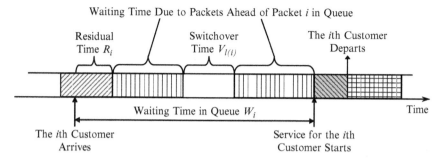

FIGURE 8.8

A gated polling system with one station.

either the station is empty or a predetermined number of customers have been served. A special case is the limited-1 system, where at most one customer is served whenever the server visits a station. In a *gated service* system, the server continuously serves a station until either the station is empty or all customers that arrived before the station is polled are served. In an *exhaustive service* system, the server serves all customers in a station continuously and leaves the station only when the station buffer is empty.

In the following, we demonstrate how to analyze a polling system by considering a single station system [300]. Figure 8.8 shows a gated service system with $m = 1$, where the blocks with different shades represent the service times of customers. Assume the arrival process is Poisson with rate λ. Let X denote the service time of a customer with mean $E[X] = 1/\mu$. Therefore, the system utilization is $\rho = \lambda/\mu$. Let V_j denote the duration of the jth switchover time. Consider the ith customer arriving at the system. Before it can be served, it has to wait in the queue until all customers that arrived earlier are served. The waiting time in the queue includes the residual time R_i, the service time for the N_i customers in the queue, and the next switchover time $V_{l(i)}$. R_i is the time to finish the current customer service or switchover time when customer i arrives. Some of the N_i customers may be served during the current service period (if they had arrived before the current service period started), and the rest, including customer i, will be served after $V_{l(i)}$. The expected waiting time for customer i can therefore be given by

$$E[W_i] = E[R_i] + \frac{E[N_i]}{\mu} + E[V_{l(i)}],\qquad (8.18)$$

where $E[R_i] = \lambda E[X^2]/2 + (1 - \rho)E[V^2]/2E[V]$, $\lim_{i\to\infty} E[N_i]/\mu = \rho W$, and $\lim_{i\to\infty} E[V_{l(i)}] = E[V]$. We therefore have the expected waiting time in queue for the single-user gated service system as

$$W = \frac{\lambda E[X^2]}{2(1 - \rho)} + \frac{E[V^2]}{2E[V]} + \frac{E[V]}{1 - \rho}.\qquad (8.19)$$

When the switchover time is a constant V, we have

$$W = \frac{\lambda E[X^2]}{2(1-\rho)} + \frac{V(3-\rho)}{2(1-\rho)}. \tag{8.20}$$

Analysis of a general polling system is quite involved. We refer interested readers to [300–302] for more details. Polling is incorporated in both IEEE 802.11 and Bluetooth for wireless LANs.

8.6.2 ALOHA and Slotted ALOHA

The original ALOHA (also called *pure ALOHA*) was the earliest MAC scheme developed for packet radio networks. With pure ALOHA, a station transmits a packet whenever it wants to. After the transmission, the station listens for an amount of time equal to the maximum round-trip propagation delay in the network plus a small fixed time increment. If no acknowledgment is received by the sender, which indicates a collision, the frame is sent again. To reduce the probability of another collision when all frames involved in the previous collision are retransmitted right away simultaneously, each station retransmits after a random amount of backoff time. A station gives up and drops the frame after several failed retransmissions.

Pure ALOHA is a very simple multiple access protocol, but its throughput is pretty low. Assume all packets have the same length and the transmission time of a packet is τ seconds. After a round-trip time $R \gg \tau$, a collision may be detected. After a random scheduling delay D, a retransmission is tried. The exact scheduling process of ALOHA is complicated and hard to analyze. To simplify the analysis, it is assumed that all transmitted packets, including retransmissions, follow a Poisson process with rate g packets/s. If a node sends a packet (either new or retransmitted) at time t and there is no other transmission during $[t - \tau, t + \tau]$, the packet will be successfully received. Since the arrival process is Poisson, the number of transmissions in $[t - \tau, t + \tau]$ has a Poisson distribution with parameter $2g\tau$. Therefore, the successful rate, denoted s, is given as

$$s = g \times \Pr\{\text{no other transmissions in} (-\tau, \tau)\} = ge^{-2g\tau}. \tag{8.21}$$

Let $G = g\tau$ be the normalized offered load; the throughput of pure ALOHA is

$$S = s\tau = g\tau e^{-2g\tau} = Ge^{-2G}. \tag{8.22}$$

By differentiating S with respect to G and setting the derivative to 0, we find the maximum throughput of pure ALOHA as

$$S^* = \frac{1}{2e} \approx 0.184, \quad \text{when } G^* = \frac{1}{2}. \tag{8.23}$$

From Equation (8.21), 2τ can be interpreted as the "vulnerable time" of pure ALOHA. If on average there is only one packet transmission during the vulnerable time (i.e., when $G = 1/2$), pure ALOHA achieves its maximum throughput.

Otherwise, the channel is either underloaded (i.e., channel bandwidth not fully exploited) or overloaded (i.e., waste of channel bandwidth due to collision).

Slotted ALOHA is an extension of pure ALOHA for improved throughput. In slotted ALOHA, time is divided into slots, and the length of one time slot is equal to the packet transmission time τ. Assume all nodes are synchronized. When a node has a packet to send, it starts to send it at the beginning of the next time slot. If no other transmissions are in the same time slot, the transmission succeeds. Otherwise, a collision occurs and the packet will be retransmitted after a random delay.

Similar to the pure ALOHA analysis, assume that all transmitted packets, including retransmissions, form a Poisson stream with rate g packets/s. If a node sends a packet (either new or retransmitted) in time slot t and there is no other transmission during the same time slot (with duration τ), the packet will be successfully received. Assume the number of packets transmitted in a time slot has a Poisson distribution with parameter $G = g\tau$. Therefore, the throughput of slotted ALOHA is

$$S = G \times \Pr\{\text{no other transmissions in}(0, T)\} = Ge^{-G}. \qquad (8.24)$$

Differentiating S with respect to G and setting the derivative to 0, we find the maximum throughput of slotted ALOHA as

$$S^* = \frac{1}{e} \approx 0.368, \quad \text{when } G^* = 1. \qquad (8.25)$$

By adopting a time-slot system (and with synchronization), the throughput is doubled. Again we can interpret the length of one time slot (i.e., τ) as the vulnerable time of slotted ALOHA. When on average one packet is transmitted during the vulnerable time (i.e., when $G = 1$), slotted ALOHA achieves its maximum throughput.

8.6.3 CSMA

One major cause of ALOHA's low throughput problem is that users do not try to avoid collisions. Users start transmission whenever they want even when the channel is busy and collision is doomed. Therefore, a simple but effective enhancement is to let each user sense the medium before starting transmission. If the channel is sensed busy, the user holds its packets until the medium is free. Such an idea is incorporated in carrier sense multiple access (CSMA).

In CSMA, one of the following three strategies can be used to sense the channel. In *1-persistent* CSMA, a user with data frame waiting senses the medium and sends the frame immediately if the medium is free. In case the medium is busy, the user continuously senses the medium and starts transmitting as soon as it finds the medium free (i.e., with probability 1). This way, the medium will not be idle when a packet is waiting to be transmitted. However, when more than one user is sensing the medium (when it is busy), they will start transmitting simultaneously when the channel becomes idle, leading to a collision. In *nonpersistent* CSMA, the node waits a random amount of time before sensing again, when it senses a busy medium. In this way, it is likely that backlogged users will come back to sense the medium at a different time, and the collision rate can thus be reduced. The downside of this method is that the medium utilization is reduced, since it is possible that the medium is idle but all the backlogged users are waiting for their random backoff timers to expire. The *p-persistent* CSMA is used when time is divided into slots. A backlogged user senses the medium and starts its transmission with probability p if the medium is found free. Otherwise, it senses the medium at the beginning of the next time slot. Thus, p-persistent CSMA combines the advantages of the preceding two schemes and can reduce the collision rate and improves the medium efficiency.

It is worth noting that, although CSMA can effectively reduce the chance of collision, it cannot completely eliminate collision. See the example shown in Figure 8.9. At T_0, Node A senses an idle channel and starts transmitting. It takes a seconds (i.e., the propagation delay) for the first bit of the packet to reach Node B. Before $T_0 + a$, Node B still senses an idle channel. If Node B starts transmitting during $[T_0, T_0 + a)$, there is a collision. Therefore, the vulnerable time of CSMA is the maximum propagation delay a.

For unslotted nonpersistent CSMA, assume that there are an infinite number of stations and fixed-length packets with transmission time τ seconds. We also assume that the propagation delay for all other stations to hear a packet is a seconds. Then, a seconds after a packet is transmitted, all other stations know that the channel is busy and do not attempt to access the channel. Assume that arriving packets follow the Poisson distribution with the rate s packets/s. When a transmission attempt fails, the packet is rescheduled for retransmission. Again to simplify the analysis, we assume the overall packets to be transmitted follow a Poisson process with rate g packets/s, including both arriving packets and rescheduled packets.

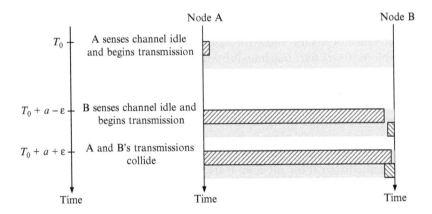

FIGURE 8.9

An example of collision in CSMA.

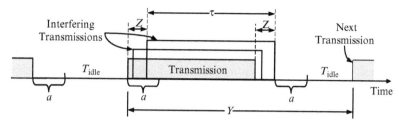

FIGURE 8.10

A typical cycle of CSMA operation.

To have a successful transmission, the channel has to be idle when the sender starts sensing, and there are no other transmissions within the next a seconds period. Therefore, we have

$$s = g \times \Pr\{\text{channel sensed idle}\} \times \Pr\{\text{no transmissions in } a \text{ seconds}\}$$
$$= g \times \Pr\{\text{channel idle}\} \times e^{-ga}. \tag{8.26}$$

It remains to derive the expression for the probability that the channel is idle. We plot a typical cycle of CSMA operation in Figure 8.10. Each cycle starts with a period of idle time, then a transmission occurs when a station with packets waiting finds the channel idle. Interfering transmissions are possible during the a seconds after the first transmission starts (i.e., the vulnerable time of CSMA), and Z is defined as the time between the beginning of the first transmission and that of the last interfering transmission. After the end of the last interfering transmission, another a seconds pass before a new cycle starts, so that all stations are aware that the current transmission is over. The length of a cycle is $Y = Z + \tau + a + T_{\text{idle}}$. From *renewal theory*, we have $\Pr\{\text{channel idle}\} = E[T_{\text{idle}}]/E[Y]$. Since the arrival process is Poisson with exponentially distributed interarrival times, we have $E[T_{\text{idle}}] = 1/g$. The probability of the channel being idle is therefore $\Pr\{\text{channel idle}\} = (1/g)/[E[z] + \tau + a + (1/g)]$, and the successful packet rate is

$$s = ge^{-ga} \times \frac{1/g}{1/g + a + \tau + E[Z]}. \tag{8.27}$$

Recall the normalized rates are $S = s\tau$ and $G = g\tau$ and let $\alpha = a/\tau$. We have

$$S = \frac{Ge^{-G\alpha}}{1 + G\alpha + G + G\alpha E[Z]/a}. \tag{8.28}$$

Note that $0 \le E[Z] \le a$, so that S is bounded as

$$\frac{Ge^{-G\alpha}}{1 + G + 2G\alpha} \le S \le \frac{Ge^{-G\alpha}}{1 + G + G\alpha}. \tag{8.29}$$

8.6.4 CSMA/CA

In wireless networks, the signal attenuates over distance following a power law behavior. If a node is transmitting, its own signal overwhelms any possible collision signal that is received. Even when the node finishes its transmission, a collision signal caused by the previous transmission may still be too weak to be detected due to power law path loss. Therefore, usually, the receiver broadcasts an acknowledgment message to notify the sender of a successful reception, while absence of acknowledgment indicates a failure and the source node will attempt a retransmission at a later time.

Collision is more costly in wireless networks than in wireline networks. When a collision occurs, bandwidth, which is limited in many wireless networks, is wasted. In addition, energy is also wasted on failed transmissions. In wired networks, the carrier sense multiple access with collision detection (CSMA/CD) scheme can be adopted to detect collision and the sender can immediately stop the transmission when collision is detected, thus avoiding wasting bandwidth and energy on transmitting the remaining part of the corrupted frames. However, CSMA/CD is not practical in wireless networks, since the wireless transceiver cannot listen for collision while transmitting; the received signal is usually overwhelmed by the transmitted signal at the transmitter. Therefore, we should try to avoid collision, rather than detect it, as in a wireline network.

In CMSA with collision avoidance (CSMA/CA), two mechanism are incorporated to avoid collision: (1) a set of delays, termed *interframe spaces* (IFS), that amounts to a priority system, and (2) a contention window and binary exponential backoff. When a station has a frame to transmit, it first senses the channel. If the channel is found to be idle, it waits for an interval of IFS to see if the channel is still idle. If so, the station starts transmission. If the channel is found busy, the backlogged station defers its transmission and keeps on sensing the channel until the current transmission is over. Then the station waits for an IFS. If the channel is still idle, the station backs off further for a random period of time. It starts transmitting only when the channel remains idle after the random backoff time. Note that during the backoff time, the backoff timer is decreased only when the channel is idle. To ensure stability, binary exponential backoff is used for handling heavy loads. With the IFS and backoff mechanism, transmissions from multiple backlogged stations are spread out over time and collision is avoided. The operation of CSMA/CA is illustrated in Figure 8.11 for both cases of a failed transmission and a successful transmission.

The *hidden terminal* and *exposed terminal* problems in wireless LANs (WLANs) are inherent from the use of wireless transmissions. Consider the scenario shown in Figure 8.12, where Nodes B and C are outside of each other's transmission range and Node A is somewhere in between and can hear both Nodes B and C's transmissions. We say Nodes B and C are *hidden* from each other with respect to Node A. Suppose Node B is transmitting to Node A and Node C also has a packet for Node A. If CSMA/CA is used, Node C senses an idle channel because it cannot hear Node B's ongoing transmission. Node C therefore starts transmitting

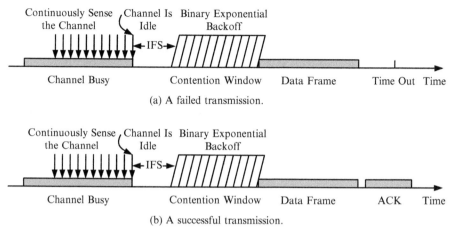

(a) A failed transmission.

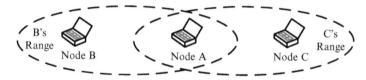

(b) A successful transmission.

FIGURE 8.11

The CSMA/CA operation.

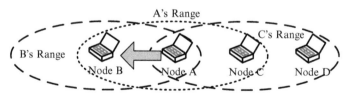

FIGURE 8.12

The hidden terminal problem, where Nodes B and C are hidden from each other with respect to Node A.

FIGURE 8.13

The exposed terminal problem, where Node C is an exposed terminal with respect to the transmission from Node A to Node B.

data to Node A and a collision occurs at Node A. This is called the *hidden terminal problem*. Now consider a different scenario shown in Figure 8.13. There are four nodes in the system. Node A is transmitting to Node B while Node C has a frame for Node D. If Node D is out of the transmission range of Node A and Node B is out of the transmission range of Node C, Node C's transmission is neither interfered with by Node A's transmission (at Node D) nor interferes with Node B's reception.

However, if CSMA/CA is used, Node C detects a busy channel and waits until Node A's transmission is over, resulting in a waste of bandwidth.

The hidden terminal problem can be solved by incorporating a request-to-send (RTS) and clear-to-send (CTS) handshake before data frame transmission. When a node has a frame to send, it first broadcasts an RTS message carrying the time needed to transmit the frame. The target node, if it is free, responds with a CTS broadcast. All other nodes that hear the RTS or the CTS mark the channel as busy for the duration of the requested transmission. Thus, collision due to hidden terminals can be avoided. However, the exposed terminal problem is not solved by this mechanism, although studies show that mitigating the exposed node problem results in more concurrent transmissions and improves network throughput performance [303].

CSMA/CA is adopted in the IEEE 802.11 MAC, which has become the most popular protocol for single- or multihop wireless networks. Xiao and Rosdahl [304] derived the *achievable maximum throughput* (MT) under the best-case scenario when (1) the channel is error free and (2) exactly one station is active (which always has a packet to send) during any transmission cycle. MT is found to be

$$MT = \frac{8L_{\text{DATA}}}{T_{D\_\text{DATA}} + T_{D\_\text{ACK}} + 2\tau + T_{\text{DIFS}} + T_{\text{SIFS}} + E[\text{CW}]}. \tag{8.30}$$

In Equation (8.30), the average backoff time is $E[\text{CW}] = \text{CW}_{\min}T_{\text{slot}}/2$, where $\text{CW}_{\min}$ is the minimum backoff window size and T_{slot} is a slot time, since the backoff time is uniformly chosen from $(0, \text{CW}_{\min} - 1)$. L_{DATA} is the payload size in bytes, τ is the propagation delay, T_{DIFS} is the *distributed interframe space*, and T_{SIFS} is the *short interframe space* [275]. The data transmission delay T_{D_DATA} and the ACK transmission delay T_{D_ACK} are given by $T_{D\_\text{DATA}} = T_P + T_{\text{PHY}} + T_{H\_\text{DATA}} + T_{\text{DATA}}$ and $T_{D\_\text{ACK}} = T_P + T_{\text{PHY}} + T_{\text{ACK}}$, where T_P is the transmission time of the physical preamble, T_{PHY} is the transmission time of the PHY header, T_{H_DATA} is the transmission time of the overhead, T_{DATA} is the transmission time for the payload, and T_{ACK} is the transmission time for the acknowledgment. Xiao and Rosdahl [304] show that, for a given payload size and a given set of overhead parameters, the *throughput upper limit* (TUL) is given as

$$\text{TUL} = \frac{8L_{\text{DATA}}}{2T_P + 2T_{\text{PHY}} + 2\tau + T_{\text{DIFS}} + T_{\text{SIFS}} + \frac{1}{2}\text{CW}_{\min}T_{\text{slot}}}. \tag{8.31}$$

An interesting observation from this analysis is that there is a finite throughput bound for the IEEE 802.11 MAC, which is independent of the link data rate. That is, even when the link rate goes to infinity, the maximum throughput is still a finite value. For example, when the payload size is 1000 bytes, the maximum achievable throughput for IEEE 802.11a is only 24.7 Mb/sec, which is about 45.7% of the 54 Mb/s nominal link capacity. The control overhead of IEEE 802.11 MAC (i.e., IFSs, random backoffs, acknowledgments, frame headers, etc.) is the main cause of the reduced throughput. The existence of such limits indicates that, by simply increasing the data rate of the wireless links without reducing overhead, the achievable throughput gain will be limited.

8.7 ROUTING, ENERGY EFFICIENCY, AND NETWORK LIFETIME

Energy efficiency is one of the most important issues in wireless networks. Since mobile nodes are powered by battery, efficient use of energy can extend the operation time of a mobile node, which in turn prolongs the time that the network can operate; that is, the *lifetime* of the network. This problem is of critical importance for wireless sensor networks, where sensor nodes may not be able to recharge their batteries.

Energy efficiency is a challenging problem and should be addressed in the design of every layer in the wireless protocol stack [305]. Among various network control and operation functions, routing has a direct impact on the energy efficiency of the underlying wireless network. In wireless networks, considerable energy is spent on transmitting and receiving data. An energy-efficient routing protocol can optimally direct the flow of data within the network, to achieve balanced depletion of battery power at the nodes [306, 307]. An important performance measure in the design of energy-efficient wireless network protocols is network lifetime. It can have various definitions, such as the network operating time until the first node depletes its energy, when a certain percentage of nodes fails, or when the network is partitioned, depending on the specific network and application under consideration.

In the rest of this section, we present an interesting routing framework to demonstrate the impact of energy efficiency on routing in wireless networks [308]. Consider a multihop wireless network illustrated in Figure 8.14, where each node operates on a limited battery, which is consumed mostly by transmitting and

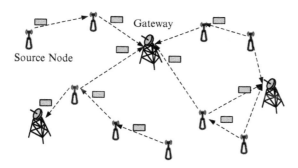

FIGURE 8.14

A wireless sensor network with sensor nodes and gateway nodes. Traffic is generated at the source nodes and delivered to the gateway nodes via multihop routes.

receiving data. At each node, some information is generated and should be delivered to the gateways via multihop routes. The transmit power is assumed to be adjusted to the minimum level appropriate for the intended receiver. This is a very general model for wireless sensor networks.

We can model the network as a directed graph $G(N, L)$, where N is the set of all nodes and L is the set of all directed links. A link is denoted as (i, j), where $i, j \in N$, and the link exists if and only if $j \in S_i$, where S_i is the set of all nodes that can be reached by Node i under a maximum transmit power. Each node has initial energy E_i. The following energy consumption model is adopted: Node i consumes energy e_{ij}^t to transmit a data unit to its neighbor Node j, while it consumes energy e_{ij}^r to receive a data unit from its neighbor Node j. Assume there are C different commodities, while each commodity c is defined by a set of source nodes and destination nodes. For each commodity $c \in C$, assume the set of original nodes is $O^{(c)}$, where data are generated at Node i with rate $Q_i^{(c)}$, and the set of destination nodes is $D^{(c)}$, each being the sink of the corresponding data. Let $q_{ij}^{(c)}$ be the transmission rate of commodity c from Node i to Node j assigned by the routing algorithm.

The lifetime of Node i under a given flow $\mathbf{q} = \{q_{ij}^{(c)}\}$ is given by

$$T_i(\mathbf{q}) = \frac{E_i}{\sum_{j \in S_i} e_{ij}^t \sum_{c \in C} q_{ij}^{(c)} + \sum_{j:i \in S_j} e_{ji}^r \sum_{c \in C} q_{ji}^{(c)}}; \qquad (8.32)$$

that is, the interval from the time when the node starts to operate (with initial energy E_i) until it depletes its energy. The network lifetime under flow $\mathbf{q}$ is defined to be the minimum lifetime among all nodes:

$$T_{sys}(\mathbf{q}) = \min_{i \in N} T_i(\mathbf{q}); \qquad (8.33)$$

that is, the interval from the time when the network starts to operate until the first node dies.

The objective of the energy-efficient routing protocol is to find the optimal flow $\mathbf{q}$ such that the network lifetime is maximized. The optimal routing problem can be formulated as

$$\text{Maximize} \quad T_{sys}(\mathbf{q}) = \min_{i \in N} T_i(\mathbf{q}) \qquad (8.34)$$

$$\text{Subject to} \quad q_{ij}^{(c)} \geq 0, \text{ for all } i \in N, j \in S_i, c \in C \qquad (8.35)$$

$$\sum_{j:i \in S_j} q_{ji}^{(c)} + Q_i^{(c)} = \sum_{j \in S_i} q_{ij}^{(c)}, \text{ for all } i \in N \backslash D^{(c)}, c \in C. \qquad (8.36)$$

The constraints are *flow conservation* constraints. Note that maximizing the system lifetime is equivalent to maximizing the amount of total information transfer given a fixed information generation rate. With some algebra, this problem can be shown to be equivalent to the following optimization problem:

$$\text{Maximize} \quad T \tag{8.37}$$

$$\text{Subject to} \quad \hat{q}_{ij}^{(c)} \geq 0, \text{ for all } i \in N, j \in S_i, c \in C \tag{8.38}$$

$$\sum_{j \in S_i} e_{ij}^t \sum_{c \in C} \hat{q}_{ij}^{(c)} + \sum_{j:i \in S_j} e_{ji}^r \sum_{c \in C} \hat{q}_{ji}^{(c)} \leq E_i, \text{ for all } i \in N \tag{8.39}$$

$$\sum_{j:i \in S_j} \hat{q}_{ji}^{(c)} + T \cdot Q_i^{(c)} = \sum_{j \in S_i} \hat{q}_{ij}^{(c)}, \ \forall \, i \in N \backslash D^{(c)}, \ c \in C \tag{8.40}$$

$$\hat{q}_{ij}^{(c)} = T \cdot q_{ij}^{(c)}. \tag{8.41}$$

An iterative *flow augmentation* algorithm is presented in [308] to solve the maximum lifetime routing problem. During each iteration, each origin node $o \in O^{(c)}$ of commodity c calculates the minimum cost path to its destination nodes in $D^{(c)}$. Then the flow is augmented by an amount $\lambda Q_i^{(c)}$ on the minimum cost path, where λ is the augmentation step size. Next the residual energy at each node and link costs are updated. With the updated link costs, the minimum cost paths are recalculated and the procedure is repeated until a node runs out of energy.

The proposed algorithm is a minimum cost–path routing algorithm where the cost of a link (i, j) is defined to be a combination of transmission and reception energy consumption (e_{ij}^t and e_{ij}^r) and the residual energy levels at both end nodes ($\underline{E}_i$ and $\underline{E}_j$); that is,

$$\text{cost}_{ij} = \left(e_{ij}^t \right)^{x_1} \cdot \underline{E}_i^{-x_2} \cdot E_i^{x_3} + \left(e_{ij}^r \right)^{x_1} \cdot \underline{E}_j^{-x_2} \cdot E_j^{x_3}, \tag{8.42}$$

where x_1, x_2, and x_3 are nonnegative weighting factors to integrate link energy consumption parameters into a single cost metric. Let the routing algorithm be FA(x_1, x_2, x_3); then we can have the following variations: (1) FA(0, 0, 0): minimum hop routing; (2) FA(1, 0, 0): minimum total energy routing; (3) FA($\cdot$, x, x): normalized residual energy is used as link cost; and (4) FA($\cdot$, $\cdot$, 0): absolute residual energy is used as link cost. Through simulation studies, the authors show that FA(1, x, x) achieves the best performance.

8.8 CONGESTION CONTROL IN WIRELESS NETWORKS

Congestion control is the key function of the transport layer. As shown in Figure 8.3, a router's buffer is shared by multiple independent sessions in a store-and-forward communication network. When the instantaneous arrival rate is higher than the output rate, the router buffer occupancy grows. Congestion occurs when the buffer is overflown, resulting in packet loss and large delay for all the sessions sharing this bottleneck router. Therefore, the main idea of congestion control is to infer (incipient) congestion and reduce the sending rates of the sources. When congestion is over, on the other hand, sources should increase their rates to fully exploit the available bandwidth in the network.

Following the *end-to-end principle*,[4] TCP implements the additive-increase-multiplicative-decrease algorithm and operates at endpoints to adjust sending rates based on network feedback. Since routers are not explicitly involved (so that they can be stateless and handle large traffic volume), TCP has to infer network congestion by timeout or duplicate acknowledgments. A fundamental assumption for such inference to work is that all packet losses are caused by congestion. Therefore, whenever there is a timeout, TCP assumes congestion in the network and starts to reduce its sending rate. This assumption, although quite reasonable for wireline networks where links are usually reliable, does not hold true for wireless networks, where packets are also dropped or lost due to transmission errors or route failures. If the sending rate is G at a wireless link with packet loss rate p, the throughput of this link is $S = G(1 - p)$. TCP should increase its sending rate G to get a large throughput if there is no congestion, rather than decreasing its rate. TCP does not distinguish between loss due to transmission errors, and due to congestion, it suffers poor performance when the end-to-end path includes wireless link(s) [309, 310].

In multihop wireless networks, TCP also suffers poor throughput performance due to frequent route failures and MAC layer contention [311–314]. The main cause of route failure is mobility; routes are broken when topology changes. Also if a frame is dropped at the MAC layer when the maximum number of retransmissions is reached (due to contention), the routing layer assumes the link is broken. In both cases, the routing engine starts to reestablish a route for the session. During this interval, TCP may take time out and reduce its sending rate. When the new path is found, it may have a different end-to-end bandwidth and delay. Therefore, TCP experiences a sudden change in round-trip time. In [312–314], the authors simulate TCP throughput using a wireless network with a chain topology. It is reported that, as the hop count increases, the end-to-end TCP throughput decreases rapidly to a very low value.

Various enhancements have been proposed in the literature to improve the TCP performance in wireless networks. For example, the class of *link layer mechanisms* adopt strong error control (e.g., forward error correction or automatic repeat request) to make wireless links more reliable. The class of *split TCP* solutions breaks down a long TCP connection into several shorter localized segments, such as a segment consisting of wireline links and another segment consisting of the last hop wireless link. The interfacing node between two adjacent segments is called a *proxy*. Independent flow and congestion control are performed on each of the segments. In the class of *explicit notification* schemes, intermediate nodes (e.g., routers or base stations) explicitly notify TCP about packet loss at wireless links.

---

[4]The end-to-end principle is one of the key design principles of the Internet. It states that, whenever possible, communications protocol operations should be implemented at the endpoints of a communications system or as close as possible to the resource being controlled.

Most of the improvements help TCP differentiate among different types of losses, and TCP takes different actions based on the type of loss detected. Interested readers can see [309] for a comparison study and [310, 311] for two excellent surveys of existing proposals.

8.9 CROSS-LAYER DESIGN AND OPTIMIZATION

So far, we demonstrated the impact of various wireless network characteristics on the design and operation of each layer in the protocol stack. Although our discussion still follows the layered structure, various factors and parameters from more than one layer are in fact implicitly considered in our discussions. For example, we jointly consider packet scheduling, a MAC layer function, with channel dynamics and power control in the physical layer. As another example, TCP suffers poor performance in wireless networks. Information from lower layers has been shown helpful for TCP to distinguish between different types of packet losses and react differently.

Cross-layer design refers to sharing information among layers for efficient use of network resources and achieving high adaptivity. In cross-layer design, each layer is characterized by a few key parameters and control knobs. The parameters are passed to other layers to help them determine the best adaptation rules for their control knobs with regard to the current network status. Cross-layer design is usually formulated as an optimization problem, with optimization variables and constraints from multiple layers. Solving the optimization problem provides the optimal values for the control knobs in the layers. Such an approach is illustrated in Figure 8.15.

Cross-layer design is especially appealing in wireless networks for the following reasons. First, the traditional architectural design approach, although highly successful in wireline networks, results in reduced search space for optimal adaptation. Unlike wireline networks, where resources are abundant, the need is compelling in wireless networks to explore a larger optimization space, including multiple layers to make the best of limited resources. Second, the existing protocol stack is defined with wireline networks in mind. It may not be suitable for wireless networks that are fundamentally different in many aspects. For example, the concept of "link" is totally different now. Connectivity between two nodes largely depends on the distance between them and their transmit powers. The unique characteristics of wireless networks require joint consideration of parameters previously located in different layers. See TCP over wireless for an example. Third, design parameters, which may locate in different layers, are more tightly coupled now than those in wireline networks, as illustrated in our earlier discussions in this chapter. It would be challenging to develop a clear-cut "separation principle" for wireless networks.

Unsurprisingly, there has been increased interest in developing wireless networking protocols and algorithms with increased interactions among various

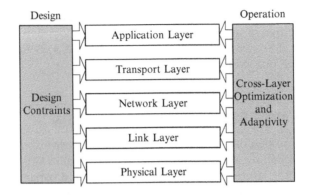

FIGURE 8.15

Cross-layer design and optimization in wireless networks, where system constraints from various layers are jointly considered for adjusting control knobs in the layers in response to network status changes.

layers. Significantly improved performance has been reported in the literature; see [315–321] for several examples.

Cross-layer design generally involves the formulation and solving optimization problems. Although it is appealing to incorporate more parameters from more layers, caution should be taken for choosing the parameters and the trade-off between performance gains and the resulting complexity. To avoid pitfalls, the following cross-layer principles have been proposed in [322]:

- **Interactions and the Law of Unintended Consequences.** Traditional architectural design allows designers to work on an isolated problem without worrying about the rest of the protocol stack. This is not true anymore in cross-layer design. Care should be taken to prevent unintended effects on other parts of the system.
- **Dependency Graph.** Cross-layer design often causes several adaptation loops that are part of different protocols. A *dependency graph* will be very helpful, where every relevant parameter is a node and a directed edge indicates the dependency relation between the parameters.
- **Timescale Separation and Stability.** If a parameter is controlled and used by two adaptation loops, stability can be proven by exploiting the notion of averaging and timescale separation. For every loop in the dependency graph consisting of interactions at similar timescales, proofs of stability are required. This is often nontrivial and requires significant analytical effort.
- **The Chaos of Unbridled Cross-Layer Design.** How do we maintain implementations of cross-layer design? The longevity and the updating cost of a cross-layer design should be carefully considered against the achievable performance gains.

8.10 CHAPTER SUMMARY

In this chapter we discuss the basic concepts of communication networks with a focus on wireless networks. After reviewing the network architecture and common building blocks, we discuss the unique characteristics of wireless communication networks, including wireless transmissions, mobility, and energy efficiency, and examine their impact on the design and optimization of various layers. The fundamental differences between wireless and wireline networks call for judicious revisiting of the existing communication network protocols (e.g., TCP), which were originally designed with wireline networks in mind. Cross-layer design and optimization provide effective means of maximizing resource utilization and optimal adaptivity in resource-limited wireless networks, but caution should be exercised to avoid the common pitfalls.

8.11 PROBLEMS

1 Error detection is incorporated in every layer of the TCP/IP protocol stack, such as frame check sequence in MAC frames and checksums in higher-layer headers. Are these functions redundant? Justify your answer.

2 Simulate the mobility of a node in a one-dimensional network in [0, 1000] following the random waypoint model without pausing. The speed is uniformly chosen from [0.1 m/s, 10 m/s]. Simulate the following two cases: Case I, start with uniformly chosen speed and initial location; Case II, start with initial speed and location drawn from the corresponding steady-state distributions. Plot the steady-state distributions (i.e., histograms) of simulated node speed and location and compare them with the analysis in Section 8.4.

3 Modify the random waypoint mobility model to simulate a node moving on a 4×4 grid of streets, with and without pausing.

4 Consider the case with pausing in the random waypoint model, where a node pauses for time t_p after reaching a destination, then moves on. Assuming t_p is chosen according to a probability density function $f_p(t)$, derive the stationary distribution that the node is pausing Pr{pausing}.

5 Jensen's inequality states that, for an integrable real-valued random variable X and a measurable convex function $f(\cdot)$, we have $f(E[X]) \leq E[f(X)]$. Follow the analysis in Section 8.5 to show that $C_{npc} \leq C_{ppc}$.

6 Consider a polling system with gated service and one station. The switchover time is $V = 1$. Consider the following two cases: Case I, the service time of each packet is a constant x; Case II, the service time of each packet is

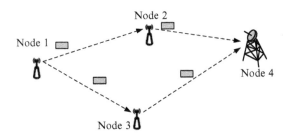

FIGURE 8.16

A four-node network with three sensor nodes and one gateway.

exponentially distributed with mean x. Use MATLAB to plot the average waiting time for offered load $\rho \in [0, 0.95]$. Compare the waiting times with an $M/G/1$ queue with the same load, assuming $x = 1$ and service time distribution for both cases.

7   Use MATLAB to plot and compare the throughput of pure ALOHA, slotted ALOHA, and CSMA (i.e., using the upper and lower bounds) for the offered load $G \in [0, 10]$. Assume $\alpha = 0.1$ for CSMA.

8   Qualitatively compare the three carrier sensing schemes: 1-persistent, non-persistent, and p-persistent sensing. Which scheme is more suitable for lightly loaded systems? Which scheme is more suitable for heavily loaded systems? Justify your answers.

9   Following the analysis in Section 8.6.4, compute and plot MT and TUL for IEEE 802.11b for a payload size increasing from 0 to 1000 bytes in steps of 100 bytes. The parameters can be found from the IEEE 802.11b specification [275].

10 Consider maximum network lifetime routing discussed in Section 8.7 for the four-node network given in Figure 8.16. Node 1 is the only source that generates traffic at $Q_1 = 10$ b/s. Therefore, only one commodity is in the network, $C = 1$. Node 4 is the only gateway, while Nodes 2 and 3 relay Node 1's bits to the gateway. All the links are of the same length with transmit energy $e^t = 50$ nJ/bit. For simplicity, we assume $e^r = 0$ nJ/bit for all the links. The initial energy at the nodes are $[E_1, E_2, E_3, E_4] = [20, 15, 20, \infty]$ J. Formulate the maximum lifetime routing problem and derive the maximum network lifetime of the network.

11 Repeat Problem 10, while the initial energy for the nodes are $[E_1, E_2, E_3, E_4] = [20, 10, 5, \infty]$ J and all other parameters remain the same.

Cognitive radio network architectures

Petri Mähönen and Janne Riihijärvi

RWTH Aachen University, Germany

9.1 INTRODUCTION

Cognitive radio research experienced a tremendous increase in volume during the last few years. It has been recognized that traditional approaches of spectrum use need to be challenged, and this lead to the emergence of dynamic spectrum access (DSA)–based cognitive radios. Another, more general definition for cognitive radio (CR) is to define CR as a context-sensitive and smart radio. This latter generalization makes CR essentially a smart, adaptive, and flexible wireless device [7]. In this chapter we consider both classes of CRs almost interchangeably.

Although the CR research produced a number of new results, it has been limited mostly to the study of cognitive radios themselves or to consider very small cooperative networks in the context of information theory, cooperative communications, or spectrum sensing (the reader should consult Chapters 7, 11, and 12 for details and further references). General cognitive radio networks, also known as *cognitive wireless networks*, have been considered by only few authors. However, our understanding of these networks is increasing rapidly. In this chapter we consider some aspects of cognitive radio network (CRN) architectures. Because there are no operational large-scale CRNs, one cannot analyze any specific architectural design. Hence, we limit our scope to describing some of the specific general features thought to be central to future CRNs. Some of these features, such as common interface specifications and agreement on various communication mechanisms between architectural entities, are essential parts of any emerging architecture. Modularity, separation of concerns into different functional components, and use of common data structures and interfaces are examples of issues any system architecture must deal with, and those of cognitive radio networks are no exception.

The general concept toward *cognitive networking* was introduced by Clark et al. in 2003 under the concept of *knowledge plane* [323]. The basic idea of a knowledge plane includes all the major components of the later cognitive networking paradigm. Clark et al. specifically state that the network with the knowledge plane (KP) has a high-level view of what its purpose is. Through a KP mechanism the network is able to adjust its operational parameters or even self-organize (reassemble) itself to more optimally serve different users and applications. This approach inherently also hints at the capability to discover problems and fix them automatically (at least partially). Furthermore, the authors introduced the central components for a new knowledge plane architecture. Specifically they introduced at an abstract level the concept that the cognitive system gathers *observations, constraints*, and *assertions*. This implies the need for some observing agents, or observation points, and mechanisms to describe assertions and constraints. Moreover, an assertion is made that a new network architecture would apply rules to generate observations and responses.

It was subsequently speculated that the use of cognitive network principles could be particularly beneficial for wireless networks [324]. The general cognitive networks, or autonomous networks, research has been starting to gain momentum, and many interesting results exist [325]. Although some of the general cognitive network research is applicable, in general CRNs also require domain-specific architectural considerations. For excellent discussion on machine learning–related architectural issues in the context of cognitive networks, we recommend to the reader Chapter 5 in [325].

9.2 COGNITIVE RADIO NETWORK ARCHITECTURES

Cognitive networks, especially in the context of cognitive radio networks, have been studied by carefully developing taxonomy and related concepts by Thomas et al. [326, 327]. The paper [326] is a highly recommended companion paper to this chapter. The concept of network knowledge representation language (NKRL) was introduced in [324], which is similar to a somewhat later suggestion by Thomas et al. under the name *cognitive specification language* (CSL) in [326]. A more specific architectural framework was later predeveloped and introduced in 2006 by Mähönen et al. [328]. Generally we agree with the referenced papers that the key issue for cognitive radio networks is their capability to perceive and model *end-to-end* goals. This may enable more efficient and global optimization. The CRNs have a great promise to enable intelligent optimization in a more flexible fashion. Cross-layer optimization should not be seen as a defining factor of CRN architectures. However, as will be seen later, various interfaces and components of the architecture will certainly support cross-layer optimization as a natural consequence. Moreover, many cross-layer optimization problems require capability to understand conflicting end-to-end goals and there is a need for a smart decision-making mechanism to resolve conflicting goals. Thus, CRN architectures probably can become an enabling technology to solve many cross-layer optimization problems without losing the benefits of modularity and some layering.

Generally the cognitive cycle introduced by Mitola [7, 182, 329], and described elsewhere in this book, describes the main abstract components of cognitive systems. Sometimes learning is defined as the key component for the system to follow cognitive architecture, but in general this is a somewhat debatable issue, if considered in the large context of artificial intelligence and CRN research. In other domains it has been common to divide cognition into three layers: *cognitive, behavioral,* and *physical* [326]. Typically, the cognitive architecture includes at least the representation of elements: memory to store goals, knowledge, and policies and all the functional processes that operate on the structures and affect the outside world.

At a more general level the CRN architecture designer must decide if the architecture is based on a centralized, distributed, or hybrid decision-making process. Most of the suggestions in the literature are based on the assumption of building either hybrid or distributed architectures. Recently, IEEE SCC41/P1900 committees have started to study general architectural building blocks for cognitive radios, mostly in the context of DSA [330]. Less abstract implementation architectures, such as recent considerations within SCC41, introduce typically separate functional entities such as *terminal reconfiguration manager* (TRM), *terminal reconfiguration controller* (TRC), and *terminal measurement collector* and their network counterparts, such as *network reconfiguration manager* (NRM) and *radio-access network reconfiguration controller* (RRC). Partially the differences are often semantic, but the controller-based architectures draw their background from classical radio resource management–based systems and may emphasize control theoretical aspects over general cognitive architectures. The reconfiguration manager terminology was originally introduced by the software-defined radio (SDR) community and places emphasis on SDR operations.

9.2.1 Cognitive Resource Manager Framework

Here we describe the cognitive resource manager (CRM)–based architecture [331], which can be seen as an example architecture for CRNs. It is introduced simply as a general architectural example with no claim to its superiority and is a sufficiently general framework to introduce the main concepts and components. Other approaches have been proposed in the literature, although most of them take a less architectural starting point and opt to describe *cognitive engines* focused on a single optimization or learning method. Perhaps the most notable exception is the recent cognitive engine work done at Virginia Tech. Another ongoing approach is to build *execution platforms* that enable one to build highly flexible DSA and SDR systems. Despite the interesting architectural issues in that domain, we leave them out from this chapter, but the reader is encouraged to consult later chapters on cognitive radio prototyping environments to learn more. The platform

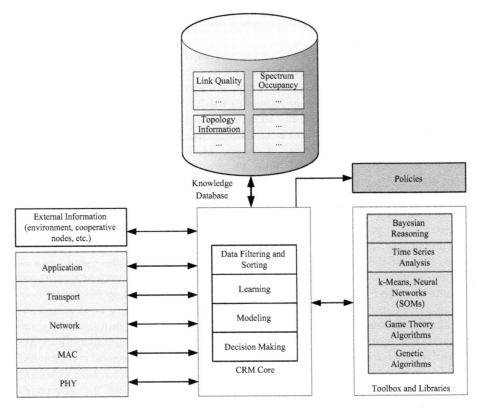

FIGURE 9.1

Main conceptual architectural components of the cognitive resource manager (Figure courtesy of M. Petrova, RWTH Aachen University).

implementation–based architectures have a very strong relation to SDR architectures, and one of the key components therein is the aforementioned reconfiguration manager.

Figure 9.1 shows the main parts of a cognitive resource manager. The core cognitive resource manager itself is a relatively small unit and partially acts like a microkernel type of operating system for cognitive radio processes. This core CRM is responsible for coordinating the information exchange between functional modules through well-defined interfaces. In this approach CRN architecture takes no specific stand on the implementation of these functional modules. Thus, protocol stack can still follow open system interconnection (OSI)-layering or be a rewirable and reconfigurable collection of different protocol entities. Similarly, functional and behavioral modules can be seen as separate "toolbox" processes and libraries. This means that different end-to-end and local optimization processes are executed independently in a coordinated fashion under the architectural framework. CRM itself is responsible for coordination and final decision making (behavior). Thus, at the general behavioral level, it makes most of the decisions. Reconfiguration managers that exist in some other more implementation-oriented architectures do not exist specifically in the CRM framework, since those would be simply functional modules with appropriate interfaces. Introducing the key components from a slightly different point of view, we have

- A CRM-core module that acts like an operating system. It is responsible for interoperability between different toolbox processes that perform modeling, optimization, and interactions. In the context of the architecture, the core should be seen partially as a distributed and networked operating entity, since it is also responsible for providing basic functionalities for different architectural blocks to communicate among themselves. Although the CRM core might not provide the actual communication protocols, it is nevertheless responsible for providing time synchronization and interface harmonization so that, for example, external policy servers and CR-centric modules can operate together.
- Toolboxes are, in principle, processes or libraries provided for optimization purposes. This modular architecture decision leaves it open for implementation purposes to decide which particular methods are actually implemented and executed in the entities of the cognitive radio networks.
- Standardized interfaces are required to ensure that different modules and protocol entities can exchange data both in a networked fashion and inside cognitive radios. At the architectural level the networked information exchange is a requirement, and it can be achieved with different ways. In [331] the authors propose that information exchange could be based on the use of *both* protocols and networked file systems.
- As will be discussed later there is the need for various databases or repositories that contain *policy information* and *models*. Although access to these repositories could be provided through independent modules, it is architecturally better to show them as standalone components due to their highly specialized and central role. Many of the repositories are often also distributed, so that part of the core policy information may be kept in the local device, but for example, spectrum rules are stored in external servers.

The CRM framework itself can work both as a standalone terminal-centric entity or a networked system with distributed decision-making capabilities. In the context of cognitive radio networks, it is natural that CRM must support networking. An important aspect of the architecture is that it includes specific interfaces. The key interfaces toward link layer, applications, and networks are briefly discussed here. Due to the wide variety of technologies that need to be covered with CRNs, it is highly advisable that the interfaces be made more flexible and open than the highly static and closed interfaces available at the present time. Finally, it is important that CRN goals and policies can be described with suitable description languages. For spectrum policy descriptions a number of languages and frameworks already exist [332, 333]. The aforementioned NKRL and CSL are required to describe different goals and utility functions of users and communicating entities in end-to-end fashion.

The CRN architectures, of course, are not entirely independent from the existing network architectures. The requirement to support legacy networks and protocol stacks makes it necessary to build modular and open interface–based information exchange. This is probably also a reason why Clark et al. [334] specifically refer to the knowledge plane as an extra architectural component instead of a new architecture. The key elements of general CRN architectures, apart from building a normal network architecture, perhaps can be summarized by the following nonexhaustive list:

1. There is a need to support the introduction of multiobjective optimization modules, which might be based in part on machine learning and heuristic algorithms.
2. CRN architecture must have a rich set of extendable interfaces to ensure that information can be exchanged efficiently between layers and agents in the network. The power of a knowledge plane is highly dependent on the existence of interfaces and protocols for efficient information exchange.
3. The architecture should support a large heterogeneity in the sense of devices, configurations, and different networking technologies.
4. CRN has a sort of self-awareness of the network so that it can do self-optimization. The self-awareness refers mainly to the capability to describe and reason about users' end-to-end goals and the architecture's ability to detect error states.
5. The architecture should provide support mechanisms and description languages to describe user and application goals and preferences, which goes beyond the present quality-of-service (QoS) parameterization mechanisms.
6. Cognitive radios and networks must have an understanding of timescales and costs. The cost of exchanging timely signalling information is required to avoid transmitting huge amounts of modeling data over the network. This is particularly important in the case of radio networks. The cost awareness can also refer to the fact that often user goals may be related to price minimization and thus the actual monetary cost of communication can become an optimization problem.

Cognitive radio networks are most probably quite heterogeneous and will have different stakeholders in their value chains. Instead of ignoring this fact, CRN architecture should take it into account and embrace it as one of the architectural principles. In the Internet architecture design this is often referred to as allowing *tussles* in the architecture [334].

A specific feature of a general CRN architecture is that it is not necessarily based on infrastructure support. Many earlier knowledge plane or autonomous computing–based proposals see a knowledge plane as a management entity of the network. In the case of CRNs several scenarios also require complete self-organization and peer-to-peer interactions. Therefore, cognitive radios themselves should have enough models and policy knowledge to handle situation awareness independently, even when there are no specific support nodes for this. Therefore, one potential challenge for CRN architectures is to embed and distribute some of the functionalities into participating nodes and not rely on network-based services.

Toolboxes and Libraries

The flexibility to build different specific cognitive systems prevents one from defining specifically which optimization components should be seen as unavoidable components. Hence, it might be better to define a general class of toolboxes and libraries that present functional components. These can then be included dynamically and in run-time with cognitive radio and network instances. A number of machine learning and heuristics-based optimization methods have been proposed as a part of cognitive radios or cognitive engines; for example, genetic algorithms, simulated annealing, and neural networks have been applied to solve a variety of problems (see also Chapter 7). These methods can be seen either as specific toolboxes to solve problems or more generally as libraries that a programmer can use to develop more complex optimization and machine learning–based solutions. This approach emphasizes that the architecture should describe as far as possible the unchangeable basic components and frameworks; and implementation- or situation-specific details should not be introduced into the architectural level.

Note that in principle one could build CRN architectures without requiring high reconfigurability from the underlying hardware components and protocol stacks. Most probably, however, both hardware and protocols will provide a lot of reconfiguration possibilities. The ideal radio hardware would be a flexible software-defined radio platform, and various recent proposals to use rewritable protocol stacks are certainly very good starting points. Regardless of the actual implementation solutions, the management of such reconfiguration requires some sort of managers and interfaces. The cognitive wireless network architecture could include these elements as specific components, but a more flexible approach is to see them as toolboxes. Thus, for example, a terminal reconfiguration manger may or may not exist in an individual terminal depending on terminal capabilities and needs.

One of the defining factors for the architecture is that it needs to support decision making and (distributed) network management. The implementation of this sort of facility is still a formidable challenge as far as one is aiming toward "cognitive"-level intelligence. There needs to be some sort of control entity that handles resource allocation and provides conflict resolution. In the preceding discussion this is referred as *core CRM*. However, this should not be understood as one monolithic control entity. The more flexible, and probable, architectural avenue is to distribute many control entities and behavioral decision-making modules into independent processes executed from toolboxes. As a specific example one could consider subsumption architectures used in robotics research. Subsumption architectures contain a number of different overlaid control entities, in a sense of decomposition. Similar modular and step-by-step instances of implementations are one likely way to build cognitive radio networks.

Unified Link Layer Applications Programmers Interface

Any CRM-like architecture clearly needs to be able to obtain information from different wireless technologies available on the given platform and be able to configure these network interfaces according to the different optimization decisions made. This creates the architectural requirement of creating a standard interface between the CRM and the wireless transceivers on the platform (which could include SDR technologies as well). Such an interface should be as technology independent as possible to ensure that architectural elements of CRN do not need to be rewritten as new wireless technologies emerge. Additional design considerations are run-time flexibility and support for key interaction paradigms, such as being able to receive event notifications upon, for example, disconnection of a link or appearance of a primary user.

The unified link layer API (ULLA) [335] is an example of a link layer interface satisfying these requirements. Wireless transceivers are represented as instances of generic classes following an object-oriented programming paradigm. These classes contain attributes yielding performance information, such as current bit rate or frame error rate, as well as methods that can be used to change the configuration. The classes form a hierarchy, starting on the highest level from a completely technology-independent representation of the available links and proceeding to more technology-specific classes yielding access to attributes that cannot be meaningfully abstracted. ULLA offers further a powerful *query functionality* using a lightweight subset of the well-known SQL language that can be used to gather information about the available links or to specify conditions for event notifications. Using such abstractions and query technologies allows creation of powerful interfaces that can remain invariant even as new technologies are developed, while still offering access to a wide variety of functionality at an abstraction level appropriate for each of the individual modules of the CRM. Perhaps one of the most important architectural indications of CRN development is that the traditional and fixed API-type interfaces may be too constrained. The requirement for a rich and flexible description of data from different technologies and layers implies the need to have lightweight parsable languages to enable queries and setup of user needs for triggers. The language-type approach, albeit not query languages, is already adapted in the domain of policy servers for DSA.

Common Application Requirement Interface and Network Knowledge Representation Language (NKRL)

Another key interface in the CRM architecture is the common application requirement interface (CAPRI). It can be used by applications (users) and protocol entities to inform the CRM about their requirements in terms of network performance. CAPRI functionality goes well beyond that offered by classical QoS service level agreements (SLAs) by allowing applications to register *utility functions* mapping network performance to concrete optimization criteria. We discuss utility-based network performance optimization as an architectural component in more detail later. For now, we remark that the utility function approach makes it possible for the CRM to reason about the interrelationships between numerous network characteristics (such as throughput, delay, packet error rate, or jitter) and their impact on application performance. Classical SLAs allow the CRM only to check if application requirements are fulfilled according to hard limits, which is not sufficient information for networkwide optimization of the use of the available resources. Information obtained on application requirements through CAPRI is a key component of NKRL by making knowledge on network performance as experienced by users of the applications visible to other elements of the architecture. The NKRL with the core CRM is a key component to be carefully implemented to have successful CRN architectures. It is clear also that especially NKRL development work requires significant standardization efforts to provide interoperability for CRNs.

Common Control Channel

Many cognitive radio architectures also introduce a common control channel (CCC) as an integral component of the architecture. In principle, control channels can be introduced as specific physical channels or logical channels, where the control information is multiplexed to the normal traffic channels. Especially in the case of DSA-based networks the physical control channel seems to be the more logical and favored approach. The principle of CCCs is no different from the use of common control and broadcast channels in the case of cellular systems. The main architectural indication is that there needs to be support to build CCCs and *standardized* channels and protocols for it. In DSA-based networks the presence of a CCC seems to be an almost inevitable approach to solve coordination issues and provide well-defined access, for example, toward spectrum policy and occupancy servers.

Knowledge Repositories and Policies

A workable architecture requires repositories and databases to store the data and knowledge that have been acquired. The storage requirement is both distributed and local, so this must be taken into account when designing interfaces and access methods. In particular, policies can be seen as a part of knowledge. However, as policies often can be quite explicit and regulatory in nature—that is, they might not require any deeper reasoning—we have chosen to show policies as a separate block in Figure 9.1. There has been a lot of work on the policy servers and languages, and a large amount of development work has been done toward implementing reference implementations in a former DARPA XG-project and other initiatives [110, 336].

The reader should note that, although spectrum policy languages have been attracting the main attention, those are not the only policies that need to be described. A lot of policies in also the commercial domain need to be described and terminal hardware and software configuration and access rights are quite natural places to use policy languages. Describing complex accountability and commercial policies especially requires further research also on the architectural level.

General Comments

Designers of CRN architecture have to also face the problem that characteristic timescales between different layers and processes are highly different. The timescale of physical layer control processes that fight against the fading effects are orders of magnitudes shorter than the typical timescales at the transport and application layers. This means that the architectural components must be able not only to handle different timescales but also allocate resources and facilitate resolution of conflicts in a highly dynamic environment. Some sort of real-time sharing and synchronization functionality is also required, since many processes need to be synchronized. If channel allocation changes in the DSA domain were done in unsynchronized fashion, this would lead to catastrophe. A similar example can be made in the case of dynamically changing a protocol stack or reconfiguring software-defined radio; receivers and transmitters must agree on a synchronized fashion on these changes, otherwise the optimization benefits can be significantly reduced and in the most severe case the communication link lost.

In the classical telecommunications standardization it is common to separate operations into user, control, and management planes. This is a rather general separation of different functionalities. The cognitive radio networks are clearly introducing a need to define new planes, as some of the aforementioned functionalities do not fit well into this previous u-, c-, and m-plane differentiation. In particular, it is quite clear that different knowledge-related operations and repositories cannot be seen strictly as a part of "just" the management plane, but rather form a new plane. Part of reasoning and optimization operation of cognitive radios, including decision making, can be seen as components of the aforementioned knowledge plane proposed by Clark et al. [323]. However, in our opinion these functionalities are better described by introducing the core CRM, since the knowledge plane is a somewhat more general abstraction. One could also see the core CRM as a sort of integrated *optimization and reasoning plane*.

9.2.2 Architectures for Spectrum Sensing

While cognitive radio networks are not only about dynamic spectrum access, providing support to DSA certainly plays a major role in any future CRN architecture. We therefore now discuss various alternatives proposed in the literature, together with some of our own work, on the specific architectures for spectrum sensing. These techniques can be used either for DSA in a traditional primary-/secondary-user context or in unlicensed bands to improve the utilization of the spectrum. We divide our discussion into two parts, first briefly commenting on technologies that individual cognitive radios can use for measuring spectrum usage at their location, then discussing the problem of fusing these individual measurements on the network level.

Spectrum Sensing on the Node Level

A single node can try to infer spectrum occupancy using a large number of techniques. We do not go into the details of the algorithms and physical layer designs involved (for those we refer the reader to Chapter 4 of this book), but rather focus on pointing out the architectural implications involved in choosing the techniques used. The simplest approach, already realizable with cheap commodity hardware, would be to use a single transceiver for both spectrum sensing and actual communications. This has two major drawbacks. First, communications are interrupted for the time used to sense the spectrum. Second, at the present time this usually limits the available spectrum sensing techniques to simple energy detection [337]. The latter problem can obviously be overcome by extending the transceiver design to support feature detection or detection of pilot signals, if used by the technology being sensed [115, 338–341]. The severity of the problem of interrupted communications depends on both the reconfiguration time of the transceiver and the behavior of the primary user. Detecting continuous transmissions could be realized in a short time, but for detecting bursty activity, longer sensing times obviously become necessary.

These problems in use of a single transceiver brought about considerable interest in architectures that include an additional *spectrum sniffer* receiver unit. The role of the sniffer is to gather information about spectrum usage in the various frequency bands potentially available for communications using a deterministic or randomized sampling scheme. While such a design would allow monitoring spectrum usage while actively communicating, the drawbacks are (minor) added complexity in the node design and increased energy consumption.

There are some architectural consequences of having spectrum sniffer or using primary transceiver for monitoring, although mostly the consequences are more practical implementation and interface related. One of the architectural issues is how the spectrum measurement information is shared and what kind of representation is given for these data. Another issue ultimately left for a system designer is how the spectrum occupancy information should be processed. A simplistic way would be to do just a carrier-sensing type of instantaneous monitoring to decide on spectrum opportunities. However, the true CRN architecture should not stop on this, but it should process, and probably store for relatively long time periods, occupancy information to generate *spectrum usage models* to learn to use spectrum more efficiently.

Spectrum Sensing on the Network Level

The fundamental problem that no spectrum sensing design operating on a level of a single node can solve is that the nature of the wireless channel might prevent it from detecting even a nearby primary user. Severe shadowing, fading, or simply the hidden terminal problem can prevent the reception of a signal from a primary user whose client would be interfered with, should the cognitive radio choose to transmit on the frequency band in question. As a solution to this problem, cooperative sensing has been proposed and studied by several research groups [342–344].

The rationale behind cooperative sensing is that, for cognitive radios that are sufficiently dispersed in space, it is highly unlikely that all of them experience deep fades or other aforementioned hindrances simultaneously. The distance scales required to reach significant gains from cooperation depend on the correlations in spectrum occupancy in different locations, a topic to which we return in greater detail later. Nevertheless, it has been clearly established that significant gains in sensing efficiency are achievable through cooperation. Lately the research focus has also started to tackle problems such as how to optimally fuse multiple spectrum occupancy measurements to obtain the most reliable results. For examples of this recent interesting line of work, we refer the reader to [345, 346].

Networking between cognitive radios also can be used to enhance spectrum sensing in ways not directly related to the cooperation gains. Policy information, models of existing wireless technologies, and regional measurement data on spectrum usage can all be made available through the network and used to further optimize the sensing process. A highly interesting example proposal in this space is the radio environment map work from Virginia Tech [347, 348].

All these forms of collaboration impose various architectural requirements. Most obvious of these is the need to communicate between the individual cognitive radios. This requires coordination in the form of a control channel, either established using a dedicated technology or in logical fashion on the main communications channel. Agreement on the representation used and the information collected and exchanged is also needed. This involves some challenging problems, such as the estimation of the accuracy of the measurements or trustworthiness of the information obtained from the network. Spectrum sensing can also generate vast amounts of data, and blind dissemination of these data throughout the network would result in a significant waste of resources. Deciding on the type of preprocessing applied to the measurements and the type of data actually exchanged are challenging problems for the architecture design of CRNs. We discuss some of these problems in more detail.

Distributed spectrum sensing can rely on three basic architectural concepts. The spectrum sensing can be done in peer-to-peer fashion, where the spectrum sensing is done by actual cognitive radios, which then share their information either directly among peers or through some spectrum servers or databases. Another approach is to introduce dedicated spectrum monitoring devices, "observatories," which are distributed as a service around the country (see, e.g., Chapter 2 of this book). This approach requires infrastructure deployment but could solve many obvious issues on the sensitivity, reliability of the data, and privacy. A third possibility is a hybrid solution between these two architectural selections. At the abstract level, the implementations of the different deployment strategies are not very different, but in practice there are differences, especially if one is considering the value chains and business models required. Distributed spectrum sensing generally shares some of the problems with wireless sensor network architectures, where we have to find the right deployment models and reasonable use of resources to achieve sufficient accuracy. This indicates also that cognitive wireless networks need to be location and environment aware when deciding how many spectrum monitors are required and where those should be located. This is discussed later in this chapter, where we discuss topology-aware CRNs.

9.2.3 Network Optimization through Utilities

Optimization of resource usage and user satisfaction are key objectives any cognitive radio architecture must be designed to support. This clearly brings in the requirement of specifying performance objectives of applications in a manner that lends itself directly to optimization techniques. NKRL needs to describe goals and objectives of different entities in the network both in the local and end-to-end contexts. Although some of the goals can be described with policies or constraints, some of the objectives must be expressed in a form that is quantitative and can be used for mathematical optimization. One of the most general ways to accomplish this is to associate to each application a *utility function*, mapping the measurable *attributes* of the network connection the application is using to a numerical value. Common examples of key attributes of network connections include throughput, delay, cost, power consumption, and various error rates. For some applications throughput is the most important attribute, and the corresponding application utility function is usually taken to have the form

$$U(\text{throughput}) = A \log(\text{throughput}/B), \tag{9.1}$$

where A and B are application-specific constants. The logarithmic dependency of the utility on network throughput is in line with the intuition that utility is proportional to the price the user would be willing to pay on the connectivity with those attributes. An example of an application with a more complicated utility function would be interactive videoconferencing sensitive to both throughput and delay.

Utility functions are simple but general tools for formalizing application requirements. Employing them in a cognitive radio network architecture requires means for both expressing utility functions of different levels of generality and mechanisms to associate applications (in most operating systems, meaning either processes or threads) with utility functions. In the CRM architecture discussed previously both these functionalities are offered through the common application requirement interface. For CAPRI a utility function is given as a string containing the description of the function in a language basically forming a subset of the textual notation for mathematical expressions typically used in modern computer algebra systems. The key elements are the fixed set of attributes represented by strings of one or more letters (t for throughput, d for delay, etc.). These can be combined with integers and real numbers (in the commonly used dotted decimal or scientific notations) through the usual basic arithmetic operations (+ for addition, − for subtraction/negation, * for multiplication, / for division, and ^ for exponentiation) together with parentheses. Finally, some predefined functions are introduced to facilitate the expression of common types of utility functions. Examples of these are $\log()$ for logarithms (as seen already, logarithms appear frequently in specifications of utilities) and the Iverson bracket notation $[\langle condition \rangle]$ evaluating to 1 if the condition is satisfied and to 0 otherwise [349]. For expressing the conditions the usual expressions $=$, $!=$, $<$, $<=$, $>$, and $>=$ are used. The value of the Iverson notation is that it makes it easy to express collections of step functions.

Naturally, the CAPRI-based approach is explained here only as a representative example. There are different ways to implement utility-based descriptions, but it is most likely that any advanced CRN architecture should be able to support this sort of description. One of the architectural consequences is that a utility type of description would allow better handling of competing goals and it requires the capability to interpret utility descriptions in the various network elements.

Specification of the optimization targets in the form of application utilities is only one component in utility-based optimization. Two additional key problems to be tackled are the aggregation of application utilities in the usual case of multiple applications or users sharing resources and actually solving the resulting optimization problem. Several design options are available for formulating the common optimization problem from individual application utilities. First is to apply techniques from multiobjective optimization, interpreting each application utility as an individual quantity to be maximized. Second would be to specify a function aggregating individual utilities to the overall utility function of the user. The simplest such function would simply sum together the individual utilities, a choice that can easily be extended by adding weights or various fairness criteria. Both of these options are again available when forming the network-wide optimization problem in architectures designed to tackle directly with global optimization.

After the relevant optimization problem(s) have been formulated, the remaining task of the network is to arrive at a solution to the problem. Unfortunately, many of the classical approaches for finding maxima for the aggregated utility function do not directly yield useful algorithms and protocols, since the attributes cannot be changed directly. Instead, each node in the network has a number of *parameters* that can be changed, ranging from choices of protocol parameters to selection of the configuration of the available radios. All these parameters influence the attributes and thus utilities, but in general the transfer function involved is unknown and is further influenced by externalities such as interference levels, which can be difficult to quantify directly. Metaheuristics, such as genetic algorithms or simulated annealing, are well suited for solving such black-box problems, but applying them blindly typically causes the network to be in an unusable state for some time as the solution space is being explored. An alternative is to combine historical information with theoretical models to estimate the responses of attributes to parameter changes and apply these models in the optimization process. In the case that the models yield smooth-enough functions, classical optimization methods again become relevant. Otherwise, metaheuristics can be applied together with the models (yielding *hybrid* optimization techniques) to improve the convergence and quality of the intermediate solutions. It should be pointed out that such combined approaches are inherently probabilistic in nature, resulting in concepts such as *expected utility*, defined as the expectation value of the utility following an action with an uncertain outcome.

Utility-based optimization has also been used to obtain architectural insight into various layered and modular networked systems. For a very readable survey on these activities, we refer the reader to [350]. Here we confine ourselves to giving a single example on how various TCP protocol flavors can be understood as solving a utility optimization problem. Given that the utility is a function of source rate x_s only (a common albeit somewhat restrictive assumption in earlier work), the basic network utility maximization problem due to Kelley et al. [351] requires maximization of $\sum_s U_s(x_s)$ subject to the constraint $Rx \leq c$, where R is the routing matrix, x is the vector of source rates, and c is the vector of link capacities. The Lagrangian dual problem [352] now is to maximize

$$D(\lambda) \equiv \sum_s \max_{x_s \geq 0} \left[U_s(x_s) - x_s \sum_l R_{ls}\lambda_l \right] + \sum_l c_l\lambda_l, \tag{9.2}$$

where λ_l are the Lagrange multipliers. It turns out that λ_l have a direct interpretation as congestion measures and that different TCP variants make different choices regarding the computation of λ_l. For example, TCP Vegas estimates the congestion by queueing delays. Also the utility functions used by different TCPs vary slightly, but most of them belong to the one-parameter family.

$$U_s(x_s) = \begin{cases} w_s \log x_s & \alpha_s = 1 \\ w_s(1 - \alpha_s)^{-1}x_s^{1-\alpha_s}, & \alpha_s \neq 1, \end{cases} \tag{9.3}$$

where w_s is a positive weight [353].

Apart from supporting utility function–based descriptions, the future CRN architectures must provide tools and descriptions to build utility functions. This is not trivial in general, as finding the suitable utility functions can be difficult. It is even more challenging if the utility function should describe, for example, the end-to-end monetary value of connectivity over different networks. The stakeholders of the current value chain are not inclined to share sensitive cost and price information with external entities even if we provide mechanisms for that.

9.2.4 Value of Perfect Information

While utility-based optimization can be applied by individual nodes in a black-box fashion, the gathering of historical data and exchange of information between the nodes should result in better models and in turn better solutions to the optimization problems at hand. However, gathering such information cannot be accomplished without cost. For local measurements the used resources are usually energy consumed by additional sensing and possibly the down time in communications if a single transceiver is used for both communications and measurements of the state of the wireless channel. If nodes further exchange information with each other, communication resources are used in greater and greater quantities and the larger the range of information exchange becomes. We can formalize the arising trade-off as the question of the *value of perfect information* (VPI) or, equivalently, as *price of ignorance* [354, 355], as follows.

Let us denote the expected value of the utility function U after the best possible action is taken based on knowledge K as $EU(K)$. Here we take best possible action to mean any change in configuration yielding maximum expected utility based on the probabilistic models formed based on K. Suppose additionally that random variable X can be observed. The improvement in the optimization result can be quantified by studying the difference $EU(K, X) - EU(K)$ in the expected utilities. Of course, when deciding whether collecting information about X is beneficial, its value is not yet known, so $EU(K, X)$ should be interpreted as the expectation over the different realizations of X as well. With this interpretation $EU(K, X) - EU(K)$ becomes precisely the classical definition of the VPI. Using this difference a cognitive radio network can quantitatively estimate whether it is beneficial to obtain more information before making a decision, or if the cost of obtaining such information no longer is justifiable by the expected gain in utility.

The architectural implication is that architecture and its components need to be aware of resource costs. This implies that all optimization and decision-making processes cannot be independent and that coordination is required on deciding when certain actions, communications, and feedback loops are efficient to use. In the CRM framework this is handled by the core CRM, but in general the take-home lesson is that the optimization modules need to be coordinated and cost/benefit analysis needs to be done in a more centralized fashion.

The costs related to exchanging optimization and signaling data are an issue that sometimes have been overlooked with different cognitive communication architecture proposals, particularly in the fixed networking domain. However, the cost issue needs to be taken seriously in the case of wireless networks. As a more specific recommendation, we believe that instead of measuring too much the cognitive architectures should also support more intelligent *estimation* of different operational states and parameters.

9.2.5 Policy Support as a Part of the Architecture

Especially in dynamic spectrum access, policy issues play a major role and should be carefully considered in any CRN architecture. Even in seemingly unregulated frequency bands, such as the 2.4 GHz ISM band, there are significant regional differences on allowable frequencies, transmit powers, and the like. This creates a need for the individual CRs to be able to obtain and reason about different local policies in a flexible manner, as fixed sets of policies reduce the evolvability of the system significantly. The related architectural requirements involve having a common way to represent policies and an infrastructure from which different policies can be obtained. Usually solutions based on the use of dedicated *policy languages* are proposed in the literature, often combined with specific *policy servers* that CRs can contact to obtain relevant policies. For an overview on such architectural issues specifically in the cognitive networks context we refer the reader to [332]. Policy issues should also be carefully considered, since earlier experiences indicate that the involved architectural components tend to become very complicated over time, the most obvious example of this being the border gateway protocol (BGP) [356] used to connect autonomous systems to form the Internet. Proliferation of different policies and unforeseen interactions of different policy statements made BGP very difficult to operate and deploy correctly but also made it a valuable case study in designing policy-based systems. CRNs and CRs will use policies very commonly, therefore it is advisable to provide architectural mechanisms to prune, aggregate, and verify rules automatically and use different reasoning methods to modify rules when necessary to avoid the known policy table problems of BGP.

9.2.6 Spectrum Brokering Services

In previous sections we primarily discuss architectural issues for networks in which dynamic spectrum access takes place based on local or distributed measurements, possibly in accordance with certain policies. A complementary approach, explored at length for example by Buddhikot et al. [357, 358], is to use dedicated *spectrum brokers* to control access to the spectrum.

Spectrum brokers distribute leases to individual cognitive radio networks to use "parts" of the spectrum, which can be defined as sets of frequency blocks over certain time and space but also in terms of other multiplexing techniques. The decisions of the brokers can in principle be based on a number of criteria, although by far the most common proposal has been to employ market mechanisms to allow stakeholders such as operators or users of CRNs to purchase or partic-ipate in an auction on spectrum resources. The proposed DIMSUMnet (dynamic intelligent management of spectrum for ubiquitous mobile access network) [357] architecture is a good example of spectrum brokering subarchitecture of CRNs. The authors introduce a hierarchical spectrum brokerage service that could very effi-ciently open spectrum markets, and presumably spectrum brokers could provide other (spectrum) policy information for devices that participate in the DIMSUMnet infrastructure. DIMUSUMnet is a cellular architecture (see Figure 9.2) that includes spectrum information and management broker(s) and new modified radio access network managers that are DIMSUMnet compatible. The base stations and termi-nals also need to be DIMSUMnet architecture capable to benefit from the system. However, there is probably no architectural limitation to include legacy system support to the system through suitable extra components, which would allow DIMSUMnet architecture and legacy cellular networks to be aware of each other and have at least a limited amount of cooperation.

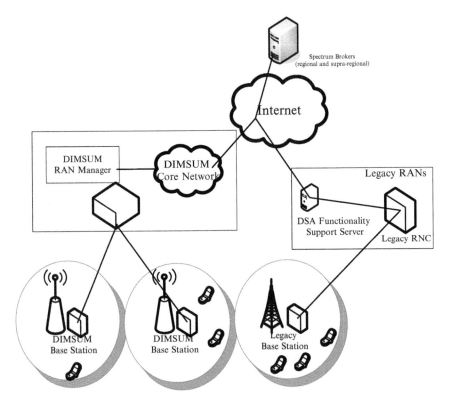

FIGURE 9.2

Modified DIMSUMnet architecture concept, partially adapted from [357].

We go into no further details of the design alternatives here, since such decisions are more policy and market than technology driven. From the point of view of the present chapter, architectural implications and requirements related to spectrum brokers are very similar to the policy servers discussed previously.

9.2.7 Information Modeling

We see that cognitive radio networks may have to deal with very diverse types of information. Because of this, we briefly outline some of the possible ways that different kinds of information can be modeled, stored, and exchanged. In the cognitive radio domain the first integrated attempt at information modeling was the radio knowledge representation language of Mitola and Maguire [329]. It arose as an effort to remove the limitations and integrate the strengths of earlier modeling languages, such as SDL, UML, and KQML [359], each of which targeted specific categories of information modeling, such as ontologies or relationships between objects. Mitola later extended his work toward the more expressive radio XML [182]. This work yielded a flexible and rich framework for organizing information, although at a price of rather high verbosity, as is usual for XML-based designs. In several areas more dedicated information models and structures might be more appropriate. Examples of such areas are various Bayesian and belief-based models, using which cognitive radios can encode uncertain information and later reason about such information or knowledge.

An interesting class of information models that appears to be particularly appropriate for optimization problems in cognitive radio networks is that of *graphical models* [360]. Knowledge or belief of relationships between observations and input parameters, interpreted as random variables, is represented in terms of a graph, with vertices corresponding to the different variables and edges encoding relationships. Bayesian networks and factor graphs are particularly common but by no means the only examples of graphical models with applications in cognitive radio networks. In our own work, for example, we applied Bayesian networks to reason about the relationships between parameters and attributes in utility-based optimization [361].

Some of the information relevant to cognitive radio networks can be rather complicated in nature. For example, when evaluating the value of perfect information, the network has to store rather rich statistical information about the different observables, actions, and their projected consequences. These models might include probability distributions on the space of utility functions (when estimating the expected behavior of other radios of which the users' utility functions are not known) or spatial statistics with uncertainties (see the next section for examples). There is clearly room for research on suitable models for such information, especially given the restrictions in complexity imposed by the wireless environment. Hence, we believe that NKRL is required to deal with the information modeling and provide framework for architectural descriptions. Naturally the actual implementation of NKRL can take diverse forms, but as an architectural component it seems inevitable for CRNs.

A problem related to the information models employed is the actual storage and transport of the information in the network. Exchanging large amounts of information models, spectrum knowledge, and measurement data over a wireless medium could easily make CRNs prohibitively expensive and inefficient. Therefore, there remains the architectural challenge to minimize the exchange of data to the absolute minimum needed to make good decisions.

9.3 TOPOLOGY-AWARE CRN ARCHITECTURES

Much of the decision making and optimization in cognitive radio networks involves reasoning about the spatial structure of the networks involved. For example, question of dynamic spectrum reuse fundamentally involves reasoning about interference probabilities, which are closely related to questions on proximities of nodes and the coverage and interference ranges of their respective transceivers. Often the spatial nature of these problems is ignored on the architectural design level, and the decisions involved are made directly on the indicator functions of events corresponding to causing interference or by representing the dependencies of channel capacities in different areas of the network by relational models. While simple, the drawback of such an approach is that it is very difficult to exploit prior information available on the spatial structure of networks and spectrum utilization in different regions.

In this section we argue that it is both possible and worthwhile to consider CRN architectures that are explicitly *topology aware*, which we take to mean awareness of spatial structure of the networks and the underlying phenomena involved. Such topology awareness can be thought of as an extension of the radio environment map concept [347] intended to provide cognitive radios increased awareness of their surroundings. The utilization of location information in CRN has also been discussed by Celebi and Arslan [362]. Here we do not go into architectural issues of using exact locations of devices, as these are covered by aforementioned article. Our focus here is to ask if the architecture should also characterize CRN and the location of its components at the macro-level more statistically. We believe that such statistical characterizations of network topology can be used to significantly improve the performance of cognitive radio networks with limited overhead. We begin by demonstrating that both spectrum usage and network structure can indeed be characterized by means of spatial statistics, after which we discuss how these techniques can actually be used in CRNs.

9.3.1 Statistical Characterization of Node Locations

Locations of nodes, whether nodes of a cognitive radio network, primary-user transmitters, or clients of the primary users can be described statistically by interpreting them as a realization of a *point process*. The operational definition of a point process is essentially that of a random variable N, the values of which are collections of points in $D \subseteq \mathbb{R}^n$. The *distribution* of N assigns probabilities for different point configurations, and it can be summarized using various statistics based on interpoint distances, correlations between node locations, and so on. For a comprehensive introduction to point processes we refer the reader to the excellent monographs [363, 364]. Typical applications of point processes related to CRNs include the following [365–367]:

- Estimating statistics of location distributions and using those to estimate other network characteristics.
- Constructing models of node locations for reasoning or simulation.
- Forming spectrum occupation models based on fitted point process models.

We consider each of these in turn. Throughout this section we use the data set depicted in Figure 9.3 as our example case study. It consists of the locations of

FIGURE 9.3

Excerpt from the FCC database on FM radio station transmitter sites.

15, 535 radio station transmitter sites obtained from the FCC database. We chose this particular data set due to its relevance for DSA applications, with frequency bands for broadcasting often being cited as prime candidates for spectrum reuse. We use this data set to introduce various statistical characterizations of point processes, followed by a more detailed look at fitting synthetic models to the data for reasoning and simulation purposes.

Point processes can be characterized using a wide variety of statistical tools. The most common examples of such statistics are various *distance-based* and *second-order statistics*. For distance-based statistics we typically either consider the distribution of the distance from a randomly selected point of U into the nearest point of N or the distribution of the distance from a randomly selected point of N to its nearest neighbor. The former yields the *spherical contact distribution function* $H_s(r)$, whereas the latter gives the *nearest-neighbor distance distribution function* $D(r)$. Being distribution functions, both of these vanish at origin and approach unity as r becomes large. Another common distance-based statistic building on these two distribution functions is the J statistic [368], defined by

$$J(r) \equiv \frac{1 - D(r)}{1 - H_s(r)}. \tag{9.4}$$

For totally random distribution of points (still commonly assumed in simulation and analysis work) we obviously have $J(r) \equiv 1$, whereas lower and higher numbers indicate spatial clustering or regularity, respectively. The J function for the FM radio station transmitter sites is illustrated in Figure 9.4. The rapid decay of $\hat{J}(r)$ is an indication of heavy clustering, in part induced by the underlying population distribution.

While the distance-based statistics just introduced are certainly expressive, their use is not without problems. They are not as robust as second-order statistics

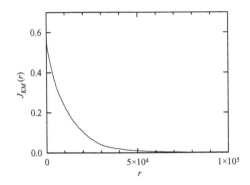

FIGURE 9.4

Estimate of the J function for the FM radio station transmitter sites.

and their power in modeling applications is limited. The cause for this is their myopic nature; both the nearest-neighbor distribution and the spherical contact distribution function are determined by the distances to the nearest point only, ignoring all the distance information to farther away neighbors. Because of this, we briefly introduce the basic second- and higher-order statistics that can be used to describe point patterns. Perhaps the most fundamental second-order statistic is the *pair-correlation function* $\xi(r)$, defined by the joint probability density

$$dP = v^2\left[1 + \xi(r)\right] dA_1 \, dA_2 \tag{9.5}$$

of finding one point in each of the two area elements dA_1 and dA_2 (v is the *intensity* of the process, giving the mean number of points per unit area). Clearly if $\xi \equiv 0$ no obvious correlation is present as one would expect for, say, the Poisson point process with complete spatial randomness. The definition of the pair-correlation function can, of course, be extended to the case of n-point correlations. For example applications of pair correlation functions in studying wireless networks in general and for the accompanying discussion, see [366, 367].

Another commonly used second-order statistic related to the pair-correlation function is Ripley's K function, defined by

$$\xi(r) \equiv \frac{1}{db_a r^{d-1}} \frac{dK(r)}{dr} - 1, \tag{9.6}$$

where b_d is the volume of the d-dimensional unit ball. This normalization is chosen to have $K(r) = \pi r^2$ for the spatially random Poisson case. Figure 9.5 shows the estimate of the K function of our data set compared to the Poisson expectation, further illustrating the clustered nature of the data.

Apart from using spatial statistics to describe in the compact-form radio environments, we could use them to provide pseudo-models for different situations, something that is particularly important for model-driven cognitive systems. Based

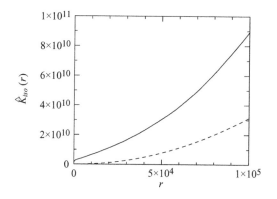

FIGURE 9.5

Estimate of Ripley's K function for the data set of FM radio station transmitter sites. The dashed line shows the theoretical value for the case of complete spatial independence (Poisson case).

on the statistical characterizations just described it is possible for cognitive systems to automatically fit point process models to data and use those models in subsequent optimization problems. For an introduction to the applicable fitting techniques with examples in the cognitive radio domain, see [365].

9.3.2 Spatial Statistics of Spectrum Usage

The theory of point processes yields a firm foundation for analyzing and modeling location distributions of primary users and nodes of cognitive radio networks. However, they are not appropriate for all spatial reasoning in topology-aware CRNs. As an example we can consider the analysis and modeling of spectrum usage, defined as the mean power spectral density (PSD) values measured for a given frequency band at different locations [369, 370]. We can treat such measurement results as a realization of some unknown *random field*. In general, random fields can be thought of as extensions of the usual theory of random processes from one-dimensional *time* to multidimensional *space*. In our case this space will be some domain D in either $\mathbb{R}^2$ or $\mathbb{R}^3$, depending on the setup of measurements and the models in question. As usual, the probabilities involved can either be interpreted as "true" randomness or simply as a model for the incompleteness of our knowledge on all the details of the system under study (Bayesian viewpoint).

Random fields can be studied in a very similar fashion to the node locations discussed earlier. Statistical characterization, quantifying for example the correlations of spectrum usage at different locations in space, can be carried out and exploited in prediction and estimation tasks. Random fields can also be modeled based on these statistics, allowing for generation of simulated instances of the mean PSD fields with realistic characteristics to be used, for example, in "what-if" analysis.

Such models are also valuable in knowledge representation and storage, as they can be used to create in a sense compressed descriptions of spectrum usage. They can also help alleviate privacy concerns as sharing detailed information on spectrum usage could be problematic for some of the stakeholders in the DSA domain, for example. Statistical models abstract away many of the potentially sensitive details while still containing the essential information from which cognitive radio networks can benefit. For a detailed introduction to the use of these techniques we refer the reader to [369, 371].

9.3.3 Applications and Discussion

Cognitive radio networks can benefit from topology information in several ways. Both point process models and random fields can be used to reason about interference probabilities as well as various statistics related to network connectivity. Location distribution can also be estimated depending on the envisaged application for both primary users and for the CRN nodes themselves. Technically such estimation is not difficult, as standard collaborative localization techniques can be applied. Spectrum models can be used to enhance both spectrum sensing and spectrum usage, given that suitable propagation models are used as well. All these applications have close connections to the required models for information representation and exchange and cannot therefore be considered in isolation in architecture design. They also require methods for CRN nodes to obtain and share the relevant information. All these considerations manifest themselves in the design process for the architecture as there is the need to have the relevant "toolboxes" in place.

9.4 PUBLISH-SUBSCRIBE CRN ARCHITECTURE

While cognitive radio techniques discussed here can certainly be used to optimize performance on the node level, we expect that practically all arising future CRN architectures will support direct communications among CRN nodes. We now briefly discuss the architectural design decisions especially related to the support of signaling or control plane interactions. Perhaps the most significant design decision on the architectural level is the choice of the overall communication paradigm. The Internet and most deployed wireless networks utilize inherently node-centric communications, in which packets are explicitly routed between nodes or *locators* of topological significance. To send data to a node in the network the source has to either know the address of the recipient or employ a *resolution service* mapping names and other types of identifiers to usable addresses or locators. While this paradigm has been very successful in present-day networks, it might not be the most natural one for signaling architecture in CRNs, where the addresses of the individual nodes would appear less important than the *type of the data* communicated.

An interesting alternative of a more data-centric nature would be to use publish-subscribe communications in the signaling architecture. In *publish-subscribe* (pub-sub) networks nodes explicitly inform the network of the kind of data they are interested in receiving; that is, they *subscribe* to the data. This subscription might be *topic based*, making distinction only between different types of data, or *content based*, allowing for additional filtering based on the actual content of the data as well. When sending or *publishing* data to the network nodes no knowledge of the potential receivers is required. In the case of a topic-based system nodes simply inform the network on the type of data they are sending, and the network protocols are then responsible for delivering the data to all subscribers. There are numerous variants of this basic concept (see, for example, [372] for an overview), but the basic feature in all of them is the decoupling of senders and receivers by focusing on the nature of the data instead of addresses or identities of the individual nodes.

9.5 CHAPTER SUMMARY

Cognitive radio networks are an exciting new development in networking, in many ways calling for new architectural solutions and concepts. In this chapter we discuss some of the key concepts, such as flexible control architectures and utility-based optimization, that we believe are central in any future CRN architecture. We also cover some slightly more speculative issues, such as using statistical topology information in designing the optimization processes involved. From these examples it appears that the greatest paradigm shift CRNs bring is the inherent need for flexibility and the requirement for architectures to allow major parts of the operational functions to be composed at runtime. It is unlikely that solutions such as classical radio resource management designs are sufficient to fully unleash the potential of CRNs. Rather, the crucial issue is to arrive at a common understanding of key interfaces and abstractions that allow the most flexibility in the actual instantiations of the architecture while guaranteeing interoperability among the different instances.

9.6 PROBLEMS

1   What are the main architectural components of cognitive radio networks?

2   How does the proposed cognitive resource manager framework provide modularity in its architecture?

3   What are the main architecture-level benefits on introducing utility functions to cognitive radio networks?

4   Why is the concept of value of perfect information, or price of ignorance, important when designing cognitive radio network architectures?

5 Give examples of the different ways one might choose to measure the "cost" of exchanging signaling information among CRN nodes.

6 In addition to the *vertical* decomposition into layers discussed in the text we often choose to decompose layers *horizontally* into various domains. Discuss why this is done and give at least three examples of different layers of horizontal decompositions taking place in the current Internet. How might these decompositions change after widespread adoption of CRN technologies?

7 Is cross-layer optimization the defining factor of cognitive radio networks?

8 Is there in general a difference in trying to detect the presence of a broadcast transmitter (such as radio or TV station) compared to a packet radio transmitter? Which type of transmitter does UMTS Node B resemble most?

9 We discuss the correlation structure in the location of radio transmitters. To what extent would you expect these correlations to arise from the relation to the underlying population distribution (which obviously is not uniform)? Would your answer be different if short-range technologies or bidirectional communications were considered?

10 How can one efficiently parameterize the clustering of cognitive radios?

User cooperative communications

Elsheikh Elsheikh[1], Kai-Kit Wong[1], Yangyang Zhang[1], and Tiejun Cui[2]

[1]University College London, United Kingdom
[2]Southeast University, China

10.1 INTRODUCTION

User cooperative communication is a form of communication in which users work together to deliver their data. Cooperative diversity is accomplished by having a node act as a relay to forward the received information from the source. The signal received directly from the source is combined at the destination with that forwarded by the relay to retrieve the transmitted message. Recently, the application of user cooperation techniques in wireless systems has received much attention.

The subject of providing diversity in reception to remedy channel impairments has been investigated for decades. Diversity can be obtained over space, time, and frequency. In fading channels, diversity reduces the risk of being in a deep fade by offering a number of independent copies of reception (see, e.g., [373]). Relaying is one way to obtain the benefits of space diversity, in which the independence of the channel responses from the source and the relay nodes to the destination node is exploited.

Relay channels have been recognized since the 1960s. A simple single-relay channel models a class of three-terminal communication channels originally examined by van der Meulen in [374, 375]. Cover and El Gamal considered discrete relay channels with additive white Gaussian noise (AWGN) and derived the achievable rate based on some random coding techniques [376]. More specifically, they also derived the capacity for the physically degraded class of channels. Later, more discussions on capacity and capacity-achieving codes appeared in [377–381].

While developing a proper model, deriving capacity, and designing coding strategies were the main concerns of previous studies, more recent work has also taken consideration of other issues and extensions, such as multiple relays and multihop transmission [382–390], resource allocation [390–395], multiuser relay networks [396], relay selection [397, 398], coding [399–401], and cross-layer issues [390, 402].

In multiuser systems, user cooperative transmission offers a flexible and dynamic alternative to obtain some of the advantages of spatial diversity, especially when hardware restrictions prevent the use of multiple antennas. This is more pronounced in self-organized networks, formed by a number of mobile terminals with no preexisting infrastructure (e.g., mobile ad hoc network, MANET). Studies by King [403], Carleial [404], and Willems et al. [405–408] examining multiple-access channels with generalized feedback can be related to the cooperative model [409]. Arguably, the work by Sendonaris et al. brought user cooperative communications to attention and renewed the interest in relay channels. In their two-part papers [410, 411] they presented an extensive set of simulation results demonstrating the great potential of cooperative diversity and discussed some implementation issues. The work by Laneman et al. in [412] is another significant contribution. They developed and analyzed low-complexity user cooperative diversity protocols, based on amplify-and-forward (AF) and decode-and-forward (DF) signaling, for delay-constrained wireless channels. Later, Nosratinia and Hunter proposed coded cooperation in which cooperation operates through channel coding in the spatial domain [413]. Instead of repeating the received bits (as in AF and DF), the cooperating node sends an incremental redundancy for its partner. Hunter et al. studied outage probability for the coded cooperation in [414]. Recently, an explosive amount of work has been done in regard to relay channels and user cooperative communication, for example, [393, 394, 412–421]. Due to a mutual relationship, the emerging of user cooperative communication revitalized researchers' interest in relay channels. In fact, most of the recent work on relay channels is motivated by user cooperative communication.

10.1.1 Diversity

It is known that detection error probability decays exponentially in received signal-to-noise ratio (SNR) in a single AWGN channel, while it decays only inversely with the SNR in fading channels [373]. That means increasing transmission power is inefficient in combating the effect of fading. Diversity techniques, on the other hand, are very effective on tackling the problem. Probability of error can be reduced by combining several copies of the signal transmitted over independent fading paths.

Diversity can be obtained over time, frequency, space, or any combination of these dimensions. *Temporal diversity* is obtained by retransmitting data packets in time intervals greater than channel *coherence time* (time over which the channel changes significantly). It can also be achieved via coding and interleaving, where information bits are dispersed over time in different coherence periods so that different parts of the code word experience independent fades. Analogously, in *frequency-selective* channels, diversity is achieved by dividing the channel into a set of orthogonal subcarriers each experiencing narrowband *frequency-nonselective* (or *flat*) fading. Information bits are then repeated or interleaved across subbands.

Spatial diversity, on the other hand, is achieved by transmitting data streams over multiple independent paths. *Multiple-antenna diversity* is one application that exploits spatial diversity. In cellular systems, for example, multiple antennas spaced sufficiently apart are employed at base stations. *Macro-diversity* is also obtained by combining signals received at different base stations. Other systems with single-antenna devices can acquire spatial diversity through relayed transmission. A relay node offers an alternate transmission path. Diversity obtained via relaying is known as *cooperative diversity*. Alternatively, single-antenna users can attain diversity via *user cooperation* where users help each other by relaying their partners' data. *User cooperative diversity* is the focus of this chapter.

In fading channels, as explained previously, diversity is used as means to improve reliability by repeating the same signal over parallel independent channels. Alternatively, by transmitting independent information streams over these independent subchannels, the data rate is increased. Both types of gains can be simultaneously obtained for a given channel, but there is a fundamental trade-off between how much of each any coding scheme can get. This is a fundamental trade-off known as *diversity-multiplexing trade-off* [422, 423].

10.1.2 User Cooperation and Cognitive Systems

Cooperative communication is emerging as a technology to enhance link connectivity, which would lead to saving in available resources (power and spectrum). Cooperative users offering alternative paths for their partners can significantly improve link reliability and increase throughput without consuming extra resources.

In the context of cognitive radio, secondary users opportunistically exploit the existence of *spectrum holes* to improve spectrum utilization. A spectrum hole is a licensed frequency band that is not being used for some time period usually in a specific area [11]. Moreover, secondary users practicing cooperation can enhance reliability and increase throughput for a given spectrum hole. In a new direction, cooperation can be performed not only between secondary users, but between secondary and primary users. By helping primary users transmit data in less time, more spectrum holes are created for secondary users [424]. In this chapter we are more concerned about the first scenario. Specifically, we look into how secondary users cooperate to improve the performance using available resources.

In a different setting secondary users are allowed to transmit at the same time as primary users given that the interference made by secondary users can be dealt with by primary users. The spectrum hole in this case is described as "gray." To avoid harmful interference to primary users, secondary users are restricted in transmission power when trying to access gray spectrum holes. A pair of secondary nodes might not be able to establish a connection if they are so distant from each other that the minimum transmission power required is larger than the maximum power allowed due to interference restrictions. In that case secondary users can cooperate to convey the message through multihop transmission. Instead of transmitting the message directly to the destination, the source sends the message to a neighboring node, which forwards it to another node and so on until it gets to the destination. *Cooperative multihop relaying* is discussed in Section 10.4.

10.1.3 Chapter Preview

The main purpose of this chapter is to introduce readers to relay channels and cooperative transmission with more emphasis on wireless applications. As these are recently active research topics, material in this chapter also reflects the ongoing research activity. In Section 10.2, some models for the general three-node relay channel are presented followed by discussions on relaying and coding strategies, optimization, relaying performance, and other issues. The wireless relay channel is tackled as a special case of the general relay channel. Section 10.3 shows how the three-node relay model can be used as a basic building block to develop user-cooperative schemes for multiuser networks. First, a two-user-cooperative setup is discussed and applied to an arbitrary multiuser wireless network. Performance results are presented demonstrating the gain from cooperation. Section 10.4 deals with multihop relay channels. Multihopping is necessary for cognitive systems exploiting a gray hole where only low-power transmission is allowed for secondary users. This chapter is summarized in Section 10.5. Finally, suggestions for additional reading are given in Section 10.5 as well as some problems to aid in understanding the material in Section 10.6.

10.2 RELAY CHANNELS

10.2.1 Introduction

In cooperative systems, users help each other enhancing the transmission rate and improving channel reliability. This is achieved by having users act as relays to their partners. A relay channel is thus the basic building block for user cooperative systems, the understanding of which is hence essential for developing effective cooperation schemes. A relay channel consists of a single source node, a single destination node, and at least one relay node assisting the source on transmission to the destination. Relays can also be applied to multiple-access channels and broadcast channels (see Figure 10.1).

The use of relays in transmission started in 1940s in the United States [425]. It was considered an efficient way to extend the range of transmission over long distances. As explained in Figure 10.2, relaying was proposed mainly to tackle

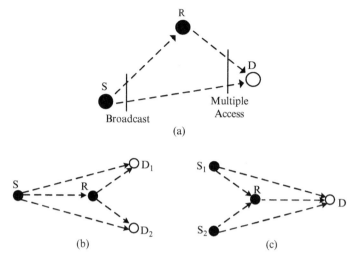

FIGURE 10.1

Graphical representations of channels with a single relay: (a) single point-to-point channel, (b) broadcast channel, and (c) multiple-access channel.

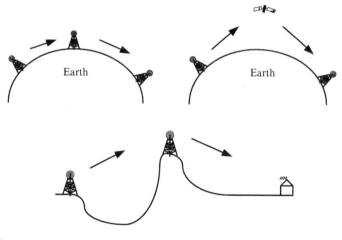

FIGURE 10.2

Example of early applications of relays in transmission over long distances.

problems like the Earth's curvature, path loss, and irregular terrains. The use of satellites as relays for Earth stations in communications added to the importance of this mode of transmission. Recently, relays are also seen as means to increase the transmission rate and save energy.

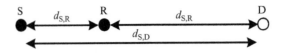

FIGURE 10.3

A linear network with a source, a relay, and a destination node.

Motivating Example

As an example, consider the linear geometry as depicted in Figure 10.3. A source node S is transmitting to a destination node D, and a relay node R is positioned somewhere along the line connecting S and D. In particular, for wireless channels, the received signal power, P_{rx}, at distance d away from the transmitter is a function of the transmitted power, P_{tx}, and is given by [373]

$$P_{rx} = \frac{kP_{tx}}{d^\alpha}, \tag{10.1}$$

where α denotes the path loss exponent (typically ranging from 2 to 6) and k is an appropriate constant that accounts for the antenna pattern in the direction of transmission and other hardware losses. In this example, if a minimum received power, say P_{min}, is required for reliable decoding at the destination, then the transmitted power from the source, through the direct path, $P_{S,direct}$, must be at least

$$P_{S,direct} = \frac{P_{min}d_{S,D}^\alpha}{k}. \tag{10.2}$$

If the relay R is employed, then less power from the source is needed, and in this case,

$$P_{S,relay} = \frac{P_{min}d_{S,R}^\alpha}{k}, \tag{10.3}$$

where $d_{S,R} < d_{S,D}$. The relay transmits at

$$P_R = \frac{P_{min}d_{R,D}^\alpha}{k}, \tag{10.4}$$

to achieve the required performance at the destination. It is interesting to note that the total transmit power, P_{relay}, through the relay path is also smaller than that of the direct path; that is

$$
\begin{aligned}
P_{relay} &\triangleq P_{S,relay} + P_R \\
&= \frac{P_{min}\left(d_{S,R}^\alpha + d_{R,D}^\alpha\right)}{k} \tag{10.5} \\
&\leq \frac{P_{min}d_{S,D}^\alpha}{k} = P_{S,direct}, \tag{10.6}
\end{aligned}
$$

in which Equation (10.6) is from the inequality $x^a + y^a \leq (x + y)^a$, for $a \geq 1$. It can be shown that the best results in terms of minimal overall transmit power can be achieved when the relay is positioned exactly in the middle between the source and the destination; that is, when $d_{S,R} = d_{R,D} = d_{S,D}/2$.

For a fixed total transmission power, a relayed transmission can also extend the transmission range by a factor of $2^{1-1/\alpha}$ compared to the direct transmission (see Problem 3b). Therefore, the use of a relay has the potential of either reducing the required transmit power, which readily extends operation lifetime in battery-operated systems, or extending transmission range. Note that the reduction in the transmit power also has the benefit of reducing the interference in wireless networks, which allows frequency reuse.

Relaying and Reliability

In the preceding example, when the relay is operated the destination ignores the direct transmission from the source. This is reasonable in path loss–dominated networks if the destination is placed farther away from the source than the relay. In most cases, however, it is more beneficial to consider a direct link as well. To illustrate, in fading channels, when the channel is in deep fade, decoding errors occur due to degraded SNR. When the source is transmitting directly to the destination, it is very likely that the channel will be in deep fade. Nevertheless, if the source sends the same information through the relay the probability of error can be reduced. This is because the probability that both channels, from the source to the destination and from the relay to the destination, are in deep fade at the same time is much less than that of the source-destination channel alone. Both received signals are processed at the destination to reveal the transmitted message. This mode of transmission is called *cooperative relaying*, in contrast to *noncooperative relaying* where there is no direct link between the source and the destination or the destination treats the source's signal as noise.

In what follows we first consider the general three-node relay channel. We model the channel and discuss relaying and coding strategies. We also study the achievable rate for both cooperative and noncooperative relay channels. Then, in the latter part of this section, we study the wireless relay channel. A model is given as well as expressions for the achievable rate and outage probability. Finally, maximizing the performance by optimally allocating available time is considered.

10.2.2 A General Three-Node Relay Channel

Theoretical study of the relay channel goes back to 1960s. The use of satellite systems in telecommunications in the 1970s motivated extensive work on the channel [425]. Van der Meulen was the first to introduce the three-node model consisting of a source node, a destination node, and a relay node in [374]. That model was investigated by Cover and El-Gamal in [376]. They proposed some relaying strategies, derived the achievable rate regions, and provided an upper bound to the capacity for the general relay channel. Although a lot of work was carried out later, the capacity of the relay channel remained an unsolved problem.

General Relay Channel: Model

Figure 10.4 shows a block diagram of a general network with a single relay. The encoder at the transmitter side encodes the message w into a sequence of channel input symbols X. Z and Z_R are random variables representing noise produced at the destination and the relay, respectively, while $h_{i,j}$ captures channel effects such as path loss, shadowing, and fading between nodes i and j. For nonfading channels, $h_{i,j}$ are constants.

The source intends to send w to the destination node through relaying. More specifically, S first broadcasts w to both R and D. At any time instant and given channel realization, the received signal at the relay is

$$y_R = h_{S,R}x + z_R. \tag{10.7}$$

After successful decoding, the relay node retransmits w to the destination. That means two independent copies of the transmitted signal are received by the destination. The received signal at the destination is

$$y = h_{S,D}x + h_{R,D}x_R + z, \tag{10.8}$$

where z_R and z are zero-mean normally distributed random variables with variance N_0.

General Relay Channel: Relaying Strategy

The relay node can make use of the signal it receives from the source node in several ways. One approach is AF relaying, where the relay simply sends a scaled copy of the received noisy signal. Another method is DF relaying, where the relay transmits a reencoded copy to the destination, if the relay can successfully decode the transmitted message. AF is easy to implement, but the noise at the relay may be amplified, which makes it unsuitable for multihop scenarios. On the other hand, DF provides a more reliable solution at cost of increased complexity. Due to the

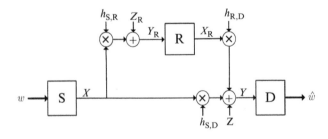

FIGURE 10.4

The three-node relay channel.

repetitive nature, a common disadvantage of AF and DF relaying is the inefficient use of the available degree of freedom. That greatly affects the performance of the relay channel especially in systems with half-duplex constraint on relays, as we explain later. Unless otherwise specified, in this chapter we always assume that the relay operates in a DF manner.

General Relay Channel: Coding

If we assume a perfect channel between the source and the relay with unlimited capacity and zero delay, the relay is able to reliably decode the signal sent by the source at any transmission rate. The relay channel, in that case, assembles a multiple-input, single-output (MISO) channel. The source and the relay allocate their transmit power and signals according to channel conditions. At the receiver side, signals add constructively to maximizing SNR. We call this type of cooperative transmission *beamforming*. Beamforming requires the source and the relay to be fully synchronized, which is another idealized assumption.

More practical strategies were proposed for real-life relay systems. Particularly for the DF, three coding strategies are known to achieve the maximum transmission rate possible. Using the descriptive names used in [386], these strategies are *irregular encoding/successive decoding*, *regular encoding/backward decoding*, and *regular encoding/sliding-window decoding*. All three strategies are based on dividing the message into blocks before transmission. The first strategy was used by Cover and El Gamal in [376], the second one introduced by Willems [405], while the last one was developed by Carleial [404] (originally for multiple-access channels). Although all three strategies achieve the maximum rate, regular encoding/sliding-window decoding is currently the preferred coding strategy for DF relay channels, since it is the simplest of the three and causes limited delay. In the following, and with reference to Figure 10.5, we give a simplified description on how the regular encoding/sliding-window decoding strategy works.

In regular encoding/sliding-window decoding, the message w is divided into B blocks $w_1, w_2, \ldots, w_B$. The whole transmission is performed in $B + 1$ time slots. In the first time slot, the source transmits $x(w_1)$, the sequence corresponds to the first block. Meanwhile, the relay sends $x_R(1)$, which is a constant sequence known to the destination. The relay starts decoding w_1 at the end of the first time slot. In the second time slot, the source transmits $x(w_2)$ while the relay sends $x_R(w_1)$.

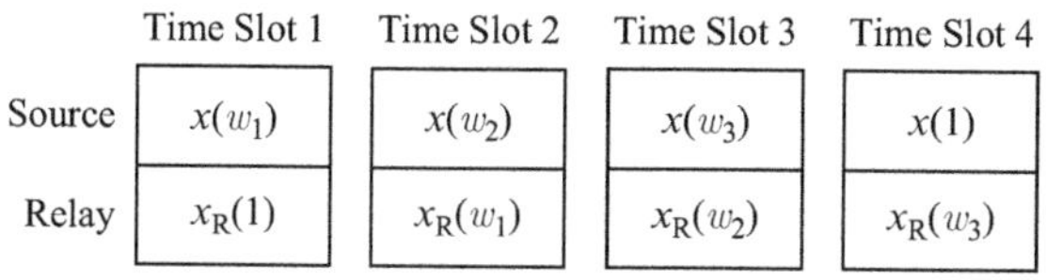

	Time Slot 1	Time Slot 2	Time Slot 3	Time Slot 4
Source	$x(w_1)$	$x(w_2)$	$x(w_3)$	$x(1)$
Relay	$x_R(1)$	$x_R(w_1)$	$x_R(w_2)$	$x_R(w_3)$

FIGURE 10.5

A regular encoding/sliding-window decoding scheme.

Generally, in time slot i the source sends $x(w_i)$ and the relay sends $x_R(w_{i-1})$. In time slot $B + 1$ the source sends a constant sequence $x(1)$ and the relay sends $x_R(w_B)$. An example of this encoding scheme is shown in Figure 10.5 with $B = 3$.

The decoding process is carried out as follows. The receiver at the destination starts decoding w_1 after the second time slot, using the first two blocks. As transmission proceeds, after receiving time slot i, the receiver uses blocks $i - 1$ and i to retrieve w_{i-1}. In other words, the receiver uses a *window* of size two blocks for decoding and shifts (*slides*) the window each time a new block is received. Note that there is a decoding delay of two blocks at the destination. Similarly, the relay uses a sliding window of size one.

The advantages of this encoding/decoding scheme are that it is simple, achieves the maximum rate, can be easily extended to multihop relaying, and has tolerable decoding delay.

General Relay Channel: Achievable Rate
Channel Capacity

Channel capacity is the basic information theoretic performance measure for a communication channel. In information theory, a mathematical representation of a point-to-point *discrete memoryless channel* (DMC) consists of two random variables, X and Y, corresponding to the input and output sequences and a set of conditional probability mass functions (pmf), $p(y|x)$, for each $x \in \mathcal{X}$ and $y \in \mathcal{Y}$, where $\mathcal{X}$ and $\mathcal{Y}$ are finite sets of channel input and output alphabets, respectively (see Figure 10.6). As first introduced by Claude Shannon in [426], *channel capacity* is the maximal rate at which information can be sent over the channel with arbitrary low probability of error [297]. Channel capacity also equals the maximum mutual information between X and Y, maximized over all possible input distributions, $p(x)$ [179],

$$C = \max_{p(x)} \mathcal{I}(X; Y) \text{ bit/channel use.} \tag{10.9}$$

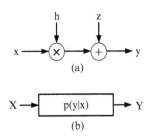

FIGURE 10.6

Single point-to-point channel: (a) operational (physical) representation and (b) information theoretic (probabilistic) representation.

Mutual information between X and Y, denoted $\mathcal{I}(X; Y)$, is a quantity that measures their mutual dependence. In other words, it tells us how much, on average, information we have about X given Y or vice versa [297]. That is expressed by

$$\mathcal{I}(X; Y) = \sum_{x \in \mathcal{X}} \sum_{y \in \mathcal{Y}} p(x, y) \log_2 \frac{p(x, y)}{p(x)p(y)}, \tag{10.10}$$

where $p(x)$, $p(y)$, and $p(x, y)$ are the marginal pmf for X, the marginal pmf for Y, and the joint pmf for X and Y, respectively. In this chapter *transmission rate* and *information transfer rate* are two other terms that refer to the mutual information. DMC capacity in Equation (10.9) can be extended to the case when X and Y are continuous. For the single channel in Figure 10.6, the received signal at the destination is

$$y = hx + z, \tag{10.11}$$

where h is a constant accounting for the channel gain. Z accounts for the noise and is assumed to be normally distributed with mean equals zero and variance N_0. This AWGN channel has a capacity of

$$C = \tfrac{1}{2} W \log_2 \left(1 + \frac{P|h|^2}{N_0} \right) \text{ b/s}, \tag{10.12}$$

where W is the bandwidth of the channel and P is the transmission average power constraint given by

$$\frac{1}{n} \sum_{t=1}^{n} E \left[|x_t|^2 \right] \leq P, \tag{10.13}$$

for any code word $(x_1, x_2, \ldots, x_n)$. C in Equation (10.12) is achievable when X is normally distributed with zero mean and variance P.

When h is randomly changing, the channel is known as a fading channel. Goldsmith and Varaiya studied the point-to-point fading channel in [427]. They showed that capacity is achievable by means of adaptive power allocation techniques where transmission power is allocated according to the channel state. That requires full knowledge of the channel at both the transmitter and receiver. The allocated power, $P(h)$, is subject to the long-term power constraint

$$\int_0^\infty P(h) f_{|h|^2}(h) dh \leq \bar{P}, \tag{10.14}$$

where $|h|^2$ is the channel power gain with the probability distribution function (pdf) $f_{|h|^2}$. The optimal power allocation that achieves capacity is found as

$$P(h) = \begin{cases} WN_0 \left(\frac{1}{H_{\text{th}}} - \frac{1}{|h|^2} \right), & |h|^2 \geq H_{\text{th}} \\ 0, & |h|^2 < H_{\text{th}}, \end{cases} \tag{10.15}$$

for some cut-off value H_{th}, which can be obtained by numerically solving

$$\int_{H_{\text{th}}}^\infty \frac{WN_0}{\bar{P}} \left(\frac{1}{H_{\text{th}}} - \frac{1}{h} \right) f_{|h|^2}(h) dh = 1. \tag{10.16}$$

The corresponding capacity is thus

$$C = \int_{H_{th}}^{\infty} W \log_2 \left(\frac{b}{H_{th}} \right) f_{|b|^2}(b) db \text{ b/s.} \tag{10.17}$$

This kind of power allocation is known as *water filling* in time. Water filling can also be in frequency for frequency-selective channels or in space for MIMO systems.

Channel capacity is in general an uneasy-to-find quantity. For most channels, there is no closed-form solution. Although it has been investigated for a long time, capacity for the general relay channel remains a challenge, even for the simple three-node case. An upper bound was obtained by Cover and El Gamal in [376] by an application of the cut-set theorem. Instead, we focus on the achievable rate for the DF relay channel. In what follows we use $\mathcal{I}_{A,B}$ to refer to the transmission rate between any pair of nodes A and B. When relaying takes place, $\mathcal{I}_{A,B...M}$ refers to information transfer rate between nodes A and M with nodes B, C, ... as relays.

Achievable Rate of the Three-Node Relay Channel

Earlier in this section we showed a coding strategy that achieves the maximum rate for DF relaying. Here we find what that maximum rate is. A three-node relay channel is modeled as a channel with two random inputs X and X_R for the source and the relay, respectively; two random outputs Y and Y_R for the destination and the relay, respectively; and a set of pmf $p(y, y_R | x, x_R)$ for each $(x, x_R, y, y_R) \in \mathcal{X} \times \mathcal{X}_R \times \mathcal{Y} \times \mathcal{Y}_R$.

For the simple case of noncooperative relaying—that is, when there is no direct link between the source and the relay—the relay channel is represented by two consecutive point-to-point channels as shown in Figure 10.7(a). This channel can achieve any rate as long as it can be supported by both source-relay and relay-destination subchannels. The achievable rate is thus bounded by the minimum of

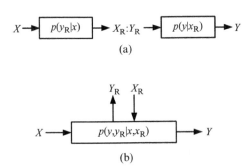

(a)

(b)

FIGURE 10.7

Information theoretic representation of a (a) noncooperative relay channel and (b) cooperative relay channel.

the two subchannels, or

$$\mathcal{I}_{S,D} = \min\left\{\mathcal{I}_{S,R}, \mathcal{I}_{R,D}\right\} \text{ b/s,} \qquad (10.18)$$

where, for any channel realization,

$$\mathcal{I}_{S,R} = W\log(1 + \gamma_{S,R}) \text{ b/s,} \qquad (10.19)$$

and

$$\mathcal{I}_{R,D} = W\log(1 + \gamma_{R,D}) \text{ b/s,} \qquad (10.20)$$

where

$$\gamma_{i,j} = \frac{P_i|b_{i,j}|^2}{N_0} \qquad (10.21)$$

is the received SNR at node j from node i.

Cooperative relaying, on the other hand, is possible only if the network is fully connected. Further, the destination combines both received signals. The relay channel can be viewed as a combination of a broadcast channel (from the sender to the relays and destination) and multiple-access channel (from the relays to the destination) [297] as illustrated in Figure 10.1(a).

With the aid of the relay, the destination receives two copies of the transmitted signal. If the same bandwidth is allocated for both direct transmission and relaying, then the optimum receiver at the destination can be achieved by maximally combining the two signal copies according to their SNRs. The maximum achievable rate at the destination is

$$\mathcal{I}_0 = \log_2(1 + \gamma_{S,D} + \gamma_{R,D}) \text{ b/s.} \qquad (10.22)$$

Comparing Equations (10.22) and (10.20) we notice the power gain advantage of combining both signals. $\mathcal{I}_0$ can be achieved at the destination only if the source-relay channel can support that rate. As a result, the maximum achievable rate by the relay channel is given by

$$\mathcal{I}_{S,R,D} = \begin{cases} \mathcal{I}_0, & \text{if } \mathcal{I}_{S,R} \geq \mathcal{I}_0 \\ \mathcal{I}_{S,R}, & \text{if } \mathcal{I}_{S,R} < \mathcal{I}_0 \end{cases}$$

$$= \min\left\{\mathcal{I}_{S,R}, \mathcal{I}_0\right\}, \qquad (10.23)$$

where $\mathcal{I}_{S,R}$ is again given by Equation (10.19). Equation (10.23) is equivalent to Equation (10.18). In Equation (10.23) as well as Equation (10.18), the source-relay channel limits the performance. This disadvantage demonstrates the importance of relay selection issues in DF relay systems.

10.2.3 **Wireless Relay Channel**

Designing a wireless communication system proves to be a challenge. Wireless systems have their unique attributes. That is due to the nature of propagation in the wireless environment and the hardware limitations on wireless devices. A telecommunication engineer needs to take into consideration these unique aspects of the wireless channel and find ways to tackle or use them if possible. Three key features that differentiate wireless channels and wireless networks from other types of channels and networks are as follows.

- **Wireless broadcast property (WBP).** A signal transmitted by a node is received by all nodes located within the coverage area of that node. The wireless network is thus described as *fully connected*. As a result, if two nodes transmit at the same time, then their signals will interfere. To avoid interference, orthogonality between signals can be achieved by techniques such as *time division multiple access* (TDMA), *frequency division multiple access* (FDMA), or *code division multiple access* (CDMA). Cooperative wireless systems exploit this broadcast property. Nearby nodes are able to receive the source's signal, with no extra cost, and these nodes then act as relays to help the destination revealing the transmitted message.
- **Half-duplex constraint.** Unlike their wireline counterparts, wireless devices are often incapable of transmitting and receiving at the same time and frequency band. Although a full-duplex device is possible, it is expensive and may be ineffective, because the dynamic range of incoming and outgoing signals can easily go beyond the supported range. A half-duplex relay is therefore more realistic. In this case transmission time or bandwidth is shared between the source and the relay. Although theoretically both techniques achieve the same rate, time sharing has implementation advantages over frequency sharing. The relay listens only when the source node is transmitting, and the source node remains silent during the relaying period [394] (see Figure 10.8).

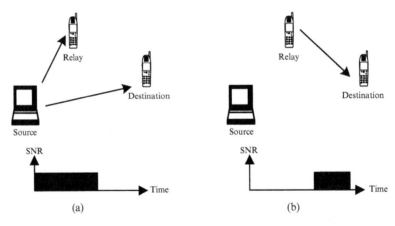

FIGURE 10.8

A wireless relay channel with half-duplex constraint. Transmission is carried out in two stages: (a) the source node broadcasts the message to both the relay and destination nodes, while (b) the relay node forwards the transmission in the second stage.

Obviously, the half-duplex constraint at the relay terminal causes a loss in the degree of freedom.

■ **Channel behavior.** In addition to path loss, communication over the wireless medium suffers from large-scale and small-scale fading caused by shadowing and the multipath propagation of the electromagnetic signal through the wireless medium. Small-scale fading is even harder to deal with. Fading channels tend to change rapidly and unpredictably. Fundamentally, fading causes poor performance in communication systems. The probability of experiencing a fade on the channel (and associated bit errors) is a limiting factor in the link's performance. The effects of fading can be combated by using diversity techniques, where the coded message is sent over multiple channels, experiencing independent fading, and coherently combining received signal at the receiver to retrieve the message. MIMO systems are an example of using spatial diversity to combat fading. When mounting multiple antennas is unfeasible, spatial diversity can be obtained through relaying. In multiuser networks, users can cooperate by relaying each other's data.

Wireless Relay Channel: Model and Signaling

Figure 10.9 shows a block diagram of the wireless relay channel with a single half-duplex constraint relay (compare with Figure 10.4). Communication between S and D takes place in two time instances as shown in Figure 10.8. First, S broadcasts w to both R and D for a duration of $(1 - \tau)$ units of time. Then, R retransmits w to D in the second period of time τ (if w is correctly decoded at R). Denoting the source's transmitted symbol x, the received signal y_R at the relay is given by

$$y_R = h_{S,R}x + z_R, \tag{10.24}$$

where z_R is the noise following Gaussian distribution with zero mean and variances N_0 and $h_{S,R}$ denote the random channel gain between S and R. Generally, we use $\{h_{ij}\}$ to represent the random channel gain between node i and node j with

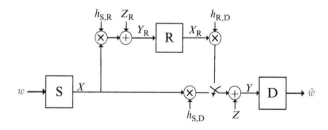

FIGURE 10.9

The wireless channel with single relay.

$i, j \in \{S, R, D\}$. Channel gain coefficients encompass the effects of path loss, shadowing, and fading.

On the other hand, D has two independent copies of the received signals, one from S and another from R. During the direct transmission phase, the received signal at D can be expressed as

$$y_D^{(1)} = b_{S,D}x + z_D^{(1)}, \tag{10.25}$$

where the channel and the noise are defined in a similar way. The received signal at D during the relaying phase is given by

$$y_D^{(2)} = b_{R,D}x_R + z_D^{(2)}. \tag{10.26}$$

Wireless Relay Channel: Achievable Rate

The detection at D is performed by maximum-likelihood (ML) decoding over the two copies of received signals. With τ being the amount of time allocated for relaying, we have

$$\mathcal{I}_{S,R}(\tau) = (1 - \tau)\log_2(1 + \gamma_{S,R}). \tag{10.27}$$

The achievable rate at the destination due to combining signals from the two transmission phases, $\mathcal{I}_0(\tau)$, is

$$\mathcal{I}_0(\tau) = (1 - \tau)\log_2\left[\gamma_{S,D} + \left(1 + \gamma_{R,D}\right)^{\frac{\tau}{1-\tau}}\right]. \tag{10.28}$$

$\mathcal{I}_0(\tau)$ in Equation (10.28) differs from that in Equation (10.22), because the source and the relay are assumed not to be transmitting at the same period of time.

As before, the transmission rate is limited by the source-relay channel. $\mathcal{I}_0(\tau)$ is an achievable subject to $\mathcal{I}_0(\tau) \leq \mathcal{I}_{S,R}(\tau)$. Furthermore, when $\tau = 0$, the relay is listening all the time and is not allowed to relay. In that case the source can ignore the relay altogether and the channel reverts to a direct transmission achieving rate up to

$$\mathcal{I}_{S,D} = \log_2\left(1 + \gamma_{S,D}\right). \tag{10.29}$$

As a result, the achievable rate for the wireless relay channel with half-duplex, as a function of the relaying time, is given by

$$\mathcal{I}_{S,R,D}(\tau) = \begin{cases} \min\{\mathcal{I}_{S,R}(\tau), \mathcal{I}_0(\tau)\}, & \text{if } 0 < \tau < 1 \\ \mathcal{I}_{S,D}, & \text{if } \tau = 0, \end{cases} \tag{10.30}$$

where, $\mathcal{I}_{S,R}(\tau)$, $\mathcal{I}_0(\tau)$, and $\mathcal{I}_{S,D}$ are given by Equations (10.27), (10.28), and (10.29), respectively.

The value of τ can also be considered a measure of the degree of cooperation. Where $\tau = 0$ and $\tau = 1$ are the special cases of no cooperation (or direct transmision) and full cooperation, respectively. Note that full cooperation achieves 0 b/s. The term $\tau = 1/2$ is another special case where time is allocated equally for the source and the relay. An *equal time allocation* policy restricts relaying gain.

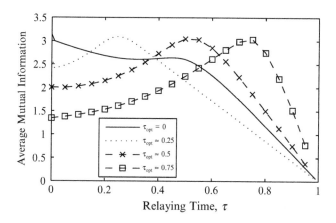

FIGURE 10.10

Mutual information of the three-node relay channel as a function of τ for different channel configurations and averaged over many channel realizations.

Figure 10.10 shows the average mutual information performance of the relay channel for $\tau \in (0, 1)$. Each curve refers to a different setup and the average is taken over many channel realizations assuming independent and identical Rayleigh fading on all channels. It can be seen that a judicious choice of τ could greatly boost the average capacity between S and D.

Maximizing Transmission Rate

We seek the optimal time allocation, τ_{opt}, that maximizes the transmission rate for the three-node wireless relay channel. The problem of interest is to solve

$$\mathbb{P}: \max_{0 \leq \tau < 1} \mathcal{I}_{S,R,D}(\tau). \tag{10.31}$$

In Figure 10.11, the mutual information for the three-node DF relay channel is plotted as a function of τ for some arbitrary setting. Subject to the channel realization, there could be a discontinuity of $\mathcal{I}_{S,R,D}$ at $\tau = \varsigma$, where ς is given by

$$\varsigma = \frac{\log_2 \left(1 + \gamma_{S,R} - \gamma_{S,D} \right)}{\log_2 \left(1 + \gamma_{S,R} - \gamma_{S,D} \right) + \log_2 \left(1 + \gamma_{R,D} \right)}. \tag{10.32}$$

From Equation (10.32), ς exists only if $\gamma_{S,R} > \gamma_{S,D}$. By considering $\mathcal{I}_{S,R}(\tau)$ and $\mathcal{I}_0(\tau)$ in Equation (10.30) separately, it can be shown that both functions are convex. Moreover, $\mathcal{I}_{S,R}(\tau)$ is a line and $\mathcal{I}_{S,R}(\tau_1) \geq \mathcal{I}_{S,R}(\tau_2)$ for all $\tau_1 \leq \tau_2$. The optimal time allocation, τ_{opt}, is therefore

$$\tau_{opt} = \begin{cases} \varsigma, & \text{if } \gamma_{S,D} < \left(1 + \gamma_{S,R} \right)^{1-\varsigma} - 1, \\ 0, & \text{if } \gamma_{S,D} \geq \left(1 + \gamma_{S,R} \right)^{1-\varsigma} - 1 \text{ or } \varsigma \text{ undefined.} \end{cases} \tag{10.33}$$

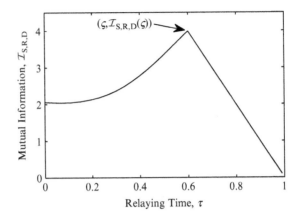

FIGURE 10.11

Mutual information of the three-node relay channel as a function of τ for a particular channel realization.

In words, Equation (10.33) tell us that direct transmission is optimal unless there exists $\tau = \varsigma$ given by Equation (10.32), and

$$\gamma_{S,D} < \left(1 + \gamma_{S,R}\right)^{1-\varsigma} - 1. \qquad (10.34)$$

In that case, $\tau = \varsigma$ is the optimal time allocation. Maximization of mutual information corresponding to Equation (10.33) is given by

$$
\mathcal{I}_{S,R,D}(\tau_{opt}) = \mathcal{I}_{S,R,D}(\varsigma)
$$

$$
= \begin{cases}
\mathcal{I}_{S,R}(\varsigma), & \text{if } \gamma_{S,D} < \left(1 + \gamma_{S,R}\right)^{1-\varsigma} - 1 \\
\mathcal{I}_{S,D}, & \text{if } \gamma_{S,D} \geq \left(1 + \gamma_{S,R}\right)^{1-\varsigma} - 1 \text{ or } \varsigma \text{ undefined.}
\end{cases} \qquad (10.35)
$$

We saw earlier that mutual information for the DF relay channel is always bounded by the *interuser* channel; that is, the source-relay channel. In many cases direct transmission is favored if the interuser channel is not good enough. Condition (10.34) is particularly important for adaptive relaying where the transmission strategy changes according to channel conditions. For example, in a modified version of the *selection relaying* scheme in [412], assuming a *channel state information transmitter* (CSIT), Equation (10.34) tells the source and the relay when to cooperate and when to switch to direct transmission. We also show in the next section that Equation (10.34) can be useful in partner selection.

Maximizing the mutual information requires CSIT. CSIT may be estimated in different ways [373]. In time division duplex (TDD) systems, the channel state can be estimated directly at the transmitter's receiving period. For frequency division duplex (FDD) systems, CSI is estimated at the receivers and fed back to the transmitter.

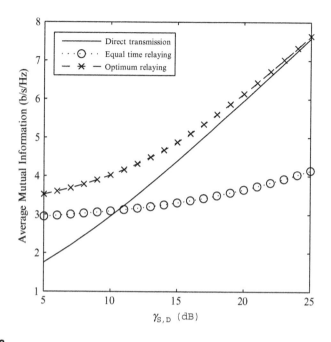

FIGURE 10.12

Comparing throughput performance of different transmission techniques. The source transmission power is increased while relay transmission power is fixed such that the relay-destination SNR is 20 dB.

To see the effectiveness of relaying with optimum time allocation, a comparison is made with direct transmission and equal time allocation. Figure 10.12 provides the achievable rates when $\alpha = 3$, $d_{S,D} = 7$, $d_{S,R} = 1$, and $d_{R,D} = 6.5$. In the results, we also assumed that $P_S = E[|x|^2] = 1$ and $P_R = E[|x_R|^2] = 1$. Results demonstrate that optimized time allocation is important and can significantly improve the achievable rates, while at high SNR from the source node to the destination node, direct transmission is asymptotic optimal, meaning that relaying is not quite needed in this case.

Outage Probability

In delay-constraint fixed-rate applications (e.g., voice applications), channel capacity and achievable rate are not of much interest. It is sufficient to know if the channel can support the required rate. In a fading channel, that question takes a probabilistic form. The question of interest is, How likely (or how often) will the channel be able to support the required rate? This leads to the principle of outage probability and outage capacity. *Outage probability*, $\mathcal{P}$, is the probability that a given rate, R, is not supported by the channel. When channel information is available at the receiver only,

$$\mathcal{P} \triangleq \Pr\{\mathcal{I} < R\}. \tag{10.36}$$

Outage probability is closely related to the probability of error. On the other hand, *outage capacity*, C_ϵ, is the maximum rate at which information can be transmitted over the channel such that $\mathcal{P} < \epsilon$,

$$C_\epsilon \triangleq \max_{p(x)} \arg \mathcal{P}(R), \quad \text{subject to } \mathcal{P}(R) < \epsilon. \tag{10.37}$$

For the wireless relay channel and a given time allocation τ,

$$\mathcal{P}(R) = \Pr\{\mathcal{I}_{S,R,D}(\tau) < R\} = \Pr\{\min\left[\mathcal{I}_{S,R}(\tau), \mathcal{I}_0(\tau)\right] < R\}, \tag{10.38}$$

which can be reduced to [72]

$$\begin{aligned}
\mathcal{P}(R) &= 1 - \Pr\{\mathcal{I}_{S,R} \geq R\}\Pr\{\mathcal{I}_0 \geq R\} \\
&= 1 - \left[1 - \mathcal{P}_{\mathcal{I}_{S,R}}(R)\right]\left[1 - \mathcal{P}_{\mathcal{I}_0}(R)\right],
\end{aligned} \tag{10.39}$$

where $\mathcal{P}_{\mathcal{I}}(x) \triangleq \Pr\{\mathcal{I} < x\}$. For Rayleigh fading channels, the received signal power, $\gamma_{i,j}$, is exponentially distributed with the cumulative distribution function

$$\mathcal{F}_{\gamma_{i,j}}(\gamma) = \begin{cases} 1 - e^{-\frac{\gamma}{\Gamma_{i,j}}}, & \text{for } \gamma \geq 0 \\ 0, & \text{for } \gamma < 0, \end{cases} \tag{10.40}$$

where $E\left[\gamma_{i,j}\right] = \Gamma_{i,j}$. That immediately yields

$$\mathcal{P}_{\mathcal{I}_{S,R}}(R) = 1 - e^{-\frac{2^{\frac{R}{1-\tau}} - 1}{\Gamma_{S,R}}}. \tag{10.41}$$

$\mathcal{P}_{\mathcal{I}_0}$ is more evolved. First, take $\Xi_{R,D} = \left(1 + \gamma_{R,D}\right)^{\frac{\tau}{1-\tau}}$. The term $\Xi_{R,D}$ is a Weibull distributed random variable with the probability density function

$$f_{\Xi}(x) = \begin{cases} \frac{1}{\Gamma_{R,D}} \frac{1-\tau}{\tau} (x+1)^{\frac{1-\tau}{2\tau}} e^{-\frac{(x+1)^{\frac{1-\tau}{\tau}}}{\Gamma_{R,D}}}, & \text{for } x \geq 1 \\ 0, & \text{for } x < 1. \end{cases} \tag{10.42}$$

Now define $\Omega \triangleq \gamma_{S,D} + \Xi_{R,D}$. Then for independent channels and $z \geq 0$, we have the cumulative distribution function for Ω as

$$\mathcal{F}_\Omega(z) = \int_{-\infty}^{z} \int_{-\infty}^{z-y} f_{\Xi}(x) f_{\gamma_{S,D}}(y) dx dy. \tag{10.43}$$

As a result, we have

$$\begin{aligned}
\mathcal{P}_{\mathcal{I}_0}(R) = 1 - e^{-\frac{2^{\frac{R}{1-\tau}} - 1}{\Gamma_{S,D}}} \\
+ \frac{1}{\Gamma_{S,D}} \int_{0}^{2^{\frac{R}{1-\tau}} - 1} \exp\left[-\frac{\left(2^{\frac{R}{1-\tau}} - x\right)^{\frac{1-\tau}{\tau}} - 1}{\Gamma_{S,R}} - \frac{x}{\Gamma_{S,D}}\right] dx.
\end{aligned} \tag{10.44}$$

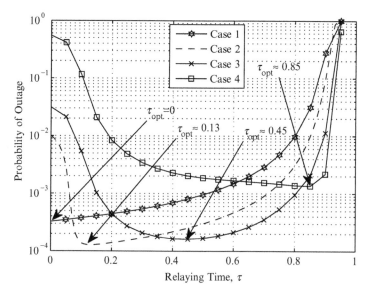

FIGURE 10.13

Outage probability examples for two-hop relay channels for arbitrary settings.

Substituting in Equation (10.39), the outage probability for the wireless relay channel is

$$P(R) = 1 - e^{-\left(\frac{1}{\Gamma_{S,R}} + \frac{1}{\Gamma_{S,D}}\right)\left(2^{\frac{R}{1-\tau}} - 1\right)}$$

$$+ \frac{e^{-\frac{2^{\frac{R}{1-\tau}} - 1}{\Gamma_{S,R}}}}{\Gamma_{S,D}} \int_0^{2^{\frac{R}{1-\tau}} - 1} \exp\left(-\frac{\left(2^{\frac{R}{1-\tau}} - x\right)^{\frac{1-\tau}{\tau}} - 1}{\Gamma_{S,R}} - \frac{x}{\Gamma_{S,D}}\right) dx. \qquad (10.45)$$

It appears that no known result is available to further simplify Equation (10.45) for arbitrary time allocation. Therefore, in this case, the outage probability (and outage capacity) can be computed only numerically. In Figure 10.13, outage probability is plotted for various τ for different cases. Again we see that optimum operation can be achieved by choosing the τ that minimizes the outage probability. The outage probability expression in Equation (10.45), however, is also too complicated to determine the optimum time allocation analytically, though it may be achieved numerically. As shown in [394], it is also possible to derive tight bounds on the outage probability for the optimization. The results in Figure 10.14 compares the outage probability performance for optimum time allocation, equal time allocation, and direct transmission for wireless three-node channels. To

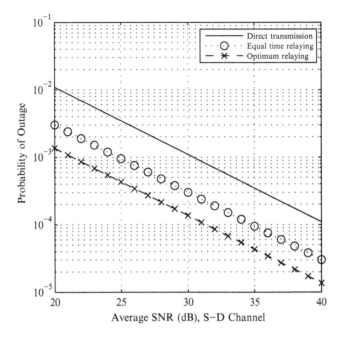

FIGURE 10.14

Comparing the outage probability performance of different transmission techniques. The source transmission power is increased while the relay transmission power is fixed such that the the relay-destination SNR is 35 dB.

generate the numerical results, we considered Rayleigh fading and that $d_{\text{SD}} = 10$, $d_{\text{RD}} = 8$, $d_{\text{SR}} = 3$, and $\alpha = 2$. The results show that the optimal time allocation can substantially reduce the outage probability as compared to the other two schemes. This is more apparent in lower transmission power.

Summary

In this section we present models for the three-node relay channel. A general three-node channel is treated first, followed by the wireless relay channel as a special case. Expressions for the achievable rate (for a given set of channel realization) and outage probability (for statistical channel information) are presented. The optimal time allocation to maximize performance is discussed. That also helps establishing, based on channel realization, the condition for relaying to achieve better performance than direct transmission. All these results are carried over to the next section to see how user cooperative networks can be built based on the model of the relay channel.

10.3 USER COOPERATION IN WIRELESS NETWORKS

10.3.1 Introduction

In a wireless cooperative system, a signal transmitted by one user is overheard by other neighboring users. These neighboring users can then process and retransmit the overheard information to the intended destination. This creates spatial diversity and can lead to increased throughput and channel reliability.

Diversity techniques are known to be a powerful means to combat the effect of multipath fading on wireless systems. Multiple antennas at base stations have become de facto in cellular systems. However, multiple antennas at mobile phones are infeasible due to hardware limitations. Instead, single-antenna users can gain diversity by cooperating with other users in the network. This mode of operation is called *user cooperative diversity*. User cooperative diversity also helps improve resource utilization in the wireless network. For example, the cost of transmission (measured in joule/bit or watt/bit) is reduced when users cooperate. Spectrum efficiency may be improved as well by user cooperation. Two networks, primary and secondary, can occupy the same band at the same time if users in one network or both cooperate. User cooperative diversity is particularly important for MANET. The noncentralized setup and changing topology necessitate cooperation among network users.

Cooperative networks are based on the classical relay channel. In the previous section, we considered the scenario in which a relay node is spending its resources (e.g., bandwidth and power) selflessly to assist another node. In multiuser networks, users act as relays forwarding their partners' data. If cooperation is done properly, gain in the network capacity is anticipated. The fact that *each of the cooperating users has its own data to transmit* is the key difference between relay channels and cooperative networks. Therefore, individual users' constraints have to be considered when cooperation is performed.

User cooperative diversity can be categorized in different ways based on the forwarding scheme [428]. On the one hand, cooperative transmission can be either *repetitive* or *coded*. Relaying schemes based on repeating partners' transmission are simple and popular. AF and DF are two examples. Coded cooperation, in contrast, is more complicated. Instead of repeating a partner's signal, a cooperating user sends an overhead that may help the destination correctly decode the signal. Coded cooperation has the advantage of using the available degree of freedom more efficiently. On the other hand, cooperative transmission can either be *fixed* or *adaptive* [428]. In fixed relaying, the relay continuously forwards the signal received from the source. On the other hand, adaptive relaying means that the relay remains in listening mode and relays only when necessary. A decision is made either based on channel conditions (*selection relaying*) or feedback from the destination (*incremental relaying*).

From Relaying to User Cooperative Diversity

Relayed transmission can take many different forms. There exist several relaying strategies, several coding schemes, and several ways to process the received signal at the destination. As well as in single channels (i.e., single source and single destination), relaying can take place in networks with multiple sources (multiple-access relay channels) and networks with multiple sinks (broadcast relay channels). Further, multiple relays could be arranged such that transmission happens over two or more hops. Therefore, in multiuser networks, users can cooperate through a large number of methods. Any cooperation scheme must take into account all network features and characteristics. These may include the nature and value of the information transfered, network resources, constraints and limitations, and the purpose behind cooperation. For a given network, the optimal cooperation strategy is, very likely, nontrivial. Moreover, added complexity from cooperation might overweigh its gains. Therefore, suboptimal solutions may be considered.

In this section, we take advantage of the material presented in the previous section to show how the three-node relay channel can be used as a prototype for the two-user cooperative network. Particularly, we want to see how a four-node network, of two sources and two destinations, can be arranged into a cooperative network. We focus on the achievable rate (assuming CSIT) after cooperation for each source-destination pair. Power and time allocation is discussed as well as a number of optimization problems relevant to this setup. Later in this section an arbitrary multiuser network is considered. A two-user network is used as a model for cooperation. Several issues arise, such as partner selection and issues related to upper layers. Some results, and discussion on these results, are also presented.

10.3.2 Two-User Cooperative Network

In light of Section 10.2, we here consider a partnership of two cooperative users in a four-node network with two source nodes and two destination nodes.

System Model

Consider two users, A and B, in partnership, as shown in Figure 10.15. User A (B) wants to send a message w_A (w_B) to a destination node D_A (D_B), with the help of user B (A) serving as a relay. Hence, we have two relay channels, (A,B,D_A) and (B,A,D_B). For the channel (A,B,D_A), user A transmits w_A for a duration $1 - \tau_A$ with power P_1^A. Upon successful decoding, user B retransmits w_A to D_A for a duration τ_A with power P_2^B (assuming that one unit of time is allocated for conveying w_A to D_A). Similarly, we have τ_B, P_1^B and P_2^A for the channel (B,A,D_B).

The maximum achievable rate for user A is therefore given by

$$\mathcal{I}_{A,B,\,D_A}(\tau_A) = \begin{cases} \min\left\{\mathcal{I}_{A,B}(\tau_A), \mathcal{I}_0^A(\tau_A)\right\}, & \text{if } 0 < \tau_A \leq 1 \\ \mathcal{I}_{A,\,D_A}, & \text{if } \tau_A = 0, \end{cases} \tag{10.46}$$

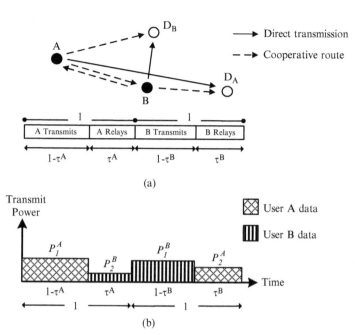

FIGURE 10.15

(a) A two-user cooperative network and (b) the associated time and power allocation.

where

$$
\begin{cases}
\mathcal{I}_{A,B}(\tau_A) = (1 - \tau_A) \log_2(1 + \gamma_{A,B}) \\[2mm]
\mathcal{I}_0^A(\tau_A) = (1 - \tau_A) \log_2 \left[\gamma_{A,D_A} + (1 + \gamma_{B,D_A})^{\frac{\tau_A}{1-\tau_A}} \right] \\[2mm]
\mathcal{I}_{A,D_A} = \log_2(1 + \gamma_{A,D_A}).
\end{cases}
\tag{10.47}
$$

Equation (10.46) is copied from Equation (10.30) with the triple (S,R,D) is replaced by (A,B,D$_A$). For channel B,A,D$_B$, we similarly have

$$
\mathcal{I}_{B,A,D_B}(\tau_B) =
\begin{cases}
\min \left\{ \mathcal{I}_{B,A}(\tau_B), \mathcal{I}_0^B(\tau_B) \right\}, & \text{if } 0 < \tau_B \le 1 \\[2mm]
\mathcal{I}_{B,D_B}, & \text{if } \tau_B = 0,
\end{cases}
\tag{10.48}
$$

where

$$
\begin{cases}
\mathcal{I}_{B,A}(\tau_B) = (1 - \tau_B) \log_2(1 + \gamma_{B,A}), \\[2mm]
\mathcal{I}_0^B(\tau_B) = (1 - \tau_B) \log_2 \left[\gamma_{B,D_B} + (1 + \gamma_{A,D_B})^{\frac{\tau_B}{1-\tau_B}} \right], \\[2mm]
\mathcal{I}_{B,D_B} = \log_2(1 + \gamma_{B,D_B}).
\end{cases}
\tag{10.49}
$$

Also, in Equations (10.46)–(10.49),

$$\gamma_{i,j} \triangleq \frac{P_k^i}{d_{i,j}^\alpha}|b_{i,j}|^2, \quad \text{for } i \in \{A, B\}, j \in \{A, B, D_A, D_B\} \text{ and } k = 1, 2. \tag{10.50}$$

The term P_k^i is user's i transmit power during the direct transmission phase ($k = 1$) or relaying phase ($k = 2$); α, $d_{i,j}$, and $b_{i,j}$ are as defined as in the previous section. Note that we assumed $N_0 = 1$.

System Constraints

The way user A and user B cooperate depends on the resources available, constraints that govern the use of these resources, and any other constraints of the system. Regarding the transmission power, cooperating users may operate under one of the following constraints (but are not limited to them):

- **Fixed uniform peak transmit power.** The same peak power is imposed for direct transmission and relaying, regardless of time allocation; that is,

$$P_1^i = P_2^i = \bar{P}_i, \text{ for } i \in \{A, B\}. \tag{10.51}$$

  Thus, the time allocation, τ, is the only cooperation parameter.

- **Average transmit power per user.** Each user is constrained to a maximum average power for the direct transmission and cooperation. Specifically,

$$P_1^A(1 - \tau_A) + P_2^A \tau_B = \bar{P}_A, \tag{10.52}$$

  and

$$P_1^B(1 - \tau_B) + P_2^B \tau_A = \bar{P}_B. \tag{10.53}$$

  The decision to make for cooperation at each node therefore includes both time and power allocation. We define the power sharing factors $\{\phi_j\}$,

$$\phi_A \triangleq \frac{P_2^A \tau_B}{\bar{P}_A}, \tag{10.54a}$$

$$\phi_B \triangleq \frac{P_2^B \tau_A}{\bar{P}_B}, \tag{10.54b}$$

  to denote the proportion of power to be used by each user to help its partner. Thus, we have cooperation parameters, τ_A, τ_B, ϕ_A, and ϕ_B.

- **Total power.** Users may share a total power constrained by

$$P_1^A(1 - \tau_A) + P_2^A \tau_B + P_1^B(1 - \tau_B) + P_2^B \tau_A = \bar{P}. \tag{10.55}$$

We can also define $\{\phi_j^i\}$ in a similar way as to Equation (10.54) so that we have

$$\phi_1^A = \frac{P_1^A(1 - \tau_A)}{\bar{P}}, \qquad (10.56a)$$

$$\phi_2^A = \frac{P_2^A \tau_B}{\bar{P}}, \qquad (10.56b)$$

$$\phi_1^B = \frac{P_1^B(1 - \tau_B)}{\bar{P}}, \qquad (10.56c)$$

$$\phi_2^B = \frac{P_2^B \tau_A}{\bar{P}}. \qquad (10.56d)$$

In this case, we have

$$\sum_{\substack{i \in \{A,B\} \\ j \in \{1,2\}}} \phi_j^i = 1. \qquad (10.57)$$

Optimizing Performance

Here, we aim to find the optimum sharing strategy of resources (power and degree of freedom) for the two-user cooperative network shown in Figure 10.15. This four-node system can be viewed as two three-node relay channels. It is important to allocate their resources appropriately to make sure that the partnership is beneficial to both users. In particular, we need an answer to the question, What is the appropriate choice of ϕ_A, ϕ_B, τ_A, and τ_B?

Before formalizing the optimization problems, we define $\phi \triangleq (\phi_A, \phi_B)$ in the case of a per-user average power constraint or $\phi \triangleq \left(\phi_1^A, \phi_2^A, \phi_1^B, \phi_2^B\right)$ in the case of total power constraint and $\tau \triangleq (\tau_A, \tau_B)$. To guide the user cooperation with the appropriate time and power allocation constraints, the following optimization problems are commonly considered:

- **Maximizing total throughput.** When the sum of achievable rates for both users is concerned, the aim is to

$$\max_{\phi, \tau} \; \mathcal{I}_{A,B,D_A}(\phi, \tau) + \mathcal{I}_{B,A,D_B}(\phi, \tau). \qquad (10.58)$$

While this strategy will surely maximize the spectrum utilization in terms of achievable rates, the performance of individual users is not necessarily improved. In particular, it tends to enhance the strongest user's performance but sacrifice the weakest user's performance after cooperation.

- **Maximizing minimum rate.** To improve fairness for user cooperation, one may like to

$$\max_{\phi, \tau} \; \min \left\{ \mathcal{I}_{A,B,D_A}(\phi, \tau), \mathcal{I}_{B,A,D_B}(\phi, \tau) \right\}. \qquad (10.59)$$

This strategy takes advantage of user cooperation to make sure that the achievable rate for the weaker user is maximized.

- **Cooperation with rate constraint.** Cooperation can also be performed to maximize a user's rate while making sure that the cooperating user is getting a minimum required rate, say R_B, for itself; that is,

$$\max_{\phi,\tau} \mathcal{I}_{A,B,D_A}(\phi,\tau) \text{ subject to } \mathcal{I}_{B,A,D_B}(\phi,\tau) \geq R_B. \tag{10.60}$$

- **Fair cooperation.** Fairness hardly can be imposed by the construction of the problem. In particular, an equal-rate constraint can be added to the problem, if this is appropriate. That is,

$$\max_{\phi,\tau} \mathcal{I}_{A,B,D_A}(\phi,\tau) \text{ or } \max_{\phi,\tau} \mathcal{I}_{B,A,D_B}(\phi,\tau)$$

$$\text{subject to } \mathcal{I}_{A,B,D_A}(\phi,\tau) = \mathcal{I}_{B,A,D_B}(\phi,\tau). \tag{10.61}$$

Solving these problems requires advanced optimization techniques and separate chapters to address them properly. In this chapter, if required, they are solved by numerical approaches. In particular, stochastic optimization, such as a cross-entropy method or Markov chain Monte Carlo, is found to be useful [429, 430].

10.3.3 Cooperative Wireless Network

In general, a wireless network has multiple source and multiple destination terminals and any number of nodes. As described previously, cooperation can be performed in different ways for the benefit of the network, individual users, or both. Based on the motivation behind cooperation, any partnership selection must be made under certain conditions determined by the network and individual users. Cooperation policies are also subject to devices and network attributes and constraints such as access techniques, rate and reliability requirements, information availability at different network entities, the nature of the network topology, the power constraint, and so on. In this section, we look at several issues specific to cooperative communication raised in the context of multiuser networks. To do that, a specific network model is assumed.

Network Model and Rules for Cooperation

We consider a wireless network, where nodes are randomly positioned and half the nodes are data sources and half are sinks. Each source is assigned a unique destination. Network topology is assumed to be changing slowly. Typically, it remains fixed for an entire session of communication. All nodes are assumed to be subject to a fixed peak power constraint and obey the half-duplex constraint. We assume that a source node is aware of all other nodes within the range to its destination node.

To improve network performance, users are permitted to work in pairs cooperatively to help each other. In particular, users are allowed to choose their partners on the basis that cooperation increases the mutual information for both users. As such, for a given user, all other users in the network can be categorized as either *useful* or *harmful*. To illustrate this, user B is deemed a *useful user* for user A if user A, with user B as a relay, can achieve a higher rate than that achievable by direct transmission; otherwise, user B is deemed to be a *harmful user*. To further this idea, it also makes sense to consider constructive-only cooperation, meaning that the cooperative users are a useful user pair. This is formally addressed in the next subsection.

Partner Selection

This subsection formalizes the partnership selection for user cooperation based on the constructiveness or mutual usefulness.

Useful User

We now consider the scenario in which user A in the network is looking for a single partner among the other users, labeled B, C, D, ... Taking user B as an example, user B is not considered for the partnership unless it is classified by user A as a useful user; that is, such that user A achieves a rate higher than that achievable by direct transmission. With reference to the previous section, it is found that direct transmission achieves the maximum rate, unless

$$\gamma_{A,D_A} < \left(1 + \gamma_{A,B}\right)^{1-\varsigma_A}, \tag{10.62}$$

where $\varsigma_A \in (0, 1)$ is given by

$$\varsigma_A = \frac{\log\left(1 + \gamma_{A,B} - \gamma_{A,D_A}\right)}{\log\left(1 + \gamma_{A,B} - \gamma_{A,D_A}\right) + \log\left(1 + \gamma_{B,D_A}\right)}, \tag{10.63}$$

For ς_A to exist we must have

$$\gamma_{A,D_A} < \gamma_{A,B}, \tag{10.64}$$

which is a looser bound than that in inequality (10.62).

Thus, in regard to user A, inequality (10.64) is a necessary condition for user B to be a useful user (and hence a possible candidate for partnership) and inequality (10.62) is a sufficient condition. Both conditions can be combined as

$$\gamma_{A,D_A} < (1 + \gamma_{A,B})^{(1-\varsigma_A)} - 1 < \gamma_{A,B}. \tag{10.65}$$

Useful Set

Generally, user A with user B as a relay can achieve any rate $\mathcal{I}_{A,D_A} < R < \mathcal{I}_{A,B,D_A}(\varsigma_A)$. Although ς_A gives user A the maximum possible advantage from cooperating with user B, it might not be acceptable for cooperation given the mutual usefulness criterion. It is therefore more useful to find all possible $\tau_A \in (0, 1)$ such that $\mathcal{I}_{A,B,D_A}(\tau_A) > \mathcal{I}_{A,D_A}$. With reference to Figure 10.16, we can observe that, as long as $\tau_A \in \left(\tau_L^A, \tau_U^A\right)$, user B is a useful partner. We refer to the set $\Lambda_{A,B,D_A} \triangleq \left(\tau_L^A, \tau_U^A\right)$ as the *useful set*.

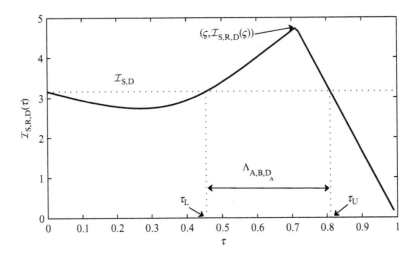

FIGURE 10.16

Mutual information for various τ, where it shows that Λ_{A,B,D_A} includes all τ that makes user B a useful partner for user A.

Constructive Partnership and Constructive Set

The previous discussion shows the perspective seen by user A. From the viewpoint of user B, we also can have the useful set Λ_{B,A,D_B}, with τ_L^B and τ_U^B defined in the same way. In what follows, for user A and user B to form a constructive partnership, it is necessary to have

$$
\begin{cases}
\Lambda_{A,B,D_A} \neq \emptyset, \\
\Lambda_{B,A,D_B} \neq \emptyset.
\end{cases}
\tag{10.66}
$$

Generally, $\tau_A \neq \tau_B$. To simplify our analysis, however, we assume that cooperation is symmetry such that $\tau_A = \tau_B$. With this symmetry condition, for the cooperation to be constructive, we require that

$$
\Lambda_{A+B} = \Lambda_{A,B,D_A} \bigcap \Lambda_{B,A,D_B}
\tag{10.67a}
$$

$$
= \left\{ \tau : \max\left(\tau_L^A, \tau_L^B\right) < \tau < \min\left(\tau_U^A, \tau_U^B\right) \right\} \neq \{\} .
\tag{10.67b}
$$

Inequality (10.67b) is a sufficient condition for constructive cooperation between A and B. Λ_{A+B} is thus regarded as the *constructive set*, which contains all possible τ that permit constructive cooperation between A and B. Figure 10.17 shows the relationship between Λ_{A,B,D_A}, Λ_{B,A,D_B}, and Λ_{A+B}. It can be shown that Equation (10.67b) is satisfied if the following is true:

$$
\begin{cases}
\mathcal{I}_0^A\left(\tau_U^B\right) > \mathcal{I}_{A,D_A} \\
\mathcal{I}_0^B\left(\tau_U^A\right) > \mathcal{I}_{B,D_B}.
\end{cases}
\tag{10.68}
$$

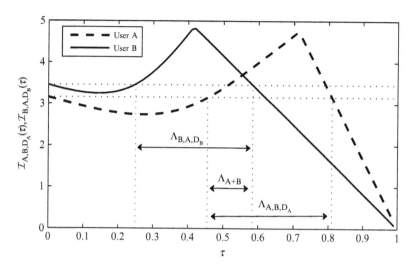

FIGURE 10.17

The constructive set, Λ_{A,B,D_A}, which is the set of all τ that allows user A and user B to form a constructive partnership.

We conclude that any two users in the network can establish a cooperative partnership if they satisfy Equations (10.62) and (10.68).

Data Link Layer

The previous sections established different conditions that need to be satisfied before a pair of users is allowed to cooperate. For networks with many users, upper layers of the communication model have to be considered. For the given model, we establish here a data link layer protocol that facilitates the partnership selection. A possible procedure for user A is described as follows:

1. Initially, user A knows γ_{A,D_A} and all users in the network such that

$$\gamma_{A,i} > \gamma_{A,D_A}, \tag{10.69}$$

   which is a necessary condition for any user i to be a useful user to user A (equivalent to Equation (10.62)).
2. User A sends a cooperation request packet (CRP) to all users satisfying the preceding condition. The CRP contains also the ID of D_A, γ_{A,D_A}, and $\gamma_{A,B}$.
3. Users who receive the CRP, check condition (10.62). Equation (10.62) is a sufficient condition for any user to be a useful user for user A.
4. Each useful user replies with a cooperation request reply (CRR) containing the ID of its destination, D_i, as well as γ_{i,D_i} and $\gamma_{i,A}$.
5. User A checks condition (10.68). All successful users are now promoted to be constructive partners by user A.

6. For each constructive partner, user A finds τ_{opt}, which solves

$$\max_{\tau \in \Lambda_{A+i}} \quad \min \left\{ \mathcal{I}_{A,i,D_A}(\tau), \mathcal{I}_{i,A,D_i}(\tau) \right\}. \tag{10.70}$$

This is the "maximize minimum rate" optimization policy discussed in Section 10.3.2.

7. If more than one constructive partner is available from Step 5, user A chooses one partner such that

$$\max_{i \in \{B,C,D,\dots\}} \quad \mathcal{I}_{A,i,D_A}(\tau_{\text{opt}}^i) + \mathcal{I}_{i,A,D_i}(\tau_{\text{opt}}^i), \tag{10.71}$$

to maximize the sum rate given τ_{opt}^i from Step 6.

8. User A sends a cooperation acceptance packet (CAP) to the selected user containing the corresponding optimal time allocation τ_{opt}.

A flowchart showing these steps is given in Figure 10.18. In Figure 10.18 user A negotiates a cooperation partnership with other users in the network. User B is assumed to be the successful partner.

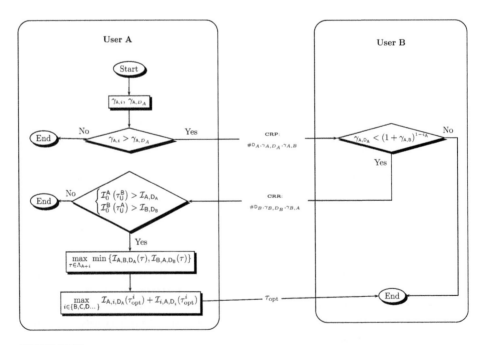

FIGURE 10.18

How user B is chosen by user A for cooperation.

Network Performance

In this subsection, we assess network performance of user cooperative systems using computer simulations.

Generating a Random Network

A network with N nodes is considered. Node position (x_i, y_i) for any node i is randomly generated with X and Y zero-mean normally distributed random variables. Nodes are arranged into $N/2$ source-destination pairs and each is given an ID. Users are arbitrarily ordered, and when the network is online, users begin to look for their possible partners starting with the user with the least ID. Once a partnership is formed for a pair of users, they no longer are available for cooperation. If at the end one user fails to find a partner, direct transmission takes place.

Since nodes are randomly positioned, the distance between each pair of nodes is random and so is the corresponding SNR. To have an indicative measure, we consider the average SNR,

$$\bar{\gamma} = E\left[\frac{P_i}{N_0 d_{i,j}^{\alpha}}\right] = \frac{P}{N_0 E\left[d_{i,j}^{\alpha}\right]}, \tag{10.72}$$

as the cost of transmission, in which the average transmit power for every node is assumed to be P. It can be shown that, for a Gaussian distributed network, we have

$$E\left[d_{i,j}^{\alpha}\right] = (2\sigma)^{\alpha} \Gamma\left(1 + \frac{\alpha}{2}\right), \tag{10.73}$$

where $\Gamma(.)$ denotes the gamma function.

Network Performance

The first important question to be answered is how easy it is to find a constructive partner. Figure 10.19 shows the percentage of transmitters that succeed to build a constructive partnership. Results indicate that the denser the network, the easier users find partners. Another observation is that user cooperation seems to be more useful when the average SNR is low. This can be explained as, in the case of low SNRs, the achievable rate for direct transmission can be easily exceeded by a partner.

Figure 10.20 shows the gain in the network mutual information resulting from cooperation. This gain is calculated using

$$G = \frac{\bar{\mathcal{I}}_{\text{coop}} - \bar{\mathcal{I}}}{\bar{\mathcal{I}}} \times 100\%, \tag{10.74}$$

where

$$\bar{\mathcal{I}} = \frac{2}{N} \sum_i \mathcal{I}_{i,D_i} \tag{10.75}$$

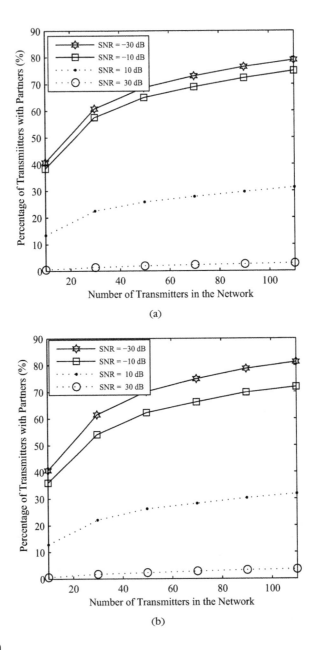

FIGURE 10.19

The percentage of transmitters with partners versus the total number of transmitters in the network, for (a) $\alpha = 2$ and (b) $\alpha = 4$.

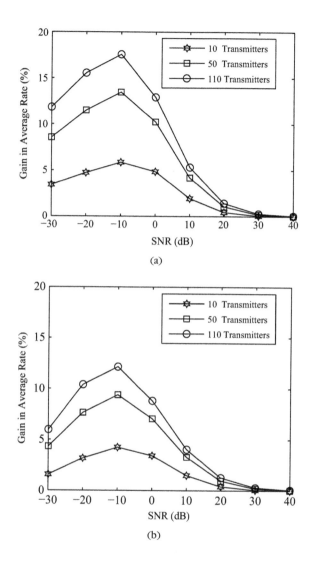

FIGURE 10.20

The percentage of transmitters with partners versus the total number of transmitters in the network, for (a) $\alpha = 2$ and (b) $\alpha = 4$.

is the average rate per link without user cooperation, and

$$\overline{\mathcal{I}}_{\text{coop}} = \frac{2}{N} \sum_{i,j(i)} \mathcal{I}_{i,j(i),D_i}\left(\tau^i_{\text{opt}}\right) \tag{10.76}$$

is the average rate per link with user cooperation, where $j(i)$ denotes the user's index for user i's partner. The results in this figure demonstrate that, as the network

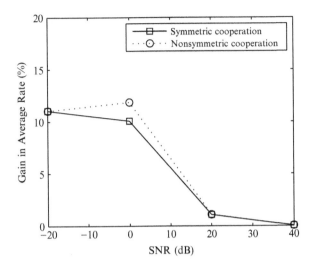

FIGURE 10.21

Performance comparison between symmetric and nonsymmetric cooperation for a network of 55 transmitters and $\alpha = 2$.

density increases, the gain from cooperation becomes more significant. Results however also illustrate that at high SNR user cooperation is less important, which aligns with the results in Figure 10.19.

Thus far, results for user cooperation are provided for symmetric cooperation, meaning that the time allocation for relaying is the same for both cooperative users in any given partnership, say $\tau_A = \tau_B$. This certainly reduces the degree of freedom when partnership is sought and therefore limits network performance. This is observed from the results in Figure 10.21, where results for both symmetric and nonsymmetric cooperation are provided. Other assumptions that affect obtained results include the order at which users select their partners, number of cooperating users, and number of allowed hops. Relaxing these assumptions is expected to further improve the results. A globally optimal cooperative protocol is still under investigation.

Results for nonsymmetric cooperation in Figure 10.21 are produced numerically. Optimization is carried out using a population-based optimization method called *particle swarm optimization* [431]. It should be noted that the results for the nonsymmetric cooperation are given for comparison purposes.

Summary

In this section we develop a two-user cooperative model based on the three-node relay channel studied in Section 10.2. A multiuser network is then considered for cooperation. Each user is given a chance to choose a partner according to a set of predefined rules. Several relevant issues are discussed, including partner selection and upper-layer protocol. Simulation results are shown to confirm substantial gains from cooperation.

10.4 MULTIHOP RELAY CHANNEL

Earlier in this chapter, we examined the scenario of a single relay. In this section we consider a single source-destination pair served by multiple relays. More specifically, we are interested in the case where relays are arranged for multihop transmission such that data are routed to the destination node over multiple hops. Multiple relays could also be arranged for two-hop transmission. In that case relays use space-time codes to convey the message, as described in [409]. Generally, multiple relays can be arranged into different groups where nodes at the same level decode the message at the same time and cooperatively retransmit it during the same time period.

Historically, multihopping was in use long time before modern telecommunication. Smoke and fire beacons used to send important messages over long distances during ancient times. In modern telecommunication, the first multihop link connected New York and Boston in the United States in 1947. More intercontinental links were established later in the United States [425]. The purpose of these links was to transmit over long distances where the Earth's curvature and natural obstacles made direct transmission unattainable. Recently, relay channels are also considered as a means to increase transmission rate and reliability. Increased throughput means more efficient usage of power and spectrum.

Multiple-User Cooperative Communication

In Section 10.3, a two-user cooperative model was developed based on the three-node relay channel studied earlier. In fact, a cooperative partnership could take different forms and more users. Specifically, network users may be grouped into *clusters* in which users cooperatively deliver their data (see Figure 10.22). Further, more flexibility is added to the way cooperative partnership is formed if nonreciprocal partnership is allowed. That means if user A helps user B, user B has the freedom to help user A or any other user. The *reciprocity* condition assumed in the previous section (in the sense that if user A helps user B, then user B must help user A) is one way to ensure some degree of cooperation fairness in the system. Multiple-user cooperation is expected to offer extra benefits at the cost of added complexity.

In this section we consider a source-destination pair assisted by a number of other users. Cooperating users are arranged to help in a multihop fashion. The aim here is to extend and generalize the results of the single relay channel presented in Section 10.2 to channels with multiple relays.

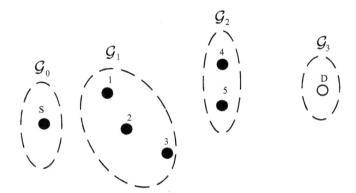

FIGURE 10.22

A general multiple relay channel. In this particular case, five relays aid transmission between the source and the destination. All nodes are arranged into four levels. Nodes on the same level decode the message at the same time and retransmit it cooperatively during the same time period.

Multihopping to Exploit Gray Holes

In the paradigm of cognitive radio, the spectrum of licensed channels can be categorized as [11]

1. **White holes.** The spectrum where the primary network is inactive.
2. **Gray holes.** The spectrum where the primary network is operating at low power.
3. **Black holes.** The spectrum where the primary network is operating at high power.

Secondary users have the right to fully exploit white holes but are banned from black holes. They are allowed to operate in gray holes as well, if interference caused at the primary receivers is tolerable. In regard to the setting in Figure 10.23, the secondary source cannot exceed a maximum transmission power P_{max}; otherwise connection between primary users is disturbed. On the other hand, a minimum power P_{min} is needed to establish a connection with the secondary destination. If $P_{min} \leq P_{max}$, the secondary source is allowed to use the hole; otherwise, it has to keep silent unless nodes in the network are willing to help in a multihopping fashion. The disadvantage of multihopping is that transmission delay increases as the number of hops increases.

Multihop Relay Channel: Model

A general multirelay channel consists of a number of relays arranged into different groups or levels, where relays on the same level cooperatively decode and transmit the received information [390]. In the classical multihop relay channel, information

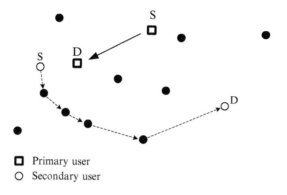

☐ Primary user
○ Secondary user

FIGURE 10.23

Secondary users are allowed to use licensed bands if those bands are unused or if they can make sure that no harmful interference is caused to primary receivers. Multihop relaying is helpful in the sense that transmission power can be kept low to reduce interference to primary users.

is sent in a consecutive fashion: first from the source to the first relay, then from the first relay to the second relay, and so on until the message reaches the destination terminal. This mode of transmission is also called *noncooperative relaying*. Apart from the source and the destination, noncooperative transmission assumes that each node is connected to one node downstream and another node upstream. In contrast, in cooperative relaying, each node combines all the signals received from all nodes upstream. It is therefore necessary to have a fully connected network, such as the one in Figure 10.24, to carry out cooperative transmission.

We are interested in the multihop relay channel with a single relay per hop. We further consider the wireless medium where the network is fully connected

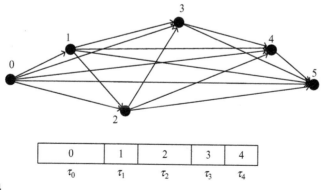

0	1	2	3	4
τ_0	τ_1	τ_2	τ_3	τ_4

FIGURE 10.24

Multihop wireless relay channel and time allocation. Here cooperative transmission is possible, since the network is fully connected.

and relays have half-duplex constraint. We consider that an M-hop relay channel comprises a source node, labeled as node 0; a destination node, labeled as node M; and $M - 1$ relay nodes, labeled accordingly by $\{1, 2, \ldots, M - 1\}$. Due to the half-duplex constraint, only one node can transmit at a time. Available time is thus shared by all transmitting nodes. Node i listens during time periods $\tau_0, \tau_1, \ldots, \tau_{i-1}$, transmits during τ_i, and remains idle during periods $\tau_{i+1}, \tau_{i+2}, \ldots, \tau_{M-1}$. An example is given in Figure 10.24 for a five-hop relay channel. Figure 10.25 further

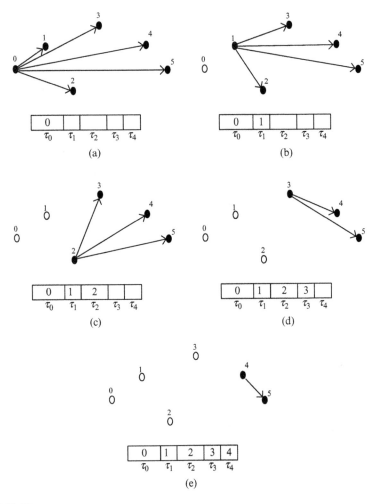

FIGURE 10.25

Transmission from node 0 (source node) to node 5 (destination node) over five hops. (a)–(e) Nodes 0–4 transmit/forward the signal using $\tau_0, \ldots, \tau_4$ of the available time in each hop.

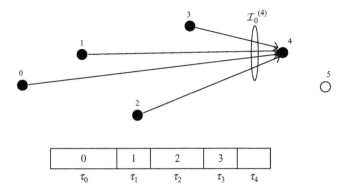

FIGURE 10.26

The accumulated rate at node 4 after four hops, $\mathcal{I}_0^{(4)}$.

describes transmission from a source node, 0, to a destination node, 5. First node 0 broadcasts the message to all nodes $1, 2, \ldots, 5$. Node 1 starts decoding the message immediately at the end of phase 1 and retransmits the message in the second phase. Nodes $2, 3$, and 4 follow. Each node combines all the received copies of the signal to decode the message before transmission. For example, as shown in Figure 10.26, node 4 combines signals transmitted during time periods $\tau_0, \tau_1, \ldots, \tau_3$ and retransmits the message during τ_4.

Multihop Relay Channel: Achievable Rate

Using a DF strategy, for a relay to be useful, it must be able to decode the source's message by combining signals received from previous transmissions. Therefore, the achievable rate at the destination is bounded by the minimum of all rates achievable at relays (i.e., nodes $1, 2, \ldots, M - 1$). Transmission rate can hence be up to

$$\mathcal{I}_M = \min_{j \in \{1,2,\ldots,M\}} \mathcal{I}_0^{(j)}, \tag{10.77}$$

where $\mathcal{I}_0^{(j)}$ the accumulated rate at node j after combining signals received from nodes $0, 1, \ldots, j - 1$. $\mathcal{I}_0^{(j)}$ is given by

$$\mathcal{I}_0^{(j)} = \tau_0 \log_2 \left[\sum_{i=0}^{j-1} \left(1 + \gamma_{i,j}\right)^{\frac{\tau_i}{\tau_0}} - j + 1 \right], \quad j = 1, 2, \ldots, M, \tag{10.78}$$

and $\tau \triangleq \{\tau_0, \tau_1, \ldots, \tau_{m-1}\}$ is the corresponding time allocation vector. Node i is allocated a proportion τ_i of the available time for transmission.

Redundant Relays

In the case of a three-node relay channel, we saw how direct transmission sometimes can be better than relaying. Further, Equation (10.34) states a condition, based on channel realization, for when relaying can achieve a rate higher than direct transmission. In a similar way, in a multihop relay channel, some relays could in fact degrade the performance. Using techniques similar to those in Section 10.2, these redundant relays could be realized and removed from the chain. Taking the first three consecutive nodes (i.e., nodes 0, 1, and 2), node 1 is useful for transmission only if the following condition is satisfied:

$$\gamma_{0,2} < \left(1 + \gamma_{0,1}\right)^{\frac{\tau_0}{\tau_0 + \tau_1}} - 1. \tag{10.79}$$

Proceeding and taking three consecutive nodes every time, node $j \in \{1, 2, \ldots, M - 1\}$ is useful for transmission given nodes i, j, and k in order; we have

$$\gamma_{i,k} < \left(1 + \gamma_{i,j}\right)^{\frac{\tau_i}{\tau_i + \tau_j}} - 1. \tag{10.80}$$

This way the resulting route is free of redundant relays. It should be noted, however, that this is not necessarily the optimum route.

Optimal Time Allocation

After removing redundant relays, performance could be further improved by optimizing time allocation. For a given route, it can be shown that the optimal time allocation $\tau_{\text{opt}} = (\tau_0, \tau_1, \ldots, \tau_{m-1})$ is the one that satisfies

$$\frac{\tau_i}{\tau_0} = \frac{\log_2\left(1 + i + \gamma_{0,1} - \sum_{j=0}^{i-1}\left(1 + \gamma_{j,i+1}\right)^{\frac{\tau_j}{\tau_0}}\right)}{\log_2\left(1 + \gamma_{i,i+1}\right)}, \quad \text{for } i = 1, 2, \ldots, m - 1, \tag{10.81}$$

and

$$\sum_{i=0}^{m-1} \tau_i = 1, \tag{10.82}$$

assuming a unit of time is available for transmission.

Multihop Relay Channel: Outage Probability

Assuming Rayleigh fading channels and CSIR, the outage probability at the destination for a M-hop relay channel can be expressed as

$$\mathcal{P}_M(R) \triangleq \Pr\{\mathcal{I}_M < R\} = 1 - \prod_{j=1}^{M}\left[1 - \mathcal{P}_{\mathcal{I}_0^{(j)}}(R)\right], \tag{10.83}$$

where $\mathcal{P}_{\mathcal{I}_0^{(j)}}(R) \triangleq \Pr\left\{\mathcal{I}_0^{(j)} < R\right\}$. To derive $\mathcal{P}_M(R)$, we find it useful to write

$$\mathcal{I}_0^{(j)} = \tau_0 \log_2\left(\sum_{i=0}^{j-1} \Xi_{i,j} + 1\right), \quad j = 1, 2, \ldots, M, \tag{10.84}$$

where

$$\Xi_{i,j} = \left(1 + \gamma_{i,j}\right)^{\frac{\tau_i}{\tau_0}} - 1. \tag{10.85}$$

For Rayleigh fading channels, $\gamma_{i,j}$ is exponentially distributed with a cumulative distribution function:

$$\mathcal{F}_{\gamma_{i,j}}(\gamma) = \begin{cases} 1 - e^{-\frac{\gamma}{\Gamma_{i,j}}}, & \text{for } \gamma \geq 0, \\ 0, & \text{for } \gamma < 0, \end{cases} \tag{10.86}$$

where $\mathrm{E}\left[\gamma_{i,j}\right] = \Gamma_{i,j}$. Additionally, it can be shown that $\Xi_{i,j}$ is a Weibull-distributed random variable with the probability density function

$$f_{\Xi_{i,j}}(x) = \begin{cases} \dfrac{1}{\Gamma_{i,j}} \dfrac{\tau_0}{\tau_i} (x+1)^{\frac{\tau_0 - \tau_i}{\tau_i}} e^{-\dfrac{\left[(x+1)^{\frac{\tau_0}{\tau_i}} - 1\right]}{\Gamma_{i,j}}}, & \text{for } x \geq 0 \\ 0, & \text{for } x < 0. \end{cases} \tag{10.87}$$

If we now define $\Omega_0^{(j)} \triangleq \sum_{i=0}^{j-1} \Xi_{i,j}$, then for independent channels and $z \geq 0$, we have the cumulative distribution function for $\Omega_0^{(j)}$ as

$$\mathcal{F}_{\Omega_0^{(j)}}(z) = \int_\infty^z \int_\infty^{z-x_{0,j}} \cdots \int_{-\infty}^{z-\sum_{i=0}^{j-2} x_{i,j}} \prod_{i=0}^{j-1} f_{\Xi_{i,j}}\left(x_{i,j}\right) dx_{0,j} dx_{1,j} \cdots dx_{j-1,j}. \tag{10.88}$$

As a result, we have

$$\mathcal{P}_{\mathcal{I}_0^{(j)}}(\kappa) = \mathcal{F}_{\Omega_0^{(j)}}\left(2^{\frac{\kappa}{\tau_0}} - 1\right)$$

$$= \int_\infty^{2^{\frac{\kappa}{\tau_0}} - 1} \int_\infty^{2^{\frac{\kappa}{\tau_0}} - 1 - x_{0,j}} \cdots \int_{-\infty}^{2^{\frac{\kappa}{\tau_0}} - 1 - \sum_{i=0}^{j-2} x_{i,j}}$$

$$\prod_{i=0}^{j-1} f_{\Xi_{i,j}}\left(x_{i,j}\right) dx_{0,j} dx_{1,j} \cdots dx_{j-1,j}. \tag{10.89}$$

As expected, no closed form is obtained. Equation (10.89) is used to find $\mathcal{P}_{\mathcal{I}_0^{(1)}}(R)$, $\mathcal{P}_{\mathcal{I}_0^{(2)}}(R), \ldots, \mathcal{P}_{\mathcal{I}_0^{(M)}}(R)$ and substitute them in Equation (10.83) to get $\mathcal{P}_M(R)$. $M = 1$ and $M = 2$ are the special cases of direct transmission and two-hop relay channel studied in Section 10.2, respectively.

Summary

In this section we extend the three-node relay model discussed in Section 10.2 to an M-hop one. Specifically, we consider a channel with a single source, a single destination, and $M - 1$ relay nodes with half-duplex constraint. A model was presented, in addition to expressions for the mutual information and outage probability. For more on multiple-relay channels please refer to [390,393,409,432].

10.5 CHAPTER SUMMARY AND FURTHER READINGS

In this chapter, cooperative diversity obtained via relaying is introduced. The relay channel is a basic element for user cooperative diversity systems. Multihop relaying is also studied as a natural extension of the three-node relay channel. The wireless case is considered as a special case of the general relay channel. Expressions for the achievable rate as well as outage probability are presented. Other relevant issues such as optimization and partner selection are discussed.

For elementary reading on wireless communications, we refer readers to [179, 373]. Fundamental concepts of cooperative communications can be found in [409, 412,415,416,432]. Topics in relaying and wireless networks are studied by Kumar et al. in [269,384,385,433,434] and by Gastpar and Vetterli in [387,388]. A tutorial paper on cooperative communications is given in [435].

Traditionally, space diversity is obtained by employing multiple receiver antennas for independent reception. Recent advanced multiantenna technologies such as multiple-input, multiple-output antennas also are widely acknowledged. As the work in MIMO systems is related, we refer interested readers to [373,422,436–441] for results and fundamental concepts in the area. For the use of MIMO in relay channels, readers may refer to [442–444].

10.6 PROBLEMS

1 Explain how diversity can be obtained over time, frequency, or space.

2 What are the similarities and differences between the following diversity schemes:
   (a) Multiple antenna diversity.
   (b) Cooperative diversity.
   (c) User cooperative diversity.

3 (a) Show that the best position for the relay in the linear channel of Figure 10.3 to minimize power consumption is in the middle between the source and the destination.
   (b) For the linear relay network shown in Figure 10.3, find the maximum transmission range if β of the transmission power is allocated for the relay and the remaining $1 - \beta$ is used by the source. Show that $\beta = 1/2$ is optimum and achieves a maximum gain on range of $2^{1-1/\alpha}$. Find a general expression for the maximum range for an M-hop link.

4 (a) The outage probability for a channel can be defined as the probability that the channel power gain, G, falls bellow a threshold value H_{th}. For Rayleigh fading channels, G is exponentially distributed. As a function of H_{th}, the probability of outage is given by

$$\mathcal{P}(H_{\text{th}}) = 1 - e^{-\frac{H_{\text{th}}}{E[G]}}. \tag{10.90}$$

Use MATLAB plot $\mathcal{P}$ as a function of H_{th} using Equation (10.90) and assuming $E[G] = 1$.

   (b) In a similar way, the outage probability for m parallel channels, $\mathcal{P}_m$, is defined as the probability that G for all channels is below H_{th}. Find an expression for $\mathcal{P}_m(H_{\text{th}})$ assuming independent and identically distributed Rayleigh fading channels. Plot $\mathcal{P}_m(H_{\text{th}})$ for $m = 2, 4$, and 8 assuming $E[G] = 1$ for all channels. Discuss the resulting graph. Explain how diversity could improve channel reliability.

5 What is the difference between cooperative and noncooperative relaying?

6 What three features make transmission over a wireless medium differ from that of a wired medium?

7 List four optimization problems where power and time are allocated to achieve optimum performance. Discuss different scenarios where these optimization problems could be applicable.

8 The network generated in Section 10.3.3 has a Gaussian distribution in the sense that the x-coordinate and the y-coordinate that mark a node position are generated randomly according to a normal distribution. $E[d_{i,j}^{\alpha}]$ in Equation (10.73) is obtained assuming zero-mean and the same variance for both the x-coordinate and the y-coordinate for all nodes. Find an expression for $E[d_{i,j}^{\alpha}]$ when the variance used to generate source nodes' positions is different than the one used to generate the destination nodes' positions.

9 In a system where up to three users are allowed to cooperate, discuss different ways in which users can be arranged for cooperation.

10 What are the three types of spectrum holes? Explain how multihop relaying could be useful for secondary users exploiting a gray hole.

Information theoretical limits on cognitive radio networks

11

Natasha Devroye

University of Illinois at Chicago, United States

11.1 INTRODUCTION

In Chapter 10, the benefits, from a theoretical perspective, of cooperative communications were introduced. In this chapter we expand upon cooperative communications by allowing a subset of the nodes to be cognitive radios with their own data to transmit (and not merely relay), thereby forming cognitive radio networks. Cognitive networks, for our purposes, are wireless networks that consist of two types of users:

- **Primary users.** These wireless devices are the primary license holders of the spectrum band of interest. In general, they have priority access to the spectrum and are subject to certain quality-of-service (QoS) constraints that must be guaranteed.
- **Secondary users.** These users may access the spectrum, which is licensed to the primary users. They are thus secondary users of the wireless spectrum and are often envisioned to be cognitive radios. For the rest of this chapter, we assume the secondary users are cognitive radios (and the primary users are not) and use the terms interchangeably. These cognitive users employ their "cognitive" abilities to communicate while ensuring the communication of primary users is kept at an acceptable level.[1]

The study of cognitive networks is relatively new and many questions and aspects must be tackled before cognitive radios can seamlessly and opportunistically

---

[1]*Acceptable* may mean a number of things. Different mathematical models may be useful for different situations.

Doi: 10.1016/B978-0-12-374715-0.00011-3

employ spectrum licensed to primary user(s). Of both theoretical and practical importance is the question, What are the fundamental limits of communication in a cognitive network? Information theory provides an ideal framework for analyzing this question, as it encompasses a number of tools and metrics suited to such fundamental studies. The limits obtained provide benchmarks for the operation of cognitive networks, where researchers may gauge the efficiency of any practical network as well as draw inspiration as to which direction to pursue in their design.

In this chapter, we outline recent information theoretic advances pertaining to the limits of cognitive networks. We emphasize and explore the impact of *cognition*, defined as extra information (or side information) the cognitive radio nodes have about their wireless environment, on the fundamental limits. We first briefly describe why cognitive networks are of intense contemporary interest before outlining several types of *cognition* that cognitive networks could exploit.

11.1.1 The Rise and Importance of Cognitive Networks

Cognitive networks are motivated by the apparent lack of spectrum under the current spectrum management policies. The right to use the wireless spectrum in the United States is controlled by the Federal Communications Commission (FCC) [445]. Most of the frequency bands useful to wireless communication have already been licensed by the FCC (see Chapter 2). A few bands have however been designated by the FCC to be unlicensed bands, most notably the industrial, scientific, and medical bands (ISM bands), over which the immensely popular WiFi devices transmit. These bands are filling up fast, and despite their popularity, the vast majority of the wireless spectrum is in fact licensed. Currently, the primary license holders obtain from the FCC the exclusive right to transmit over their spectral bands. As most of the bands have been licensed out and the unlicensed bands are rapidly filling up, it would appear that we are approaching a spectral crisis. This, however, is far from the case. Recent measurements, such as those presented in Chapter 2, have shown that, for as much as 90% of the time, large portions of the *licensed* bands remain unused. As licensed bands are difficult to reclaim and release, the FCC is considering *dynamic* and *secondary* spectrum licensing [60, 446] as an alternative to reduce the amount of unused spectrum. Bands licensed to primary users could, under certain negotiable conditions, be shared with nonprimary users without having the primary licensee release its own license. Whether the primary users would be willing to share their spectrum depends on a number of factors, including the impact on their own communication.

Cognitive radios, wireless devices with reconfigurable hardware and software (including transmission parameters and protocols) [6], are capable of delivering what these secondary devices would need: the ability to intelligently sense and adapt to their spectral environment. Along with this newfound flexibility comes the challenge of understanding the limits and designing protocols and transmission schemes to fully exploit these cognitive capabilities. In particular, to design practical and efficient protocols, the theoretical limits must be well understood. We next describe different scenarios, assumptions, and corresponding types of cognitive behavior for which information theoretic limits have been considered.

11.1.2 Types of Cognitive Behavior

Networks that contain cognitive radios should intuitively be able to achieve better performance than networks in which they are absent. We are being intentionally vague at this point as to what *performance* and *fundamental limits* mean as they vary among applications. Different information theoretic metrics are discussed in the next section. Cognitive networks should achieve better performance, as they are able to (1) exploit their cognitive abilities—that is, sense and adapt to their wireless environment, and (2) often (but not necessarily) exploit new policies in secondary spectrum licensing scenarios in which the agile cognitive radios are permitted to share the spectrum with primary users. Naturally, the extent to which the performance of the network can be improved depends on what the cognitive radios know about their spectral environment and, consequently, how they adapt to this. We depart from the assumption that the secondary users are cognitive radios wishing to share the primary users' spectrum. Cognitive behavior, or how the secondary cognitive users employ the primary spectrum, may be grouped into three categories, as is done with slight variations in [447–450], each of which exploits varying degrees of knowledge of the wireless environment at the secondary user(s). Another implicit assumption is that the burden of guaranteeing primary-user communication at a predetermined level is borne by the secondary users. That is, the legacy primary system does not necessarily (but may) adapt to the cognitive users, while the cognitive users definitely adapt to the primary, often legacy, system. The three types of cognitive behavior we consider are:

- **Interference-avoiding behavior (spectrum interweave).** The secondary users employ the primary spectrum without interfering with the primary users whatsoever. The primary and secondary signals may be thought of as orthogonal to each other: They may access the spectrum in a time division multiple access (TDMA) fashion, in a frequency division multiple access (FDMA) fashion, or in any fashion that ensures that the primary and secondary signals do not interfere with each other. The cognition required by the secondary users to accomplish this is knowledge of the spectral gaps (for example, in time or frequency) of the primary system. The secondary users may then fill in these spectral gaps. Notice that this form of behavior is referred to as *spectrum overlay* in Chapter 5; unfortunately, information theory uses a different definition of spectrum overlay, as given later.

■ **Interference-controlling behavior (spectrum underlay).** The secondary users transmit over the same spectrum as the primary users but do so in a way that the interference seen by the primary users from the cognitive users is controlled to an acceptable level. This acceptable level is captured by primary QoS constraints.[2] This is termed *underlay*, as often the cognitive radios transmit in such a fashion that they appear to be noise under the primary signals. The cognition required is knowledge of the "acceptable levels" of interference at primary users in a cognitive user's transmission range as well as knowledge of the effect of the cognitive transmission at the primary receiver. This last assumption boils down, in classic wireless channels, to knowledge of the channel(s) between the cognitive transmitter(s) and the primary receiver(s). In Chapter 5, spectrum underlay is described in a similar manner, with an emphasis on how underlay techniques are achieved: using spread-spectrum techniques.

■ **Interference-mitigating behavior (spectrum overlay).** The secondary users transmit over the same spectrum as the primary users, but in addition to knowledge of the channels between primary and secondary users (nature), the cognitive nodes have information about the primary system and its operation. Examples are knowledge of the primary users' code books, allowing secondary users to decode the primary users' transmissions, or in certain cases even knowledge of the primary users' message. In Section 11.5 we discuss why these assumptions may be plausible and realizable in cognitive networks. *Spectrum overlay* in Chapter 5 refers to the spectrum interweave concept; there is presently no analogous concept of overlay in dynamic spectrum access systems in agile transmission techniques.

To illustrate the effect of different types of cognition, in this chapter we take as an example a simple channel in which a primary transmitter-receiver pair (white, $\mathcal{P}_{tx}, \mathcal{P}_{rx}$) and a cognitive transmitter-receiver pair (gray, $\mathcal{S}_{tx}, \mathcal{S}_{rx}$) share the same spectrum, shown in Figures 11.1 and 11.2. We derive fundamental limits on the communication possible under each type of cognition. One information theoretic metric that lends itself well to illustrative purposes and is central to many studies is the capacity region of the channel in Figure 11.2. Under Gaussian noise, we illustrate different examples of cognitive behavior and in Sections 11.3–11.5 we build up to the right illustration in Figure 11.2, which corresponds to the rates achieved under different levels of cognition.

The basic and natural conclusion is that, the higher the level of cognition at the cognitive terminals, the higher the achievable rates. However, increased cognition often translates into increased complexity. At what level of cognition future secondary spectrum licensing systems will operate depends on the available side information and network design constraints. We next outline the chapter in more detail.

---

[2]What constitutes an acceptable level is described later and may vary from system to system.

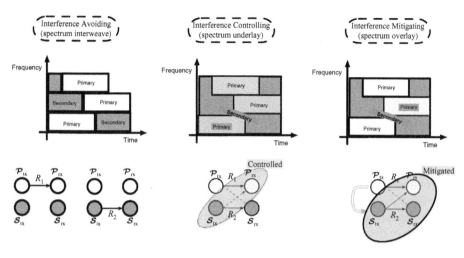

FIGURE 11.1

Three types of cognitive behavior.

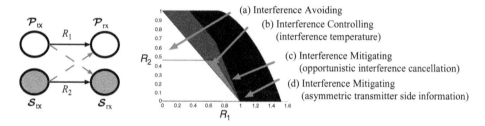

(a) Interference Avoiding

(b) Interference Controlling
(interference temperature)

(c) Interference Mitigating
(opportunistic interference cancellation)

(d) Interference Mitigating
(asymmetric transmitter side information)

FIGURE 11.2

The primary users (white) and secondary users (gray) wish to transmit over the same channel. Solid lines denote desired transmission, dotted lines denote interference. The achievable rate regions under four cognitive assumptions and transmission schemes are shown on the right. Strategies (a)–(d) are in order of increasing cognitive abilities.

11.1.3 Chapter Preview

We start the study of information theoretic limits of cognitive networks in Section 11.2, where we define classic information theoretic channels and measures of interest. In Section 11.3 we outline challenges and trends in interference-avoiding cognitive behavior. In Section 11.4 we first outline the limits of some spectrum underlay techniques in small networks then pursue throughput scaling laws in large cognitive underlay networks. In Section 11.5 we explore the communication possibilities when the cognitive radios are able to decode some of the primary users' messages, allowing for behavior that borders on and overlaps *cooperative* behavior. Of particular interest in cognitive networks is the lack of symmetry in the cooperation. These sections progressively build the reader up to the achievable rate regions of Figure 11.2, which illustrates the main point: Different levels of cognition result in different fundamental limits.

11.2 INFORMATION THEORETIC BASICS

One of information theory's main contributions is the characterization of fundamental limits of communication. We first define two of the most common types of communication channels: the discrete memoryless channel and additive white Gaussian noise channel. We then outline two information theoretic metrics of interest: the capacity (region) and the sum-throughput scaling law, which are examined for cognitive networks. We then briefly outline classic information theoretic channels that form a nice basis from which to explore channels with primary and secondary users.

11.2.1 Communications Channels

A channel is modeled as a set of conditional probability density functions relating the inputs and outputs of the channel. Communication over this channel takes place through the use of an encoder, which may be viewed as a function that maps a message into an encoded channel *input* (*sequence*) and a decoder, which may be viewed as a function that takes the channel *output* (*sequence*) and tries to recover the sent message. Two of the most common information theoretic channels are discrete memoryless channels (DMCs) and additive white Gaussian noise (AWGN) channels. While we outline information theoretic results for both types of channels, the latter lends itself well to examples and illustrations and so is focused on more heavily.

Discrete Memoryless Channels

Communication over a discrete memoryless channel takes place in a discrete number of "channel uses," indexed by the natural number $n \in \mathbb{N}$. We illustrate concepts using the simple two-transmitter, two-receiver channel shown in Figure 11.3. The primary (secondary) transmitter $\mathcal{P}_{tx}$ ($\mathcal{S}_{tx}$) wishes to communicate a message to a single primary (secondary) receiver $\mathcal{P}_{rx}$ ($\mathcal{S}_{rx}$). The transmitters communicate their messages by transmitting *codewords*, which span n channel uses (one input symbol per channel use). The receivers independently decode the received signals, often corrupted by noise according to the statistical channel model, to obtain the desired message. One quantity of fundamental interest in such communication is the maximal *rate*, typically cited in bits/channel use at which communication can take place. Most information theoretic results of interest are asymptotic in the number of channel uses; that is, they hold in the limit as $n \to \infty$.

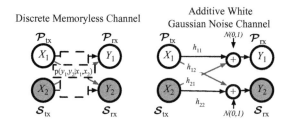

FIGURE 11.3

The primary users (white, $\mathcal{P}_{tx} \to \mathcal{P}_{rx}$) and secondary users (gray, $\mathcal{S}_{tx} \to \mathcal{S}_{rx}$) wish to transmit over the same channel. We show a discrete memoryless channel, described by $p(y_1, y_2 | x_1, x_2)$ and an additive white Gaussian noise channel with channel coefficients $h_{11}, h_{12}, h_{21}, h_{22}$.

A *discrete channel* has finite input alphabets and output alphabets $\mathcal{X}_1, \mathcal{X}_2$ and $\mathcal{Y}_1, \mathcal{Y}_2$, respectively, which are related though a collection of conditional probability mass functions $p(y_1, y_2 | x_1, x_2)$. These conditional distributions define the DMC. Transmitter $\mathcal{P}_{tx}$ ($\mathcal{S}_{tx}$) wishes to send a message $W_1 \in \{1, 2, \ldots, 2^{nR_1}\}$ ($W_2 \in \{1, 2, \ldots, 2^{nR_2}\}$), where R_1 (R_2) are the transmission rates, encoded as the n-sequence $\mathbf{x}_1^n$ ($\mathbf{x}_2^n$) to its receivers in n channel uses. The received signal, which the decoder uses to obtain the transmitted signal, is y_1^n (y_2^n). We consistently use the notation x as an instance of the random variable X, which takes on value in the alphabet $\mathcal{X}$. Vectors and matrices are denoted using bold fonts, and we omit super- and subscripts n when it is clear from context.

Additive White Gaussian Noise Channels

The additive white Gaussian noise channel is typically considered the most important continuous alphabet channel [297]. For the purpose of this chapter, at each channel use, we assume that outputs at the primary and cognitive receivers, Y_1 and Y_2, respectively, are related to the inputs at the primary and cognitive transmitters X_1 and X_2, respectively, as shown in Figure 11.3, as

$$Y_1 = X_1 + h_{21}X_2 + N_1, \quad N_1 \sim \mathcal{N}(0, 1)$$

$$Y_2 = h_{12}X_1 + X_2 + N_2, \quad N_2 \sim \mathcal{N}(0, 1).$$

Here h_{12}, h_{21} are the quasistatic [451] fading coefficients assumed to be known to all transmitters and receivers. By normalizing we may assume without loss of generality that $h_{11} = h_{22} = 1$. The rate achieved by the primary and cognitive tx-rx pairs are R_1 and R_2, respectively, measured in bits/channel use. In large cognitive networks, we assume that each transmitter-receiver pair in the wireless network sees independent additive white Gaussian noise and, when applicable, independent fading. In networks, we refer to the *sum-rate* or *sum-throughput* as the sum of all the rates of the different transmitter-receiver pairs in the network that may be simultaneously achieved (we define what it means for a rate to be achieved shortly). For more extended definitions of standard information theoretic quantities, see [297, 452, 453].

Multiple-Input, Multiple-Output Channels

Wireless channels in which transmitters and receivers employ multiple antennas, or multiple-input, multiple-output (MIMO) channels, have been extensively studied in recent years due to their ability to combat fading, increase data rates, or allow one to beamform signals to desired users or spatial subspaces [436, 439]. In cognitive radio networks, the use of MIMO adds another dimension to the problem and could, for example, allow MIMO cognitive users to transmit in the null space of the primary channels.

11.2.2 Information Theoretic Metrics of Interest

Given a probabilistic characterization of a channel, the fundamental limits of communication over the channel may be expressed in terms of a number of metrics. In this section, we consider the following two commonly considered and powerful metrics:

1. **Capacity/capacity regions.** Largest rate/rate tuples at which reliable communication may be ensured.
2. **Sum-throughput scaling.** How the sum-rate of the network scales with the number of nodes n, as $n \to \infty$.

Capacity/Capacity Regions

The capacity of a point-to-point channel (single transmitter, single receiver) is defined as the supremum over all rates (expressed in bits/channel use) for which reliable communication may take place. Reliable communication is achieved when the probability of decoding error may be made arbitrarily small, and is usually achieved in an asymptotic sense as the number of channel uses tends to infinity. Shannon's pioneering work [426] proved that, for a simple discrete memoryless point-to-point channel with inputs x of the input alphabet $\mathcal{X}$ to the outputs y of the output alphabet $\mathcal{Y}$, the capacity C is given by the supremum over all input distributions $p(x)$ of the mutual information:

$$I(X; Y) = \sum_{x,y} p(x,y) \log_2 \left[\frac{p(x,y)}{p(x)p(y)} \right].$$

Naturally, C depends on the conditional distributions $p(y|x)$, which define the DMC. The mutual information $I(X; Y)$ intuitively measures how much information the variables X and Y share; that is, how much one can tell you about the other. One of the most challenging aspects in obtaining the capacity of a channel is determining what input distribution $p(x)$ maximizes the mutual information.

In the point-to-point AWGN channel, the output Y is related to the input X according to $Y = hX + N$, where h is a fading coefficient (often modeled as a Gaussian random variable), and N is the noise, which is $N \sim \mathcal{N}(0,1)$. Under an average input power constraint $E[|X|^2] \leq P$, it is known that the optimal input distribution is Gaussian as well, allowing one to obtain the well-known capacity

$$C = \frac{1}{2}\log_2\left(1 + |h|^2 P\right) = \frac{1}{2}\log_2\left(1 + \text{SINR}\right) = \mathcal{C}(\text{SINR})\text{(bits/channel use)}.$$

Here SINR is the received signal-to-interference-plus-noise ratio, and $\mathcal{C}(x) = \frac{1}{2}\log_2(1+x)$. Gaussian noise channels have the computationally convenient property that the optimal capacity-achieving input distribution $p(x)$ is often Gaussian as well. Thus, in Gaussian noise channels, even when the capacity-achieving input distribution of the channel is unknown, achievable rate regions are often computed assuming Gaussian input distributions.

Capacity as defined above is a single number. When multiple data streams are to be communicated, for example, in a channel that consists of two users transmitting at rates R_1 and R_2, respectively, the *capacity region*, in this case a two-dimensional region, is the generalization of the capacity to multiple dimensions. While capacity is central to many information theoretic studies, it is often challenging to determine. Inner bounds, or achievable rates, as well as outer bounds to the capacity (or capacity region) may be more readily available. Inner bounds (or inner bound regions when there are multiple simultaneous data streams) lie inside the capacity (capacity region) and are obtained by suggesting a coding scheme and proving that it achieves asymptotically small error as the block length $n \to \infty$. The suggested scheme may not be the best transmission scheme, and a larger achievable rate (achievable rate regions) may exist. The capacity (or capacity region for multiple data streams) is the supremum over all achievable rates (rate regions, respectively). An outer bound to the capacity is a rate above which reliable communication (in the sense of the probability of error $\to 0$ as the block length $n \to \infty$) may be shown to be impossible. If one can determine an outer bound on the capacity region and show that a particular scheme achieves all points on the outer bound of a channel, then that encoding scheme is said to be *capacity achieving*.

Capacity regions have been particularly difficult to obtain for channels in which multiple transmitters and multiple receivers simultaneously wish to communicate. Indeed, *multiuser information theory* or *network information theory* is a challenging field with a plethora of open questions. As an example, one of the central and simplest of multiuser channels is the information theoretic *interference channel*. This channel, introduced over 30 years ago [454, 455], consists of two independent transmitters that wish to communicate independent messages to two independent receivers, much like the simple channel depicted in Figure 11.3. Although the channel capacity region[3] is known in certain cases, the general

---

[3]Capacity regions and achievable rate regions are natural extensions of the notions of capacity and achievable rate to higher dimensions.

capacity region, despite promising recent advances [456–459], is still an open problem. At the crux of this lies the information theory community's lack of understanding of how to deal with interference and overheard, undesired information.

Sum-Throughput Scaling Laws

When we have a network of nodes, the exact capacity region of the network is currently out of reach. The sum-throughput scaling law of a network is a more tractable asymptotic approximation commonly used in describing wireless channels. The study of throughput scaling laws was initiated by the work of Gupta and Kumar [269] and expanded to consider a variety of wireless channel models and communication protocol assumptions [433, 460–470]. One typically assumes n (transmitter-receiver) pairs of randomly located devices wish to communicate and asks how their sum-rate scales as a function of n. The number of nodes n is allowed to grow to infinity by either letting the density of nodes stay fixed and the area increase with n (extended network) or fixing the network area and letting the density increase with n (dense network). Due to node limitations such as power constraints, multiple hops may be needed for a specific message to reach a distant destination.

As expected, the throughput scaling in ad hoc networks depends greatly on the node distribution and the physical layer processing capability, more specifically the ability to cooperate among nodes. Some specific examples are

- In the interference-limited regime, in which no cooperation is allowed (except simple decode and forward), and all nodes treat other signals as interference, the per-node throughput scales at most as $1/\sqrt{n}$ [269].
- If the nodes are uniformly distributed, a simple nearest-neighbor forwarding scheme achieves a $1/[n\log(n)]$ per-node throughput [269].
- When the nodes are distributed according to a Poisson point process, a backbone-based routing scheme achieves the per-node scaling of $1/\sqrt{n}$ [468], meeting the upper bound.
- When nodes are able to cooperate in a MIMO-like fashion, a novel hierarchical scheme can achieve *linear* growth in the sum rate, corresponding to a constant per-node throughput [470].

The development of these scaling laws show that the assumptions about the network and the nodes' signal processing capability are crucial to the scaling law. We later outline scaling law results for *cognitive networks* in which primary and secondary devices coexist. This scenario differs from existing scaling law results in that the network is no longer homogeneous and the cognitive nodes must transmit in a way that is acceptable to the primary nodes.

11.2.3 Classic Channels

We briefly outline four classic multiuser channels, illustrated in Figure 11.4, that are relevant to the study of cognitive networks.

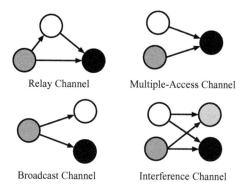

FIGURE 11.4

Four classic information theoretic multiuser channels. Contemporary cognitive channels encompass and may use the clever techniques used in these well-studied channels.

Relay Channels

In a relay channel [376] (see also Chapter 10), the communication between a source (gray) and a destination (black) may be aided through the use of a relay node (white), which has no information of its own to convey. Despite its simplicity, the capacity of this channel remains unknown in general. A comprehensive survey of relay channels may be found in [386], while interesting approximation results for relay networks are presented in [471]. The notion of having one node relay information for another is a fundamental one, and techniques used in relay channels (for example, block Markov coding) may be used in cognitive channels. In addition, cognitive nodes may act as relays for other cognitive or even primary nodes.

Multiple-Access Channels

In the classic multiple-access channel (MAC) two independent nodes wish to communicate to a single common receiver. The capacity of this channel and techniques employed such as superposition coding and successive decoding are well understood [297, 452, 453] and form the building blocks for achievable rate regions of other closely related channels including the interference channel. Interesting aspects and properties of the capacity region of MAC may be found in [404, 405, 472, 473].

Broadcast Channels

In the classic broadcast channel a single transmitter wishes to communicate independent messages[4] to two independent receivers. Its capacity region is in general unknown. For the DMC, some of the best achievable rate regions and outer bounds

---

[4]In general, the transmitter may wish to transmit a common message destined to both receivers as well as independent ones to each receiver.

may be found in [474, 475]. The capacity region of the Gaussian MIMO broadcast channel was recently unveiled [476, 477], where it was shown that encoders using Gaussian code books with successive dirty paper coding [174, 478] are capacity achieving.

Interference Channels

In the classic interference channel [454, 455] two independent transmitters wish to transmit two independent messages to two independent receivers. The capacity region of this channel is in general unknown, though there are capacity results for a few special cases [479, 480]. The best inner bounds for discrete memoryless interference channels may be found in [459, 481] and tight outer bounds are obtained in [482]. Some of the main techniques used in interference channels include rate splitting, superposition coding, and binning, all of which are useful for cognitive channels. In Gaussian noise, recent advances are bringing us close to the capacity [456–458, 483]. The degrees of freedom in MIMO interference channels may be found in [456, 484, 485].

Cognitive Channels

We now turn to a novel class of channels: *cognitive channels*. In these channels a subset of the nodes may be cognitive radios that wish to access the spectrum licensed out to primary user(s). A simple example is shown in Figure 11.3. As stated before, cognitive channels seek to exploit the cognition enabled by the cognitive radios; that is, exploit the fact that cognitive radios are able to sense their spectral environment and adapt to it. Various types of cognition or side information may be available to the cognitive radios. Their behavior varies depending on this side information. We now use some of the tools, metrics, and channels outlined in this section to explore three types of cognitive behavior: interference avoiding, interference controlling, and interference mitigating.

11.3 INTERFERENCE-AVOIDING BEHAVIOR: SPECTRUM INTERWEAVE

Secondary spectrum licensing cognitive radio was arguably conceived with the goal and intent of implementing interference-avoiding behavior [7, 11]. Indeed, in this intuitive approach to secondary spectrum licensing, cognitive radios sense the spatial, temporal, or spectral voids and adjust their transmission to fill in the sensed *white spaces*. Cognition in this setting corresponds to the ability to accurately detect the presence of other wireless devices; the cognitive side information is knowledge of the spatial, temporal, and spectral gaps a particular cognitive tx-rx pair would experience. The cognitive radios adjust their transmission to fill in the spectral (or spatial/temporal) void, as illustrated in Figure 11.1, with the potential to drastically increase the spectral efficiency of wireless systems.

The cognition required for this type of behavior is knowledge of the spectral gaps. In a realistic system the secondary transmitter spends some of its time sensing the channel to determine the presence of the primary user. As an illustrative example and idealization, we assume that knowledge of the spectral gaps is perfect: When primary communication is present the cognitive devices are able to precisely determine it, instantaneously. While such assumptions may be valid for the purpose of theoretical study and provide outer bounds on what can be realistically achieved, practical methods for detecting primary signals have also been of great recent interest. A theoretical framework for determining the limits of communication as a function of the sensed cognitive transmitter and receiver gaps is formulated in [486,487]. Studies on how detection errors may affect the cognitive and primary systems are found in [115,488,489]. Because current secondary spectrum licensing proposals demand detection guarantees of primary users at extremely low levels in harsh fading environments, a number of authors have suggested improving detection capabilities through allowing multiple cognitive radios to collaboratively detect the primary transmissions [344,490–492] (see also Chapter 4).

Under our idealized assumptions, the rates R_1 of the primary tx-rx pair and R_2 of the cognitive tx-rx pair achieved through ideal white space filling are shown as the inner white triangle of Figure 11.2. When a single user transmits the entire time in an interference-free environment, the axes intersection points are attained. The convex hull of these two interference-free points may be achieved by time sharing (TDMA fashion). Where on this line a system operates depends on how often the primary user occupies the specific band. If the primary and secondary power constraints are P_1 and P_2, respectively, then the white space filling rate region may be described as

White space filling region (a)

$$= \left\{ (R_1, R_2) | 0 \leq R_1 \leq \eta \mathcal{C}(P_1), 0 \leq R_2 \leq (1 - \eta)\mathcal{C}(P_2), 0 \leq \eta \leq 1 \right\}.$$

Interference Avoidance through MIMO

In addition to detecting the spectral white spaces, interference at the primary user may be avoided or controlled if the cognitive user is equipped with multiple antennas and is able to place its transmit signal in the null space of the primary users' receive channel. In this scenario, the exact channel between the secondary transmit antennas and the primary receive antennas must be known. In [493], the authors study the fundamental trade-off a cognitive transmitter faces between maximizing its own transmit throughput and minimizing the amount of interference it produces at each primary receiver. They address this from an information theoretic perspective by characterizing the secondary user's channel capacity under both its own transmit power constraint and a set of interference power constraints, each imposed at one of the primary receivers. In particular, this paper exploits multiantennas at the secondary transmitter to effectively balance between spatial multiplexing for the secondary transmission and interference avoidance at the primary receivers. In [494] a sum-rate maximization problem for single-input, multiple-output multiple-access channels (SIMO-MAC) under interference constraints for the primary users as well as a peak transmission power constraint for each secondary user is considered. The authors wish to maximize the rate of the secondary users subject to interference constraints on the primary users as well as peak power constraints for the secondary users. Other works that similarly exploit multiple antennas to avoid or control the interference seen by primary users may be found in [495–497]. The scenarios considered in these papers can be considered an *interference-avoiding* scheme if the tolerable interference at the primary receivers is set to zero; otherwise, it falls under the *interference-controlled* paradigm we look at in the next section.

11.4 INTERFERENCE-CONTROLLED BEHAVIOR: SPECTRUM UNDERLAY

When the interference caused by the secondary users on the primary users is permitted when below a certain level or while guaranteeing a certain level of quality of service, the more flexible *interference-controlled behavior* emerges. We look at spectrum underlay techniques and an example of the resulting achievable rate region in small networks, as well as the resulting throughput scaling laws in two types of large networks. We note that this type of interference-controlled behavior covers a large spectrum of cognitive behavior and highlight only three examples, while referring to only a small subset of all the possible references. In Figure 11.1 spectrum underlay is graphically depicted as having primary users slightly gray, as opposed to the interference-free white color illustrated in interference avoiding.

11.4.1 Underlay in Small Networks: Achievable Rates

Interference Temperature

Rather than detecting white spaces, in spectrum underlay, a cognitive radio simultaneously transmits with the primary user(s) while using its cognitive abilities to control the amount of harm it inflicts upon them. While the definition of harm may be formulated mathematically in a number of ways, one common definition involves the notion of *interference temperature*, a term first introduced by the FCC [498] to denote the average level of interference power seen at a primary receiver. In secondary spectrum licensing scenarios, the primary receiver's interference temperature should be kept at a level that will satisfy the primary user's desired quality of service. That is, primary transmission schemes may be designed to withstand a certain level of interference, which cognitive radios or secondary nodes may exploit for their own transmission. Provided the cognitive user knows (1) the maximal interference temperature for the surrounding primary receivers, (2) the current interference temperature level, and (3) how its own transmit power translates to received power at the primary receiver, then the cognitive radio may adjust its own transmission power to satisfy any interference temperature constraint the primary user(s) may have. The work in [499–502] all consider the capacity of cognitive systems under various receive-power (or interference temperature–like) constraints.

As an illustrative example, we consider a very simple interference temperature–based cognitive transmission scheme. Assume, in the channel model of Figure 11.3, that each receiver treats the other user's signal as noise, a lower bound to what may be achieved using more sophisticated decoders [503]. The rate region obtained is shown as the light-gray region (b) of Figure 11.2. This region is obtained as follows: We assume the primary transmitter communicates using a Gaussian code book of constant average power P_1. We assume the secondary transmitter allows its power to lie in the range $[0, P_2]$ for P_2 some maximal average power constraint. The rate region obtained may be expressed as

Simultaneous-transmission rate region (b)

$$= \left\{ (R_1, R_2) | 0 \le R_1 \le C \left(\frac{P_1}{b_{21}^2 P_2^* + 1} \right), \right. \tag{11.1}$$

$$\left. 0 \le R_2 \le C \left(\frac{P_2^*}{b_{12}^2 P_1 + 1} \right), \ 0 \le P_2^* \le P_2 \right\}.$$

The actual value of P_2^* chosen by the cognitive radio depends on the interference temperature or received power constraints at the primary receiver.

11.4.2 Underlay in Large Networks: Scaling Laws

Information theoretic limits of interference-controlled behavior have also been investigated for large networks; that is, networks of which the number of nodes $n \to \infty$. We illustrate two types of networks: single-hop networks and multi-hop networks. In the former, secondary nodes transmit the subject to outage probability–like constraints on the primary network. In the latter, the multihop secondary network is permitted to operate as long as the scaling law of the primary network is kept the same as in the absence of the cognitive network.

Single-Hop Cognitive Networks

The planar network model considered in [504] is depicted in Figure 11.5, where multiple primary and secondary users coexist in a network of radius R (the number of nodes grows to infinity as $R \to \infty$). Around each receiver, either primary or cognitive, we assume a protected circle of radius $\epsilon_c > 0$, in which no interfering transmitter may operate. Other than the receiver-protected regions, the primary

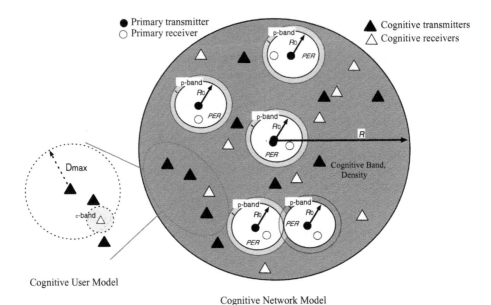

FIGURE 11.5

Network model: A cognitive network consists of multiple primary and cognitive users. The primary users locations are arbitrary with a minimum distance R_0 between any two primary transmitters. The cognitive transmitters are distributed randomly and uniformly with density λ. Cognitive user model: Each cognitive transmitter wishes to transmit to a single cognitive receiver that lies within a distance $\leq D_{max}$. Each cognitive receiver has a protected circle of radius $\epsilon_c > 0$, in which no interfering transmitter may operate.

transmitter and receiver locations are arbitrary, subject to a minimum distance R_0 between any two primary transmitters. This scenario corresponds to a broadcast network, such as TV or the cellular networks, in which the primary transmitters are base stations. The cognitive transmitters, on the other hand, are uniformly and randomly distributed with constant density λ. We assume that each cognitive receiver is within a D_{max} distance from its transmitter, as shown in the cognitive user model of Figure 11.5, and transmits with constant power P. We assume that the channel gains are path loss dependent only (no fading or shadowing) and each user treats unwanted signals from all other users as noise.

The quality-of-service guarantee of the primary users is in the form Pr[primary user's rate $\leq C_0] \leq \beta$. That is, the secondary users must transmit so as to guarantee that the probability that the primary users' rates fall below C_0 is less than a desired amount β. This may be done by appropriate selection of the network parameters $P, \epsilon_c, R_0, \lambda$, as done in [505]. The questions answered in [504] and [505] that relate to this single-hop cognitive network setting may be summarized as

- **What is the scaling law of the secondary network?** By showing that the average interference to the cognitive users remains bounded due to the finite transmission ranges D_{max} of the cognitive users and R_0 of the primary users, one can show that the lower and upper bounds to each user's average transmission rate are constant and thus the *average* network throughput grows linearly with the number of users [506].
- **How should the network parameters be chosen to guarantee Pr[primary user's rate $\leq C_0] \leq \beta$?** This interesting question is addressed in [505, 506] and omitted here for brevity.
- **May the cognitive nodes scale their power depending on the distance from the primary network?** Yes, under the assumption that a single primary network is at the center of the cognitive network, the further away the cognitive users are from the center, the larger their transmit power may be (or alternatively the larger the distance D_{max} may get). Specifically, suppose that a cognitive user at distance r transmits with power $P = P_c r^\gamma$, where P_c is a constant. Then, provided that $0 \leq \gamma < \alpha - 2$, the total interference from the cognitive users to the primary user is still bounded, making the power scaling an attractive option for the cognitive users. With the power scaling, the maximum distance D_{max} between a cognitive tx and rx can now grow with the network size as $D_{max} \leq K_d r^{\gamma/\alpha} < K_d r^{1-2/\alpha}$, where r again is the distance from the cognitive transmitter to the primary transmitter and K_d is a constant. Therefore, depending on the path loss α, the cognitive tx-rx distance can grow with an exponent of up to $1 - 2/\alpha$. For a large α, this growth is almost at the same rate as the network.

Multihop Cognitive Networks

We now consider a cognitive network consisting of multiple primary and multiple cognitive users, where there is no restriction on the maximum cognitive tx-rx distance. We assume tx-rx pairs are selected randomly, as in a classic [269] standalone ad hoc network. Both types of users are ad hoc, randomly distributed according to Poisson point processes with different densities. Here the quality-of-service guarantee to the primary users states that the scaling law of the primary ad hoc network does not diminish in the presence of the secondary network.

In [507] it is shown that, provided that the cognitive node density is higher than the primary node density, using multihop routing, *both* types of users, primary and cognitive, can achieve a throughput scaling as if the other type of users were not present. Specifically, the throughput of the m primary users scales as $\sqrt{m/\log m}$, and that of the n cognitive users as $\sqrt{n/\log n}$.

What is of particular interest in this result is that, to achieve these throughput scalings, the primary network need not change anything in its protocols; it is *oblivious* to the secondary network's presence. The cognitive users, on the other hand, rely on their higher density and a clever routing technique (in the form of *preservation regions* [507]) to avoid interfering with the primary users.

11.5 INTERFERENCE-MITIGATING BEHAVIOR: SPECTRUM OVERLAY

Thus far, the side information available to the cognitive radios has been (a) knowledge of the primary spectral gaps and (b) knowledge of the primary interference constraints and secondary to primary channel gains. In this section we increase the level of cognition even further. In interference-mitigating cognitive behavior, the cognitive user transmits over the same spectrum as the primary user but uses this additional cognition to mitigate (1) the interference it causes to the primary receiver and (2) the interference the cognitive receiver experiences from the primary transmitter.

To mitigate interference, the cognitive nodes must have the primary system's *code books*. This allows the cognitive transmitter or receiver to opportunistically decode the primary users' messages, which in turn may lead to gains for both the primary and secondary users, as we will see. We consider two types of interference-mitigating behavior in this section:

1. **Opportunistic interference cancellation.** The cognitive nodes have the code books of the primary users. The cognitive receivers opportunistically decode the primary users' messages that they pull off their received signal, increasing the secondary channel's transmission rates.
2. **Asymmetrically cooperating cognitive radio channels.** The cognitive nodes have the code books of the primary users, and the cognitive transmitter(s) has knowledge of the primary user's message. The cognitive transmitter may use this message knowledge to carefully mitigate interference at the cognitive receiver as well as cooperate with the primary users in boosting its signal at its receiver.

11.5.1 Opportunistic Interference Cancellation

We assume the cognitive link has the same knowledge as in the interference temperature case (b) and some additional information about the primary link's communication: the primary user's *code book*. Primary code book knowledge translates to being able to decode primary transmissions; we suggest a scheme that exploits this extra knowledge next.

In *opportunistic interference cancellation*, as first outlined in [508], the cognitive receiver opportunistically decodes the primary user's message, which it then subtracts off its received signal. This intuitively cleans up the channel for the cognitive pair's own transmission. The primary user is assumed to be oblivious to the cognitive user's operation and so continues transmitting at power P_1 and rate R_1. When the rate of the primary user is low enough relative to the primary signal power at the cognitive receiver (or $R_1 \leq C\left(h_{12}^2 P_1\right)$) to be decoded by $\mathcal{S}_{\text{rx}}$, the channel $(\mathcal{P}_{tx}, \mathcal{S}_{tx} \to \mathcal{S}_{rx})$ forms an information theoretic multiple-access channel, the capacity region of which is well known [297]. In this case, the cognitive receiver first decodes the primary's message, subtracts it off its received signal, and proceeds to decode its own. When the cognitive radio cannot decode the primary's message, the latter is treated as noise.

The region (c) of Figure 11.2 illustrates the gains opportunistic decoding may provide over the former two strategies. It is becoming apparent that higher rates are achievable when a higher level of cognition in the network is properly exploited. What type of cognition is valid to assume naturally depends on the system and application.

11.5.2 Asymmetrically Cooperating Cognitive Radio Channels

We increase the cognition even further and assume the cognitive node(s) has the primary code books as well as the message to be transmitted by the primary sender(s). For simplicity of presentation we consider again the two-transmitter, two-receiver channel shown in Figures 11.2 and 11.6. This additional message knowledge allows for a form of *asymmetric cooperation* between the primary and cognitive transmitters. This asymmetric form of transmitter cooperation, first introduced in [175, 509], can be motivated in a cognitive setting in a number of ways.

- Depending on the device capabilities, as well as the geometry and channel gains between the various nodes, certain cognitive nodes may be able to hear or obtain the messages to be transmitted by other nodes. For example, if $\mathcal{S}_{tx}$ is geographically close to $\mathcal{P}_{tx}$ (relative to $\mathcal{P}_{rx}$), then the wireless channel $(\mathcal{P}_{tx} \to \mathcal{S}_{tx})$ could be of much higher capacity than the channel $(\mathcal{P}_{tx} \to \mathcal{P}_{rx})$. Thus, in a fraction of the transmission time, $\mathcal{S}_{tx}$ could listen to and obtain the message transmitted by $\mathcal{P}_{tx}$. These messages would need to be obtained in real time and could exploit the geometric gains between cooperating transmitters relative to receivers in, for example, a two-phase protocol [175].

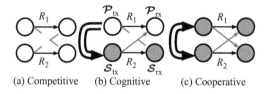

(a) Competitive   (b) Cognitive   (c) Cooperative

FIGURE 11.6

Three types of behavior depending on the amount and type of side information at the secondary transmitter. (a) Competitive: The secondary terminals have no additional side information. (b) Cognitive: The secondary transmitter has knowledge of the primary user's message and code book. (c) Cooperative: Both transmitters know each others' messages. The double line denotes noncausal message knowledge.

- In an automatic repeat request (ARQ) system, a cognitive transmitter, under suitable channel conditions (if it has a better channel to the primary transmitting node than the primary receiver), could decode the primary user's transmitted message during an initial transmission attempt. In the event that the primary receiver was not able to correctly decode the message and it must be retransmitted, the cognitive user would already have the to-be-transmitted message, or asymmetric side information, at no extra cost (in terms of overhead in obtaining the message).
- The authors in [510] consider a network of wireless sensors in which a sensor S_2 has a better sensing capability than another sensor S_1 and thus is able to sense two events, while S_1 is able to sense only one. Therefore, when they wish to transmit, they must do so under an asymmetric side information assumption: Sensor S_2 has two messages, and the other has just one.

The main question that information theory helps in answering is, How can the cognitive system best exploit this extra level of cognition; that is, knowledge of the primary user's message?

Background: Exploiting Transmitter Side Information

A key idea behind achieving high data rates in an environment where two senders share a common channel is interference cancelation or mitigation. The capacity of a discrete memoryless channel $p(y|x, s)$ when side information s (which may be thought of as interference) is known noncausally at the transmitter but not the receiver was first considered by Gel'fand and Pinsker [511]. They showed that the capacity of this discrete memoryless channel is given by

$$C = \max_{p(u|s)p(x|u,s)} I(U; Y) - I(U; S), \tag{11.2}$$

for an auxiliary random variable U distributed jointly with X and S.

The result of Gel'fand and Pinsker was later generalized by Costa [174] to real alphabets in his well-known paper entitled "Writing on Dirty Paper." There, he showed that, in a Gaussian noise channel with noise N of power Q, input X power constraint $E[|X|^2] \leq P$, and additive interference S of arbitrary power known noncausally to the *transmitter* but not the receiver,

$$Y = X + S + N, \quad E[|X|]^2 \leq P, \quad N \sim \mathcal{N}(0, Q)$$

the capacity is that of an interference-free channel (see Figure 11.7), or

$$C = \max_{p(u|s)p(x|u,s)} I(U; Y) - I(U; S) \tag{11.3}$$

$$= \frac{1}{2} \log_2 \left(1 + \frac{P}{Q}\right). \tag{11.4}$$

This remarkable and surprising result has found its application in numerous domains including data storage [512, 513], watermarking/steganography [514], and most recently, *dirty paper coding* has been shown to be the capacity-achieving

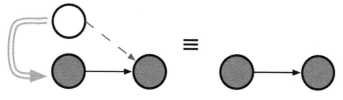

FIGURE 11.7

A channel with noncausal knowledge of the interference has, in Gaussian noise, a capacity equivalent to an interference-free channel.

technique in Gaussian MIMO broadcast channels [476, 478]. We now apply dirty paper coding techniques to the Gaussian cognitive channel.

Bounds on the Capacity of Cognitive Radio Channels

Although in practice the primary message must be obtained causally, as a first step, numerous works have idealized the concept of message knowledge: Whenever the cognitive node S_{tx} is able to hear and decode the message of the primary node P_{tx}, it is assumed to have full a priori knowledge.[5] The one-way double arrow in Figure 11.6 indicates that S_{tx} knows P_{tx}'s message but not vice versa. This is the simplest form of asymmetric noncausal cooperation at the transmitters. The term *cognitive* is used to emphasize the need for S_{tx} to be a device capable of obtaining the message of the first user and altering its transmission strategy accordingly.

This asymmetric transmitter cooperation present in the *cognitive* channel has elements in common with the *competitive* channel and the *cooperative* channels of Figure 11.6, which may be explained as follows:

1. **Competitive behavior/channel.** The two transmitters transmit independent messages. There is no cooperation in sending the messages, and therefore the two users *compete* for the channel. This is the same channel as the two-sender, two-receiver interference channel [454]. The largest to-date known general region for the interference channel is that described in [481], which is stated more compactly in [459]. Many of the results on the cognitive channel, which contains an interference channel if the noncausal side information is ignored, use a similar rate-splitting approach to derive large rate regions [175, 515, 516].
2. **Cognitive behavior/channel.** Asymmetric cooperation is possible between the transmitters. This asymmetric cooperation is a result of S_{tx} knowing P_{tx}'s message but not vice versa. We discuss this actively researched channel in this section.

---

[5]This assumption is often called the *genie assumption*, as these messages could have been given to the appropriate transmitters by a genie.

3. **Cooperative behavior/channel.** The two transmitters know each others' messages (two-way double arrows) and can therefore fully and symmetrically cooperate in their transmission. The channel pictured in Figure 11.6(c) may be thought of as a two-antenna sender, two single-antenna receivers broadcast channel, where, in Gaussian MIMO channels, dirty paper coding was recently shown to be capacity achieving [476, 478].

Cognitive behavior may be modeled as an interference channel with asymmetric, noncausal transmitter cooperation. This channel was first introduced and studied in [175, 509].[6] Since then, a flurry of results, including capacity results in specific scenarios, of this channel have been obtained. When the interference to the primary user is weak ($b_{21} < 1$), the rate region (d) has been shown to be the capacity region in Gaussian noise [517] and in related discrete memoryless channels [510]. In channels where interference at both receivers is strong, both receivers may decode and cancel out the interference, or where the cognitive decoder wishes to decode both messages, capacity is also known [515, 518, 519]. However, the most general capacity region remains an open question for both the Gaussian noise and discrete memoryless channel cases.

When using an encoding strategy that properly exploits this asymmetric message knowledge at the transmitters, the region (d) of Figure 11.2 is achievable in AWGN and the weak interference regime ($b_{21} < 1$ in AWGN) corresponds to the capacity region of this channel [517, 520]. The encoding strategy used assumes both transmitters use random Gaussian code books. The primary transmitter continues to transmit its message of average power P_1. The secondary transmitter splits its transmit power P_2 into two portions, $P_2 = \psi P_2 + (1 - \psi)P_2$ for $0 \leq \psi \leq 1$. Part of its power, ψP_2, is spent in a *selfless* manner: on relaying the message of $\mathcal{P}_{tx}$ to $\mathcal{P}_{rx}$. The remainder of its power, $(1 - \psi)P_2$, is spent in a *selfish* manner on transmitting its own message using the interference-mitigating technique of dirty paper coding. This strategy may be thought of as selfish, as power spent on dirty paper coding may harm the primary receiver (and is indeed treated as noise at $\mathcal{P}_{rx}$). The rate region (d) may be expressed as [517, 521]

Asymmetric cooperation rate region (d)

$$= \left\{ (R_1, R_2) | 0 \leq R_1 \leq C \left[\frac{(\sqrt{P_1} + b_{12}\sqrt{\psi P_2})^2}{b_{12}^2(1 - \psi)P_2 + 1} \right], \right. \tag{11.5}$$

$$\left. 0 \leq R_2 \leq C((1 - \psi)P_2), \ 0 \leq \psi \leq 1 \right\}.$$

By varying ψ, we can smoothly interpolate between strictly selfless behavior to strictly selfish behavior. Of particular interest from a secondary spectrum licensing perspective is the fact that the primary user's rate R_1 may be strictly increased

---

[6]It was first called the *cognitive radio channel* and is also known as the *interference channel with degraded message sets.*

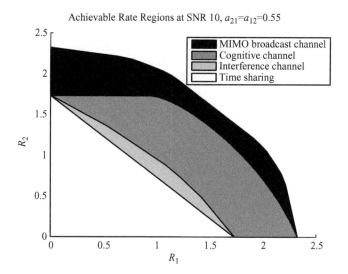

FIGURE 11.8

Capacity region of the Gaussian 2 × 1 MIMO two-receiver broadcast channel (outer), cognitive channel (middle), achievable region of the interference channel (second smallest), and time-sharing (innermost) region for Gaussian noise powers $N_1 = N_2 = 1$, power constraints $P_1 = P_2 = 10$ at the two transmitters, and channel parameters $h_{12} = 0.55$, $h_{21} = 0.55$.

with respect to all other three cases (i.e., the *x*-intercept is now to the right of all other three cases). That is, by having the secondary user possibly relay the primary's message in a selfless manner, the system essentially becomes a 2 × 1 multiple-input, single-output system that sees all the associated capacity gains over noncooperating transmitters or antennas. This increased primary could serve as a motivation for having the primary share its code book and message with the secondary user.

While Figure 11.2 shows the impact of increasing cognition (or side information at the cognitive nodes) on the achievable rate regions corresponding to protocols that use this side information, Figure 11.8 shows the impact of transmitter coopera-tion. In the figure, the region achieved through asymmetric transmitter cooperation (*cognitive behavior*) is compared to (1) the Gaussian MIMO broadcast channel region (in which the two transmitters may cooperate, *cooperative behavior*), (2) the achievable rate region for the interference channel region obtained in [481] (the largest known to date for the Gaussian noise case, *competitive behavior*),[7] and

---

[7] The achievable rate region of [481] used in these figures (as the "interference channel" achievable region) assumes the same Gaussian input distribution as in [175] and is omitted for brevity.

(3) the time-sharing region where the two transmitters take turns using the channel (*interference-avoiding behavior*). The capacity region for the Gaussian MIMO broadcast channel with two single-antenna receivers and one transmitter with two antennas subject to per-antenna power constraints of P_1 and P_2, respectively, is given by Equation (11.6), which may be obtained from the general formulation in [476, 478]:

MIMO BC region = Convex hull of

$$
\left\{
\begin{array}{c}
(R_1, R_2): \\[4pt]
\begin{array}{cc}
R_1 \le \frac{1}{2}\log_2\left(D_1''\right) = R_1(\pi_{12}) & R_1 \le \frac{1}{2}\log_2\left(D_1'\right) = R_1(\pi_{21}) \\[4pt]
R_2 \le \frac{1}{2}\log_2\left(D_2'\right) = R_2(\pi_{12}) \quad \bigcup \quad & R_2 \le \frac{1}{2}\log_2\left(D_2''\right) = R_2(\pi_{21})
\end{array} \\[10pt]
B_1, B_2 \succeq 0, \quad B_1 = \begin{bmatrix} b_{11} & b_{12} \\ b_{12} & b_{22} \end{bmatrix}, \quad B_2 = \begin{bmatrix} c_{11} & c_{12} \\ c_{12} & c_{22} \end{bmatrix}, \quad B_1 + B_2 \preceq \begin{bmatrix} P_1 & z \\ z & P_2 \end{bmatrix}
\end{array}
\right\},
$$

(11.6)

where

$$
D_1' = \frac{H_1(B_1)H_1^\dagger + Q_1}{Q_1},
$$
(11.7)

$$
D_1'' = \frac{H_1(B_1 + B_2)H_1^\dagger + Q_1}{H_1(B_2)H_1^\dagger + Q_1},
$$
(11.8)

$$
D_2' = \frac{H_2(B_2)H_2^\dagger + Q_2}{Q_2},
$$
(11.9)

and

$$
D_2'' = \frac{H_2(B_1 + B_2)H_2^\dagger + Q_2}{H_2(B_1)H_2^\dagger + Q_2}.
$$
(11.10)

Here $X \succeq 0$ denotes that the matrix X is positive semi-definite, and we define $H_1 = [1\ b_{21}]$ and $H_2 = [b_{12}\ 1]$. The framework for this Gaussian MIMO broadcast channel region may also be used to express an achievable rate region for the Gaussian asymmetrically cooperating channel [521]. We notice the similarity with the MIMO broadcast region: The differences lie in the fact that only one dirty paper coding order is permitted and the transmit covariance matrix B_2 corresponding to the cognitive user's message is constrained, reflecting the asymmetry of the cooperation. We note that this region is equivalent to Equation 11.5.

Cognitive region = Convex hull of

$$
\left\{
\begin{array}{c}
(R_1, R_2): \\[2mm]
R_1 \leq \tfrac{1}{2} \log_2 \left(D_1'' \right) = R_1(\pi_{12}) \\[1mm]
R_2 \leq \tfrac{1}{2} \log_2 \left(D_2' \right) = R_2(\pi_{12}) \\[3mm]
B_1, B_2 \succeq 0, \quad B_1 = \begin{bmatrix} b_{11} & b_{12} \\ b_{12} & b_{22} \end{bmatrix}, \quad B_2 = \begin{bmatrix} 0 & 0 \\ 0 & c_{22} \end{bmatrix}, \\[3mm]
B_1 + B_2 \preceq \begin{bmatrix} P_1 & z \\ z & P_2 \end{bmatrix}, \quad z^2 \leq P_1 P_2
\end{array}
\right\}
$$

This 2×2, noncausal cognitive radio channel has been extended in a number of ways. The effect of generalized feedback has been studied in [522] and that of partial message knowledge in [515]. While this channel assumes noncausal message knowledge, a variety of two-phase half-duplex causal schemes have been presented in [175, 523], while a full-duplex rate region was studied in [524]. Many achievable rate regions are derived by having the primary transmitter exploit knowledge of the *exact* interference seen at the receivers (e.g., dirty paper coding in AWGN channels). The performance of dirty paper coding when this assumption breaks down has been studied in the context of a compound channel in [525] and in a channel in which the interference is partially known [526].

Cognitive channels have also been explored in the context of multiple nodes and antennas. Extensions to channels in which both the primary and secondary networks form classic multiple-access channels has been considered in [527, 528]. Cognitive versions of the X channel [529] have been considered in [530, 531], while cognitive transmissions using multiple antennas, without asymmetric transmitter cooperation, have been considered in [493].

Finally, while we outline some results on the exact rate regions for cognitive radio channels, how these rates scale at a high signal-to-noise ratio (SNR $\to \infty$) is also a measure of interest. The multiplexing gain, m, of a cognitive network[8] defines how the sum-rate of a network, $C_{\text{sum}}[\log(\text{SNR})]$, grows as a function of SNR; that is $C_{\text{sum}}(\text{SNR}) = m \log(\text{SNR}) + o[\log(\text{SNR})]$ as SNR $\to \infty$. The multiplexing gain is of particular interest in networks in which exact capacity expressions are lacking and may be thought of intuitively as the number of independent streams of information that may be simultaneously transmitted at high SNR. Great strides have been made in characterizing the degrees of freedom of interference networks [484], cognitive and X channels [530, 531], and wireless networks in which cooperation is causally enabled [532, 533].

---

[8]*Multiplexing gain, degrees of freedom,* and *prelog* are all terms used interchangeably in the literature.

11.6 CHAPTER SUMMARY

In this chapter, we outline recent results on the information theoretic limits of cognitive networks. Two main metrics are used: achievable rate/capacity regions for small networks and throughput scaling laws, as the number of nodes $n \to \infty$ for large networks. The general conclusion has been that increasing the amount of cognition or side information available to the cognitive transmitters and receivers increases the amount and quality of the communication. In interference-avoiding behavior, the cognitive nodes transmit in an orthogonal fashion to the primary users, thereby avoiding any mutual interference. Spectral efficiency may, however, be increased if cognitive nodes transmit over the same spectrum as the primary nodes, as done in the interference-controlling and -mitigating cognitive behaviors. In the former, cognitive transmitters require knowledge of the impact their transmission will have on the primary system and control their transmissions to stay within the acceptable limits for the primary user. When the secondary users further obtain the code books and possibly messages of the primary users, interference-mitigating behavior may be accomplished. The receivers may either opportunistically cancel the primary transmitter's interference or, at the cognitive transmitters, may judiciously select their power levels to either amplify or mitigate the primary user's signal. When building a cognitive network, the issue of obtaining these different types of side information, as well as realistically exploiting them, becomes crucial. It will be up to the individual applications to judge whether the promised gains by increasing levels of cognition are worth the effort and cost in obtaining and properly using them.

11.7 PROBLEMS

1 What are three types of cognitive behavior? Can you think of a fourth? How does it relate to the others?

2 Into which types of cognitive behavior does code division, multiple access fit? Why?

3 Formally define what it means for a
   (a) Rate R to be *achievable* in a discrete memoryless channel.
   (b) Rate tuple (R_1, R_2) to be *achievable* in a multiuser discrete memoryless channel.

4 State an inner and an outer bound to the capacity region of the relay channel. Under what conditions are the inner or outer bounds tight?

5 State the capacity region of the two- and three-user discrete memoryless multiple-access channels.

6   State an inner and outer bound to the capacity region of the discrete memoryless broadcast channel.

7   State the capacity region of a two-user Gaussian broadcast channel, where all nodes have single antennas.

8   State the best-known inner and outer bounds to the capacity region of the discrete memoryless interference channel.

9   In spectrum interweave cognitive behavior the sensing is idealized: The cognitive transmitters and receivers are able to sense the channel perfectly and instantaneously. Determine the impact on the achievable rate region for spectrum interweave cognitive behavior if perfect sensing of the primary user requires a finite duration of time T and the cognitive transmitter subsequently transmits for a time period T'.

10 What are different types of beamforming? What types are most useful to interference avoidance in cognitive networks with multiple antennas at transmitters and receivers?

11 In spectrum underlay systems in Gaussian noise, the region (b) is achievable. If it is acceptable that the primary user's rate drops to half its interference-free rate when a cognitive user is present, determine the power $P_{2}*$ at which the cognitive user may transmit. This $P_{2}*$ is necessarily a function of the channel gains, the noise, and the primary user power.

12 From the description of opportunistic interference cancellation, mathematically describe the rate region (c) in terms of the channel gains, power constraints, noise power, as well as interference temperature constraints of the primary user.

13 Determine two nonidentical (and noninclusive) inner bounds to the capacity region of the discrete memoryless cognitive channel with asymmetric cooperation.

14 What is the best-known outer bound to the capacity region of the discrete memoryless cognitive channel with asymmetric cooperation?

15 Using MATLAB, plot the equivalent of Figure 11.8 for the following channel gains:
   (a) $h_{12} = 0.2, h_{21} = 0.8$.
   (b) $h_{12} = 0.8, h_{12} = 0.2$.
   (c) $h_{12} = 1.2, h_{21} = 0.55$.
   (d) $h_{21} = 1.2, h_{12} = 0.55$.

Cross-layer optimization for multihop cognitive radio networks

12

Yi Shi and Y. Thomas Hou

Virginia Polytechnic Institute and State University, United States

12.1 INTRODUCTION

From a wireless networking perspective, cognitive radio (CR) offers a whole new set of research problems in algorithm design and protocol implementation. To appreciate such opportunity, we compare CR with a closely related wireless networking technology called *multi-channel multi-radio* (MC-MR), which has also been under intensive research in recent years (see e.g., [534–538] and the references therein). First, MC-MR employs traditional hardware-based radio technology and hence each radio can operate on only a single channel at a time and there is no switching of channels on a per-packet basis. Therefore, the number of concurrent channels that can be used at a wireless node is limited by the number of radios. In contrast, the radio technology in a CR is software based; a software radio is capable of switching frequency bands on a per-packet basis. As a result, the number of concurrent frequency bands that can be used by a CR is typically much larger than that which can be supported by MC-MR. Second, a common assumption for MC-MR is that a set of "common channels" is available at every node in the network; each channel typically has the same bandwidth. Such assumption is hardly true for CR networks, in which each node may have a different set of available frequency bands, each of which may be of unequal size. A CR node is capable of working on a set of "heterogeneous" channels that are scattered over widely separated slices of the frequency spectrum with different bandwidths. An even more profound advance in CR technology is that CR can work on noncontiguous channels for transmission/reception. These important differences between MC-MR and CR warrant that algorithm design for future CR networks is substantially more complex than that

Doi: 10.1016/B978-0-12-374715-0.00012-5

for current MC-MR wireless networks. In some sense, an MC-MR wireless network can be considered a special case of CR network. Thus, algorithms designed for future CR networks can be tailored to address current MC-MR networks while the converse is not true.

Our goal is to optimize network-level performance of a multihop CR network. It is now well understood that network performance for such a network is tightly coupled with lower-layer behaviors [539]. For example, to determine the amount of flow that can be transported between two nodes, we need to compute this link's capacity under a particular power control and scheduling. Then we can determine how to fully utilize this link capacity by optimally assigning a flow rate on this link. For scheduling, before we decide if a link should be active on certain frequency bands, we should confirm that this link is indeed used in upper-layer routing. For those links that are not used in routing, we should not consider them for scheduling. For power control, before we determine transmission power at a node on a frequency band for a transmission, we should confirm that this transmission is indeed needed in routing and this band is indeed scheduled to be active on this node. Due to these interdependencies among the layers, a cross-layer design is essential to achieve optimal CR network performance.

Following a cross-layer optimization approach, in this chapter, we show how to jointly consider power control, scheduling, and routing at multiple layers (see Table 12.1). At the physical layer, we determine the transmission power at each CR node on each of its available frequency bands. We note that, although a larger transmission power can increase the transmission rate at this node, it also produces a larger interference at other nodes and may degrade the network performance. Therefore, we cannot simply set the transmission power to its maximum. We use the physical model [269] to characterize the power-rate relationship. Under the physical model, a transmission is successful if and only if the signal-to-interference-plus-noise-ratio (SINR) at the intended receiver exceeds a threshold so that the transmitted signal can be decoded with an acceptable bit error rate (BER). Further, capacity calculation is based on SINR (via Shannon's formula), which takes into account interference due to simultaneous transmission at other nodes. At the link layer, we arrange transmissions on different frequency bands such that each transmission has an SINR that exceeds a required threshold. We note that, although

Table 12.1 Mechanism Considered in This Chapter at Each Layer in the Protocol Stack

Layer in Protocol Stack	Mechanism Considered in This Chapter
Network layer	Multipath multihop flow routing
Link layer	Frequency scheduling
Physical layer	Per-node-based power control

activating multiple frequency bands on a link can increase the transmission rate on this link, it also produces interference to other nodes on the same band. Therefore, we need an optimal scheduling such that all transmission rates on all nodes can be maximized. At the network layer, we allow flow splitting and multihop routing to achieve the best performance. When we determine flow rates on each link, we need to consider the following two constraints. The first constraint is flow balance; that is, at each node (except the source and destination nodes) the total incoming data rates should be equal to the total outgoing data rates. The second constraint is link capacity constraint; that is, the total flow rates on each link cannot be more than the achievable capacity under power control and scheduling. We develop mathematical models for these complex relationships among power control, scheduling, and routing. These models are general and can be used for many related cross-layer problems for multihop CR networks.

As a case study, we apply our models to a specific cross-layer optimization problem. We consider how to maximize throughput capacity for CR networks. We assume there are multiple user communication sessions (source-destination pairs), each with a minimum rate requirement. We aim to maximize a common scaling factor of these minimum rate requirements via joint optimization of power control, scheduling, and routing. By applying our mathematical models, we formulate this problem as a mixed integer nonlinear program (MINLP). Although this problem is difficult to solve, we show how to develop a centralized solution to this complex optimization problem based on the so-called *branch-and-bound* (BB) framework and *reformulation-linearization technique* (RLT) [540]. The basic idea of branch and bound is divide and conquer. We apply this solution procedure on several network instances and show its efficacy.

We organize this chapter as follows. In Section 12.2, we present mathematical models for power control, scheduling, and routing. In Section 12.3, as a case study, we study the throughput maximization problem and give a solution based on the branch-and-bound framework. In Section 12.4, we present some numerical results for the proposed solution procedure. Section 12.5 concludes this chapter. Section 12.6 offers some problems as a review for this chapter.

12.2 MATHEMATICAL MODELS AT MULTIPLE LAYERS

We consider a CR-based ad hoc network with a set of nodes $\mathcal{N}$. For a node $i \in \mathcal{N}$, the set of available frequency bands $\mathcal{M}_i$ depends on its location and may not be identical to the available frequency bands at other nodes. We assume that the bandwidth of each frequency band (channel) is W. Denote by $\mathcal{M}$ the set of all frequency bands present in the network; that is, $\mathcal{M} = \bigcup_{i \in \mathcal{N}} \mathcal{M}_i$. Denote $\mathcal{M}_{ij} = \mathcal{M}_i \cap \mathcal{M}_j$, which is the set of frequency bands that is common on both nodes i and j and thus can be used for transmission between these two nodes. In the rest of this section, we give mathematical models characterizing the interrelationships among different layers. Table 12.2 lists all notation in this chapter.

12.2.1 Scheduling and Power Control

In a multihop CR network, it is likely that there are multiple simultaneous transmissions in the network. On the same band, simultaneous transmission and reception are prohibited, as the transmission overwhelms the reception (also known as *self-interference*). For such self-interference, we must schedule transmission/reception either on different frequency bands or in different time slots. Also, within the network, transmissions at some nodes may interfere with other nonintended receiving nodes if they are on the same band and close to each other. Therefore, it is necessary to schedule the interfering transmission/reception either on a different band or in a different time slot.

Scheduling for transmission at each node in the network can be done either in the time domain or the frequency domain. In this chapter, we consider scheduling in the frequency domain in the form of assigning frequency bands (channels). Concurrent transmissions within the same channel are allowed as long as the interference level is acceptable.

Denote

$$x_{ij}^m = \begin{cases} 1 & \text{if node } i \text{ transmits data to node } j \text{ on band } m \\ 0 & \text{otherwise.} \end{cases} \tag{12.1}$$

To simplify the model, we assume that node i cannot use a band $m \in \mathcal{M}_i$ for transmitting different data to multiple nodes or for receiving different data from multiple nodes; that is,

$$\sum_{j \in \mathcal{T}_i^m} x_{ij}^m \leq 1, \tag{12.2}$$

$$\sum_{k \in \mathcal{T}_i^m} x_{ki}^m \leq 1, \tag{12.3}$$

where $\mathcal{T}_i^m$ is the set of nodes to which node i can transmit (and receive) on band m in the network. Further, due to self-interference, node i cannot use it for both transmission and reception; that is,

$$x_{ki}^m + x_{ij}^m \leq 1 (j, k \in \mathcal{T}_i^m, j \neq k). \tag{12.4}$$

Combining Equations (12.2), (12.3), and (12.4), we have

$$\sum_{k \in \mathcal{T}_i^m} x_{ki}^m + \sum_{j \in \mathcal{T}_i^m} x_{ij}^m \leq 1. \tag{12.5}$$

Figure 12.1 illustrates the scheduling constraint, where node i is receiving from node k and transmitting to nodes j and h. Then, by Equation (12.5), node i needs to use three different bands for these transmissions/receptions.

Table 12.2 Notation Used in This Chapter

Symbol	Definition
$\mathcal{N}$	The set of nodes in the network
$\mathcal{M}_i$	The set of available bands at node $i \in \mathcal{N}$
$\mathcal{M}$	The set of frequency bands in the network, $\mathcal{M} = \sum_{i \in \mathcal{N}} \mathcal{M}_i$
$\mathcal{M}_{ij}$	The set of available bands on link $i \rightarrow j$, $\mathcal{M}_{ij} = \mathcal{M}_i \bigcap \mathcal{M}_j$
W	Bandwidth of a frequency band
$\mathcal{L}$	The set of active user communication sessions in the network
$s(l), d(l)$	Source and destination nodes of session $l \in \mathcal{L}$
$r(l)$	Minimum rate requirement of session l
K	Rate scaling factor for all sessions
$P_{\max}$	The maximum transmission power at a transmitter
η	Ambient Gaussian noise density
g_{ij}	Propagation gain from node i to node j
α	The minimum required SINR
$\mathcal{T}_i^m$	The set of nodes that node i can transmit to (and receive from) on band m
$\mathcal{T}_i$	The set of nodes that node i can transmit to (and receive from) $\mathcal{T}_i = \bigcup_{m \in \mathcal{M}_i} \mathcal{T}_i^m$
$\mathcal{I}_j^m$	The set of nodes that may produce interference on band m at node j
x_{ij}^m	Binary indicator to mark whether or not band m is used by link $i \rightarrow j$
$f_{ij}(l)$	Data rate attributed to session l on link $i \rightarrow j$
Q	The number of transmission power levels at a transmitter
q_{ij}^m	The transmission power level from node i to node j on band m
t_i^m	The transmission power level at node i on band m, $t_i^m = \sum_{j \in \mathcal{T}_i^m} q_{ij}^m$
s_{ij}^m	The SINR from node i to node j on band m
ε	A small positive constant reflecting the desired accuracy
Ω_z	The set of all possible values of $(\mathbf{x}, \mathbf{q})$ in problem z
LB_z, UB_z	The lower and upper bounds of problem z
ψ_z	The solution obtained by local search for problem z
LB, UB	The minimum lower and upper bounds among all problems
ψ_ε	A $(1 - \varepsilon)$ optimal solution

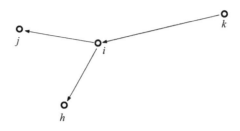

FIGURE 12.1

An example of scheduling constraint at node i. Each transmission/reception at node i needs a different band.

Now we consider power control and its relationship with scheduling. At each CR node, the node's transmission power on each band is bounded by $P_{\max}$. Since the interference from this link to other links depends on transmission power, it is necessary to determine the optimal transmission power for each node on each band.

For power control, it is reasonable to assume that the transmission power at a node is limited to a finite number of discrete levels between 0 and $P_{\max}$. To model this discrete power control mathematically, we introduce an integer parameter Q that represents the total number of power levels to which a transmitter can be adjusted; that is, $0, \frac{1}{Q}P_{\max}, \frac{2}{Q}P_{\max}, \ldots, P_{\max}$. Denote $q_{ij}^m \in \{0, 1, 2, \ldots, Q\}$ the integer power level. Clearly, when node i does not transmit data to node j on band m, q_{ij}^m is 0. Under the maximum allowed transmission power level Q, we have

$$q_{ij}^m \begin{cases} \le Q & \text{if } x_{ij}^m = 1 \\ = 0 & \text{otherwise.} \end{cases}$$

With joint consideration of x_{ij}^m and q_{ij}^m, this relationship can be rewritten as

$$q_{ij}^m \le Q x_{ij}^m. \tag{12.6}$$

As discussed earlier, concurrent transmissions by different nodes on the same band are allowed as long as the interference level is kept under control. Under the physical model [269], a transmission is successful if and only if the SINR at the receiving node exceeds a certain threshold, say α. We now formulate this constraint. For a transmission from node i to node j on band m, when there is interference from concurrent transmissions on the same band, the SINR at node j (denoted as s_{ij}^m) is

$$s_{ij}^m = \frac{g_{ij}\frac{q_{ij}^m}{Q}P_{\max}}{\eta W + \sum_{k \in \mathcal{N}}^{k \ne i,j} \sum_{b \in T_k^m}^{b \ne i,j} g_{kj}\frac{q_{kb}^m}{Q}P_{\max}}$$

$$= \frac{g_{ij}q_{ij}^m}{\frac{\eta W Q}{P_{\max}} + \sum_{k \in \mathcal{N}}^{k \ne i,j} \sum_{b \in T_k^m}^{b \ne i,j} g_{kj}q_{kb}^m},$$

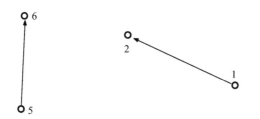

FIGURE 12.2

Three simultaneous transmissions on the same frequency band. At receiving node 2, transmissions from nodes 3 and 5 are considered interference, as node 2 is not the intended receiver for these transmissions.

where η is the ambient Gaussian noise density and g_{ij} is the propagation gain from node i to node j. As an example, suppose three transmissions, $1 \rightarrow 2$, $3 \rightarrow 4$, and $5 \rightarrow 6$, are on band m at the same time (see Figure 12.2). At node 2, there is interference from nodes 3 and 5. Therefore, the SINR at node 2 is

$$s_{12}^m = \frac{g_{12}q_{12}^m}{\frac{\eta W Q}{P_{\max}} + g_{31}q_{34}^m + g_{51}q_{56}^m}.$$

To get a more compact constraint, note that, when there is a transmission from node i to node j on band m, we have $x_{ij}^m = 1$. Then, by Equation (12.5), $x_{ki}^m = 0$ for $k \in T_i^m$ and $x_{kj}^m = 0$ for $k \in T_j^m$. Hence, by Equation (12.6), $q_{ki}^m = 0$ and $q_{kj}^m = 0$. Then we have

$$s_{ij}^m = \frac{g_{ij}q_{ij}^m}{\frac{\eta W Q}{P_{\max}} + \sum_{k \in \mathcal{N}}^{k \neq i,j} \sum_{b \in T_k^m} g_{kj}q_{kb}^m}.$$

Denote $t_k^m = \sum_{b \in T_k^m} q_{kb}^m$. We have

$$s_{ij}^m = \frac{g_{ij}q_{ij}^m}{\frac{\eta W Q}{P_{\max}} + \sum_{k \in \mathcal{N}}^{k \neq i,j} g_{kj}t_k^m}. \qquad (12.7)$$

Note that this SINR computation also holds when $q_{ij}^m = 0$; that is, when there is no transmission from node i to node j on band m.

Recall that, under the physical model, a transmission from node i to node j on band m is successful if and only if the SINR at node j exceeds a threshold α; that is, $s_{ij}^m \geq \alpha$. This is the necessary and sufficient condition for successful transmission under the physical model. Then, by Equation (12.1), we have

$$x_{ij}^m = \begin{cases} 1 & \text{if } s_{ij}^m \geq \alpha \\ 0 & \text{otherwise.} \end{cases}$$

This can be written into the following compact inequality:

$$s_{ij}^m \geq \alpha x_{ij}^m.$$

12.2.2 Routing

In an ad hoc network, consider a set of $\mathcal{L}$ active user communication (unicast) sessions. Denote by $s(l)$ and $d(l)$ the source and destination nodes of session $l \in \mathcal{L}$ and $r(l)$ the minimum rate requirement (in bits/second) of session l. To route these flows from their respective source nodes to destination nodes, it is necessary to employ multihop due to the limited transmission range of a node. Further, to have more flexibility, it is desirable to allow flow splitting and multipath routing. That is, the data flow from a source can be split into subflows and each subflow can traverse different paths to the flow's destination. This is because a single path is overly restrictive and may not yield an optimal solution. In routing constraints, we need to ensure that flow balance holds at each node (except source and destination nodes).

In the case study in Section 12.3, we consider how to maximize a rate scaling factor K for all session rates. That is, for each session $l \in \mathcal{L}$, $r(l)K$ amount of data rate is to be sent from $s(l)$ to $d(l)$. In this context, the routing constraints can be modeled as follows.

Denote by $f_{ij}(l)$ the data rate on link (i,j) attributed to session l, where $i \in \mathcal{N}, j \in \mathcal{T}_i = \bigcup_{m \in \mathcal{M}_i} \mathcal{T}_i^m$. If node i is the source node of session l—that is, $i = s(l)$—then flow balance at node i must hold. That is,

$$\sum_{j \in \mathcal{T}_i} f_{ij}(l) = r(l)K. \tag{12.8}$$

If node i is an intermediate relay node for session l—that is, $i \neq s(l)$ and $i \neq d(l)$—then

$$\sum_{\substack{j \neq s(l) \\ j \in \mathcal{T}_i}} f_{ij}(l) = \sum_{\substack{k \neq d(l) \\ k \in \mathcal{T}_i}} f_{ki}(l). \tag{12.9}$$

If node i is the destination node of session l—that is, $i = d(l)$—then

$$\sum_{k \in \mathcal{T}_i} f_{ki}(l) = r(l)K. \tag{12.10}$$

It can be easily verified that, once Equations (12.8) and (12.9) are satisfied, Equation (12.10) is also satisfied. As a result, it is sufficient to list only Equations (12.8) and (12.9) in the formulation.

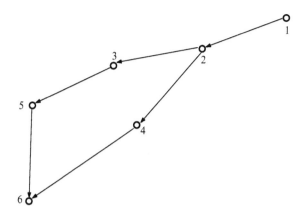

FIGURE 12.3

Multipath routing for a session l with $s(l) = 1$ and $d(l) = 6$.

An example of multipath routing for a session l is shown in Figure 12.3, where node 1 is the source and node 6 is the destination. At the source node 1, we have

$$f_{12}(l) = r(l)K. \tag{12.11}$$

At intermediate relay node 2, the data received from node 1 are split into two flows, and we have

$$f_{23}(l) + f_{24}(l) = f_{12}(l). \tag{12.12}$$

At intermediate relay node 3, the data received from node 2 are sent to node 5, and we have

$$f_{35}(l) = f_{23}(l). \tag{12.13}$$

Similarly, we have the following constraints at nodes 4 and 5:

$$f_{46}(l) = f_{24}(l), \tag{12.14}$$

$$f_{56}(l) = f_{35}(l). \tag{12.15}$$

At destination 6, we have

$$f_{46}(l) + f_{56}(l) = r(l)K. \tag{12.16}$$

Note that, by taking the sum of the left and right sides in Equations (12.11), (12.12), (12.13), (12.14), and (12.15), we have

$$f_{12}(l) + [f_{23}(l) + f_{24}(l)] + f_{35}(l) + f_{46}(l) + f_{56}(l)$$
$$= r(l)K + f_{12}(l) + f_{23}(l) + f_{24}(l) + f_{35}(l).$$

After cancelation of common terms on both sides, we have

$$f_{46}(l) + f_{56}(l) = r(l)K,$$

which is precisely Equation (12.16). Therefore, it is not necessary to include the flow constraint at destination nodes (i.e., Equation (12.10)) once we have Equation (12.8) at source nodes and Equation (12.9) at relay nodes.

In addition to these flow constraints at each node $i \in \mathcal{N}$ for session $l \in \mathcal{L}$, the aggregated flow rates on each radio link cannot exceed that link's capacity. That is, on a link $i \rightarrow j$, we must have

$$\sum_{\substack{l \in \mathcal{L} \\ s(l) \neq j, d(l) \neq i}} f_{ij}(l) \leq \sum_{m \in \mathcal{M}_{ij}} W \log_2(1 + s_{ij}^m). \tag{12.17}$$

The constraint in Equation (12.17) illustrates the coupling relationship among flow routing (via $f_{ij}(l)$), power control (embedded in s_{ij}^m), and scheduling (embedded in s_{ij}^m).

12.3 A CASE STUDY: THE THROUGHPUT MAXIMIZATION PROBLEM

As a case study, we study the throughput maximization problem for a multihop CR network. We consider multihop, multipath routing to transmit data from each source to their corresponding destination. We show how to maximize a rate scaling factor K for all active sessions. That is, for each active session $l \in \mathcal{L}$, what is the maximum K we can have while $r(l)K$ amount of data rate can be transmitted from $s(l)$ to $d(l)$?

12.3.1 Problem Formulation

By applying the mathematical models in Section 12.2 and putting together all the constraints for scheduling, power control, and flow routing, we have the following mathematical formulation:

$$\text{Max} \qquad K \tag{12.18}$$

$$\text{s.t.} \qquad \sum_{i \in T_k^m} x_{ki}^m + \sum_{j \in T_i^m} x_{ij}^m \leq 1 \quad (i \in \mathcal{N}, m \in \mathcal{M}_i)$$

$$q_{ij}^m - Q x_{ij}^m \leq 0 \quad (i \in \mathcal{N}, m \in \mathcal{M}_i, j \in T_i^m) \tag{12.19}$$

$$\sum_{j \in T_i^m} q_{ij}^m - t_i^m = 0. \quad (i \in \mathcal{N}, m \in \mathcal{M}_i) \tag{12.20}$$

$$\frac{\eta WQ}{P_{\max}}s_{ij}^m + \sum_{k \in \mathcal{N}}^{k \neq i,j} g_{kj}t_k^m s_{ij}^m - g_{ij}q_{ij}^m = 0 \quad (i \in \mathcal{N}, m \in \mathcal{M}_i, j \in T_i^m) \qquad (12.21)$$

$$\alpha x_{ij}^m - s_{ij}^m \leq 0 \quad (i \in \mathcal{N}, m \in \mathcal{M}_i, j \in T_i^m) \qquad (12.22)$$

$$\sum_{l \in \mathcal{L}}^{s(l) \neq j, d(l) \neq i} f_{ij}(l) - \sum_{m \in \mathcal{M}_{ij}} W \log_2(1 + s_{ij}^m) \leq 0 \quad (i \in \mathcal{N}, j \in T_i)$$

$$\sum_{j \in T_i} f_{ij}(l) - r(l)K = 0 \quad (l \in \mathcal{L}, i = s(l))$$

$$\sum_{j \in T_i}^{j \neq s(l)} f_{ij}(l) - \sum_{i \in T_k}^{k \neq d(l)} f_{ki}(l) = 0 \quad (l \in \mathcal{L}, i \in \mathcal{N}, i \neq s(l), d(l))$$

$$x_{ij}^m \in \{0,1\}, q_{ij}^m \in \{0,1,2,\cdots,Q\}, t_i^m, s_{ij}^m \geq 0 \quad (i \in \mathcal{N}, m \in \mathcal{M}_i, j \in T_i^m)$$

$$K, f_{ij}(l) \geq 0 \quad (l \in \mathcal{L}, i \in \mathcal{N}, i \neq d(l), j \in T_i, j \neq s(l)),$$

where $Q, \eta, W, \alpha, P_{\max}, g_{ij}$, and $r(l)$ are all constants and $K, x_{ij}^m, q_{ij}^m, t_i^m, s_{ij}^m$, and $f_{ij}(l)$ are all optimization variables. This formulation is a mixed-integer nonlinear program, which is NP-hard in general [541].

12.3.2 Solution Overview

For the complex MINLP problem, we employ the so-called *branch-and-bound* framework [542] to develop a solution. Under branch and bound, we aim to provide a $(1 - \varepsilon)$ optimal solution, where ε is a small positive constant reflecting our desired accuracy in the final solution.

Initially, branch and bound analyzes the value sets for each partition variable; that is, all discrete variables and all variables in a nonlinear term. For our problem, these variables include all $x_{ij}^m, q_{ij}^m, t_i^m$, and s_{ij}^m variables. The value sets for each partition variable are $x_{ij}^m \in \{0,1\}, q_{ij}^m \in \{0,1,2,\ldots,Q\}, t_i^m \in \{0,1,2,\ldots,Q\}$, and $s_{ij}^m \in \left[0, g_{ij}P_{\max}/\eta W\right]$. By using some *relaxation* techniques—that is, replacing a discrete variable by a continuous variable and replacing a nonlinear constraint by several linear constraints—branch and bound obtains a linear relaxation for the original problem based on the value sets for each partition variable. The solution to this relaxed problem provides an upper bound (UB) to the objective function. As we shall show shortly, this critical step is made possible by the convex hull relaxation for nonlinear discrete terms. We call the approximation errors caused by relaxation *relaxation errors*. Due to these relaxation errors, the solution to the relaxed problem usually is infeasible to the original problem. To obtain a feasible solution to the original problem, branch and bound uses a *local search* algorithm and the relaxation solution as the starting point. The obtained feasible solution provides a lower bound (LB) for the objective function (see Figure 12.4(a) for an example). If the obtained lower and upper bounds are close to each other within

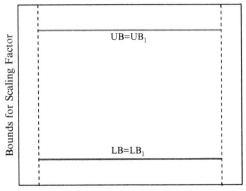

(a) Iteration 1. We can estimate UB and LB for the original problem.

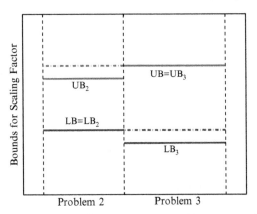

(b) Iteration 2. By dividing the original problem into two new problems, we can develop tighter bounds on each problem (UB_2, LB_2 for Problem 2 and UB_3, LB_3 for Problem 3). Thus, bounds ($UB = UB_3$ and $LB = LB_2$) for the original problem also become tighter.

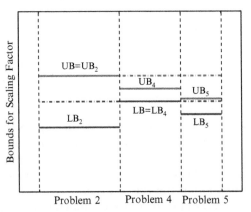

(c) Iteration 3. By dividing Problem 3 (with the largest upper bound) into two new problems, we can obtain tighter upper and lower bounds to Problem 3. Thus, bounds for the original problem can be further improved as $UB = UB_2$ and $LB = LB_4$.

FIGURE 12.4

Illustration of branch-and-bound solution procedure.

a factor of ε—that is, $LB \geq (1 - \varepsilon)UB$—then the current feasible solution is $(1 - \varepsilon)$ optimal and we are done.

If the relaxation errors for nonlinear terms are not small, then the gap between the upper bound UB and the lower bound LB could be large. To close this gap, we must have a tighter linear relaxation; that is, with smaller relaxation errors. This could be achieved by further narrowing down the value sets of partition variables. Specifically, branch and bound selects a partition variable with the maximum relaxation error and divides its value set into two sets by its value in the relaxation solution. Then the original Problem 1 is divided into two new Problems 2 and 3 (see Figure 12.4(b)). Again, branch and bound performs relaxation and local search on these two new problems. Now we have upper bounds UB_2 and UB_3 for Problems 2 and 3, respectively. We also have feasible solutions that provide lower bounds LB_2 and LB_3 for Problems 2 and 3, respectively. Since the relaxations in Problems 2 and 3 are both tighter than in Problem 1, we have $\max\{UB_2, UB_3\} \leq UB_1$ and $\max\{LB_2, LB_3\} \geq LB_1$. For a maximization problem, the upper bound of the original problem is updated from $UB = UB_1$ to $UB = \max\{UB_2, UB_3\}$. Also, the best feasible solution to the original problem is the solution with a larger LB_i. Then the lower bound of the original problem is updated from $LB = LB_1$ to $LB = \max\{LB_2, LB_3\}$. As a result, we now have a smaller gap between UB and LB. Then we either have a $(1-\varepsilon)$-optimal solution (if $LB \geq (1-\varepsilon)$ UB), or choose a problem with the maximum upper bound (Problem 3 in Figure 12.4(b)), and perform partition on this problem.

An important property of branch and bound is that we may remove some problems from further consideration before we solve it completely. During the iteration process for branch and bound, if we find a Problem z with $LB \geq (1-\varepsilon)UB_z$, then we conclude that this problem cannot provide much improvement on LB, see Problem 4 in Figure 12.4(c). That is, further branching on this problem will not yield much improvement so we can remove this problem from further consideration.

It has been shown that, under very general conditions, a branch-and-bound solution procedure always converges [540]. Although the worst-case complexity of such a procedure is exponential, the actual running time, based on our experience, is reasonably fast if all partition variables are discrete variables (e.g., the problem considered in our case study).

Figure 12.5 shows the general framework of the branch-and-bound procedure. Note that the key components in the branch-and-bound framework are problem specific and must be carefully designed to make it work. These include (1) how to obtain a linear relaxation, (2) how to perform a local search, and (3) how to choose a variable for partition. The details of these components are presented in Sections 12.3.3, 12.3.4, and 12.3.5, respectively.

12.3.3 Linear Relaxation

During each iteration of the branch-and-bound procedure, we need a linear relaxation technique to obtain an upper bound on the objective function (see line 5 in Figure 12.5).

Branch-and-Bound Procedure

1. Initialization:
2.   Let the initial best solution $\psi_\varepsilon = \emptyset$ and the initial lower bound LB $= -\infty$.
3.   Determine initial value set for each partition variable.
4.   Let the initial problem list include only the original problem, denoted as Problem 1.
5.   Build a linear relaxation and obtain the relaxation solution $\hat{\psi}_1$.
6.   The objective value of $\hat{\psi}_1$ is an upper bound UB$_1$ to Problem 1.
7. Iteration:
8.   Select Problem z that has the maximal UB$_z$ among all problems in the problem list.
9.   Update upper bound UB $=$ UB$_z$.
10.   Find a feasible solution ψ_z from $\hat{\psi}_z$ via a local search algorithm and denote its objective value as LB$_z$.
11.   If (LB$_z$ > LB) {
12.   Update $\psi_\varepsilon = \psi_z$ and LB $=$ LB$_z$.
13.   If LB $\geq (1 - \varepsilon)$UB, we stop with the $(1 - \varepsilon)$ optimal solution ψ_ε.
14.   Otherwise, remove all Problems z' with LB $\geq (1 - \varepsilon)$UB$_{z'}$ from the problem list. }
15.   Select a variable with the maximum relaxation error and divide its value set into two sets by its value in $\hat{\psi}_z$.
16.   Create two new Problems z_1 and z_2 based on these two sets.
17.   Remove Problem z from the problem list.
18.   Obtain UB$_{z1}$ and UB$_{z2}$ for Problems z_1 and z_2 via their linear relaxations.
19.   If LB $< (1 - \varepsilon)$UB$_{z1}$, add Problem z_1 into the problem list.
20.   If LB $< (1 - \varepsilon)$UB$_{z2}$, add Problem z_2 into the problem list.
21.   If the problem list is empty, we stop with a $(1 - \varepsilon)$ optimal solution ψ_ε.
22.   Otherwise, go to the next iteration.

FIGURE 12.5

Pseudocode for the branch-and-bound solution procedure.

For the polynomial term $t_k^m s_{ij}^m$ in the problem formulation, we can apply a method called *reformulation-linearization technique* [540]. For a nonlinear term $t_k^m s_{ij}^m$, we introduce a new variable u_{ijk}^m; replace $t_k^m s_{ij}^m$ with u_{ijk}^m; and add RLT constraints on these variables. Suppose t_k^m and s_{ij}^m are bounded by $(t_k^m)_L \leq t_k^m \leq (t_k^m)_U$ and $(s_{ij}^m)_L \leq s_{ij}^m \leq (s_{ij}^m)_U$, respectively. Hence, we have $[t_k^m - (t_k^m)_L] \cdot [s_{ij}^m - (s_{ij}^m)_L] \geq 0$, $[t_k^m - (t_K^m)_L] \cdot [(s_{ij}^m)_U - s_{ij}^m] \geq 0$, $[(t_k^m)_U - t_k^m] \cdot [s_{ij}^m - (s_{ij}^m)_L] \geq 0$, and $[(t_k^m)_U - t_k^m] \cdot [(s_{ij}^m)_U - s_{ij}^m] \geq 0$. From these relationships and substituting $u_{ijk}^m = t_k^m s_{ij}^m$, we have the following RLT constraints for u_{ijk}^m:

$$(t_k^m)_L \cdot s_{ij}^m + (s_{ij}^m)_L \cdot t_k^m - u_{ijk}^m \leq (t_k^m)_L \cdot (s_{ij}^m)_L,$$

$$(t_k^m)_U \cdot s_{ij}^m + (s_{ij}^m)_L \cdot t_k^m - u_{ijk}^m \geq (t_k^m)_U \cdot (s_{ij}^m)_L,$$

$$(t_k^m)_L \cdot s_{ij}^m + (s_{ij}^m)_U \cdot t_k^m - u_{ijk}^m \geq (t_k^m)_L \cdot (s_{ij}^m)_U,$$

$$(t_k^m)_U \cdot s_{ij}^m + (s_{ij}^m)_U \cdot t_k^m - u_{ijk}^m \leq (t_k^m)_U \cdot (s_{ij}^m)_U.$$

For the log term, we propose to employ three tangential supports as an approximation (see Figure 12.6). These three tangential segments form a convex hull linear relaxation. We first analyze the bounds for $1 + s_{ij}^m$. Then, we introduce a variable

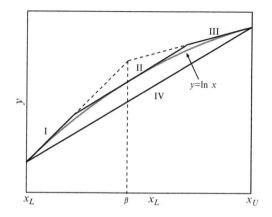

FIGURE 12.6

A convex hull for $y = \ln x$ to obtain its linear relaxation, where segments I, II, and III define upper bounds and segment IV defines a lower bound.

$c_{ij}^m = \ln(1 + s_{ij}^m)$ and consider how to get a linear relaxation for $y = \ln x$ over $x_L \leq x \leq x_U$. This function can be bounded by four segments (or a convex hull), where segments I, II, and III are tangential supports and segment IV is the chord (see Figure 12.6). In particular, three tangent segments are at $(x_L, \ln x_L)$, $(\beta, \ln \beta)$, and $(x_U, \ln x_U)$, where $\beta = [x_L \cdot x_U \cdot (\ln x_U - \ln x_L)]/(x_U - x_L)$ is the horizontal location for the point intersects extended tangent segments I and III; segment IV is the segment that joins points $(x_L, \ln x_L)$ and $(x_U, \ln x_U)$. The convex region defined by the four segments can be described by the following four linear constraints:

$$x_L \cdot y - x \leq x_L(\ln x_L - 1),$$

$$\beta \cdot y - x \leq \beta(\ln \beta - 1),$$

$$x_U \cdot y - x \leq x_U(\ln x_U - 1),$$

$$(x_U - x_L)y + (\ln x_L - \ln x_U)x \geq x_U \cdot \ln x_L - x_L \cdot \ln x_U.$$

As a result, the non-polynomial (log) term can also be relaxed into linear constraints.

Denote $\mathbf{x}, \mathbf{q}, \mathbf{t}$, and $\mathbf{s}$ as the vectors for variables $x_{ij}^m, q_{ij}^m, t_i^m$, and s_{ij}^m, respectively. We have the following linear relaxation for Problem z:

$$
\begin{aligned}
&\text{Max} && K \\
&\text{subject to} && \sum_{i \in T_k^m} x_{ki}^m + \sum_{j \in T_i^m} x_{ij}^m \leq 1 && (i \in \mathcal{N}, m \in \mathcal{M}_i) \\
& && q_{ij}^m - Q x_{ij}^m \leq 0 && (i \in \mathcal{N}, m \in \mathcal{M}_i, j \in T_i^m) \quad (12.23) \\
& && \sum_{j \in T_i^m} q_{ij}^m - t_i^m = 0 && (i \in \mathcal{N}, m \in \mathcal{M}_i)
\end{aligned}
$$

$$\frac{\eta W Q}{P_{\max}} s_{ij}^m + \sum_{k \in \mathcal{N}}^{k \neq i,j} g_{kj} u_{ijk}^m - g_{ij} q_{ij}^m = 0 \quad (i \in \mathcal{N}, m \in \mathcal{M}_i, j \in T_i^m)$$

$$\text{RLT constraints for } u_{ijk}^m \qquad (i, k \in \mathcal{N}, m \in \mathcal{M}_i, j \in T_i^m, k \neq i, j)$$

$$\alpha x_{ij}^m - s_{ij}^m \leq 0 \qquad (i \in \mathcal{N}, m \in \mathcal{M}_i, j \in T_i^m)$$

$$\sum_{l \in \mathcal{L}}^{s(l) \neq j, d(l) \neq i} f_{ij}(l) - \sum_{m \in \mathcal{M}_{ij}} \frac{W}{\ln 2} c_{ij}^m \leq 0 \quad (i \in \mathcal{N}, j \in T_i)$$

$$\text{Convex hull constraints for } c_{ij}^m \qquad (i \in \mathcal{N}, m \in \mathcal{M}_i, j \in T_i^m)$$

$$\sum_{j \in T_i} f_{ij}(l) - r(l)K = 0 \qquad (l \in \mathcal{L}, i = s(l))$$

$$\sum_{j \in T_i}^{j \neq s(l)} f_{ij}(l) - \sum_{i \in T_k}^{k \neq d(l)} f_{ki}(l) = 0 \qquad (l \in \mathcal{L}, i \in \mathcal{N}, i \neq s(l), d(l))$$

$$c_{ij}^m, u_{ijk}^m \geq 0 \qquad (i, k \in \mathcal{N}, m \in \mathcal{M}_i, j \in T_i^m, k \neq i, j)$$

$$K, f_{ij}(l) \geq 0 \qquad (l \in \mathcal{L}, i \in \mathcal{N}, i \neq d(l), j \in T_i, j \neq s(l))$$

$$(\mathbf{x}, \mathbf{q}, \mathbf{t}, \mathbf{s}) \in \Omega_z,$$

where Ω_z is defined as $\Omega_z = \{(\mathbf{x}, \mathbf{q}, \mathbf{t}, \mathbf{s}) : (x_{ij}^m)_L \leq x_{ij}^m \leq (x_{ij}^m)_U, (q_{ij}^m)_L \leq q_{ij}^m \leq (q_{ij}^m)_U, (t_i^m)_L \leq t_i^m \leq (t_i^m)_U, (s_{ij}^m)_L \leq s_{ij}^m \leq (s_{ij}^m)_U\}$, which is the set of all possible values of $(\mathbf{x}, \mathbf{q}, \mathbf{t}, \mathbf{s})$ in Problem z, where $(x_{ij}^m)_L, (x_{ij}^m)_U, (q_{ij}^m)_L, (q_{ij}^m)_U, (s_{ij}^m)_L$, and $(s_{ij}^m)_U$ are constant bounds. For example, Ω_1 for the original Problem 1 is

$$\{(\mathbf{x}, \mathbf{q}, \mathbf{t}, \mathbf{s}) : 0 \leq x_{ij}^m \leq 1, 0 \leq q_{ij}^m \leq Q, 0 \leq t_i^m \leq Q, 0 \leq s_{ij}^m \leq \tfrac{g_{ij} P_{\max}}{\eta W}\}.$$

12.3.4 Local Search Algorithm

A linear relaxation for a Problem z as discussed in Equation (12.23) can be solved in polynomial time. Denote the relaxation solution as $\hat{\psi}_z$, which provides an upper bound to Problem z but may not be feasible to the original problem (12.18) due to relaxation errors. We now show how to obtain a feasible solution ψ_z to the original problem based on relaxed solution $\hat{\psi}_z$ (see line 10 in Figure 12.5).

To obtain a feasible solution, we need to determine the integer values for $\mathbf{x}$ and $\mathbf{q}$ in solution ψ_z such that Equations (12.5), (12.19), and (12.22) hold. All other variables are based on $\mathbf{x}, \mathbf{q}$. Initially, each q_{ij}^m is set to the smallest value $(q_{ij}^m)_L$ in its value set and x_{ij}^m is fixed to 0 or 1 if its value set has only one element 0 or 1, respectively. Based on these values of q_{ij}^m, we can compute the capacity $\sum_{m \in \mathcal{M}_{ij}} W \cdot \log_2 \left[1 + g_{ij} q_{ij}^m / (\eta W Q / P_{\max} + \sum_{k \in \mathcal{N}}^{k \neq i,j} g_{kj} t_k^m) \right]$ for each link $i \rightarrow j$. The requirement on a link $i \rightarrow j$ is $\sum_{l \in \mathcal{L}}^{s(l) \neq j, d(l) \neq i} \hat{f}_{ij}(l)$. Thus, we can compute k_{ij}, the ratio between the capacity and the requirement. The objective value for the

Local Search Algorithm
1. Initialization:
2.   Set $q_{ij}^m = (q_{ij}^m)_L$ and x_{ij}^m as 0 or 1 if its value set only has one element 0 or 1, respectively.
3.   Compute the requirement $\sum_{l \in \mathcal{L}}^{s(l) \neq j, d(l) \neq i} \hat{f}_{ij}(l)$.
4. Iteration:
5.   Compute capacity $\sum_{m \in \mathcal{M}_{ij}} W \log_2 \left(1 + \dfrac{g_{ij} q_{ij}^m}{\eta WQ/P_{max} + \sum_{k \in \mathcal{N}}^{k \neq i,j} g_{kj} t_k^m} \right)$ and the ratio k_{ij}
between the capacity and the requirement for each link $i \to j$.
6.   Suppose link $i \to j$ has the smallest k_{ij}. Try to increase its capacity as follows.
7.   If q_{ij}^m can be increased on an already used band {
8.   Suppose band m has the largest $\hat{q}_{ij}^m$ among these bands.
9.   Increase q_{ij}^m under the constraints of $q_{ij}^m \leq (q_{ij}^m)_U$ and the corresponding $k_{ij} \leq 1$. }
10.   else, if q_{ij}^m can be increased on an available but currently unused band {
11.   Suppose band m has the largest $\hat{q}_{ij}^m$ among these bands.
12.   Increase q_{ij}^m under the constraints of $q_{ij}^m \leq (q_{ij}^m)_U$ and the corresponding $k_{ij} \leq 1$.
13.   Set $x_{ij}^m = 1$, $x_{ib}^m = 0$ for $b \in T_i, b \neq j$, $x_{ki}^m = 0$ for $k \in T_i$. }
14.   else the iteration terminates.

FIGURE 12.7

Pseudocode of a local search algorithm to find a feasible solution.

current $\mathbf{x}$ and $\mathbf{q}$ is $K \cdot \min\{k_{ij} : i \in \mathcal{N}, j \in T_i\}$. Therefore, we aim to increase the minimum k_{ij}. For the link with the smallest k_{ij}, we try to increase some q_{ij}^m under the constraint of $q_{ij}^m \leq (q_{ij}^m)_U$. When we cannot further increase the smallest k_{ij}, we are done. The details of this local search algorithm are shown in Figure 12.7.

12.3.5 Selection of Partition Variables

If the relaxation error for a Problem z is not small, the gap between its lower and upper bounds may be large. To obtain a small gap, we generate two new subproblems z_1 and z_2 from Problem z. We hope that these two new problems will have smaller relaxation errors. Then the bounds for them can be tighter than the bounds for z. Therefore, we identify a variable based on its relaxation error in line 15, Figure 12.5.

Note that the choice of a partition variable affects the convergence speed. Here, the candidate variables for partitioning are based on their impacts on the objective value, variables in $\mathbf{x}$ are more significant than variables in $\mathbf{q}$. Hence, we should first select one of $\mathbf{x}$ variables as the branch variable. In particular, for the relaxation solution $\hat{\psi}_z$, the relaxation error of a discrete variable x_{ij}^m is $\min\{\hat{x}_{ij}^m, 1 - \hat{x}_{ij}^m\}$, where $\hat{x}_{ij}^m$ is the value of variable x_{ij}^m in solution $\hat{\psi}_z$. We choose an x_{ij}^m with the maximum relaxation error among all $\mathbf{x}$ variables and set its value to $\{0\}$ and $\{1\}$ in Problems z_1 and z_2, respectively. Since the value set for this x_{ij}^m has only one element, this x_{ij}^m can be replaced by a constant in the new problem. As a result, some constraints may also be removed.

It should be noted that we may pose more limitations on other variables based on the new value set of x_{ij}^m. That is, if x_{ij}^m is 0, then we have $q_{ij}^m = 0$ based on Equation (12.19). If x_{ij}^m is 1, then we have $x_{ib}^m = 0$ for $b \in T_i, b \neq j$ and $x_{ki}^m = 0$ for $k \in T_i$ based on Equation (12.5).

When none of the **x** variables can be partitioned (i.e., all x variables are already set to 0 or 1), we select one of **q** variables for partitioning. In particular, for the relaxation solution $\hat{\psi}_z$, the relaxation error of a discrete variable q_{ij}^m is $\min\{\hat{q}_{ij}^m - \lfloor \hat{q}_{ij}^m \rfloor, \lfloor \hat{q}_{ij}^m \rfloor + 1 - \hat{q}_{ij}^m\}$, where $\hat{q}_{ij}^m$ is the value of variable q_{ij}^m in solution $\hat{\psi}_z$. Assuming the value set of q_{ij}^m in Problem z is $\{q_0, q_1, \ldots, q_K\}$, its value set in Problems z_1 and z_2 will be $\{q_0, q_1, \ldots, \lfloor \hat{q}_{ij}^m \rfloor\}$ and $\{\lfloor \hat{q}_{ij}^m \rfloor + 1, \lfloor \hat{q}_{ij}^m \rfloor + 2, \ldots, q_K\}$, respectively. Again, we may pose more limitations on other variables based on the new value set of q_{ij}^m. In particular, if q_{ij}^m is 0, then we have $x_{ij}^m = 0$ based on Equation (12.22). If the new value set of q_{ij}^m does not include 0, then we have $x_{ij}^m = 1$ based on Equation (12.19).

Note that when all possible partition variables in **x** and **q** can no longer be partitioned (i.e., all values are assigned), the other variables can be solved via a linear program (LP).

12.4 NUMERICAL RESULTS FOR THE THROUGHPUT MAXIMIZATION PROBLEM

In this section, we present some numerical results for the case study in Section 12.3. The purpose of this effort is to offer quantitative understanding on the joint optimization at different layers.

12.4.1 Simulation Setting

For ease of exposition, we normalize all units for distance, bandwidth, rate, and power based on Equation (12.17) with appropriate dimensions. We consider 20-, 30-, and 50-node CR networks with each node randomly located in a 50×50 area (see later Figures 12.8, 12.10, and 12.12). We assume there are $|\mathcal{M}| = 10$ frequency bands in the network and each band has a bandwidth of $W = 50$. At each CR node, only a subset of these bands is available. For the 20- and 30-node networks, we assume there are five user communication sessions, with source node and destination node randomly selected and the minimum rate requirement of each session randomly generated within $[1, 10]$ (see later Tables 12.4 and 12.6). For the 50-node network, the number of user communication sessions is 10 (see later Table 12.8).

We assume that the propagation gain model is $g_{ij} = d_{ij}^{-4}$ and the SINR threshold is $\alpha = 3$ [543]. The maximum transmission power at each node is $P_{\max} = 4.8 \cdot 10^5$ mW. We assume that the number of power control levels is $Q = 10$.

For our proposed branch-and-bound solution procedure, we set ε to 0.1, which guarantees that the obtained solution is 90% optimal.

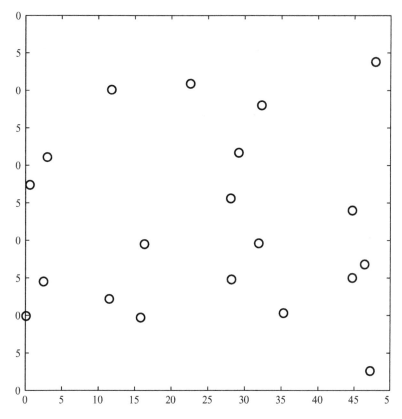

FIGURE 12.8

A 20-node network.

12.4.2 **Results and Observations**

A 20-Node Network

For the 20-node network in Figure 12.8, the location and available bands at each node are given in Table 12.3. There are five sessions. The source node, destination node, and minimum rate requirement of each session are shown in Table 12.4. The transmission power levels on their respective frequency bands in the final solution are:

Band 1: $q^1_{7,3} = 1$, $q^1_{16,12} = 7$
Band 2: $q^2_{8,2} = 2$
Band 3: $q^3_{13,14} = 2$
Band 4: $q^4_{1,7} = 7$, $q^4_{2,10} = 2$
Band 5: $q^5_{11,10} = 1$
Band 6: $q^6_{15,19} = 9$

Table 12.3 Location and Available Frequency Bands at Each Node for a 20-Node Network

Node	Location	Available Bands	Node	Location	Available Bands
1	(0.1, 9.9)	1, 2, 3, 4, 7, 8, 9, 10	11	(28.1, 25.6)	1, 2, 3, 4, 5, 6, 7, 8, 9, 10
2	(29.2, 31.7)	1, 2, 3, 4, 5, 7, 8, 10	12	(32.3, 38)	1, 8, 9, 10
3	(3, 31.1)	1, 4, 5, 6	13	(47.2, 2.6)	3, 5, 10
4	(11.8, 40.1)	1, 2, 3, 4, 6, 9, 10	14	(44.7, 15)	2, 3, 6, 7, 8
5	(15.8, 9.7)	1, 2, 3, 5, 6, 8, 9	15	(44.7, 24)	1, 2, 3, 4, 5, 6, 7, 8, 9, 10
6	(16.3, 19.5)	3, 5, 6, 8, 9	16	(47.9, 43.8)	1, 3
7	(0.6, 27.4)	1, 4, 8, 9, 10	17	(46.4, 16.8)	1, 7, 9
8	(22.6, 40.9)	1, 2, 3, 5, 7, 9, 10	18	(11.5, 12.2)	2, 5, 6, 10
9	(35.3, 10.3)	2, 9	19	(28.2, 14.8)	4, 5, 6, 7, 8, 9, 10
10	(31.9, 19.6)	1, 2, 3, 4, 5, 6, 7, 8, 9, 10	20	(2.5, 14.5)	1, 7, 10

Table 12.4 Source Node, Destination Node, and Minimum Rate Requirement of Each Session in the 20-Node Network

Session l	Source Node $s(l)$	Dest. Node $d(l)$	Min Rate Req. $r(l)$
1	16	10	9
2	18	3	1
3	12	11	4
4	13	17	3
5	15	6	2

Band 7: $q^7_{14,17} = 1$, $q^7_{20,1} = 1$
Band 8: $q^8_{12,11} = 3$
Band 9: $q^9_{12,8} = 1$, $q^9_{19,6} = 3$
Band 10: $q^{10}_{18,20} = 1$

Note that the same frequency band may be used by concurrent transmissions; for example, both node 7 (to 3) and node 16 (to 12) are transmitting on band 1. To minimize interference, our solution places these concurrent transmissions sufficiently apart and sets the optimal transmission power less than the maximum.

Figure 12.9 shows the flow routing topology in the final solution. The flow rates are:

Session 1: $f_{2,10}(1) = 103.30$, $f_{8,2}(1) = 103.30$, $f_{11,10}(1) = 15.86$, $f_{12,8}(1) = 103.30$, $f_{12,11}(1) = 15.86$, $f_{16,12}(1) = 119.16$
Session 2: $f_{1,7}(2) = 13.24$, $f_{7,3}(2) = 13.24$, $f_{18,20}(2) = 13.24$, $f_{20,1}(2) = 13.24$
Session 3: $f_{12,11}(3) = 52.96$

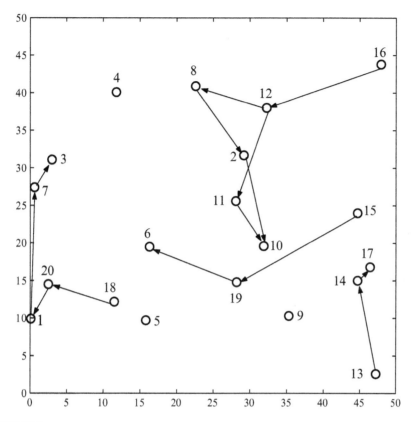

FIGURE 12.9

The flow routing topology in the final solution for the 20-node network.

Session 4: $f_{13,14}(4) = 39.72, f_{14,17}(4) = 39.72$
Session 5: $f_{15,19}(5) = 26.48, f_{19,6}(5) = 26.48$

We can see that, to maximize the achieved capacity, multipath routing is used (e.g., for session 1).

Under this solution, the achieved data rates for sessions 1 to 5 are 119.16, 13.24, 52.96, 39.72, and 26.48, respectively, and the achieved rate scaling factor is 13.24.

A 30-Node Network

For the 30-node network in Figure 12.10, the location and available bands at each node are given in Table 12.5. There are five sessions. The source node, destination node, and minimum rate requirement of each session are shown in Table 12.6. The transmission power levels on their respective frequency bands in the final solution are

Band 1: $q_{4,1}^1 = 1, q_{21,28}^1 = 1$
Band 2: $q_{28,13}^2 = 3$
Band 3: $q_{19,29}^3 = 9$
Band 4: $q_{23,29}^4 = 4$
Band 5: $q_{26,29}^5 = 1$
Band 6: $q_{13,11}^6 = 2$
Band 7: $q_{16,21}^7 = 4$
Band 8: $q_{16,13}^8 = 2$
Band 9: $q_{26,22}^9 = 7$
Band 11: $q_{17,16}^{11} = 2$
Band 12: $q_{19,23}^{12} = 1$
Band 14: $q_{22,15}^{14} = 1$

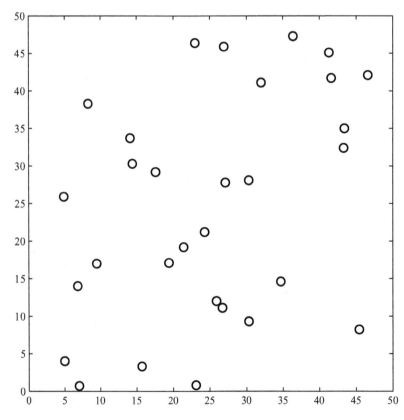

FIGURE 12.10

A 30-node network.

Table 12.5 Location and Available Frequency Bands at Each Node for a 30-Node Network

Node	Location	Available Bands	Node	Location	Available Bands
1	(7, 0.7)	1, 2, 6, 7, 16, 17, 19, 20	16	(30.3, 28.1)	7, 8, 11, 16, 17, 19, 20
2	(5, 4)	3, 5, 9, 12, 14, 15	17	(32, 41.1)	7, 11, 16, 17, 19, 20
3	(6.8, 14)	1, 2, 6, 7, 8, 11, 16, 17, 19, 20	18	(14.1, 33.7)	3, 4, 5
4	(15.7, 3.3)	1, 2, 7, 16, 20	19	(23, 46.4)	3, 12, 15
5	(9.5, 17)	3, 4, 5, 9, 12	20	(30.3, 9.3)	5, 9
6	(19.4, 17.1)	1, 2, 6, 7, 8, 16, 19, 20	21	(17.6, 29.2)	1, 2, 6, 7, 8, 11, 16, 17, 19, 20
7	(34.7, 14.6)	3, 4, 5, 9, 12, 14	22	(27.1, 27.8)	9, 12, 14, 15
8	(4.9, 25.9)	3, 4, 12	23	(26.9, 45.9)	3, 4, 5, 9, 10, 12, 13, 14, 15, 17
9	(46.6, 42.1)	10, 18	24	(43.3, 32.4)	1, 2, 11, 16, 17, 20
10	(8.3, 38.3)	3, 4, 5, 9, 14	25	(45.4, 8.2)	3, 4, 5, 9, 12, 14
11	(26.7, 11.1)	1, 6, 7, 8, 11, 16, 17, 19, 20	26	(43.4, 35)	3, 5, 9, 15
12	(36.4, 47.3)	10, 13, 18	27	(41.3, 45.1)	1, 16, 20
13	(24.3, 21.2)	1, 2, 6, 8, 11, 19	28	(14.4, 30.3)	1, 2, 6, 7, 8, 11, 16, 17, 20
14	(23.1, 0.8)	3, 5, 9, 14	29	(41.6, 41.7)	3, 4, 5, 9, 10, 12, 14, 15, 18
15	(21.4, 19.2)	4, 9, 12, 14	30	(25.9, 12)	1, 2, 6, 7, 8, 11, 16, 17, 19, 20

Table 12.6 Source Node, Destination Node, and Minimum Rate Requirement of Each Session in the 30-Node Network

Session l	Source Node $s(l)$	Dest. Node $d(l)$	Min Rate Req. $r(l)$
1	16	28	4
2	24	11	7
3	13	1	1
4	19	29	8
5	26	15	1

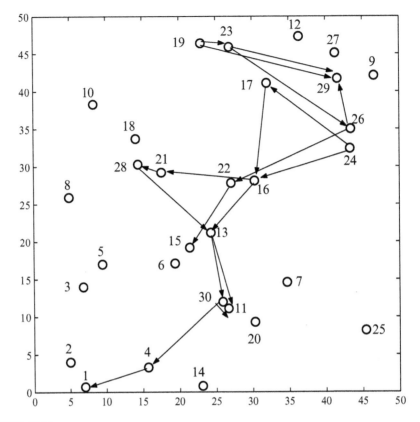

FIGURE 12.11

The flow routing topology in the final solution for the 30-node network.

Band 15: $q_{23,26}^{15} = 10$
Band 16: $q_{24,17}^{16} = 3, q_{30,11}^{16} = 1$
Band 17: $q_{24,16}^{17} = 7$
Band 19: $q_{13,30}^{19} = 1$
Band 20: $q_{30,4}^{20} = 4$

Note that the same frequency band may be used by concurrent transmissions; for example, both node 4 (to 1) and node 21 (to 28) are transmitting on band 1. To minimize interference, our solution places these concurrent transmissions sufficiently apart and sets the optimal transmission power less than the maximum.

Figure 12.11 shows the flow routing topology in the final solution.

Session 1: $f_{16,21}(1) = 124.72, f_{21,28}(1) = 124.72$
Session 2: $f_{13,11}(2) = 160.58, f_{13,30}(2) = 57.68, f_{16,21}(2) = 24.22, f_{16,13}(2) = 194.04, f_{17,16}(2) = 104.36, f_{21,28}(2) = 24.22, f_{24,16}(2) = 113.90, f_{24,17}(2) = 104.36, f_{28,13}(2) = 24.22, f_{30,11}(2) = 57.68$
Session 3: $f_{13,30}(4) = 31.18, f_{30,4}(3) = 31.18, f_{4,1}(3) = 31.18$
Session 4: $f_{19,23}(4) = 211.19, f_{19,29}(4) = 103.33, f_{23,26}(4) = 102.46, f_{23,29}(4) = 108.73, f_{26,29}(4) = 102.46$
Session 5: $f_{22,15}(5) = 39.32, f_{26,22}(5) = 39.32$

We can see that, to maximize the achieved capacity, multipath routing is used (e.g., for session 2).

Under this solution, the achieved data rates for sessions 1 to 5 are 124.72, 218.26, 31.18, 314.52, and 39.32, respectively, and the achieved rate scaling factor is 31.18.

A 50-Node Network

For the 50-node network in Figure 12.12, the location and available bands at each node are given in Table 12.7. There are 10 sessions. The source node, destination

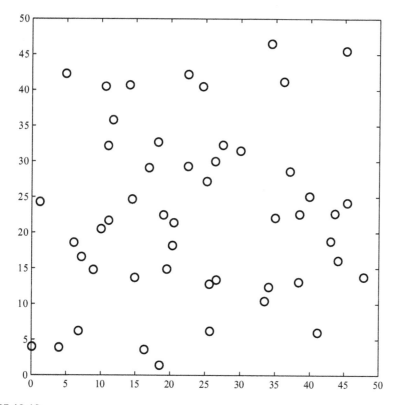

FIGURE 12.12

A 50-node network.

Table 12.7 Location and Available Frequency Bands at Each Node for a 50-Node Network

Node	Location	Available Bands	Node	Location	Available Bands
1	(11.1, 21.7)	2, 3, 4, 8, 25	26	(25.2, 27.2)	10, 14, 20, 24, 26
2	(0.1, 4)	6, 7, 10, 13, 14, 20, 23, 24, 26, 28	27	(22.5, 42.2)	5, 9, 12, 16, 18, 27, 29, 30
3	(7.2, 16.6)	6, 10, 14, 20, 23, 24, 26	28	(30, 31.5)	6, 13, 24, 26, 28
4	(11, 32.2)	6, 7, 10, 13, 14, 20, 23, 24, 26, 28	29	(35, 22.1)	6, 10
5	(16.3, 3.6)	10, 13, 14, 20, 23	30	(25.7, 6.2)	5, 9, 12, 17, 18, 22, 27, 29, 30
6	(14.5, 24.7)	8, 11, 25	31	(34.1, 12.4)	9, 12, 16, 17, 30
7	(14.9, 13.7)	5, 9, 12, 16, 17, 18, 22, 27, 29, 30	32	(26.4, 30)	5, 9, 12, 16, 17, 18, 22, 27, 29, 30
8	(19.5, 14.9)	7, 24, 28	33	(14.1, 40.7)	1, 2, 25
9	(26.6, 13.4)	1, 19, 21, 25	34	(34.4, 46.5)	9, 17, 18, 30
10	(22.5, 29.3)	1, 3, 4, 8, 11, 15, 19	35	(19, 22.5)	1, 6, 7, 10, 13, 14, 20, 23, 24, 28
11	(24.6, 40.5)	3, 8, 25	36	(39.9, 25.1)	6, 13, 14, 20, 23, 24, 26, 28
12	(38.4, 13.1)	2, 8, 11, 15	37	(20.3, 18.2)	1, 2, 3, 4, 8, 11, 15, 19, 21, 27
13	(4, 3.9)	9, 12, 16, 22, 27, 29, 30	38	(10, 20.5)	6, 7, 10, 13, 14, 20, 23, 24, 26, 28
14	(6.1, 18.6)	9, 12, 16, 17, 18, 22, 27, 30	39	(20.5, 21.4)	1, 2, 3, 4, 8, 11, 15, 19, 21, 25
15	(38.5, 22.6)	2, 4, 11, 15, 19, 21, 25	40	(37.1, 28.6)	7, 10, 13, 14, 20, 23, 24, 26
16	(1.2, 24.3)	5, 9, 12, 17, 22, 29, 30	41	(44.1, 16.1)	1, 15, 21
17	(4.9, 42.3)	5, 27	42	(41.1, 6)	9, 29
18	(18.5, 1.4)	5, 9, 12, 17, 18, 27, 30	43	(43, 18.8)	5, 9, 12, 16, 18, 22
19	(16.9, 29.1)	3, 4, 10, 11, 12, 15	44	(45.4, 24.2)	9, 12, 16, 17, 18, 30
20	(33.5, 10.4)	7, 13, 14, 20, 23, 24, 26, 28	45	(36.2, 41.2)	5, 9, 17, 27, 29, 30
21	(25.6, 12.8)	6, 7, 20, 23, 24, 28	46	(27.5, 32.3)	12, 16, 17, 18, 29, 30
22	(45.2, 45.5)	2, 8, 15, 19	47	(47.8, 13.8)	22, 27, 29, 30
23	(43.6, 22.7)	1, 2, 3, 4, 11, 15, 19, 21	48	(8.9, 14.8)	5, 30
24	(10.6, 40.5)	4, 15, 19, 21, 25	49	(6.8, 6.2)	5, 9, 12, 16, 17, 27, 30
25	(18.2, 32.7)	9, 12, 18, 22, 27	50	(11.7, 35.8)	1, 2, 3, 4, 8, 11, 15, 19, 21, 25

Table 12.8 Source Node, Destination Node, and Minimum Rate Requirement of Each Session in the 50-Node Network

Session *l*	Source Node *s(l)*	Dest. Node *d(l)*	Min Rate Req. *r(l)*
1	21	4	4
2	5	26	7
3	19	20	6
4	33	6	10
5	37	10	9
6	23	11	2
7	25	46	3
8	42	43	9
9	44	27	8
10	47	30	1

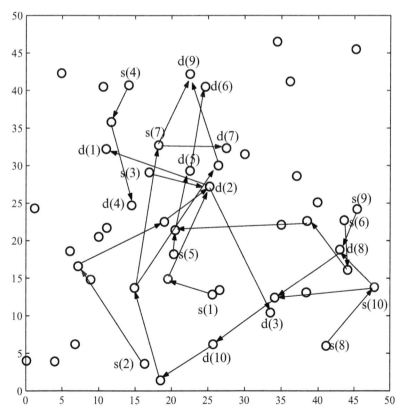

FIGURE 12.13

The flow routing topology in the final solution for the 50-node network.

node, and minimum rate requirement of each session are shown in Table 12.8. The routing topology is shown in Figure 12.13. The detailed transmission power levels and flow rates in this solution are omitted to conserve space. We can see that it is necessary to employ multipath routing for sessions 9 and 10. The scaling factor in the final solution is 13.36.

12.5 CHAPTER SUMMARY

In this chapter, we study cross-layer optimization for multihop CR networks. We give joint consideration of power control at the physical layer, scheduling at the link layer, and flow routing at the network layer. We present mathematical models to characterize the interdependency among power control, scheduling, and routing. Our models are general and can be applied to many cross-layer optimization problems for multihop CR networks. As a case study, we apply our models to a throughput maximization problem for a multihop CR network. We develop a centralized solution procedure for this optimization problem based on a branch-and-bound framework. Using numerical results, we offer quantitative understanding on the joint optimization at different layers.

12.6 PROBLEMS

1 Compare CR and MC-MR wireless networks.

2 Discuss the impact of power control on scheduling and flow routing.

3 Consider a CR network where each node uses peak transmission power P_{max} (i.e., no power control).
   (a) Formulate the constraints for scheduling and successful transmission.
   (b) Comparing solutions with and without power control, which provides better performance (larger scaling factor)? Explain.

4 Discuss the impact of scheduling on power control and flow routing.

5 Discuss the impact of flow routing on power control and scheduling.

6 Suppose flow splitting and multipath routing are not allowed.
   (a) What are the routing constraints?
   (b) Compare the single-path solutions and the multipath solutions and explain which achieve better performance.

7 Suppose the objective is to maximize the total data rate utility for all user sessions, where utility is $u(l) = \ln r(l)$ for session l. Write down the problem formulation in this case.

8 The problem considered in this chapter aims to maximize an objective. The branch-and-bound procedure to solve this problem is shown in Figure 12.5.

For another problem that aims to minimize an objective, can we use the same branch-and-bound procedure? If not, how should we change the procedure to solve a minimization problem?

9   Why do we need linear relaxation in the branch-and-bound procedure?

10   Why do we need a local search in the branch-and-bound procedure?

11   When we select a partition variable, why should we select an x variable even if the relaxation error caused by this variable is less than that by a q variable?

12   After we obtain the values for all x and q variables, how do we solve the optimal values for other variables?

13   By a branch-and-bound procedure, can we get an optimal solution for the problem considered in this chapter? If yes, how should we change the procedure in Figure 12.5? Discuss the pros and cons to get an optimal solution and a $(1 - \varepsilon)$-optimal solution.

Applications, standards, and implementations of cognitive radio

3

Defining cognitive radio

13

Przemysław Pawełczak[1] and Rangarao Venkatesha Prasad[2]

[1]University of California, Los Angeles, United States
[2]Delft University of Technology, The Netherlands

13.1 INTRODUCTION

Since the early 1990s we observed increasing interest around the world in *cognitive radio* (CR) technology. For example in Figure 13.1, we present the result of a simple web crawl, which shows a monotonic increase in the amount of Internet web pages containing phrases like *cognitive radio*, *dynamic spectrum access*, and *opportunistic spectrum access*. Yet another indication of CR popularity is that, in the past five years, more than 20 books were published, more than 20 journal special issues were arranged, and over 40 conferences organized—all solely dedicated to CR and related concepts (for more detailed information we refer the reader to www.scc41.org/crinfo). However, by closely looking into a random set of recently published papers having CR in their title, we observe that each author has his or her own definition of the topic. For example, hardware specialists have a notion of CR that interrelates to software-defined radio (SDR), while physical layer researchers look at CR with an eye on the bit pipe that can be achieved from the information theory (IT) point of view. Finally, protocol designers look into it from the optimization and implementation perspectives. Not only has there been a lot of ambiguity in CR terminology, almost all interrelated concepts of CR are hitherto defined based on the context. It is then imperative to devote a chapter of this book to clearly understanding all the interpretations of CR and the related concepts.

This chapter is structured as follows. In Section 13.2 we present related terminology that is important to CR. Later, in Section 13.3, we briefly describe the efforts of IEEE to standardize concepts related to CR. The chapter concludes in Section 13.4, while Section 13.5 presents some open questions to the inquisitive reader.

© 2010 Elsevier Inc. All rights reserved.
Doi: 10.1016/B978-0-12-374715-0.00013-7

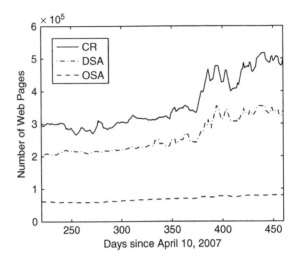

FIGURE 13.1

Statistics of the Google search engine responses for cognitive radio (CR), dynamic spectrum access (DSA), and opportunistic spectrum access (OSA) phrases in terms of number of WWW pages found; see also [544].

13.2 DEFINING CR: HISTORY, APPLICATIONS, AND RELATED CONCEPTS

We can state that the emergence of CR is strongly correlated with contemporary spectrum management that resulted in highly inefficient radio spectrum use. At the core of this problem lies archaic spectrum licensing and management. Such static spectrum assignment, applied to radio frequencies for almost a century, lead to a so-called quasi-scarcity of the spectrum. Therefore, it would be a welcome step if resource-starved users with no license are allowed to utilize dynamically (opportunistically) licensed frequencies when they are free (to minimize interference) at a specific place and time. Such an approach would increase overall frequency reuse and boost the throughput for applications that opportunistically use the empty frequencies.

Many successful attempts have been made in the past to liberalize spectrum access. Before going further with the definition of CR, let us briefly discuss the history of the nonconservation approach to spectrum management.

13.2.1 A Brief History of Elastic Spectrum Management

Looking at the history of radio regulations (especially in the United States), we can find many examples of attempts to liberalize spectrum market. Here, by

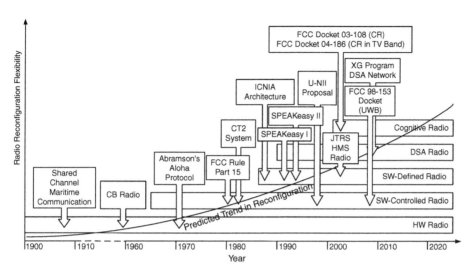

FIGURE 13.2

History of dynamic spectrum access systems and their relation to the implementation platforms, with the view to the future [545]; HW: hardware, SW: software (for the explanation of all other terms the reader is referred to Section 13.2.7).

liberalization, we mean maintaining a set of radio channels to be used by the public without ownership or license. Such maintenance would be completely and spatially distributed (using specific "radio layer management protocols") or supported by a spectrum regulator.

One of the first communication systems with shared radio resources was maritime communication developed in the early 1920s, see Figure 13.2. A 2.182 kHz band was used as an emergency and control channel on which all ships could listen for someone intending to communicate by broadcasting a working carrier identifier for further communication. After World War II, around 1960, the U.S. Federal Communications Commission (FCC) allowed using shared channels in land mobile communication, where one trunking channel could be used by many parties. Basically, hardware extensions like the tone-coded squelch or listen before talk (LBT) were used to share the frequency band. The fact that most transmitted messages were short made the shared channel communication very efficient. In the mid-1970s, the FCC allowed sharing channels at the 27 MHz band (so-called citizen band, CB) on a first-come, first-served basis. The only restriction to the users of the CB bands was to adhere to the maximum transmit power limit.

With innovations in wireless data communication, more flexible ways of spectrum management were possible. Abramson's Aloha protocol, presented first in 1970, was a solution to use radio channels for wireless data communication without a centralized coordinating entity. The ideas of random access were later extended to Packet Radio Networks. This, indirectly, lead to a FCC Rule Part 15, which described the ways of coexistence of low-power wireless devices in the industrial, scientific, and medical (ISM) band. Adopted in 1985, it initially described the methods for wireless devices using spread spectrum as a communication technique. Later, the Part 15 rule was changed to specify any modulation technique that met required power limits, was flexible enough, and contained no "strong spectral lines." Neither etiquette nor LBT protocols were defined in the FCC ruling. Its huge success was later legitimized by FCC's acceptance of the Apple Corp. proposal in 1995 to allow everyone to use the 5 GHz band (called unlicensed-national information infrastructure, U-NII) with no prior license or permission. Currently, U-NII is used with a higher success rate for wireless packet-based communication.

The British cordless telephone second generation (CT2) system, standardized in the mid-1980s, was another example of a successful distributed channel management technique. The 40 MHz band divided into 40 channels was managed by a base station (BS) that could monitor the level of interference on all channels and choose one that possessed minimum interference. CT2 systems were very popular in Hong Kong and Singapore.

George Gilder, in "Auctioning the Airwaves" (published in *Forbes* on April 11, 1994), envisioned that in the future "the wireless systems ... will offer bandwidth on demand and send packets wherever there is room." In parallel, Eli Noam from University of Columbia, proposed, in 1995, an "open spectrum access" paradigm [546], in which interested parties would pay for bandwidth whenever there is demand. Although both proposals addressed no technical issues and were mainly aiming at packet data communication, it was a sign for radio regulators that real steps in liberalizing the spectrum market is the order of the day; that is, it was clearly visible that it might be better to promote licensed parties that share their under- and nonutilized resources. Therefore, in 2002, FCC issued the 98-153 docket, permitting many users to transmit using low-power communication based on ultra wideband (UWB) communication. Recently released FCC docket 03-122 revisited Rule 15, allowing wireless data users to share channels with radar systems on an LBT basis. Finally the FCC realized that CR techniques are the future substrate that stimulate full growth of an "open spectrum" (see FCC Docket 03-108 on CR techniques and FCC Docket 04-186 on CR in TV spectrum). In the second report, FCC 08-260 finally approved allowing the use of TV white spaces in November 2008 (see [63]).

We note that some probes of radio channel liberalization were not so spectacular, mainly due to inflexible rules of operation given by the regulator. Examples of such systems were the radio common carrier (RCC) issued in the mid-1970s, the 800 MHz channel air ground telephone service (AGTS) from the 1990s, the unlicensed personal communications service (UPCS), and the large-scale low-earth orbit satellite system (called *Big LEOS*) with shared code division multiple access (CDMA) channels (early 1990s). First, the RCC could operate only when multiple service providers decided how to share common channels, which was not so financially attractive due to competition among all interested parties. Second, the AGTS was not popular due to the many rules of operation that FCC imposed. Third, the UPCS specification by FCC also included many restrictions to the operation of potential systems. Moreover, it had to share channels with microwave point-to-point links and often spatial separation was necessary between different UPCS devices. Finally, Big LEOS failure was mainly due to the financial problems of the service providers because of licensing fees.

The general idea of CR, as described already, then started to attract lots of attention. Since the introduction of this concept formally in 1999 by Mitola [6], a massive amount of literature has been published on that topic.

A brief illustration of this discussion is given in Figure 13.2. More information on the historical developments in dynamic spectrum management can be found in [547]. Now, given the knowledge of past inflexible spectrum management, we obviously need to look at the future.

13.2.2 A View of Wireless Network Futurists

In the late 1990s, in parallel to what had been happening over the last 100 years in radio spectrum management, a community of researchers, visionaries, futurists, and alike started to think about combining flexible spectrum access concepts with intelligent radio hardware platforms and smart networks. In this framework, emerging paradigms of dynamic spectrum access were related to cognitive communications.[1] The computation abilities of current electronic devices as well as recent developments in computer science and artificial intelligence allowed researchers to start thinking of introducing cognition into wireless networks and devices. This functionality would allow wireless systems in general to become more flexible, inferring from the environment to adapt the internal parameters to fulfill the needs of the user in a better way by taking necessary actions. These intelligent or learning cognitive devices would per se also allow harvesting the radio spectrum more optimally, allowing more users to communicate efficiently without an additional need for licensing. The ultimate dream is to use and reuse the available spectrum to the fullest.

CR can be applied to any communication system or network that suffers from spectrum shortage. It becomes attractive, since it needs no specially designed modulation technique, coding, or the like. We can think of no currently existing network that can be upgraded with such functionality. Ad hoc, sensor, and cellular networks are the ones that might benefit from the additional spectrum capacity it can offer. Operational specific networks can also benefit from the introduction of CR [548]. Moreover, the utility of CR has been recognized by ETSI and considered one of the candidates for the future radio interface of 4G networks. The potential for CR has also been recognized by IEEE. Its growing interest in this research topic is demonstrated by starting an IEEE Communications Society Technical Committee on Cognitive Networks [549]. Furthermore, its newest standard specifies protocols

---

[1]The term *cognition* is a popular topic in psychological and social sciences, which relates to information processing, understanding, and making sense of observations and using the attained knowledge in future interactions with the environment.

for future regional access networks, called IEEE 802.22, which aim at design of a new radio interface that would harness the so-called TV white spaces. Yet another initiative of the IEEE is a standard related to reconfigurable heterogenous radio interfaces, IEEE 1900.4. More information on the standardization initiatives of IEEE within the framework of OSA is given in Section 13.3.

As pointed out in Section 13.1, during the course of research on CR, there has been a lot of ambiguity in naming certain concepts; that is, different modern approaches of spectrum management were commonly mistaken with CR. We discuss in detail the ambiguous definitions related to CR in Section 13.2.3.

13.2.3 Ambiguity in CR Definitions

Historically, CR was first described in [6, 7, 329] as a decision-making layer at which "wireless personal digital assistants and the related networks were sufficiently computationally intelligent about radio resources, and related computer-to-computer communications, to detect user needs as a function of use context, and to provide radio resources and wireless services most appropriate to those needs." This was a vision of an intelligent wireless "black box" with which a user travels. Wherever the user goes, the CR device would adapt to the new environment, allowing the user to always be connected [6]. We need to note that Mitola was not only the creator of the CR notion, he also coined the term *SDR*; see, for example, [329] (we elaborate more on hardware architectures associated with the CR in Section 13.2.7). He thought of CR as a natural extension of SDR, where software allowed the device to flexibly alter transmission and reception parameters to all layers of the communication stack. Also, he was the first one to think of including intelligence ergo cognition to the whole radio setup.

Six years after Mitola's first article on CR, Simon Haykin, in the *IEEE Journal on Selected Areas in Communications* [11], recapitulated the idea of CR. He defined CR as "inclusive of SDR, [an idea] to promote efficient use of spectrum by exploiting the existence of spectrum holes" or an "intelligent wireless communication system ... that adapt(s) to statistical variations in the input stimuli with two primary objectives in mind: highly reliable communication ... [and] efficient utilization of radio spectrum." Thus, he limited the scope of CR to the efficient spectrum utilization-oriented device. His article focuses on signal processing techniques that could be helpful particularly in managing the second goal, efficient utilization of radio spectrum. Not only did he define his own CR, but also altered the basic cognitive cycle proposed by Mitola [329]. This article [11] was the first major article to give a totally different definition of CR and at the same time introduce terminology confusion. Interestingly, according to Google Scholar, as of September 30, 2008, the original Mitola paper on CR [329] was cited 404 times, while Haykin's paper [11] was cited 669 times!

Not only researchers but also regulatory bodies and standardization institutions defined CR in their own ways, more often than not self-serving. For example, SDR Forum explains CR is "a radio that has, in some sense, 1) awareness of changes in its environment and 2) in response to these changes adapts its operating characteristics in some way to improve its performance or to minimize a loss in performance."

In contrast to these definitions, the FCC describes CR as a wireless node or network able to negotiate cooperatively with other users to enable more efficient utilization of radio resources; see FCC Docket 03-108 and 03-186 for a more detailed description. CR would be able to identify a portion of the unused spectrum and utilize it for communication purposes. Thus, just like in Haykin's paper [11], the FCC's approach is a simplified form of Mitola and Maguire's vision where only radio spectrum conditions are considered while making a decision about future transmission and reception parameters. We can thus conclude that the original CR concept of Mitola is limited in scope in the later literature, purely to a radio that efficiently utilizes spectrum, looks for spectrum opportunities, and its adaptation process is limited to the physical layer.

Yet another notion of CR can be found in the IT community. Chapter 11 of this book is mainly devoted to the discussion of CR in an IT context. There, CR is limited to analysis of capacity and throughput of tx-rx Pair 1 (in that context called *secondary users*) with tx-rx Pair 2 (called *primary users*) that interferes with Pair 1. In this context a notion of *cognitive channel* is presented; that is, a channel in which a secondary pair of nodes possesses some kind of side information on what actually the interferer is transmitting. It is clearly seen that cognition in an IT context is far different from what Mitola expressed as "cognition." We can see that many of these approaches have different ideas and goals. Using the preceding CR explanations, some important aspects are sieved and Table 13.1 tries to consolidate them.

Luckily, some institutions aim at standardizing CR-related concepts and give them back their original meaning. One of the firsts and currently the major one

Table 13.1 A Coarse Comparison of Different Interpretations of CR

Aspects	Mitola	Haykin	SDR Forum	FCC	Inf. Theory
User's needs	✓				
Context	✓				
Intellig. & contr.	✓	✓	✓		
Radio/spectr.	✓	✓	✓	✓	✓
Spectr. effic.		✓	✓	✓	✓
Primary users		✓	✓	✓	✓
SDR	✓	✓			
Cooperation				✓	
Reliability		✓			

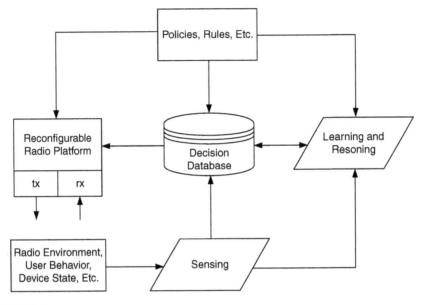

FIGURE 13.3

Components of a CR node, see also [545].

is the IEEE Standard Coordination Committee 41 (IEEE SCC41) (more information on the work of IEEE SCC41 is given in Section 13.3). One of the working groups of IEEE SCC41, IEEE 1900.1, was constituted solely for the purpose of gathering all the definitions that accompany CR. In the IEEE 1900.1 standard [550], the cognitive functionality may be spread across the layers of the communication architecture, resulting in coordination among the layers for an efficient use of the available spectrum. Figure 13.3 explains the basic functional blocks of such a CR node. Specifically, apart from a reconfigurable radio, a CR node has various components. The sensing and policies block (if available) are extensively used in deciding the availability of spectrum. These blocks also help in driving the learning and reasoning functions. The decision database, along with the input from the sensing and policies, block drive learning. The result is that the radio is configured based on input from different layers of the communication stack as well as from the environment inputs.

Moreover, recent research papers outline the possibility of extending the principle of cognition to the entire heterogeneous networks, thus defining the concept of cognitive networks (CNs) [326]. The aim of CNs is to self-adapt to changing requirements from users' applications to provide quality of service and self-management capability. Such a networking paradigm is based on the availability of software-adaptable network elements, driven and configured by a cognitive process. A cognitive process is a decision-making engine in which decisions are based on the current network conditions and involve adaptation and learning techniques. It is important to note that the concept of CN originated independently from the concept of CR.

We also need to emphasize that there is yet another ambiguity in the definition of CN, since we cannot equate CN and cognitive radio network (CRN). Referring again to Chapter 11 of this book, for example, CN is defined as a network constructed of primary and secondary users (we define these concepts later in Section 13.2.6), where secondary users are considered the cognitive ones. These users simply obtain the additional information on the activity of the primary users to employ better transmission parameters, in this context limited only to coding.

13.2.4 A Glossary of Cognitive Radio Definitions

Presented here is a *figurative* glossary of definitions—and some of them paraphrased from different sources and put together for easy comparison. This should also aid in getting a complete idea of what other researchers are thinking. We refer interested readers to the original texts and articles for more explanations and details.

Mitola [6, 7, 329]. Wireless personal digital assistants and the related networks that are sufficiently computationally intelligent about radio resources, and related computer-to-computer communications, to detect user needs as a function of use context and to provide radio resources and wireless services most appropriate to those needs.

Wikipedia. Cognitive radio is a paradigm for wireless communication in which either a network or a wireless node changes its transmission or reception parameters to communicate efficiently, avoiding interference with licensed or unlicensed users. This alteration of parameters is based on the active monitoring of several factors in the external and internal radio environment, such as radio frequency spectrum, user behavior, and network state.

IEEE 1900.1 [550]. (a) A type of radio in which communication systems are aware of their environment and internal state and can make decisions about their radio operating behavior based on that information and predefined objectives; (b) cognitive radio [as defined in item a] that uses software-defined radio, adaptive radio, and other technologies to adjust automatically its behavior or operations to achieve desired objectives.

Haykin [11]. Cognitive radio is an intelligent wireless communication system that is aware of its environment and uses the methodology of understanding by building to learn from the environment and adapt to statistical variations in the input stimuli to achieve high reliability and efficient utilization of the radio spectrum.

Scientific American [551]. Cognitive radio is an emerging smart wireless communications technology that will be able to find and connect with any nearby open radio frequency to best serve the user. Therefore, a cognitive radio should be able to switch from a band of the radio spectrum that is blocked by interference to a free one to complete a transmission link, a capability that is particularly important in an emergency.

Rondeau and Bostian [552]. Cognitive radio is a system that has a cognitive engine performing modeling, learning, and optimizing the processes to reconfigure the communication system including the radio layer by taking the information from users, radio, and the context.

13.2.5 A Generalized Definition of Cognitive Radio Network

An interesting aspect we observe with respect to CR is that it fits many fields of scientific and engineering endeavors. For example, it stimulates signal processing techniques with respect to detection and sensing; it looks like a fertile area of application for artificial intelligence (AI); it encourages estimation theory enthusiasts as well as linguists with respect to policy language; it provides new creative opportunities for systems specialists.

To conclude, here we define the CR—in layman's terms—as the concept with which the wireless nodes adapt their properties, including radio, to achieve overall efficient spectrum usage, in time and space, based on the factors such as radio, radio environment, policies, and higher-layer requirements with an inherent and constant learning to improve the spectrum usage.

13.2.6 Concepts Related to Spectrum Management

In the previous section we conclude that CR has been somehow reduced to the spectrum management layer only. Thus, all the concepts related to modern approaches in spectrum management are again mixed with CR [109, 545, 553]. To clear the ambiguity in terminology let us briefly introduce our classification in Figure 13.4.

We can consider three essential models: *exclusive spectrum management* (ESM), the *spectrum commons* (SC) sharing model, and *hierarchical spectrum management* (HSM). The ESM model still gives exclusive channel usage to each user or service provider but differs from a static assignment in the sense that the channels are allocated dynamically among possible licensees. The process of exclusive channel access is usually governed by radio regulation bodies. The differences between ESM approaches, specified in Figure 13.4, depend on the economic model, which varies from country to country. In the SC model, different users compete for the assigned frequencies on equal terms. The HSM model gives *primary (licensed) users* (PUs) more rights to use the spectrum than other *secondary (unlicensed) users* (SUs). We can distinguish two HSM approaches. In *overlay* HSM, only one user/system can use a frequency band at a particular space and time, and the SUs have to back off when a PU is present. However, when no PU is present, the SU can opportunistically use the frequency band, so this technique is also referred to as *opportunistic spectrum access*. In *underlay* HSM, an SU can transmit on an

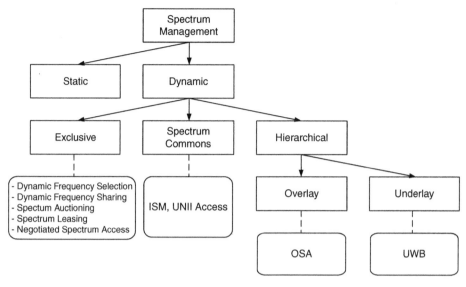

FIGURE 13.4

Modern spectrum management: classification with the application examples. See also [109, 553, 554].

already occupied band if this transmission does not increase the interference to the PU above a given threshold. A further classification of overlay HSM (not shown in Figure 13.4) involves *symmetric coexistence* (when both SU and PU networks adapt) and *asymmetric coexistence* (when only the SU network adapts, obeying the PU requirements).

13.2.7 Concepts Related to Computational Platforms

Yet another definition ambiguity comes from the CR implementation. Categories and classes of different future adaptive radio devices are listed in Table 13.2. This simplistic comparison tries to show the differences among them, since some confusion still persists in the CR community on how to classify different devices and systems. Please note that in Figure 13.2 different milestones in spectrum management flexibility have been mapped into different hardware platforms. The more flexible the given system is, the more flexible the hardware platform becomes. Certain milestones that we have to note in developing software-based radio platforms are SPEAKeasy [555], joint tactical radio system [556], DARPA XG program radios [557], and integrated communications, navigation, identification avionics (ICNIA) [558]. We can predict semi-exponential growth in hardware flexibility in the coming years (see again Figure 13.2).

Some explanations and features of the terms in Table 13.2 follow. We refer to the SCC41 1900.1 standard for very detailed descriptions of many of these terms and their interrelationships [550].

Table 13.2 Types of Adaptable Radio Devices (HW: Hardware, SW: Software)

Type of Radio	Platform	Reconfiguration	Intelligence
Hardware	HW	Minimal	None
Software	HW/SW	Automatic	Minimal
Adaptive	HW/SW	Automatic/predefined	Minimal/none
Reconfigurable	HW/SW	Manual/predefined	Minimal/none
Policy based	HW/SW	Manual (database)/ automatic	Minimal/none
Cognitive	HW/SW	Full	Artificial/ machine learning
Intelligent	HW/SW	Full	Machine learning/ predicting decision

Hardware radio. The capability of CR devices changing their radio characteristics is implemented completely in hardware. Thus, once in the field the devices will not be able to change their characteristics other than what is already built in. For example, the range of frequency programmed into the hardware always remains the same, even though the user knows that there is an opportunity to work in a different range. Therefore, the scope is limited in this case.

Software radio. The capability of CR devices changing their radio characteristics also is implemented in software. Thus, the devices are able to change their characteristics from other than what is already built in. For example, contrasting with the preceding, the range of frequency programmed into the hardware may be changed by uploading a new software patch (say, a simple configuration file).

Adaptive radio. This is the capability of CR devices where its radio characteristics are changed by mechanisms such as closed-loop or open-loop controllers. Basically, the devices adapt to the surroundings by sensing and using the preprogrammed logic and control techniques.

Reconfigurable radio. The radios in CR devices of which the functionalities can be changed manually. A hardware radio and a software radio both are reconfigurable, though in different ways and to different degrees.

Policy-based radio. The changes to the radio functionalities of CR devices are governed by the policies. The policy set usually is available as a data set (or database). For example, the frequencies used by military equipment are not allowed to be used by others under all circumstances. Basically the policy set governs the operational characteristics of the CR devices quite immaterial of whether they are capable.

Cognitive radio. It has been already defined. This includes databases, policies, learning techniques, and so forth.

Intelligent radio. This includes cognitive radios, which are also able to learn as well as predict the situations and adapt themselves. In a general and crude sense, it is a software radio. However, with respect to the previous explanation of the software radio, it just specifies the capability to work with a software control, thus an intelligent radio is much more than a simple software radio.

13.3 CR TERMINOLOGY STANDARDIZATION

In today's world of information and communication technology (ICT), specific themes require tight and efficient interaction between research and standardization efforts to enable new technologies to flourish and reach maturity. In the area of CR, IEEE SCC41 is a coordination body within the IEEE standardization organization "in the areas of dynamic spectrum access, cognitive radio, interference management, coordination of wireless systems, advanced spectrum management, and policy languages for next generation radio systems." Based on a summary of the IEEE SCC41 structure from [545, 559] we provide additional information on the recent developments within the project.

13.3.1 General Overview

The IEEE SCC41 effort was preceded by the IEEE P1900 Standards Committee, jointly sponsored by the IEEE Communications Society (ComSoc) and the IEEE Electromagnetic Compatibility (EMC) Society. The objective of P1900 was to develop supporting standards dealing with new technologies for the next-generation radio and advanced spectrum management. Keeping an eye on the broad spectrum of required technologies for the emerging wireless communications, the IEEE Standards Board constituted reorganization of the IEEE 1900 effort as IEEE SCC41 "Dynamic Spectrum Access Networks" (DySPAN). Thus, IEEE SCC41 oversaw a range of new standards that aid in the speedy development of the technology for next-generation wireless networks. Within the IEEE, SCC41 is a direct sponsor of standards, reporting directly to the IEEE Standards Board. The IEEE 1900 committee ceased to exist at the inaugural meeting of IEEE SCC41 in April 2007 during the IEEE DySPAN Conference. The work of all IEEE P1900 working groups continued under IEEE SCC41, retaining the same names as before. A brief diagrammatic representation of various working groups is represented in Figure 13.5.

The activities of IEEE SCC41 are supported by IEEE Communication Society technical committees, which represent the interface with the research community in terms of proposals for new standards and technical contributions to the internal discussions within the working groups (WGs). At the time of writing, this chapter IEEE SCC41 had six WGs, each responsible for evolving standardization processes for different aspects of DSA. WGs are identified as IEEE 1900.x, where .x represents one of the WGs. Each WG submits its evolved standard document to IEEE after

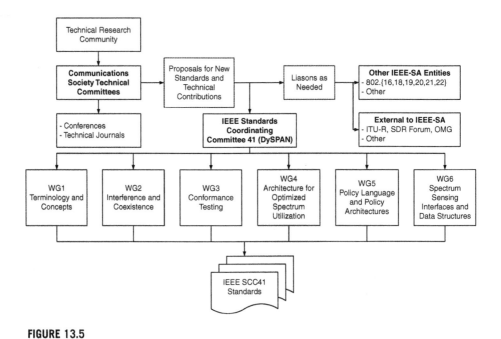

FIGURE 13.5

IEEE SCC41 organization structure and relationships with other entities. See also [545, Fig. 2].

finalizing the recommendations through discussions. The WG first prepares the document, which is open to all the IEEE Standard Association (SA) members as well as outsiders (on a fee-per-vote basis). Once the voting members are identified and the documents finalized, the documents go for ballots arranged by the IEEE SA. During the balloting, IEEE SA members vote with their comments. Negative votes are carefully considered by the working group and the issues addressed by them go into the next round of discussions, rewriting, and balloting. Thus, a fair chance is given to interested organizations and individuals to make the standards applicable to a wide spectrum of products using dynamic or opportunistic spectrum access. This process can be repeated many times, depending on the comments and attention a standard document invokes. Now we briefly explain the scope of each WG in the sequel.

13.3.2 IEEE 1900.1

As we point out in Section 13.2 many groups working on CR have defined CR and other related terms differently. Thus, IEEE decided to create 1900.1 WG, Standard Definitions and Concepts for Spectrum Management and Advanced Radio System Technologies, responsible for creating a glossary of important CR-oriented terms and concepts. It further provides explanations to germinate a coherent view of the various efforts taking place in the broad arena of CR. The key idea was to standardize and prepare technically precise definitions related to CR. In fact, 1900.1 WG acted as a glue to the other IEEE SCC41 WGs, tying them together with common definitions of CR terms. The IEEE 1900.1 has been voted by the IEEE Standard Association and is a standard now.

13.3.3 **IEEE 1900.2**

In light of new CR technology, many radio systems coexist and they try to optimize the utilization of spectrum in space and time. The accurate measurement of interference has thus become a crucial requirement for the deployment of these technologies. The mandate of the 1900.2 WG, Recommended Practice for Interference and Coexistence Analysis, was to recommend the interference analysis criteria and establish a well-thought-out framework for measuring and analyzing the interference between radio systems. New technologies, while attempting to improve spectral efficiency—by being flexible, collaborative, and adaptive—also cause disputes. Therefore, this WG established a common standard platform on which the disputing parties can present their cases and resolve them amicably.

The framework for interference analysis addresses the context of measurements and the purpose. Any new adaptive system has a trade-off between cost and gain. Therefore, the interference analysis should make this gain explicit, along with the usage model for this trade-off. Apart from the interference power measurements and the context, impact and remedies are also mentioned for analysis and comparison. Finally, parameters for analysis are derived from scenarios including the context and harmful interference thresholds. Uncertainty levels in measurements are compulsorily considered in the analysis. Just like 1900.1, the 1900.2 has been voted by IEEE Standard Association and has become a standard.

13.3.4 **IEEE 1900.3**

On the software front, IEEE 1900.3 WG, Recommended Practice for Conformance Evaluation of Software-Defined Radio Software Modules, is developing test methods for conformance evaluation of software for SDR devices. The aim is to define a set of recommendations that helps in assuring the coexistence and compliance of the software modules of CR devices before proceeding toward validation and certification of the final devices, as laid down in IEEE 1900.2. Since SDR is an important component of future CR networks, these recommended practices should help in creating high confidence in the deployed SDR devices. These devices will have multiple layers of software, each addressing different functionalities. Therefore, it is all the more essential to test the capability of SDR devices a priori to install the patches correctly over the air, assuring secure execution of intended functionalities.

As an illustration consider an implementation of the SDR device specifications into a program. This can be verified with the formal verification methods. However, formal specifications for software, mostly, do not exist. Therefore, testing in these cases becomes less formal, by focusing on only a particular subset of device operations. The aim is to design testing procedures that will comply with the semi-formal software specifications. One of the solutions is to define checkpoints (mandatory or obligatory) and assertions that will reflect the specifications. For these reasons IEEE 1900.3 WG specifies device management procedures. Since many of those exist today (e.g., Java's Mobile Device Management Server application programming interfaces), 1900.3 WG utilizes other relevant standards to achieve its goal.

13.3.5 IEEE 1900.4

Many mobile devices used today operate on multiple wireless networks. The study of network and device architectures that can help in distributed decision making for DSA is an important area, commercially as well as academically. These procedures particularly refer to reconfigurable terminals (including SDR) capable of accessing a multitude of radio access technologies. The 1900.4 WG, Coexistence Support for Reconfigurable, Heterogeneous Air Interfaces, defines the overall system architecture, splitting the functionality between terminals and the network, and the information exchange between coordinating entities. Its main goal is to increase the overall system utilization of reconfigurable terminals while increasing the perceived quality of service. All 1900.4-enabled devices should operate in an OSA or DSA manner so that they will not degrade the performance of PU radio access devices. The study of heterogeneity in wireless access technologies and multihoming of the devices—with CR capability—differentiates this WG from other WGs of SCC41. At first the 1900.4 WG looks into only the architectural and functional definitions. The corresponding protocol definitions related to the information exchange are addressed at a later stage. This standard was approved in late January 2009. After the successful work and much interest in this WG, two more projects have been assigned to 1900.4 WG:

1900.4a. Standard for Architectural Building Blocks Enabling Network-Device Distributed Decision Making for Optimized Radio Resource Usage in Heterogeneous Wireless Access Networks—Amendment: Architecture and Interfaces for Dynamic Spectrum Access Networks in White Space Frequency Bands.

1900.4.1. Standard for Interfaces and Protocols Enabling Distributed Decision Making for Optimized Radio Resource Usage in Heterogeneous Wireless Networks.

With these two projects, the scope of 1900.4 has expanded IEEE's interest in the community.

13.3.6 IEEE 1900.5

The recent WG of IEEE SCC41, started in August 2008, on Policy Language and Policy Architectures for Managing Cognitive Radio for Dynamic Spectrum Access

Applications, defines a policy language (or a set of policy languages or dialects) to specify interoperable, vendor-independent control of CR functionality and behavior for DSA resources and services. The initial work concentrates on standardizing the features necessary for a policy language to be bound to one or more policy architectures to specify and orchestrate the functionality and behavior of CR features for DSA applications (see www.scc41.org/5).

13.3.7 IEEE 1900.6

Yet another WG of IEEE SCC41 started in August 2008, on Spectrum Sensing Interfaces and Data Structures for Dynamic Spectrum Access and Other Advanced Radio Communication Systems. The intended standard defines the information exchange between spectrum sensors and their clients in radio communication systems. The logical interface and supporting data structures used for information exchange are defined abstractly without constraining the sensing technology, client design, or data link between sensors and clients (see www.scc41org/6).

13.3.8 Related Standardization Efforts

Other IEEE projects related to next-generation radio, like IEEE 802.{18, 19, 20, 22} are the sources of expertise for IEEE SCC41. As mentioned earlier, since active dialog between different standard bodies is crucial at this stage of CR development, IEEE SCC41 initiated cooperation with the FCC, Ofcom, SDR Forum, and OMG Forum, to name a few. The relation between different WGs of IEEE SCC41 as well as different standardization bodies within and outside IEEE are depicted in Figure 13.5.

IEEE 802.22 for TV White Space

To make universal broadband access a reality, allow the Internet service industry to grow, and make consumer Internet access competitive by providing more choices, the TV white space was opened for unlicensed use by the FCC [63]. Thus, the ubiquitous use of wireless access is thought of as a major deciding factor to allow more spectrum. In fact, even a fixed wireless access has many advantages over wired network access in terms of ease of setting up and use. In this direction, the white space coalition is one such initiative that influenced the FCC to open up the TV spectrum. Further, IEEE 802.22 indeed took the initiative to define a standard to use TV white space.

IEEE 802.22 is thought of as an alternative technology to WiFi with an unlicensed spectrum like that of WiFi, but a better spectrum between 54 MHz and 863 MHz. Similar to TV signals the access to Internet could be over tens of kilometers and no restrictions regarding in-building environments and the like. However, the challenges were many: identification of the primary users, listing the unused channels locally, and defining the power levels so as not to interfere with the adjacent bands. The two important entities defined here are the base station and customer premises equipment (CPE). BS controls all the CPEs, determining when to send data and the channels to use. CPEs also sense the spectrum in its vicinity, enabling distributed sensing, and send it back to the BS. With the opening up of TV white space by the FCC, this standard gained a significant role. We refer the reader to Chapter 14 of this book for a detailed description of this standard.

13.3.9 Results and Roadmap of IEEE SCC41

IEEE SCC41 and its previous avatar, IEEE P1900, successfully took steps toward the standardization process of the class of CR systems, spectrum agile systems, dynamic and opportunistic spectrum sharing systems, and their usage paradigms. The current results include an approved standard of 1900.1, 1900.2, and 1900.4.

Notwithstanding the work in these WGs, IEEE SCC41 took a new initiative to invite more projects and potentially set up new study groups (SGs). In case an SG successfully demonstrates the relevance of an issue to the goals of IEEE SCC41, it could submit a request to constitute a new WG and work toward the development of a new standard. As an example, recently, IEEE SCC41 received proposals for the standardization of requirements for disaster recovery, representation of information exchange in CR systems, and detection of primary users.

Indeed, IEEE SCC41 is actively working as an incubator and an "umbrella" for supporting, facilitating, and encouraging the efforts for standardization of the multifaceted issues of cognitive communications. For interested readers we list here some open issues for standardization that will expedite the proliferation of CR:

Regulatory. There are some issues concerning regulatory aspects that are not addressed by the current WGs. Though 1900.2 is addressing some of the dependability concerns, it is not complete at this time. Issues such as priority among various users, interference caused to primary and secondaries, measurement of interference, and detection and dependability detection are some of them.

Test procedures. It is surmised that CR devices will have to go through multiple tests. It is not just the interference caused, but an important question is how to quantify the intelligence. Hence, there is a need to standardize the metrics that would also boost efficient spectrum usage.

Protocols. Protocols with respect to the use of CR methodology need to be addressed in a prominent way. Some of the issues are integrating with OSI-layered architecture, protocols to exchange data, MAC protocols to increase the use of white space, and protocols that assure dependability.

Interoperability. Coexistence and cooperation are the hallmarks of CR devices. Many shades of CR stacks are surmised to be developed in the near future. If CR is going to enhance spectrum reuse, then these devices with different approaches need to meet at some point. They need to operate together. A standard way of doing this is needed.

Medium access control. Medium access control (MAC) protocols, particularly distributed MAC for ad hoc networks operating in the opportunistic spectrum, are not covered by any standard yet. Intelligent spectrum management such as

IEEE 802.11k covers only the unlicensed bands. There is a need for a generalized CR-MAC, defined for CR, that takes into account many of the specificities of CR. It is envisaged that many future versions of MAC inherit this generalized CR-MAC, and it requires sufficient generalization and flexibility.

Security. The standards should define the security apparatus for CRs. The enforcing agencies need such standards to nail down the devices that violate these features. It is in fact very important, since the devices are supposed to be adaptive and intelligent. The need is to develop a standard to (a) detect infringement and (b) enforce fairness. Authorization, avoiding denial-of-service attacks, and resilience to intruders are some issues important here.

13.4 CHAPTER SUMMARY

This chapter provides a terminology foundation for conceptualizing CR. It also provides discussions on difficulties in defining CR unambiguously. An explanation is given regarding the reasons why the community faces such a situation. A clear definition of CR is presented and a historical perspective of CR is also given. Moreover, most important definitions, which are interrelated to CR but hitherto used imprecisely, are also presented. Finally, this chapter discusses briefly the efforts of IEEE to standardize all concepts important to CR.

13.5 PROBLEMS

1   Are there any lessons learned from the initial "terminology mess" related to CR? Is there a way to avoid multiple interpretations of research terms? Or is it a natural research process, presumably on novel system-level concepts, mixing different notions? Discuss.

2   Give examples for each platform listed in Table 13.2.

3   Think of any other term related to CR. Does the CR community need more terms to describe its work or is the descriptive process complete? Provide justification for your statements.

4   Enhance Figure 13.2 with the help of recent literature on this topic.

5   Discuss the future of CR as a technology for the development of new applications.

6   It is envisaged that CR devices will be part of a communication stack in the near future. However, the legacy networks will not be able to cooperate and work with the new CR devices. Discuss possible ways to circumvent this problem.

7   Define *spectrum efficiency* under CR. How is spectrum efficiency dependent on the various aspects of CR?

8   Cognition is a key aspect in CRs; therefore, where would you keep "cognition"? Assuming that you are using the formalism of an OSI stack, where can cognition be placed? Can cognition be distributed?

9   Make a case study of an emergency network and explain how CR can aid the first responders.

10   Explain with reasoning why CRN will be the way to the next-generation communication networks.

Cognitive radio for broadband wireless access in TV bands: The IEEE 802.22 standards

14

Carlos Cordeiro[1], Dave Cavalcanti[2], and Saishankar Nandagopalan[3]

[1] Intel Corporation, United States
[2] Phillips Research North America, United States
[3] Broadcom Corporation, United States

14.1 INTRODUCTION

The proliferation of wireless services and devices for uses such as mobile communications, public safety, WiFi, and TV broadcast serve as the most indisputable example of how much modern society has become dependent on radio spectrum. While land and energy constituted the most precious wealth creation resource during the agricultural and industrial eras, respectively, the radio spectrum has become the most valuable resource of the modern era [17]. Notably, unlicensed bands (e.g., ISM and UNII) play a key role in this wireless ecosystem, given that many of the significant revolutions in radio spectrum usage have originated in these bands and resulted in a plethora of new applications, including last-mile broadband wireless access, health care, wireless PANs/LANs/MANs, and cordless phones. This explosive success of unlicensed operations and the many advancements in technology that resulted from it led regulatory bodies (e.g., the Federal Communications Commission, FCC, through its Spectrum Policy Task Force, SPTF [560]) to analyze the way the spectrum is currently used and, if appropriate, make recommendations on how to improve radio resource usage.

As indicated by the SPTF and numerous reports [561], the usage of the radio resource spectrum experiences significant fluctuations. For example, based on the measurements carried out in [561] for the frequency bands below 3 GHz and conducted from January 2004 to August 2005, we conclude that, on an average, only about 5.2% of the spectrum is actually in use in the United States in any given location at any given time (please refer to [561] for more detailed information). Interestingly enough, these measurements reveal that heavy spectrum utilization often takes place in unlicensed bands while licensed bands often experience low (e.g., TV bands) or medium (e.g., some cellular bands) utilization. These striking results coupled with recent advancements in radio technology led the FCC to revisit the traditional way of spectrum management. It has been realized that not only is spectrum usage very low in certain licensed bands, but also the scarcity of radio resources is becoming a crisis, hindering the development of many wireless applications, including broadband access (not only in urban/suburban areas but especially in rural/remote areas), public safety, health care, business, and leisure.

14.1.1 Cognitive Radios

Cognitive radios (CRs) [7, 60, 551] are seen as the solution to the current low usage of the radio spectrum. It is the key technology that will enable flexible, efficient, and reliable spectrum use by adapting the radio's operating characteristics to the real-time conditions of the environment. CRs have the potential to utilize the large amount of unused spectrum in an intelligent way while not interfering with other incumbent devices in frequency bands already licensed for specific uses. CRs are enabled by the rapid and significant advancements in radio technologies (e.g., software-defined radios, frequency agility, and power control) and can be characterized by the utilization of disruptive techniques such as wideband spectrum sensing, real-time spectrum allocation and acquisition, and real-time measurement dissemination (please also refer to the DARPA next-generation, XG, program request for comments (RFCs) [242] for a good overview of issues in and the potential of CRs).

With all these facts and foundations in place, the TV band Notice of Proposed Rule Making (NPRM) [60, 562] was the natural next step taken by the FCC. This NPRM, released in May 2004, proposes to allow unlicensed radios to operate in the TV broadcast bands provided no harmful interference is caused to incumbent services (e.g., TV receivers), which can be accomplished by employing CR-based technologies.

All these important events created a mindset within the IEEE that culminated in the formation of the IEEE 802.22 WG (or simply, 802.22) for wireless regional area networks (WRANs) in November 2004 [17]. This working group is chartered with the specific task of developing an air interface (i.e., physical, PHY, and medium-access control, MAC) based on CRs for unlicensed operation in the TV broadcast bands.

14.1.2 Regulatory Scenario for TV White Space

As mentioned, the 802.22 was formed in light of the TV band NPRM released by the FCC [60], which proposes to open the spectrum allocated to the TV service for unlicensed operation based on CRs. In the United States, TV stations operate from channels 2 to 69 in the VHF and UHF portions of the radio spectrum. All these channels are 6 MHz wide and span from 54–72 MHz, 76–88 MHz, 174–216 MHz, and 470–806 MHz. In addition to the TV service, also called *primary service*, other services such as wireless microphones are also allowed by the FCC to operate on vacant TV channels on a noninterfering basis (please refer to Part 74 of the FCC rules), and so are private land and commercial mobile radio services (PLMRS/CMRS) including public safety (please refer to Part 90 of the FCC rules). While it is recognized by the 802.22 WG that the FCC is yet to release the final rules for unlicensed operation in the TV broadcast bands (the rule was released on November 4, 2008[1]), there is a common feeling that these rules will not be a roadblock but rather will serve as a catalyst to the development of this new CR-based standard and promote the emergence of new markets, applications, and services.

14.1.3 Dynamic Spectrum Access Models

The two models with which spectrum access can be done are spectrum property rights and the spectrum commons approach. The *spectrum property rights* approach suggests that the spectrum can be treated like land, and private ownership of the spectrum is viable. This approach entails the holder of spectrum to have exclusive use of the spectrum portion it possesses, without the potential of harmful interference from other parties. Owners would be able to trade their spectrum in a secondary market. The use of the spectrum would be flexible, in that the authorized party could use the spectrum portion for any purpose. The FCC has chosen a partial implementation of this approach by employing spectrum auctions as a means of licensing. But this approach has efficiency issues, as seen in TV bands.

The *spectrum commons approach* allows the bands to be open to technologies as long as they follow the rules of access in the specific band; as smart technologies evolve, communicating devices will become able to avoid interference through mutual cooperation and coexistence and the spectrum will become unscarce. This phenomenon has been seen in the 2.4 GHz industrial, scientific, and medical (ISM) band as well as the 5 GHz UNII band. The emergence of cognitive and software-defined radio concepts, multiple-antenna and multicarrier techniques, as well as ultra-wideband (UWB) technologies and mesh network topologies provide a technology panacea that proponents of this approach use to support their arguments. The rules for access are governed by the FCC and device manufacturers have to comply with rules and regulations for operating in that band. Thus, even the commons regime is a form of lightly controlled shared access [563]. The approach for IEEE 802.22 WRAN is slightly different in the sense that the operation of WRAN causes no interference to the incumbents on the TV bands and there are mechanisms in the entire system design to build an efficient radio.

---

[1] Available at: http://newmobiletech.com/fcc-opens-tv-white-space-for-wireless-broadband/.

14.2 OVERVIEW OF IEEE 802.22 STANDARD

The wireless regional networks for which this standard is developed are expected to operate in lower population density areas and provide broadband access to data networks using vacant TV channels in the VHF and UHF bands in the range of frequencies between 54 MHz and 862 MHz, while avoiding interference to the broadcast incumbents on these bands. A typical application will be the coverage of the rural area around a village, as illustrated in Figure 14.1, within a radius of 17–30 km, depending on the effective isotropic radiated power (EIRP) of the base station using adaptive modulation, although the MAC could accommodate user terminals located as far as 100 km when exceptional radio frequency (RF) signal propagation conditions prevail.

As indicated in the 802.22 functional requirement document, the capacity at the user terminal is expected to be of 1.5 Mb/s in the downstream and 384 kb/s in the upstream. The service availability due to radio frequency propagation is assumed to be at 50% of locations to allow the service provider to reach subscribers in fringe areas, and 99.9% of the time to provide a reliable connection where it is possible. The average spectrum efficiency over the coverage area is expected to be around 2 bps/Hz, given the adaptive modulation parameters and the operating constraints described later; and assuming a 6 MHz TV channel bandwidth and a 40:1 oversubscription ratio resulting from the stochastic nature of the data network usage, this translates into a total of 255 user terminals that can be served by the base station per TV channel.

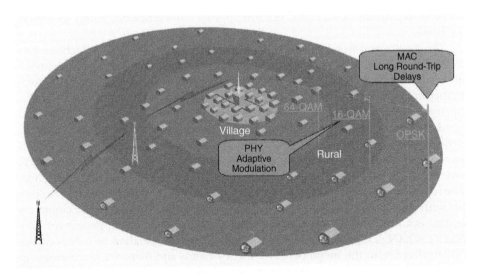

FIGURE 14.1

Typical application of the 802.22 WRAN standard [17].

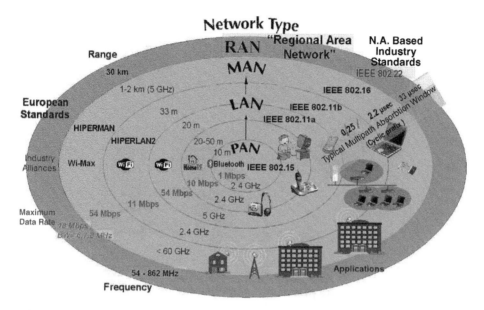

FIGURE 14.2

Characteristics of the WRAN standard relative to other wireless network standards.

Figure 14.2 illustrates the main characteristics of the 802.22 WRAN standard relative to the other existing wireless network standards.

14.2.1 Applications

The most prominent target application of 802.22 WRANs is wireless broadband access in rural and remote areas with performance comparable to those of existing fixed broadband–access technologies (e.g., DSL and cable modems) serving urban and suburban areas [564]. While the availability of broadband access may not be so critical in urban and perhaps suburban areas, the costs of which remain high, this certainly is not the case in rural and remote areas where about half of the United States and most of the developing nations (other countries located in South America, Africa, and Asia) populations are concentrated. Therefore, this has triggered the FCC to stimulate the development of new technologies (e.g., based on CRs) that increase the availability of broadband access in these underserved markets.

14.2.2 Reference Architecture

The 802.22 system specifies a fixed point-to-multipoint (PMP) wireless air interface whereby a base station (BS) manages its own cell and all associated consumer premise equipment (CPEs), as depicted in Figure 14.3.

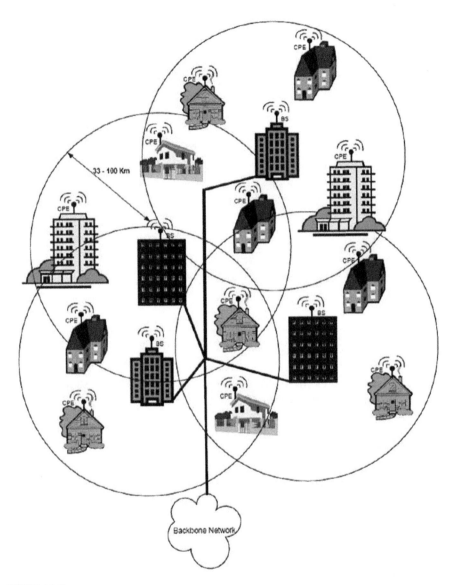

FIGURE 14.3

Reference architecture.

The BS (a professionally installed entity) controls the medium access in its cell and transmits in the downstream direction to the various CPEs, which respond back to the BS in the upstream direction. To ensure the protection of incumbent services, the 802.22 system follows a strict masters/slave relationship, wherein the BS performs the role of the master and the CPEs are the slaves. No CPE is allowed to transmit before receiving proper authorization from a BS, which also controls all the RF characteristics (e.g., modulation, coding, and frequencies of operation) used by the CPEs. In addition to the traditional role of a BS, which is to regulate data transmission in a cell, an 802.22 BS manages a unique feature of distributed sensing. This is needed to ensure proper incumbent protection and is managed by the BS, which instructs the various CPEs to perform distributed measurement activities.

14.3 IEEE 802.22 PHYSICAL LAYER

In this section we describe the physical (PHY) layer of the most recent 802.22 draft standard [17], highlighting its key characteristics. The 802.22 PHY layer is specifically designed to support a system that uses vacant TV channels to provide wireless communication access over distances of up to 100 km. The PHY specification is based on orthogonal frequency division multiple access (OFDMA) for both upstream (US) and downstream (DS) access, and its key parameters are summarized in Table 14.1. The PHY modes for a reference 6 MHz TV channel are given in Table 14.2.

14.3.1 Preamble, Control Header, and MAP Definition

Section 14.4.1 describes the basic superframe structure and frame structure, and in this section we define the key PHY features that support them. In 802.22, the

Table 14.1 IEEE 802.22 System Parameters

Parameter	Specification	Remark
Frequency range	54–862 MHz	
Bandwidth	6 MHz, 7 MHz, 8 MHz	To accommodate TV band channelization of different regulatory domains
Payload modulation	QPSK, 16-QAM, 64-QAM	BPSK used for preambles, pilots and CDMA codes
Transmit effective isotropic radiated power	Default 4 W for CPEs	Currently 4 W for BS in the United States but may vary in other regulatory domains
Multiple access	OFDMA	
FFT size	2048	
Cyclic prefix modes	1/4, 1/8, 1/16, 1/32	
Duplexing	TDD	

Table 14.2 PHY Modes the Data Rates Are Calculated Based on a CP to FTT Ratio of 1/16				
PHY Mode	**Modulation**	**Coding Rate**	**Date Rate Mbps**	**Spectral Efficiency 6 MHz Channel**
1	BPSK	Uncoded	4.54	0.76
2	QPSK	1/2	1.51	0.25
3	QPSK	1/2	4.54	0.76
4	QPSK	2/3	6.05	1.01
5	QPSK	3/4	6.81	1.13
6	QPSK	5/6	7.56	1.26
7	16-QAM	1/2	9.08	1.51
8	16-QAM	2/3	12.10	2.02
9	16-QAM	3/4	13.61	2.27
10	16-QAM	5/6	15.13	2.52
11	64-QAM	1/2	13.61	2.27
12	64-QAM	2/3	18.15	3.03
13	64-QAM	3/4	20.42	3.40
14	64-QAM	5/6	22.69	3.78

allocation of resources in the OFDMA frame can be made in terms of subchannels and symbols. A subchannel is defined as a set of 28 contiguous OFDM subcarriers (24 data and 4 pilot), and there are 60 subchannels per symbol.

In the first frame of the superframe, the first symbol is the superframe preamble, followed by a frame preamble symbol. The third symbol is the superframe control header (SCH), the fourth symbol contains the frame control header (FCH) and, when needed, the DS-MAP, US-MAP, DCD, and UCD. Due to the presence of the SCH in the first frame of the superframe, the first frame payload contains two fewer symbols than the remaining 15 frames to keep the frame length to 10 ms. The other 15 frames of the superframe contain a frame preamble; the FCH; the DS-MAP, US-MAP, DCD, UCD messages; and the data bursts.

In each frame, a transmit-receive turnaround (TTG) gap is inserted between the DS and US to allow the CPE to switch between the receive mode and transmit mode. A receive-transmit turnaround (RTG) gap is inserted at the end of each frame to allow the BS to switch between its receiving mode and transmit mode. The values of the TTG and RTG change based on the cyclic prefix and channel bandwidth under consideration, and this is indicated in Table 14.3. Note that the calculations in Table 14.3 include the number of symbols required only for the FCH, DS-MAP, US-MAP, DCD, and UCD symbols.

Table 14.3 Symbol and TTG/RTG Values per Frame

Cyclic Prefix	Number of Symbols/Frame			TTG			RTG		
Bandwidth (MHz)	6	7	8	6	7	8	6	7	8
1/4	26	30	34	210 sec			83.33 μs	190 μs	270 μs
1/8	29	33	38				46 μs	286 μs	214 μs
1/16	30	35	41				270 μs	270 μs	32 μs
1/32	31	37	42				242 μs	22 μs	88 μs

Preamble Definition

Two types of frequency domain sequences are defined to facilitate burst detection, synchronization, and channel estimation at a 802.22 receiver:

1. **Short training sequence (STS).** This sequence is formed by inserting a nonzero binary value on every fourth subcarrier. In the time domain, this results in four repetitions of a 512-sample sequence in each OFDM symbol.
2. **Long training sequence (LTS).** This sequence is formed by inserting a nonzero binary value on every second subcarrier. In the time domain, this results in two repetitions of a 1024-sample sequence in each OFDM symbol.

The STS is used to form the superframe and CBP preambles, while the LTS is used to form the frame preamble. The sequences use binary $(+1, -1)$ values in the frequency domain and are generated in an algorithmic way from m sequences to ensure a low peak-to-average power ratio (PAPR). The superframe preamble is used by the receiver for frequency and time synchronization. The superframe preamble is one OFDM symbol in duration and consists of four repetitions of the STS in the time domain preceded by a cyclic prefix. The frame preamble is one OFDM symbol in duration and consists of two repetitions of the LTS in the time domain preceded by a cyclic prefix. Finally, the CBP preamble is one OFDM symbol in duration and consists of five repetitions of the STS in the time domain. The CBP preamble is designed as to have low cross-correlation with the superframe preamble to differentiate it from the superframe preamble.

Control Header and MAP Definition

In this subsection we define the structure of the two control headers (SCH and FCH) and the MAPs (DS-MAP, US-MAP, DCD, and UCD). The SCH is transmitted using the PHY mode 1 (see Table 14.2) and TCP $= 1/4$ TFFT. It is transmitted over all data subcarriers, encoded by a rate-1/2 convolutional coder, and after interleaving, is mapped using QPSK constellation resulting in 336 QPSK symbols. To improve the robustness and make better utilization of the available subcarriers, spreading by a factor of 4 is applied to the output of the mapper, resulting in a maximum length of 42 bytes. The FCH is transmitted as part of the DS protocol data unit (PDU) in the DS subframe and uses the basic data rate mode. The length of FCH is 4 bytes and it carries, among other things, the length (in bytes) information for the DS-MAP if it exists or the length of the US-MAP. The FCH is sent in the first two subchannels of the symbol immediately following the preamble symbol. To increase the robustness of the FCH, the encoded and mapped FCH data may be retransmitted, which is indicated through the SCH. The receiver can combine corresponding symbols from the two or three OFDM slots and decode the FCH data to determine the lengths of the following fields in the frame.

These MAPs are transmitted using the base data rate mode in the logical subchannel immediately following the FCH logical subchannels. The length of the DS-MAP PDU, if present, or the US-MAP PDU is indicated in the FCH. If the DS-MAP is present, the length of the US-MAP, DCD, and UCD are specified in the DS-MAP and follow the DS-MAP in that order. The number of subchannels required to transmit these fields is determined by their lengths and could possibly exceed the number of subchannels allocated per symbol. In this case, the transmission of these PDUs will continue in the next symbol, starting with the first logical subchannel. In most situations, no more than two symbols are required to transmit the FCH and MAPs. Any unused subchannels in the second symbol can be used for data transmissions.

14.3.2 CBP Packet Format

The CBP protocol is described in detail in Section 14.4, and its packet format is depicted in Figure 14.4. The first symbol of the CBP packet is the preamble (generated as described in Section 14.3.1), followed by the CBP data payload 1 and an optional data payload 2. The length field in the first symbol enables a receiver to determine the presence or absence of the second data symbol. The CBP data symbols consist of the data and the pilot subcarriers. From the 1680 used subcarriers, 426 subcarriers are designated as pilot subcarriers and the remaining 1254 subcarriers are designated as data subcarriers.

Figure 14.5 shows a simplified block diagram of a CBP data encoder and mapper. The CBP payload is divided in to blocks of 418 bits before encoding and mapping. Each block of 418 bits is first encoded using a rate-1/2 convolutional code. The encoded bits are then mapped using QPSK constellation, which results in 418

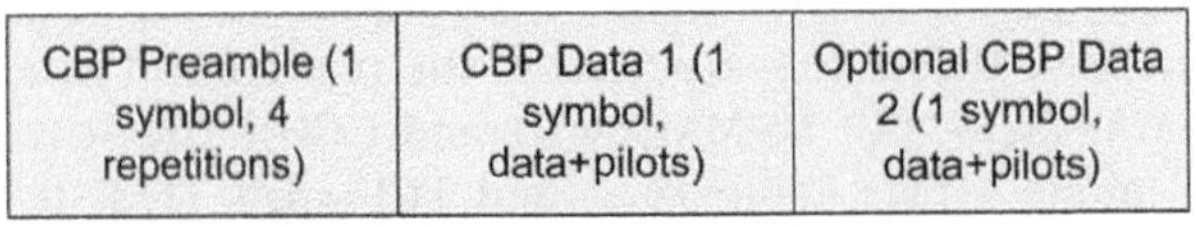

CBP Preamble (1 symbol, 4 repetitions)	CBP Data 1 (1 symbol, data+pilots)	Optional CBP Data 2 (1 symbol, data+pilots)

FIGURE 14.4

CBP packet format.

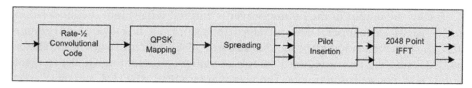

FIGURE 14.5

CBP data encoding.

symbols. Each of these QPSK symbols is transmitted on three subcarriers to provide additional frequency diversity. A simple receiver can combine the pilot symbols with the preamble symbols and perform interpolation to derive channel estimates. These channel estimates can then be used to equalize the CBP data symbols. The receiver can also use maximal ratio combining to despread the data symbols and perform the decoding using a Viterbi algorithm.

14.3.3 Channel Coding and Modulation Schemes

Figure 14.6 describes the channel coding process used in 802.22. Channel coding includes data scrambling, convolutional coding or advanced coding, puncturing, bit interleaving, and constellation mapping.

The frame payload data are first processed by the data scrambler using a pseudo-random binary sequence generator with the generator polynomial $1 + X^{14} + X^{15}$. The preamble and the control header fields of the frame are not scrambled. The forward error correction (FEC) scheme follows the data scrambler. The mandatory coding scheme in 802.22 is convolutional coding. The data burst is encoded using a rate-1/2 binary convolutional encoder. Duo-binary convolutional turbo code, low-density parity check (LDPC) codes, and shortened block turbo codes (SBTCs) are optional advanced coding schemes. For the interleaving stage, the same interleaver used for subcarrier interleaving is employed to interleave encoded bits at the output of the encoder. The interleaving algorithm used in 802.22 is described by the block size K and three integer parameters $\{p, q, j\}$. The global equation of the algorithm depends on the interleaving pattern of the previous iteration $(j - 1)$, the position index of samples (k), and two integer parameters (p, q). The parameter P gives the interleaving partition size multiple of the interleaving block size K.

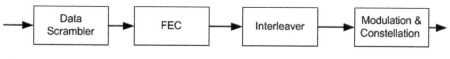

FIGURE 14.6

Channel coding in 802.22.

Finally, the output of the bit interleaver is entered serially into the modulation and constellation mapper. The input data to the mapper are first divided into groups of N_{cbpc} (two for quadrature phase-shift keying (QPSK), four for 16-QAM, and six for 64-QAM) bits and converted into complex numbers representing QPSK, 16-QAM or 64-QAM, constellation points. The mapping for QPSK, 16-QAM, and 64-QAM is performed according to gray-coding constellation mapping. The pilot subcarriers are modulated according to the BPSK modulation using a modulation-dependent normalization factor equal to 1.

14.3.4 Transmit Power Control

Transmit power control (TPC) is an important feature in 802.22, since it requires that only the minimum transmit power necessary to keep a link be used while maintaining the link quality, which further enhances incumbent protection in addition to spectrum sensing, incumbent databases, geolocation, and so on. Each regulatory domain has its own transmit power constraints, and 802.22 is defined to comply with different regulatory requirements. The BS and CPE support monotonic power-level control with accuracy of ± 0.5 dB, over a range of at least 60 dB, with a resolution (step size) of 1 dB. The EIRP of each CPE device is limited to the cap obtained from the BS, which, in turn, can obtain the power limits from an incumbent database or locally at the BS.

14.3.5 RF Mask

The RF mask to be used by 802.22-compliant BSs and CPEs has not been finalized as of the writing of this chapter. However, the limits indicated in Table 14.4 must be obeyed to protect TV receivers and wireless microphones.

Table 14.4 Emission Power Limits for BS and CPE

	If WRAN Operates	
	First adjacent channel to wireless microphone	Second adjacent channel and beyond to TV or wireless microphone
WRAN first adjacent channel limit	4.8 µV/m	200 µV/m
WRAN second adjacent channel and beyond limit	4.8 µV/m	23.5 µV/m (if orthogonal polarization is not available, this value reduces to 4.8 µV/m)

14.4 IEEE 802.22 MEDIUM-ACCESS CONTROL LAYER

Previous works [112, 565] provide an overview of 802.22 medium-access control (MAC) based on early drafts of the standard. However, only recently the 802.22 working group released its first official draft [17]. Although most of the basic features have not changed, in this section we provide an up-to-date overview of the 802.22 MAC layer with emphasis on the CR features key to support-required incumbent protection, self-coexistence among WRANs, and quality of service (QoS). The 802.22 standard defines a connection-oriented and centralized MAC layer, which borrowed some of its basic features, including resource allocation and QoS-supporting features, from the IEEE 802.16 standard for wireless metropolitan area networks (WMANs) [566]. As in 802.16 WMANs, the medium access within a cell in controlled by a base station, which uses TDM (time division multiplex) in the downstream direction and allocates resources using a DAMA (dynamic assigned multiple access) approach in the upstream direction.

14.4.1 Superframe and Frame Structures

The 802.22 MAC uses a synchronous timing structure, where frames are grouped into a superframe structure, which was introduced to allow for better incumbent protection and self-coexistence. The superframe structure, depicted in Figure 14.7, consists of 16 frames with a fixed duration of 10 ms each. The BS starts the first frame within the superframe with the superframe preamble followed by the frame preamble and the superframe control header. The superframe preamble is used for time synchronization, while the frame preamble is used for channel estimation, allowing robust decoding of the SCH and following messages. The SCH carries the

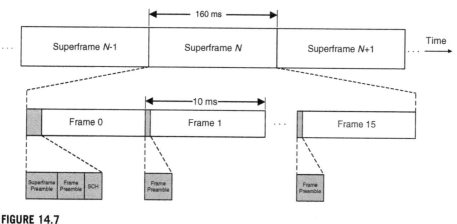

FIGURE 14.7

MAC superframe structure [17].

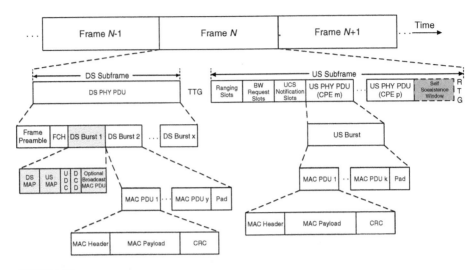

FIGURE 14.8

MAC frame structure.

BS MAC address along with the schedule of quiet periods for sensing, as well as other information about the cell [17]. The SCH is transmitted at a very robust rate to allow for successful decoding over long distances, which is important to ensure neighboring WRANs to discover each other and avoid harmful interference. After the SCH, the BS transmits the frame control header (FCH), which is followed by the messages within the first frame. The remaining 15 frames within the super-frame start with the frame preamble followed by the FCH and subsequent data messages.

The time-domain structure for a typical MAC frame is depicted in Figure 14.8. The frame is divided into DS and US subframes and the self-coexistence window (SCW), which can be scheduled by the BS at the end of the frame. The first downstream burst after the FCH is used to transmit the DS/US MAPs, the DS/US channel descriptor messages (DCD and UCD), and other MAC broadcast messages. The DS/US MAPs are broadcast messages that specify the resource allocation in the DS and US subframes, respectively. The DCD and UCD are usually transmitted by the BS at periodic intervals to define the characteristics of the DS and US physical channels, respectively. After the initial control signaling, the BS can schedule the DS bursts for data transmission using different modulation/coding schemes for each burst [17]. In the US subframe, the BS can allocate resources for contention-based access before the data bursts, which can be used for ranging, bandwidth (BW) request, and urgent coexistence situation (UCS) notification. The UCS window is another new feature in the 802.22 MAC, which can be used by CPEs to transmit an indication that an incumbent has been detected on the channel. Furthermore, the BS may also reserve up to five symbols at the end of the frame for the self-coexistence window. The SCW is used for execution of the coexistence beacon protocol (CBP), which involves transmission of coexistence beacons (or CBP packets) carrying information about the cell and specific coexistence mechanisms. The SCW and CBP packets are new cognitive radio features that allow for over-the-air coordination among neighboring 802.22 cells to facilitate incumbent protection and spectrum sharing mechanisms. They are described in more detail in the following sections.

The two-dimensional (time–frequency) MAC frame structure is shown in Figure 14.9. Interesting points to notice in this figure are the different ways that

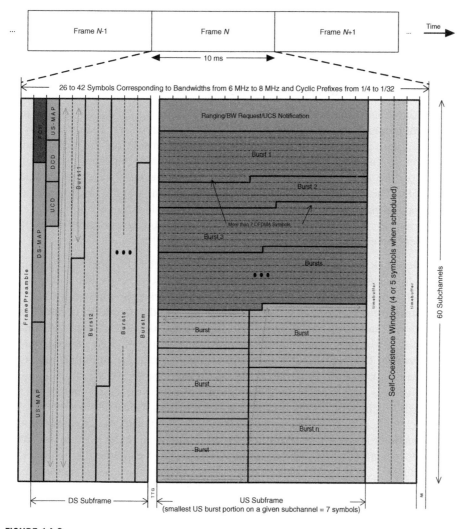

FIGURE 14.9

Time–frequency illustration of the 802.22 frame structure [17].

MAC data can be mapped into the two-dimensional orthogonal frequency division multiplexing (OFDM) structure used at the PHY layer. In the DS subframe, the MAC data are first layered vertically by subchannels then advanced into the time domain (symbol) horizontally. This structure reduces the delay for the DS data. On the other hand, two options are possible in the US subframe. First, the US MAC data can be mapped horizontally symbol by symbol in a logical subchannel and advanced to the next subchannel only when full capacity is filled in the previous one. This option with long US bursts allows the CPEs to maximize power per subcarrier, which can contribute to an increase in the communication range. The second option is to use US bursts with a maximum length of seven symbols. If this option is used, the width of the last US bursts may be between 7 and 13 symbols. This optional allocation can provide better delay performance in the US at the cost of loss of range, which may be a good trade-off for real-time multimedia applications.

14.4.2 Incumbent Detection and Notification Support

One of the design challenges for the 802.22 MAC layer is to allow for reliable protection of incumbents, while maintaining QoS for 802.22 users. This problem was addressed with incumbent detection through spectrum sensing and multichannel operation. In this section we describe how the 802.22 MAC layer supports incumbent detection and notification, which may trigger events for frequency agility operations. The PHY-related aspects of spectrum sensing and a summary of the sensing algorithm adapted in the standard are discussed later in Section 14.5.

Two important capabilities were introduced in the MAC layer to support reliable incumbent detection:

1. **Network quiet periods.** To avoid interference with spectrum sensing, which has to meet very low incumbent detection thresholds (IDTs) (e.g., −116 dBm for DTV), the BS can schedule networkwide quiet periods (QPs), during which all transmissions are suspended, and hence sensing can be performed more reliably. Without QPs for sensing, the WRAN may face a high false alarm rate, especially in areas where multiple WRANs coverage areas overlap. Two types of QPs can be scheduled: intraframe and interframe QPs (see Figure 14.10). Intraframe QPs, as the name suggests, are short-duration QPs (less than a frame) and are useful for regular sensing of in-band channels[2] without affecting the QoS for WRAN users. However, the BS can also schedule longer interframe QPs across multiple frames, in case more time is needed for sensing. Interframe QPs should be used on an on-demand basis, since it affects the QoS of the users. Overall, the BS can limit the number and duration of QPs to the minimum necessary to meet the sensing requirements in terms of probability of detection and probability of false alarm. The BS can schedule QPs by using the QP scheduling fields in the SCH or it can use

---

[2]In-band channels are the operating channel (N) and its first adjacent channels ($N − 1$ and $N + 1$).

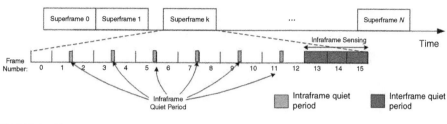

FIGURE 14.10

Intraframe and interframe quiet periods.

a specific management message, called *channel quiet request* (CHQ-REQ), to stop traffic at anytime within its cell.

2. **Channel measurement management.** In case an incumbent is detected by the BS, the BS can take the appropriate action to avoid interference, but when a CPE detects an incumbent, it has to report it to the BS. For that, the MAC layer includes channel measurement request and report messages, which allow the BS to take full control of the incumbent detection and notification process within its cell. The BS can also use management frames to request CPEs to perform other type of measurements, such as detection of other WRAN and other performance-related measurements. The BS is also responsible for allocating US resources for the CPEs to transmit their measurement reports after the sensing is completed. In case a CPE has detected the presence of incumbents but has not been allocated sufficient US bandwidth to transmit its measurements back to the BS, the CPE can use the UCS notification slots in the US subframe to inform the BS of the situation. The CPE uses a contention-based mechanism to transmit a UCS notification message to indicate an incumbent has been detected on the channel [17]. By combining the UCS notification messages with sensing results received from CPEs, the BS can react more efficiently in a timely manner to the presence of incumbents.

14.4.3 Multichannel Operation

As discussed earlier, the regulatory rules require the secondary users to vacate the channel within the channel move time (CMT) (2 sec for 802.22 [17]) once a primary user is detected. This requirement imposes an important challenge for the 802.22 network, namely, how to maintain connectivity for 802.22 users with QoS support in an incumbent detection situation. The 802.22 group tackled this issue using the concept of backup channels. During normal operation, the BS proactively maintains a list of backup channels. In case an incumbent is detected on in-band channels, the BS is responsible for triggering a switch to a backup channel within the CMT, which should occur seamlessly to maintain QoS guarantees for the 802.22 users. Obviously, the backup channel must also be clear of incumbents in order to be used right away. Therefore, incumbent detection must also be done in out-of-band channels; that is, channels that may be used as backup. However, sensing backup channels may be done during the CPE's idle time and not require QPs in the operating channel. But, if the backup channel to be sensed is occupied by another WRAN, the CPEs should use the QPs scheduled in the corresponding channel to avoid interference from the other WRAN.

To achieve seamless transition to the backup channel without service interruption, two channel management modes are specified in the 802.22 standard. In the implicit mode, the BS may use the action fields in the DCD broadcast message to signal the transition to all its CPEs. Otherwise, the BS may use a specific management message (explicit mode) to schedule a channel switch event for a specific frame in the future. The explicit approach provides the flexibility to allow channel management for a single or a group of CPEs, since the channel management message (channel switch request, CHS-REQ) can be sent as a broadcast, unicast, or multicast frame. In this way, a BS operating as multiple colocated 802.22 radios in different channels could direct a CPE to another channel and continue operation in its current channel. This feature is especially useful in case individual CPEs detect low-power incumbents (e.g., wireless microphones) in areas with limited channel availability.

14.4.4 Synchronization

Synchronization is a key factor for successful operation of 802.22 systems, and it is needed not only for communication purposes between BS and CPEs but also for incumbent protection. The BS and CPEs in a cell must be synchronized to ensure no transmissions occur during the QP for sensing. Also, neighboring WRANs sharing the same channel (N) or operating on first and second adjacent channels ($N-2, N-1, N+1, N+2$) must synchronize their QPs to avoid interference with incumbent sensing and reduce the false alarms rate. Although in-band channels include only up to first adjacent channels ($N, N+1$ and $N-1$), synchronization of QPs up to second adjacent channels ($N+2$ and $N-2$) is needed to avoid interference when sensing the first adjacent channels.

To facilitate synchronization of QPs, all the 802.22 BSs are required to be equipped with a satellite-based positioning system (e.g., GPS), which is used to derive the timing information for the superframes. Therefore, by specifying a common reference time, the standard ensures all the superframes are synchronized. Once the superframes are synchronized, the next step is to synchronize the intraframe and interframe QPs.

The interframe QPs are expected to be used for regular sensing in-band channels, since their limited duration (less than a frame) has only minimal impact on QoS. To facilitate synchronization, such quiet periods can be scheduled only at the end of each frame. Therefore, neighboring BSs need to advertise only the frames in which intraframe QPs are scheduled to occur and the QPs duration. This information is broadcasted in the SCH (using the intraframe sensing cycle frame bitmap and other supporting fields) at the beginning of every superframe and it is also included in the CBP packets transmitted by CPEs. The intraframe QP schedule can be valid for multiple superframes, as indicated by the intraframe sensing cycle field. This allows upcoming neighboring BSs to discover existing cells operating in the area and align their intraframe QPs accordingly. Also, once a BS discovers the QPs scheduled by other BSs using the same channel or adjacent channels, it has to align and adjust its intraframe QP duration (intraframe sensing duration) to the largest among the overlapping BSs. It should be noted that this synchronization is done with the information that the BS obtains via sensing other BSs in its vicinity or obtaining the report from its own CPEs in the vicinity of other BSs.

The BSs also broadcast (in the SCH) information about the interframe QPs, which includes the time to quiet period (TTQP) and duration of the quiet period (DQP). To ensure QoS, the interframe QPs should be used only when strictly needed, and hence, they are scheduled on demand. Once other neighboring BSs receive information about a new interframe QP scheduled, they perform an algorithm to decide whether they should change their own QP schedule to align with the received schedule. The mechanism is based on the following rule to reduce the ping-pong effect [17]: "A BS 1 shall only modify its inter-frame sensing quiet period schedule to synchronize with the inter-frame sensing quiet period of another nearby BS 2 if the remaining time to BS 1's next inter-frame sensing quiet period is larger than the remaining time to BS 2's next inter-frame sensing quiet period." In this case, a BS modifies its scheduled QP only if its QP is scheduled for a later time than the other BS's QP. After making the decision, the BS schedules a change in its QP. BSs can also cancel or change QPs by changing the QP schedule contained in the SCH or using specific management messages to schedule QPs. Clearly, for these procedures to be effective, the BSs should schedule QPs ahead of time, and if possible, a few superframes ahead of time to allow enough time for coordination with neighboring BSs.

14.4.5 Self-Coexistence

The unlicensed spectrum access model adopted for 802.22 systems underscores the importance of self-coexistence[3] mechanisms to ensure efficient and fair spectrum utilization. Self-coexistence also plays a key role in protecting the incumbents, due to the required coordination for reliable spectrum sensing; that is, synchronization of QPs. Proper coordination of sensing schedules also minimizes the number of required QPs and leaves more time for data communication.

The self-coexistence problem is approached in 802.22 with the following key elements:

---

[3] *Self-coexistence* refers to coexistence among 802.22 systems.

1. **Neighboring network discovery and coordination.** The nondeterministic availability of spectrum coupled with the ability to switch channels makes the 802.22 operating spectrum environment dynamic. This requires sensing not only for incumbent detection, but also for other neighboring 802.22 systems. Network discovery is part of the initialization procedures for both BSs and CPEs, but it is also continuously done during the normal network operation as well. WRANs can be discovered through the SCH transmitted by the BSs or by CBP packets, which are transmitted during the SCW window by CPEs or BSs. CPEs that discover other neighboring WRANs send this information back to their BS in the format of measurement report messages. Upon discovery of new neighboring WRANs, the BS must consider whether QP synchronization is required, as discussed in Section 14.4.4.

2. **Coexistence beacon protocol.** The CBP protocol plays a key role in enabling efficient discovery and coordination of neighboring networks. BSs and CPEs can discover other WRANs by detecting CBP packets transmitted during the SCW window. The CBP packets carry information about the cell (e.g., BS MAC address, schedule of QPs, backup channels), and BSs should open regular SCWs for transmission of CBPs to enable efficient network discovery. During normal operation, the CPE should listen to its BS's SCH to identify any changes relevant to its cell, and since superframes are synchronized across cells, CPEs may not be able to detect neighboring BSs' SCH transmissions. The CBP packets provide additional support for network discovery, since they are transmitted during the SCWs, and the BS can explicitly request CPEs to listen to the channel during the SCW to detect CBP packets from other neighboring cells. The CBP packets may also carry information required for executing several resource sharing mechanisms, which requires information exchange between cells. In other words, the CBP serves as an underlying protocol for inter-WRAN communication. As can be seen in Figure 14.9, the SCW imposes extra overhead (four or five symbols), and BSs must take into account the overhead and timing constraints for inter-WRAN communication when scheduling SCWs. The CBP protocol is useful in many scenarios, and a possible situation is illustrated in Figure 14.11. In this example, the BSs A and B are outside of the communication range of each other, but CPEs A and B in the overlap area can relay information between the two cells via CBP packets.

3. **Resource sharing mechanisms**. After network discovery, neighboring WRANs may have to consider how to share the available spectrum. Consider the example scenario in Figure 14.11 and suppose BS A and CPE A1 operate on a given channel N. When a new BS B and CPE B1 start operation, they first scan the available channels and CPE B1 eventually detects BS A's SCH or CBP packet transmitted by A1. At this point, BS B must execute the first coexistence mechanism, called *spectrum etiquette*. The main idea behind spectrum etiquette is to avoid operating cochannels

with other existing WRANs. Therefore, BS B first searches for an available channel that is not used by BS A. However, if no other channel is available, BS B can share the same channel with BS A. Although not a desirable situation, this might happen in areas with only a few channels available, and the standard provides the required support for coexistence in such cases. BSs may engage in a negotiation process to share the channel on a frame-by-frame basis. Such negotiation may be based on spectrum contention mechanisms [17]. The negotiation between BSs is carried out through CBP packet exchanges. In the example in Figure 14.11, the CPEs are used to exchange CBP packets, but in a scenario where BSs are within each other's communication range, they can transmit and receive the CBP packets themselves. It should be noted that such negotiation process between BSs may become a bottleneck as the number of neighboring BSs on the same channels increase.

Another problem that should be considered in coexistence scenarios where multiple BSs share the same channel is the potential collision of SCH transmissions from neighboring cells. Synchronization of the superframes across cells results in simultaneous transmissions of SCHs by all BSs in the beginning of every superframe, which may lead to collisions at CPEs in the overlapping areas. Due to the robust transmission mode, a CPE may be able to decode the SCH from its BS, even in presence of other SCH transmissions. However, depending on the distance between interfering BSs, SCH collisions might occur, which could prevent operation at some specific locations. This is still an open issue for the standard, but its impact might be minimal, depending on the availability of incumbent-free

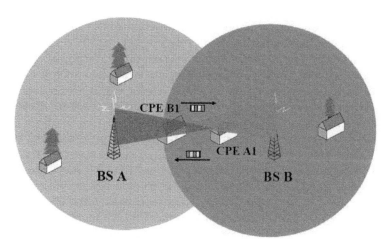

FIGURE 14.11

Example scenario of inter-WRAN communication with CBP.

channels. Neighboring BSs will always try to find empty channels, and cochannel operation with other BSs is pursued only as the last resort to maintain connectivity. Nevertheless, several proposals are being considered to address this problem. In one possible approach, called *single-frequency network* (SFN), all overlapping BSs transmit a self-coexistence mode SCH, which is the same across all BSs and carries the frame allocations for different cells, while another option is to multiplex the SCH transmissions in the time domain.

14.4.6 Quality-of-Service Support

The 802.22 MAC layer includes several mechanisms to support QoS for upstream and downstream traffic. The adopted QoS service model is based on the IEEE 802.16d standard [566], which includes the following basic concepts:

1. **Service flow QoS scheduling.** The primary purpose of the QoS support at the MAC layer is to define the transmission ordering and scheduling on-the-air interface. That is achieved by associating packets traversing the MAC layer to a service flow, which is identified by a SFID (service flow identifier). A service flow is a unidirectional flow of packets provided a particular QoS support level, which is specified by a set of QoS parameters such as latency, jitter, and throughput guarantees. Service flows may be specified by a service class, which is a name associated with a particular set of QoS parameters and can be used to facilitate specification of service flows by upper-layer protocols. Four basic scheduling services are supported: unsolicited grant service (UGS), real-time polling service (rtPS), non-real-time polling service (nrtPS), and best effort (BE). The UGS service is designed to support real-time data streams consisting of fixed-size data packets sent at periodic intervals. The rtPS is designed to support real-time data streams consisting of variable-sized data packets issued at periodic intervals, such as MPEG video. The nrtPS is designed to support delay-tolerant data streams consisting of variable-sized data packets for which a minimum data rate is required, such as FTP. The BE service is designed to support data streams for which no minimum service level is required. More details about the QoS parameter sets for these services are described in the standard [17].

2. **Activation model.** Service flows can be classified as provisioned, admitted, or active. Provisioned service flows require a negotiation between CPE and BS to be activated. During this negotiation, the CPE sends the SFID and the associated QoS parameter sets to the BS, which may authorize the flow if resources are available. To activate a service flow, the BS maps the service flows to a CID (connection identifier), which identifies a connection between the CPE and BS across which the data are delivered. Service flows can also be in a transient admitted state, where the resources are not yet completely activated. In this two-phase model, the service flows are first admitted (passing admission control, where availability of the requested

resources is verified), and the service can be activated later after the final end-to-end negotiations are finalized. This model saves network resources and is useful for applications in which a long-term reservation of resources is necessary or desirable. For instance, resources used by a voice call put on hold could be temporarily allocated to other services, but such resources should be available for resuming the call when needed.

3. **Dynamic service establishment.** The MAC layer provides a series of management messages and procedures to create (DSA), change (DSC), or delete (DSD) service flows. The DSA messages create a new service flow. The DSC messages change an existing service flow. The DSD messages delete an existing service flow. These procedures can be initiated by CPEs or BSs. Changing a service flow modifies the QoS parameters associated with the flow. The set of DSx messages provide full flexibility for the BS and CPEs to adapt the air interface resource allocation to their traffic requirements.

It must be noted that, unlike in 802.16 networks, the 802.22 system has to deal with dynamic channel availability issues, due to incumbents. Therefore, in addition to the QoS-supporting mechanisms described already, the other CR features in the 802.22 MAC layer (i.e., reliably incumbent detection, synchronization, and self-coexistence) play a key role in providing QoS support for the 802.22 users. The interaction between the CR features with underlying QoS-supporting mechanisms in 802.22 systems is not yet completely understood, and system-level evaluation in realistic deployment scenarios are yet to be done.

14.4.7 Spectrum Management Model

The cognitive radio protocols described in the previous sections provide all flexibility to the 802.22 operators to detect and share the available spectrum, as well as comply with the requirements of the DSA model for secondary operation in TV bands. In this section, we describe in detail the architecture for managing the CR capabilities and protocols. As described in [565], the 802.22 standard has adopted a spectrum management model, where each WRAN BS has a central entity, called a *spectrum manager* (SM), which is responsible for taking the key decisions and triggering the proper events and actions to ensure protection of incumbents and efficient spectrum utilization while complying with regulatory policies. The SM is shown in Figure 14.12 as part the 802.22 protocol reference architecture [567]. The architecture shown in Figure 14.12 includes a new cognitive plane and security features and it replaces the previously adapted system architecture in Draft 1.0 [17]. The data and management planes are separated from the cognitive plane, which was introduced to support the new features for spectrum sensing and management and geolocation capabilities. As can be noted, security features are included as sublayers in the three planes (data, management, and cognitive). These sublayers provide functions to verify spectrum and service availability, as well as various forms of device, data, and signal authentication; authorization;

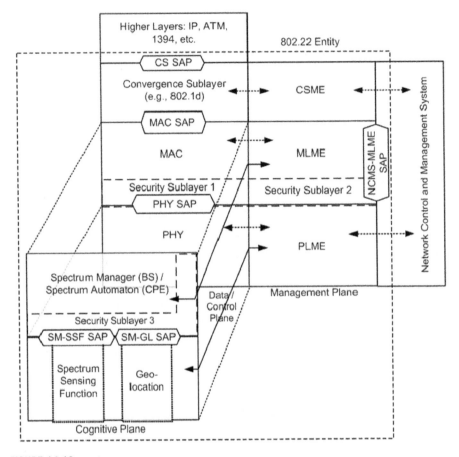

FIGURE 14.12

Protocol reference architecture for an 802.22 BS or CPE [567].

data; control and management message integrity; confidentiality; nonrepudiation; and so forth.

14.4.8 Spectrum Manager

The SM collocated with the BS can be seen as the central intelligence of the system and it is part of the cognitive plane. On the other hand, the CPEs implement a corresponding entity called *spectrum automaton* (SA), which is basically a "slave" of the SM running at the BS. The CPE SA controls the local CPE's spectrum sensing function (SSF) when it is not under the control of a BS, for example, during initialization and scanning procedures. The CPE SA can also use the SSF to perform sensing autonomously during the CPE's idle time. During normal operation the

CPE SA responds to requests from the BS's SM, for instance, to perform spectrum sensing and report the results.

Within a WRAN cell, the SM centralizes all the decisions with respect to spectrum management, which includes the following:

1. Maintain up-to-date spectrum availability information.
2. Classify, set priorities, and select channels for operation and backup.
3. Maintain association control.
4. Trigger frequency agility-related actions (i.e., channel switch).
5. Manage mechanisms for self-coexistence (interference-free scheduling, renting/offering, and spectrum contention).

Although some of the decisions about spectrum management are not specified in the standard (i.e., they are left implementation dependent to provide flexibility for manufacturers to differentiate their products), any SM implementations must comply with the regulatory rules for the operating domain (e.g., rules established by the FCC in the United States for use of TV channels). Moreover, the specific regulatory rules may differ from one domain to another, which requires a flexible standard to achieve worldwide success. Most of the regulatory rules deal with incumbent protection, and two main capabilities are mandated in the standard to support potentially different incumbent protection requirements: a spectrum sensing function and incumbent database support.

14.4.9 Spectrum Sensing Function

Every 802.22 CPE is required to implement a spectrum sensing function, which is defined in the standard as a "black box" that performs spectrum sensing. The standard specifies inputs and outputs for the SSF, as well as the performance requirements for the sensing algorithm implemented (e.g., probability of detection, incumbent detection threshold, and probability of false alarm). The specific sensing algorithm is implementation dependent, although the standard does include, for informative purposes, description and performance results of several sensing algorithms, which were proposed and evaluated during the standardization work. More details about sensing algorithms are found in Section 14.5.

14.4.10 Incumbent Database Support

Although spectrum sensing can provide the required protection to incumbents, some regulatory domains may require access to incumbent databases that store information about the presence of incumbents in a given area. Incumbent databases may reduce the overhead on the system due to spectrum sensing, especially for DTV signals. For instance, if a channel is occupied in a given area by a TV station and this information is accessible through a database, the 802.22 BS may decide to reduce the frequency in which the channel is sensed by its associated CPEs or it may even decide to perform no sensing at all on that channel. On the other hand, there are other issues related to incumbent databases, such as accuracy of the database information and maintenance costs. Also, incumbent databases may not be efficient for more dynamic incumbents, such as wireless microphone; and the cost of maintaining a database for a large number of low-power devices (i.e., wireless microphones) may be prohibitive. In fact, one of the main open questions is related to maintenance of such databases. Regulators could maintain a database infrastructure, but that would come at a certain cost. A group of service providers and incumbents could be another option to maintain such a database. At the Draft 1.0 stage in the standardization process, it is not yet clear how the implementation of incumbent databases will play along with the business models of 802.22 operators.

Nevertheless, to allow for aggregation of information from multiple available sources and support potential regulatory policies, the 802.22 standard also requires BSs to be able to connect to external databases. Access to incumbent databases is implemented through the higher-layer protocols and it is outside the scope of the standard. However, to allow for modular and interoperable implementations, the standard defines interfaces and primitives to connect to the databases [17]. If an incumbent database is not available or accessing a database is not required by regulation, the SM is still able to make decisions based on spectrum sensing information. This is an important flexibility for the success of the 802.22 standard, especially in less populated and rural areas, due to the low cost in implementing spectrum sensing, which allows operation with minimal infrastructure costs and no dependency on third parties to maintain an incumbent database.

14.5 SPECTRUM SENSING

Spectrum sensing is an important component of cognitive radio and the IEEE 802.22 standard, and this sections first derives the keep-out radius and the sensing radius to which the incumbents have to abide.

14.5.1 Incumbent Protection Radius

Before we outline how the incumbent protection radius is derived, we refer to Table 14.5, which states the incumbent protection parameters as outlined in IEEE 802.22 standard [17]. Assume that a DTV broadcasting station, is transmitting at 1 MW (90 dBm) effective radiated power (ERP) with an antenna height of 500 m. This DTV operates at 615 MHz in the UHF band. Next, consider a WRAN sensor with a 0 dBi receive antenna gain, operating at some distance from the TV transmitter. The receive power for such a sensor is plotted in Figure 14.13 as a function of distance from the DTV transmitter, assuming a signal propagation model is conforming

Table 14.5 IEEE 802.22 Incumbent Protection Parameters

Parameter	Wireless Microphones	TV Services
Incumbent detection threshold (IDT)	−107 dBm (over 200 kHz)	−116 dBm (over 6 MHz)
Probability of detection (PD)	90%	90%
Probability of false alarm (PFA)	10%	10%
Channel detection time (CDT)	≤2 sec	≤2 sec
Channel move time (CMT)	2 sec	2 sec
Channel closing transmission time (CCTT)	100 ms	100 ms

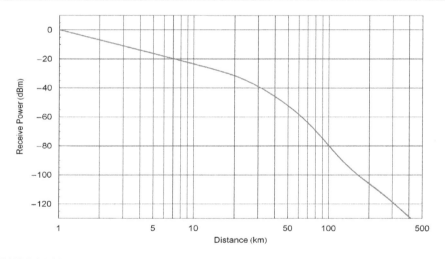

FIGURE 14.13

DTV receive power versus distance for a 0 dBi rx antenna.

to ITU-R P1546[4] [561]. The ITU-R propagation model describes not only the spatial mean pathloss, but also the expected standard deviation of the shadow fading. Hence, each sensor should be subject to the typical lognormal shadow fading with a 5.5 dB standard deviation [561]. Let P_{rx} be the instantaneous receive power

---

[4]This paper uses the propagation models specified in the ITU-R document P.1546-1, as is used in the IEEE 802.22 working group.

(in dBm) and the average receive power be represented by $\bar{P}_{rx}$ (in dBm). The spatial mean receive power is given by the ITU-R recommended practice [561] as the sum of the average power and the shadow fading, which is a zero-mean normal random variable [72] with a standard deviation of 5.5 dB. This is represented by

$$P_{rx} = \bar{P}_{rx} + F_S \quad [\text{dBm}].$$

Having constructed the propagation model, we now describe how to calculate the "keep-out region" as outlined in IEEE 802.22 WRAN system. The keep-out region is a region around the primary user (e.g., DTV transmitter) where the WRAN system is not permitted to transmit, since it may cause harmful interference to the primary user. These calculations are based on [568, 569].

Calculating the Base Station Keep-out Region

The DTV protection contour, also referred to as the *noise-limited contour*, is located at a distance where the field strength is 41 dBμ using the F(50, 90)[5] propagation curve. All TV receivers within the protection contour must be free of harmful interference. In this scenario, this contour occurs at 134.2 km from the DTV transmitter. According to the FCC NPRM for DTV the desired to undesired (D/U)[6] ratio is 23 dB assuming that the interferer (i.e., the undesired signal) is another DTV transmitter. For the purpose of our study, we assume this D/U ratio also applies when the interferer is a WRAN signal.

Next, calculate the maximum undesired field strength at the edge of the noise limited contour. From that field strength, it is possible to determine how far away the WRAN transmitter must be located to cause no harmful interference to the DTV receiver. The undesired signal's allowable field strength is given by the following formula:

$$\text{FS}_U = \text{FS}_D - D/U + F/B.$$

Here FS_U, FS_D, and F/B represent the undesired signal field strength at the noise limited contour, the desired field strength, and the antenna front-to-back ratio, respectively. The desired $F(50, 90)$ field strength at this point is 41 dBu, which is the signal that needs to be protected. Hence, the D/U ratio must exceed 23 dB. Using a front-to-back ratio of 14 dB for the DTV receive antenna [570], the following limit on the undesired field strength at the DTV receiver is derived:

$$\text{FS}_U \leq 41 - 23 + 14 = 32 \, \text{dBu}.$$

Given this upper limit on the undesired field strength we can obtain the distance between the WRAN transmitter and the DTV receiver such that the presence of the WRAN transmitter causes no harmful interference. The transmission of a WRAN

---

[5]$F(X, Y)$ represents the spatial and temporal relationship of the TV signal propagation as specified in P1546. It represents the field strength that would exceed a certain threshold at X% of locations for Y% of time [570].

[6]This is also referred to as the *signal-to-interference ratio* in wireless communications.

FIGURE 14.14

WRAN base station field strength.

base station is limited to 36 dBm EIRP. Using the ITU propagation curves, we convert from EIRP to ERP as follows:

$$ERP = EIRP - 2.2 = 33.8 \text{ dBm}.$$

We assume a base station antenna height of 75 m [17]. The distance at which the field strength of the undesired signal reaches 32 dBu is approximately 16.1 km, as can be seen in Figure 14.14. Adding 16.1 km to the DTV protection contour of 134.2 km, we obtain a keep-out region of 150.3 km from the DTV transmitter. At the edge of the keep-out region the DTV field strength using the $F(50, 90)$ curve is 36.5 dBu. The receive power assuming an isotropic sensing antenna is calculated to be −96.48 dBm.

Calculating the CPE Keep-out Region

This analysis is similar to what was done for the base station keep-out region. The primary differences are the antenna height and antenna directionality. For residential use, antenna heights are typically 10 m, and antennas are assumed to be directional with a front-to-back ratio of 14 dB. The CPE is assumed to be transmitting at 36 dBm EIRP with the directional antenna pointed toward the WRAN base station and away from the DTV receiver. If we assume a back-to-front ratio of 14 dB for the WRAN transmit antenna, then the EIRP in the direction of the DTV receiver is given by

$$EIRP = 36 - 14 = 22 \text{ dBm}.$$

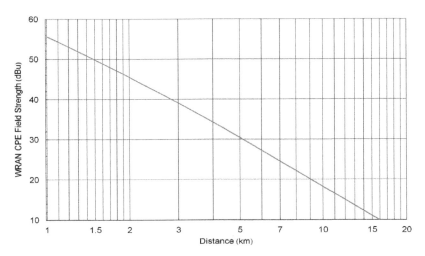

FIGURE 14.15

WRAN CPE field strength.

Using the ITU propagation curves, we convert from EIRP to ERP, giving

$$\text{ERP} = \text{EIRP} - 2.2 = 19.8 \text{ dBm}.$$

Assuming the CPE antenna height of 10 m [17] and using the $F(50, 10)$ propagation curve, we can calculate the required separation between the CPE and the noise protection contour. The distance at which the field strength of the undesired signal reaches 32 dBu is approximately 3.45 km, as seen in Figure 14.15. Adding 3.45 km to the DTV protection contour of 134.2 km gives a keep-out region of 137.6 km around the DTV transmitter.

Now consider a CPE on the opposite side of the WRAN cell to the TV protection contour; it has higher EIRP in the direction of the DTV receiver and hence it is worth calculating the distance the CPE must be placed from the DTV receiver. Its EIRP is fixed at 36 dBm, which corresponds to a distance of 6.9 km from the DTV receiver. So the CPE on the "opposite" side of the base station that is more than 3.5 km away from the protection contour causes approximately the same interference. Typically the WRAN cell diameter is larger than 10 km and CPEs that are not at the edge of the cell use transmit power control, so even though they are somewhat closer they transmit at a lower power. Hence, it is required to place the CPE at least 3.45 km from the DTV receiver so that it does not cause harmful interference.

14.5.2 Sensing Algorithms

Having determined the interference radius and hence the keep-out radius for WRAN systems to operate in the same band as a TV signal, let us now look into the typical sensing algorithms considered by the IEEE 802.22 WRAN standard [17]. Sensing algorithms can be classified as blind and nonblind. In the former, no assumption is made on signals of the incumbents, while in the latter we use the characteristics of the incumbent signal to detect it.

Power Detector

The power detector [72, 571, 572] is the simplest detector that can be constructed in practice. This detector uses very limited a priori information regarding the DTV signal. The detection is based on only the signal power. Though this detector is unlikely to be used in practice it does give a lower bound on sensing performance, since it is likely other complex detectors will outperform the power detector. The performance of more complex detectors should be measured against the power detector to evaluate if the higher complexity and sophisticated techniques result in significant performance gains. The power detector and the energy detector are very similar. The test statistic for the power detector is an estimate of the signal power, while the test statistic for the energy detector is an estimate of the signal energy during the sensing time. Since we prefer a test statistic of which the mean value does not change with the sampling duration, we choose to use the power detector. This is just a matter of convenience.

The DTV signal has a bandwidth of 6 MHz, as illustrated in Figure 14.16. Since the signal is at the baseband, the frequency ranges from -3MHz to $+3$MHz. It is assumed that the signal is sampled at the Nyquist rate resulting in a complex signal with 6 MHz sampling. If the sampling rate is much higher than the Nyquist rate, then the assumption about the independence of the noise samples is no

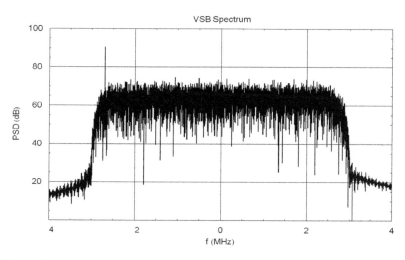

FIGURE 14.16

DTV spectrum.

longer valid. There are two hypotheses in this problem. The null hypothesis is that the channel is vacant, which is represented as H_0. The alternative hypothesis is that the channel is occupied by the DTV transmitter represented as H_1. The test statistic is an estimate of the received signal power, which is given by

$$T = \frac{1}{T_S} \sum_{n=1}^{M} y(n)y^*(n),$$

where T_S represents the sensing time, which is defined as the product of number of samples, M, with the sampling rate T_S; $y(n)$ is the sample at the nth time instant; and $y^*(n)$ represents the complex conjugate. The sampling time is give by

$$T_S = \frac{M}{f_s}.$$

We sample at the bandwidth B so that $f_s = B$. The test statistic is then

$$T = \frac{B}{M} \sum_{n=1}^{M} y(n)y^*(n).$$

The mean value of the test statistic is the signal power. Observed signals under the hypotheses take the following forms:

$$H_0: \quad y(n) = w(n)$$

$$H_1: \quad y(n) = x(n) + w(n),$$

where complex additive white Gaussian Noise is represented as

$$w(n) = w_R(n) + jw_I(n).$$

The real and imaginary parts of the noise are independent and identically distributed city-place-normal (Gaussian) random variables:

$$f_{WR}(w) = f_{WI}(w) = N(0,\ N/2).$$

Here the noise spectral density is N. We chose to use N instead of $N0$, since the receiver is likely to have a nonzero noise figure, which we would like to consider in our analysis. It is not necessary to consider multipath fading for the power detector explicitly, since the power detector collects all the energy of the signal and the time variation of the total energy over a 6 MHz bandwidth due to multipath is insignificant. In more complex detectors we have to explicitly consider multipath fading. Shadow fading is addressed by allowing the signal level to have a lognormal distribution. The test statistic for the power detector is an estimate of the signal power.

Probability of False Alarm for the Power Detector

The threshold in the power detector is based on the probability density function of the test statistic under hypothesis H_0. In [17] we see that several values of the probability of false alarm should be considered. Typical values of false alarm probabilities are 0.1 and 0.01. In this section we see how to select the detector threshold to meet the PFA requirements. The probability density function for the test statistic under H_0 is given by

$$f_T(t) = N\left(NB, \frac{(NB)^2}{M}\right).$$

If we define a new threshold, γ_0, as

$$\gamma_0 = \frac{\sqrt{M}}{NB}(\gamma - NB),$$

we have that

$$T > \gamma \quad \text{if and only if} \quad S > \gamma_0.$$

So the probability of false alarm can also be written as

$$P_{FA} = P(S > \gamma_0).$$

Since S is a zero-mean Gaussian random variable with unit variance the probability of S exceeding a threshold is given by the Q function [571]:

$$Q(x) = \frac{1}{\sqrt{2\pi}} \int_x^\infty \exp\left(\frac{-y^2}{2}\right) dy.$$

The threshold γ_0 is then given by

$$\gamma_0 = Q^{-1}(P_{FA}).$$

Therefore, the value of the detector threshold for test statistic T is

$$\gamma = NB\left[1 + \frac{Q^{-1}(P_{FA})}{\sqrt{M}}\right].$$

Probability of Misdetection for the Power Detector

Having set the detection threshold based on the required PFA, we can calculate the formula for the probability of misdetection. The probability of misdetection is simply the probability that the test statistic will not exceed the threshold under hypothesis H_1,

$$P_{MD} = P(T < \gamma) \quad \text{under } H_1.$$

Under hypothesis H_1 the probability density function of the test statistic is given as

$$f_T(t) = N\left[P + BN, \frac{(P+NB)^2}{M}\right].$$

To determine the probability of misdetection it is convenient to transform the test statistic into a zero-mean Gaussian random variable with unit variance as was done previously. Proceeding as we did for hypothesis H_0, we have P_{MD} as

$$P_{MD} = P(S < -\gamma_0) = P(S > \gamma_0) = Q(\gamma_0).$$

Therefore, the probability of misdetection is seen to be

$$P_{MD} = Q(\gamma_0) = Q \left\{ \frac{\sqrt{M}}{(P + NB)}[(P + NB) - \gamma] \right\}.$$

To illustrate these theoretical results, they have been applied to the specific DTV detection problem as described in [568]. To specify the detection threshold we must specify the noise power spectral density, the bandwidth, the probability of detection, and the number of samples observed to generate T. As in [570] we assume the receiver has an 11 dB noise figure that includes the noise figure of the RF front end and any other cable and coupling losses. Hence, the noise power spectral density (PSD) is given by

$$N = N_0 + 11 = -174 + 11 = -163 \text{ dBm/Hz}.$$

The signal bandwidth is $B = 6$ MHz, so the total noise power is

$$NB = -174 + 11 + 10\log(6 \times 10^6) = -95.2 \text{ dBm}.$$

The Power Detector with Noise Uncertainty

Previously the noise PSD was assumed to be known exactly. In this section we assume the noise PSD is known to within some tolerance. As discussed previously we assume an 11 dB receiver noise figure, which includes not only the noise figure of the LNA, but also effect of cable and coupling losses. The average noise power is the same as used previously,

$$\bar{N} = N_0 + 11 = -174 + 11 = -163 \text{ dBm/Hz}.$$

We assume the noise PSD is known to within $\pm\Delta$ dB so the formula for the system noise PSD is

$$N = \bar{N} \pm \Delta = -163 \pm \Delta \text{ dBm/Hz}.$$

The signal bandwidth is $B = 6$ MHz so the average noise power is

$$\bar{N}B = -174 + 11 + 10\log(6 \times 10^6) = -95.2 \text{ dBm}.$$

And the resulting noise power can be written as

$$NB = (\bar{N} \pm \Delta)B = -95.2 \pm \Delta \text{ dBm}.$$

The value of the noise uncertainty depends on a variety of factors, including an initial error in the noise estimate, the change in noise due to changes in temperature, the change in amplifier gain due to changes in temperature, and of course any errors due to changes in interference levels. When noise uncertainty is considered a minimum signal power can be detected. This effect is referred to as the SNR *wall* [573]. The value of this wall is given in terms of the noise uncertainty. If we define

$$\alpha = 10^{(\Delta/10)},$$

then the SNR wall is given by [573]

$$\text{SNR wall} = \frac{\alpha^2 - 1}{\alpha}.$$

Eigenvalue Sensing Technique

The other blind sensing technique considered in IEEE 802.22 is the eigenvalue sensing technique [17]. Let $y(t)$ be the continuous-time received signal of interest with central frequency f_c, bandwidth W, that is sampled at a sampling rate $f_s (\gg W)$. Let $T_s = 1/f_s$ be the sampling period. The received discrete signal is then $x(n) = y(nT_s)$. There are two hypothesizes as before: H_0: signal does not exist; and H_1: signal exists. The received signal samples under the two hypotheses are therefore as follows:

$$H_0: \_x(n) = \eta(n)$$

$$H_1: \_x(n) = s(n) + \eta(n),$$

where $s(n)$ is the transmitted signal passed through a wireless channel (including fading and multipath effect), and $\eta(n)$ is the white-noise samples. Note that $s(n)$ can be a superposition of multiple signals. The received signal is generally passed through a filter. Let $f(k), \_k = 0, 1, ..., K$ be the filter. After filtering, the received signal is turned to

$$\tilde{x}(n) = \sum_{k=0}^{K} f(k)x(n - k), \_n = 0, 1, ...$$

Let

$$\tilde{s}(n) = \sum_{k=0}^{K} f(k)s(n - k), \_n = 0, 1, ...$$

$$\tilde{\eta}(n) = \sum_{k=0}^{K} f(k)\eta(n - k), \_n = 0, 1, ...$$

Then

$$H_0: \_\tilde{x}(n) = \tilde{\eta}(n)$$

$$H_1: \_\tilde{x}(n) = \tilde{s}(n) + \tilde{\eta}(n).$$

Note that here the noise samples $\tilde{\eta}(n)$ are correlated. If the sampling rate f_s is larger than the channel bandwidth W, we can down-sample the signal. Let $M \geq 1$ be the down-sampling factor. If the signal to be detected has a much narrower bandwidth than W, it is better to choose $M \geq 1$. For notation simplicity, we still use $\tilde{x}(n)$ to denote the received signal samples after down-sampling; that is, $\tilde{x}(n) \triangleq \tilde{x}(Mn)$. Choose a smoothing factor $L > 1$ and define

$$x(n) = [\tilde{x}(n) \quad \tilde{x}(n - 1) \quad ... \quad \tilde{x}(n - L + 1)]^T,$$
$$\_n = 0, 1, ..., N_s - 1.$$

A suggested value of L is about 10. Define a $L \times [K + 1 + (L - 1)M]$ matrix as

$$
H = \begin{bmatrix}
f(0) & \cdots & \cdots & f(K) & 0 & \cdots & 0 \\
0 & \cdots & f(0) & \cdots & f(K) & \cdots & 0 \\
 & & \cdots & & \cdots & & \\
0 & \cdots & \cdots & \cdots & f(0) & \cdots & f(K)
\end{bmatrix}
$$

Let $G = HH^H$. Decompose the matrix into $G = QQ^H$, where Q is a $L \times L$ Hermitian matrix. The matrix G is not related to signal and noise and can be computed offline. If an analog filter or both an analog filter and digital filter are used, the matrix G should be revised to include the effects of all the filters. In general, G is found to be the covariance matrix of the received signal, when the input signal is white noise only. The matrix G and Q are computed only once and only Q is used in detection.

Maximum-Minimum Eigenvalue (MME) Detection

Step 1. Sample and filter the received signal as described previously.

Step 2. Choose a smoothing factor L and compute the threshold γ to meet the requirement for the probability of false alarm.

Step 3. Compute the sample covariance matrix:

$$
R(N_s) = \frac{1}{N_s} \sum_{n=0}^{N_s-1} x(n)x^H(n).
$$

Step 4. Transform the sample covariance matrix to obtain

$$
\tilde{R}(N_s) = Q^{-1}R(N_s)Q^{-H}.
$$

Step 5. Compute the maximum eigenvalue and minimum eigenvalue of the matrix $\tilde{R}(N_s)$ and denote them as λ_{max} and λ_{min}, respectively.

Step 6. Determine the presence of the signal based on the eigenvalues and the threshold: If $\lambda_{max}/\lambda_{min} > \gamma$, the signal exists; otherwise, the signal does not exist.

Energy with Minimum Eigenvalue (EME) Detection

Step 1. Sample and filter the received signal as described previously.

Step 2. Choose a smoothing factor L and compute the threshold γ to meet the requirement for the probability of false alarm.

Step 3. Compute the sample covariance matrix:

$$
R(N_s) = \frac{1}{N_s} \sum_{n=0}^{N_s-1} x(n)x^H(n).
$$

Step 4. Transform the sample covariance matrix to obtain

$$
\tilde{R}(N_s) = Q^{-1}R(N_s)Q^{-H}.
$$

Step 5. Compute the average energy of the received signal ρ and the minimum eigenvalue of the matrix $\tilde{R}(N_s)$, λ_{min}.

Step 6. Determine the presence of the signal: If $\rho/\lambda_{min} > \gamma$, the signal exists; otherwise, the signal does not exist.

Fast Fourier Transform–Based Pilot Sensing Algorithms

The FFT-based pilot sensing technique described in this section is a *nonblind* (ATSC-specific) sensing technique and is classified as a *fine* sensing technique [17]. The ATSC VSB signal has a pilot at the lower-band edge in a known location relative to the signal. It is assumed that the signal to be sensed is a bandpass signal at a low intermediate frequency (IF) of 5.38 MHz with the nominal pilot location at 2.69 MHz and is sampled at 21.52 MHz. However, the basic steps of the sensing algorithm can be implemented with suitable modifications for any IF and sampling rate. The essential features of the proposed method are as follows:

1. The signal is demodulated to the baseband by the nominal frequency offset of f_c = 2.69 MHz. Hence, if $x(t)$ is the real band pass signal at low IF, $y(t) = x(t)e^{-j2\pi f_c t}$ is the complex demodulated signal at baseband.
2. The signal $y(t)$ is filtered with a low-pass filter of bandwidth, say, 40 kHz (± 20 kHz). The filter bandwidth should be large enough to accommodate any unknown frequency offsets.
3. The output of the low-pass filter is down-sampled from 21.52 MHz to 53.8 kHz, to form the signal $z(t)$.
4. A fast Fourier transform is applied to the down-sampled signal $z(t)$. Depending on the sensing period, the length of the FFT will vary. For example, a 1 ms sensing window allows a 32-point FFT while a 5 ms window allows a 256-point FFT.
5. Then the FFT output is squared and the maximum value and location are determined.

Steps 1 and 3 are shown in Figure 14.17. Signal detection can then be done either by setting a threshold on the maximum value or by observing the location of

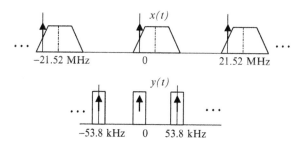

FIGURE 14.17

Frequency-domain description.

the peak over successive intervals. Instead of the FFT, other well-known spectrum estimation methods, such as the Welch periodigram, can be used in Step 5. The basic method just described can be adapted to a variety of scenarios:

1. Multiple fine sensing windows, such as 5 ms, sensing dwells every 10 ms. The 256-point FFT outputs squared from each sensing window can be averaged to form a composite statistic; also the location information from each measurement can be used to derive a detection metric.
2. If a single long sensing window, such as 10 ms, is available, a 512-point FFT or periodigram can be used to obtain better detection performance.

The parameters of the sensor can be chosen depending on the desired sensing time, complexity, probability of missed detection, and probability of false alarm. Detection based on location is robust against noise uncertainty, since the position of the pilot can be pinpointed with accuracy, even if the amplitude is low due to fading. Various combining schemes can be developed for both pilot-energy and pilot-location sensing.

ATSC Cyclostationary Sensing Technique

It has been recognized that many random time series encountered in the field of signal processing are more appropriately modeled as cyclostationary than stationary, due to the underlying periodicities in these signals [17, 571]. Another reason to use cyclostationary signal model is that random signals, such as white Gaussian noise, are not cyclostationary. Thus, cyclostationarity provides us a way to separate desired signals from noise.

Cyclostationary Feature of ATSC DTV Signals

According to [570], DTV data are vestigial sideband (VSB) modulated. Before VSB modulation, a constant of 1.25 was added to the eight-level pulse amplitude modulated (8-PAM) signal. Therefore, there is a strong pilot tone on the power spectrum density of the ATSC DTV signal. Let $s(t)$ be this pilot tone signal, which is a sinusoidal signal in the time domain, and further assume that this strong pilot tone is located at frequency f_0; that is,

$$s(t) = \sqrt{2P}\cos(2\pi f_0 t + \theta) \otimes b(t), \tag{14.1}$$

where P and θ are the power and the initial phase of the sinusoidal function, respectively. The function $b(t)$ is the channel impulse response and $\otimes$ is the convolution operator. The received signal must contain the signal

$$x(t) = s(t)e^{-j2\pi vt} + w(t), \tag{14.2}$$

where $w(t)$ is the additive white Gaussian noise and v is the amount of frequency offset in units of hertz. We assume that $w(t)$ is zero-mean with autocorrelation function $Rw(\tau) = E[w(t)w^*(t - \tau)] = \sigma^2 \delta(\tau)$. The cyclic spectrum of the received signal must contain the cyclic spectrum of $x(t)$, which is given by

$$S_x^\alpha(f) = \begin{cases} \frac{P}{2} \left[\delta(f - f_0 - v) + \delta(f + f_0 + v) \right] |H(f)|^2 + \sigma^2 & \text{for } \alpha = 0 \\ \frac{P}{2} \delta(f) H(f - f_0 - v) H^*(f + f_0 + v) & \text{for } \alpha = \pm 2(f_0 + v) \\ 0 & \text{otherwise,} \end{cases}$$

$$(14.3)$$

where $H(f)$ is the frequency response of the channel. The parameter α is the cyclic frequency. From Equation (14.3), ideally, the noise does not contribute to the cyclic spectrum of $x(t)$ when cyclic frequencies $\alpha = \pm(2f_0 + v)$. Thus, performing spectrum sensing by detecting the peaks on the cyclic spectrum of the signal should be better than using PSD.

Initial Processing of Received Signal

The RF ATSC DTV signal for a given DTV channel is first filtered and down-converted to a given intermediate frequency. The IF signals are usually sampled at a rate that is multiple times the symbol rate. The samples can be expressed as

$$y[n] = x[n] + w[n], \qquad (14.4)$$

where $x[n]$ are samples of the transmitted DTV signal. The noise $w[n]$ is assumed to be zero-mean with variance σ^2. Then, $y[n]$ is used to perform cyclostationarity-based sensing algorithms.

Test Statistic Using Cyclic Spectrum

First, we use a proper narrow bandpass filter to filter $y[n]$ and obtain a small frequency band that contains the pilot tone. Then, $y[n]$ is down-converted to have a lower central frequency. Note that we perform down-conversion multiple times. Let $z_l[n]$ denote the down-converted signal, which has a central frequency $f_{IF} + l\Delta f$. Note that Δf is chosen to be small, which depends on the sample rate and FFT size used in computation of the cyclic spectrum. We decimate $z_l[n]$ by a proper decimation ratio D to obtain $z_l^D[n]$, which has a lower sampling rate. Finally, we compute the cyclic spectrum by

$$S_z^\alpha(k) = \frac{1}{2L+1} \frac{1}{\Delta t} \sum_{l=-L}^{L} Z_l^D(k + \alpha/2) \cdot Z_l^D(k - \alpha/2), \qquad (14.5)$$

where

$$Z_l^D(k) = \sum_{n=0}^{N-1} z_l^D[n] e^{-j2\pi kn/N}. \qquad (14.6)$$

Note that, in Equation 14.5, we use a spectral smoothing method by averaging $2L + 1$ times to obtain cyclic spectrum. In Equation 14.6, N is the number of time samples used to compute short-term Fourier transform. The parameter Δt is the length of data segment, which equals $(N - 1)T_s$, where T_s is the time-sampling increment of the signal $z_l^D[n]$. Finally, we use

$$T = \max_{\alpha} \left| S_z^{\alpha}(0) \right| \tag{14.7}$$

as our decision statistic. The range of α depends on f_{IF} and frequency offset.

Sensing Procedure for TV Signals

Several to a large number of frequency components are taken in a TV channel band, depending on the required sensing accuracy. Refer to a later section on how these frequency components can be selected.

To compare values of these components with prestored information, the following two methods are applied:

1. **Correlation calculation.** To compare the shape of spectrum of the received signal with the well-known shapes of possible incumbent signals, calculate correlations with prestored values of spectral information for NTSC and DTV signals or other TV signals; if one of these correlation values is larger than the predetermined values, the judgment is that the National Television Systems Committee (NTSC) or DTV or one of other TV signals exists.
2. **Pilot detection.** To check whether a pilot signal exists or not, calculate the ratio of a pilot component to another component around the pilot after another component is picked such that this ratio is maximized; for example, if F417/F1200 > th_n, this signal is NTSC where th_n is the predetermined threshold for NTSC signals where F_n is the nth spectral component. If F103/F1200 > th_d, this signal is DTV where th_d is the predetermined threshold for DTV signals.

Frequency component values, correlation values, or ratios for several symbol periods can be averaged to have better sensing results.

Sensing Procedure for Wireless Microphone Signals

Wireless microphone systems should not be operated on the same frequency channel band as a local TV station. It means only open (unoccupied) frequencies should be used for microphones. Most microphone signals are FM modulated with a bandwidth of much less 200 kHz and at most 200 kHz. (Refer to IEEE 802.22-07/0124r0, Wireless Microphone Signal Simulation Method, March 2007.)

With these assumptions, microphone signals can be detected using the following procedure by sensing the spectral components using FFT devices:

1. For the preceding example, for every 3 kHz in a 6 MHz band a spectral component is measured and compared with other components.

2. If a 200 kHz band has many components, the judgment is that a wireless microphone is operated in that band as follows.
3. For the previous case, for example, if consecutive six components spaced equally in 200 kHz have considerable energy, it is judged that a microphone signal is detected.
4. Or if more correlations with stored information on various microphone signal spectral signatures (mainly FM signatures) than a predetermined value exist, a wireless microphone is operated in that band using a correlation calculation.

14.6 OTHER STANDARDIZATION ACTIVITIES

Within and outside the IEEE 802.22 are other ongoing industry activities that involve CR technologies.

14.6.1 IEEE 802.22.1 Standard

Within the IEEE 802.22 [17] are two task groups that are not as popular as its parent working group: the IEEE 802.22.1 and IEEE 802.22.2 task groups. The IEEE 802.22.1 task group was created to define a standard for CR devices to provide enhanced protection to low-powered licensed devices, such as wireless microphones, which also operate in the TV bands. Given the very low-power (as compared to TV transmitters) characteristics of wireless microphones, it becomes much harder to detect their presence through spectrum sensing technologies. Hence, the IEEE 802.22.1 is defining a beacon-based system that transmits at a higher power and can be triggered when a wireless microphone is to be used and thus can notify nearby 802.22 systems to vacate the current operating channel. The IEEE 802.22.2 task group is defining detailed technical guidance to installers, deployers, and operators of IEEE 802.22-compliant systems to help assure that such systems are correctly installed and deployed. Correct installation and deployment of IEEE 802.22-compliant systems are important to assure that those systems can maximally achieve their design goals in terms of system performance, reliability, and noninterference to incumbent licensed systems with which they share the TV broadcast bands.

14.6.2 Other Related Standards: IEEE 802.16h, SCC41

In addition to IEEE 802.22, the IEEE 802.16h task group [574] is also defining technologies inspired by CR concepts suitable to the IEEE 802.16 standard (commercially known as WiMax). The IEEE 802.16h task group does not, however, define some of the mechanisms that are considered to be needed to operate in the TV bands (e.g., spectrum sensing, incumbent databases), which is one of its differences from IEEE 802.22. Some of the technology options defined by 802.16h include:

1. The definition of a coexistence control channel, which allows the construction of collaborative networks that can communicate to, for example, signal quite periods for spectrum measurements.
2. Mechanisms for spectrum sharing and trading among cochannel WiMax networks.
3. Detection and avoidance of interference generated by similar (i.e., WiMax) and dissimilar (i.e., non-WiMax) systems.

Finally, the IEEE Standards Coordinating Committee (SCC) 41 [575], formerly known as IEEE P1900, is also developing standards in the area of dynamic spectrum access networks and is meant to provide coordination and information exchange among standards developing activities in the IEEE. The focus of SCC41 is on improved use of the spectrum, which includes the definition of new techniques and methods of dynamic spectrum access for managing interference, coordination of wireless technologies and network management, and information sharing. A total of six working groups are defined under the SCC41 umbrella.

14.7 CHAPTER SUMMARY AND FUTURE DIRECTIONS

The IEEE 802.22 working group defined the first worldwide air interface standard based on CR techniques. This new standard, which operates in the TV bands, uses techniques such as spectrum sensing, incumbent detection and avoidance, and spectrum management to achieve effective coexistence and radio resource sharing with existing licensed services. The 802.22 is leading the way for efficient utilization of the spectrum and, if this is successful, will lead to further development of newer technologies that can efficiently access the available spectrum.

However, to fully realize the dream of efficient spectrum utilization would require that sensing and the other methodologies outlined in [17] work well and be tested in complex environments. The business case of deploying a system based on these concepts in the band of interest with an eye on revenue generation is weak. Also research has to be done to design RF front ends that do not exhibit nonlinearity in different bands when spectrum agility is employed. Another area is the design of small form factor antennas as the frequency range is less than 1 GHz to be efficiently deployed in devices such as cell phones and PDA.

Since the FCC voted to open the TV spectrum, it provides a good lead to the existing unlicensed wireless technologies, such as cordless phone, WiFi, Bluetooth, and Zigbee, the coverage ranges of which are small compared to IEEE 802.22. In this case the sensing requirements coupled with listen before talk may well help them initially explore the rich spectrum.

14.8 **PROBLEMS**

1 Explain how the superframe structure and the SCH can be used to support reliable incumbent detection and meet the incumbent protection requirements described in Table 14.5.

2 Explain the main reasons for supporting the two different formats of uplink data burst allocations illustrated in the 802.22 frame structure (Figure 14.8).

3 What features in the 802.22 frame structure and MAC protocol enable over-the-air signaling between different 802.22 cells, and in which scenarios are these features most useful?

4 Explain why and in which situations neighboring 802.22 BSs need to synchronize their quiet periods for spectrum sensing.

Cognitive radio network security

15

Jung-Min "Jerry" Park, Kaigui Bian, and Ruiliang Chen
Virginia Polytechnic Institute and State University, United States

15.1 INTRODUCTION

Cognitive radio (CR) is a revolutionary technology that promises to alleviate the spectrum shortage problem and bring about remarkable improvements in the efficiency of spectrum utilization. However, the successful deployment of CR networks and the realization of their benefits depend on the placement of essential security mechanisms in sufficiently robust form to resist misuse of the systems. The emergence of the opportunistic spectrum sharing (OSS) paradigm and cognitive radio technology raises new security implications that have not been studied previously. Researchers have only recently started to examine the security issues specific to CR devices and networks.

Figure 15.1 provides a taxonomy of security threats to CR networks. Note that this taxonomy focuses on "active" threats unique to CR networks. Passive threats, such as eavesdropping, and threats also applicable to non-CR networks, such as jamming, are not included in the taxonomy. Figure 15.1 categorizes the threats into two broad categories: spectrum access–related security threats and radio software security threats. The former can be further classified into two subcategories: threats to incumbent coexistence and threats to self-coexistence. This taxonomy is neither the only appropriate way nor a widely accepted way of classifying threats to CR networks—it is used here for the sake of discussion. In the following subsections, we briefly describe the security threats in each category.

15.1.1 Overview of Security Threats to Incumbent Coexistence

Spectrum sharing, or coexistence, is an important attribute of CR networks. CR networks support two types of coexistence: incumbent coexistence (i.e., coexistence

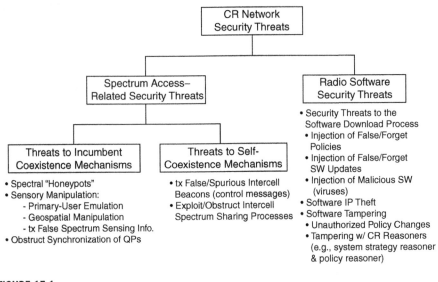

FIGURE 15.1

A taxonomy of security threats.

between primary and secondary networks) and self-coexistence (i.e., coexistence between secondary networks). Adversaries can exploit the vulnerabilities in the coexistence mechanisms to attack CR networks. In this section we limit our discussion to security threats to incumbent coexistence.

A CR needs to carry out spectrum sensing to identify fallow spectrum bands; that is, spectrum "white spaces." Ensuring the trustworthiness of the spectrum sensing process is important in the OSS paradigm, since spectrum sensing directly affects incumbent coexistence. Here, we focus our discussions on two particular security threats to incumbent coexistence: *primary-user emulation* and *transmission of false sensing results*.

In the OSS paradigm, *secondary* users equipped with CRs opportunistically utilize fallow licensed bands after identifying them via spectrum sensing. Secondary users are permitted to operate in licensed bands only on a noninterference basis to the *incumbent* (a.k.a. *primary*) users. Because the incumbent users' usage of licensed spectrum bands may be sporadic, a secondary user must constantly monitor for the presence of incumbent signals in the current operating band and candidate bands. A secondary user that detects the presence of incumbent signals in the current band must immediately switch to another band. On the other hand, if the secondary user detects the presence of a secondary user, it invokes a self-coexistence mechanism to share spectrum resources. In a *primary-user emulation* (PUE) attack, a rogue secondary user attempts to gain priority over other secondary users by transmitting signals that emulate the characteristics of the incumbent's signals. Due to the programmability of CRs, it is possible for an adversary to modify the radio software of a CR to change its emission characteristics so that they resemble those of an incumbent transmitter. The potential impact of a PUE attack depends on the legitimate secondary users' ability to distinguish the attacker's signals and actual incumbent signals while conducting spectrum sensing.

Another security threat is closely related to spectrum sensing. This security problem threatens the reliability of the distributed spectrum sensing (DSS) process in CR networks. Existing research suggests that DSS, when compared with individual spectrum sensing, enhances sensing accuracy while reducing the need for very sensitive sensing technology, which can be costly [24, 125, 576]. In DSS, individual nodes send their local sensing data to a fusion center, which processes the data to determine a sensing decision. However, DSS raises a security concern: *Byzantine failures* in DSS. In the context of DSS, Byzantine failures may be caused by either malfunctioning sensing nodes or malicious nodes launching *spectrum sensing data falsification* (SSDF) attacks. In either case, incorrect spectrum sensing data are reported to the fusion center, which can affect the accuracy of the sensing decision.

The Byzantine failure problem in the context of DSS is fundamentally different from the problem of resilient/secure data aggregation in sensor/ad hoc networks [577–581]. The existing algorithms for secure data aggregation cannot be used to address the problem of Byzantine failures. In secure data aggregation, an aggregate function (e.g., sum, count, or average) of the sensed data residing on sensor nodes needs to be computed while assuming that a subset of the sensor nodes are controlled by an adversary that is attempting to skew the final result. Secure aggregation's goal is to compute the aggregate function in such a way that an adversary that tampers with the aggregation process is unable to gain an advantage beyond what is possible with direct data injection. On the other hand, the goal of DSS is to make an accurate sensing decision after carrying out data fusion of local sensing results in the presence of Byzantine failures (which is equivalent to direct data injection attacks). Moreover, the investigation of DSS Byzantine failures involves not only the study of data fusion techniques, but also the interplay between the data fusion techniques and the spectrum sensing techniques.

15.1.2 Overview of Security Threats to Self-Coexistence

Self-coexistence mechanisms for a CR network are defined as part of the network's air interface and have features specific to the air interface. Therefore, here we focus on the self-coexistence mechanisms of IEEE 802.22 [582] to facilitate our discussion. IEEE 802.22 is the first standard for wireless access networks based on CR technology. It specifies the air interface for a wireless regional area network (WRAN) that uses fallow segments of the UHF/VHF TV bands between 54 and 862 MHz. Although we limit our discussions to threats against 802.22's self-coexistence mechanisms, the security issues discussed, in principle, are relevant to other types of CR networks as well.

Ensuring the congruous coexistence of 802.22 networks (i.e., self-coexistence) is of paramount importance. In 802.22, the self-coexistence problem is exacerbated by the fact that a base station's (BS) coverage range may be as large as 100 km. It is quite possible for a number of 802.22 cells to have overlapping coverage areas. Without proper mechanisms to handle self-coexistence in such a situation, the resulting self-interference can render the 802.22 systems useless. IEEE 802.22 addresses self-coexistence using inter-BS dynamic resource sharing and prescribes two supporting mechanisms: nonexclusive spectrum sharing and exclusive spectrum sharing. A network in need of more spectrum resources executes nonexclusive sharing as long as the maximum achievable signal-to-interference ratio on the selected channel is higher than the required threshold of the network's supported services. If nonexclusive sharing is not feasible, then an 802.22 WRAN needs to acquire spectrum resources through exclusive spectrum sharing, viz. the *on-demand spectrum contention* (ODSC) protocol [583]. The ODSC process enables a cell to acquire better channels or more channels to support the quality of service of the admitted workloads. Because direct inter BS communication is not possible in most cases, a BS collects neighboring cells' spectrum utilization information by receiving intercell control messages forwarded by consumer premise equipment (CPE) devices under its control. These control messages, called *intercell beacons*, are used by BSs to exchange spectrum utilization information.

Although intercell beacons play a crucial role in self-coexistence, they are not protected by 802.22's security sublayer. In other words, intercell beacons are vulnerable to unauthorized modification, forgery, or replay. In Section 15.4.3, we describe an attack that disrupts the exclusive spectrum sharing process by transmitting forged intercell beacons. We coin the term *beacon falsification* (BF) attack to refer to such an attack.

15.1.3 Radio Software Security Threats

The flexibility and adaptability brought by modern software, low-cost microprocessors, and smart antennas have made software-defined and cognitive radios a reality. However, the advantages of such radios can be offset by the lack of security and reliability of the underlying software, which serves as the control and command for the radio system. Consequently, the emergence of software-defined radio (SDR) and software-based CR have brought about new security threats not considered previously. In particular, the programmability of CR devices raises serious security concerns. Without proper software protection mechanisms in place, CRs are vulnerable to a host of attacks targeting the radio software. These attacks may include execution of malicious code, removal of software-based authentication or access control functions, intellectual property (IP) theft via reverse engineering, disruption of radio software reconfiguration, and the like. In Section 15.5, we discuss these and other related threats.

The rest of this chapter is organized as follows. We focus our discussion on primary-user emulation attacks and Byzantine failures in DSS in Sections 15.2 and

15.3, respectively. We then describe security threats to self-coexistence mechanisms in the context of the 802.22 air interface in Section 15.4. In Section 15.5, we discuss security issues that affect the trustworthiness of CR software.

15.2 PRIMARY-USER EMULATION ATTACKS

15.2.1 Spectrum Sensing in Hostile Environments

A CR's ability to distinguish between primary-user signals and secondary-user signals is key to the implementation of the OSS paradigm. Distinguishing the two signals is nontrivial, but it becomes especially difficult when the CRs operate in hostile environments. In a hostile environment, an attacker may modify the air interface of its own CR to mimic a primary-user signal's characteristics, thereby causing legitimate secondary users to erroneously identify the attacker as a primary user. We coin the term *PUE attack* to refer to this attack. There is a realistic possibility of PUE attacks, since CRs are highly reconfigurable due to their software-based air interface [11]. To thwart such attacks, a scheme that can reliably distinguish between legitimate primary signal transmitters and secondary signal transmitters launching PUE attacks is needed. In hostile environments, such a scheme should be integrated into the spectrum sensing mechanism to enhance the trustworthiness of the sensing result.

One of the major technical challenges in spectrum sensing is the problem of precisely distinguishing primary-user signals from secondary-user signals. To distinguish the two signals, existing spectrum sensing schemes based on energy detectors [584, 585] implicitly assume a "naive" transmitter verification scheme. When energy detection is used, a secondary user can recognize the signals of other secondary users, since they can be demodulated and decoded, but that is not true for primary-user signals. When a secondary user detects a signal that it can demodulate and decode, it assumes that the signal is that of a secondary user; otherwise, it determines that the signal is that of a primary user. Under such an overly simplistic transmitter verification scheme, a selfish or malicious secondary user (i.e., an attacker) can easily exploit the spectrum sensing process. For instance, a PUE attacker may "masquerade" as a primary user by transmitting unrecognizable signals in one of the licensed bands, thus preventing other secondary users from accessing that band.

There are alternative techniques for spectrum sensing, such as matched filter and cyclostationary feature detection [119]. Devices capable of such detection techniques are able to recognize the intrinsic characteristics of primary-user signals, thus enabling them to distinguish those signals from those of secondary users. However, such detection techniques are still not robust enough to counter PUE attacks. To defeat cyclostationary detectors, an attacker may make its transmissions indistinguishable from primary-user signals by transmitting signals that have the same cyclic spectral characteristics as primary-user signals.

The current research and standardization efforts suggest that one of the first applications of CR technology will be its use for dynamic spectrum access (DSA) of fallow TV spectrum bands. The Federal Communications Commission (FCC) is considering opening up TV bands for DSA because TV bands often experience lower and less dynamic utilization than other primary-user networks such as cellular networks [34]. We focus on a scenario in which a primary-user network is composed of TV transmission towers and receivers placed at fixed locations. In such a setting, the *location* of a given transmitter along with other factors can be utilized to determine whether the transmitter is a primary transmitter or a PUE attacker.

Estimating the location of a wireless device is a well-studied problem [586–590]. However, localization of primary transmitters in the context of DSA is not a trivial problem when one considers the requirement prescribed by the FCC [60], which states that *no modification to the incumbent system should be required to accommodate opportunistic use of the spectrum by secondary users.* For this reason, conventional approaches, such as embedding signed location information in a primary user's signal or employing an interactive protocol between a primary signal transmitter and a localization device, cannot be used.

15.2.2 Classification of PUE Attacks

In PUE attacks, the adversary transmits only in fallow bands. Hence, the aim of the attackers is not to cause interference to primary users but to preempt spectrum resources that could have been used by legitimate secondary users. Depending on the motivation behind the attack, a PUE attack can be classified as either a selfish PUE attack or a malicious PUE attack:

- **Selfish PUE attacks.** In this attack, an attacker's objective is to maximize its own spectrum usage. When selfish PUE attackers detect a fallow spectrum band, they prevent other secondary users from competing for that band by transmitting signals that emulate the signal characteristics of primary-user signals. This attack is most likely to be carried out by two selfish secondary users of which the intention is to establish a dedicated link.
- **Malicious PUE attacks.** The objective of this attack is to obstruct the DSA process of legitimate secondary users; that is, prevent legitimate secondary users from detecting and using fallow licensed spectrum bands, causing denial of service. Unlike a selfish attacker, a malicious attacker does not necessarily use fallow spectrum bands for its own communication purposes. It is quite possible for an attacker to obstruct the DSA process simultaneously in multiple bands by exploiting two DSA mechanisms implemented in every CR. The first mechanism requires a CR to wait for a certain amount of time before transmitting in the identified fallow band to make sure that the band is indeed unoccupied. Existing research shows that this time delay is nonnegligible [576, 584]. The second mechanism requires a CR to periodically sense the current operating

band to detect primary-user signals and immediately switch to another band when such signals are detected. By launching a PUE attack in multiple bands in a round-robin fashion, an attacker can effectively limit the legitimate secondary users from identifying and using fallow spectrum bands.

Both attacks could have disruptive effects on CR networks. (Their disruptive effects are studied using simulation in Section 15.2.4.) To thwart PUE attacks, one needs to first detect the attack. Next, we describe a *transmitter verification* scheme that can be integrated into a spectrum sensing scheme to detect PUE attacks under certain conditions.

15.2.3 Noninteractive Localization of Primary Signal Transmitters

Here, we describe a *transmitter verification* scheme for detecting PUE attacks. Before describing the scheme, we state some of the assumptions that form the foundation of the scheme. The primary user is assumed to be a network composed of TV signal transmitters (i.e., TV broadcast towers) and receivers. A TV tower's transmitter output power is typically hundreds of thousands of watts [591], which corresponds to a transmission range from several miles to tens of miles. We assume that the secondary users, each equipped with a hand-held CR device, form a mobile ad hoc network. Each CR is assumed to have self-localization capability and a maximum transmission output power that is within the range from a few hundred milliwatts to a few watts—this typically corresponds to a transmission range of a few hundred meters. An attacker, equipped with a CR, is capable of changing its modulation mode, frequency, and transmission output power.

Based on these assumptions, Chen and Park [592] propose a transmitter verification scheme for spectrum sensing that is appropriate for hostile environments; the transmitter verification scheme is illustrated in Figure 15.2. In the network model under consideration, the primary signal transmitters are TV broadcast towers placed at fixed locations. Hence, if a signal source's estimated location deviates from the known location of the TV towers and the signal characteristics resemble those of primary-user signals, then it is likely that the signal source is launching a PUE attack. An attacker, however, can attempt to circumvent this location-based detection approach by transmitting in the vicinity of one of the TV towers. In this case, the signal's energy level in combination with the signal source's location is used to detect PUE attacks. It would be infeasible for an attacker to mimic both the primary-user signal's transmission location and energy level since the transmission power of the attacker's CR is several orders of magnitude smaller than that of a typical TV tower. Once an instance of a PUE attack has been detected, the estimated signal location can be further used to pinpoint the attacker.

As Figure 15.2 shows, the transmitter verification scheme includes three processes: verification of signal characteristics, measurement of received signal energy level, and localization of the signal source. To date, the technical problems related to the first two processes, in the context of CR networks, have attracted a lot

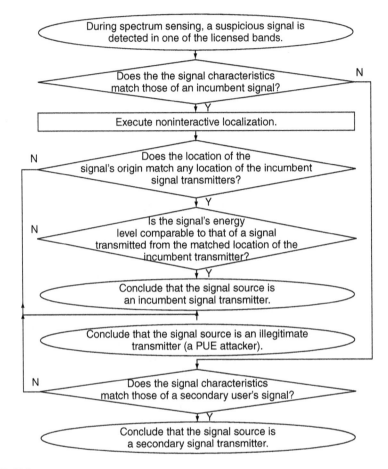

FIGURE 15.2

A flowchart of the transmitter verification scheme.

of attention [22]. In contrast, very little existing research directly addresses the third process. Therefore, in the following discussion, we focus on the problem of transmitter localization. This problem—called by various names, such as *location estimation*, *location identification*, *localization*, and *positioning*—has been studied extensively in the past. The primary signal transmitter localization problem (which is referred to as the *PST localization problem* hereafter), however, is more challenging for two reasons. First, the following requirement must be met: No modification should be made to primary users to accommodate the DSA of licensed spectrum. Because of this requirement, including location information in a primary user's signal, this is not a viable solution. The requirement also excludes the possibility of using a localization protocol that involves the interaction between a primary user and the localization device(s). Thus, the PST localization problem becomes a *noninteractive* localization problem. Second, it is the transmitter (and not the receiver) that needs to be localized. When a receiver is localized, one need not consider the existence of other receivers. However, the existence of multiple transmitters adds to the difficulty of transmitter localization.

Existing Localization Techniques

Before introducing the proposed localization system, we first summarize conventional localization techniques in wireless networks and discuss how these techniques should be improved to address the PST localization problem in CR networks.

The conventional localization approaches are based on one or several of the following techniques: time of arrival (TOA), time difference of arrival (TDOA), angle of arrival (AOA), and resolved signal strength (RSS).

GPS [590] is a typical localization system based on TOA. A mobile node receives signals from satellites that contain their location and time information. Based on the information, the node can calculate its own position.

TDOA is a passive localization technique that utilizes the difference between the arrival times of pulses transmitted by a transmitter but does not rely on any knowledge of the pulse transmission time. The technique measures the time differences at multiple receivers with known locations and subsequently computes a location estimate [586].

In the AOA technique, a receiver measures the angel of arrival from two or more transmitters. If the locations of the transmitters are known, the receiver can calculate its own location using triangulation [589]. Using the same principle, angle of arrival information to multiple receivers can be used to determine the transmitter's location.

RSS-based localization techniques arise from the strong correlation between the distance of a wireless link and RSS [588, 593]. Specifically, given a transmitter-receiver pair, RSS can be modeled as a function of transmitted power and transmitter-receiver distance. Therefore, if a correct model is used and multiple observers take RSS measurements from a transmitter, then the transmitter location can be estimated using the model. For example, one of the techniques for radio location in Wireless E911 [594] is to use "location signatures." The location signature scheme stores and matches multipath patterns (fingerprints) that mobile phone signals are known to exhibit at different locations in each cell.

TOA is a receiver-localization technique and needs to be modified to support transmitter localization. Such a modification is not trivial, especially when one considers the possibility of malicious transmitters that may obstruct the localization. Both TDOA and AOA techniques can be used for transmitter localization and have relatively high localization precision. When applying TDOA and AOA to the PST localization problem, one needs to consider the scenario in which a number of transmitters are equipped with a directional antenna. The common drawback of both techniques is the use of expensive hardware. In contrast, RSS-based techniques require no such costly hardware. However, they also have drawbacks, such as difficulty in handling multiple transmitters and the innate inaccuracy of RSS measurements. In the following discussions, we show that these issues can be addressed by taking many RSS measurements and properly processing them.

Architecture of the Localization System

The basic idea of the proposed localization system uses the fact that the magnitude of an RSS value typically decreases as the distance between the signal transmitter and the receiver increases [587]. Therefore, if one is able to collect a sufficient number of RSS measurements from a group of receivers spread throughout a large network, the location with the peak RSS value is likely to be the location of a transmitter. The advantage of this technique is twofold when it is used for the PST localization problem in CR networks: It both obviates modification of primary users and supports localizing multiple transmitters that transmit signals simultaneously.

The requirement to collect RSS distribution in a network naturally leads us to resorting to an underlying wireless sensor network (WSN) that can help collect RSS measurements across the network. It should be noted that the idea of using an underlying WSN to facilitate the operation of a CR network is not new. For example, in [576], it was proposed that a spectrum-aware sensor network be used for distributed spectrum sensing, so that the sensor network can provide secondary users with information about spectrum opportunities throughout a network. If sensor nodes in a WSN have the capability to measure RSS and are aware of their positions [587], they can be used to solve the PST localization problem. However, two problems need to be addressed for the aforementioned approach to be viable.

First, path fading may change over time and a PUE attacker may constantly change its location or vary its transmission power to evade localization, thus causing RSS measurements to fluctuate drastically within a short period of time. This problem cannot be mitigated by taking the average of measurements taken at different times, since the RSS values measured at a given position at different times have different distributions. A possible solution to this problem is to take a "snapshot" of the RSS distribution in a given network; that is, require the sensors of a WSN to take synchronized RSS measurements in a given band.

The second problem arises from the fact that RSS usually varies by a large magnitude (30 dB to 40 dB) [593] over short distances. This makes it very challenging to decide the location of primary users just by reading the raw data in a snapshot of RSS distribution. We conducted a simulation experiment to illustrate this problem. A 2000 m × 2000 m network with two transmitters located at (800 m, 1800 m) and (1300 m, 550 m) was simulated. Each transmitter's transmission power was 500 mW, working at the UHF frequency of 617 MHz. The phase shift between the two transmitters was randomly chosen. A statistical log-loss signal propagation model, which was shown to be appropriate for modeling signal propagation behavior in many situations [595], was employed in the simulation. In this model, the expected RSS in decibels is given by

$$\mu = p + \beta_0 + \beta_1 \ln s, \tag{15.1}$$

where s is the transmitter-receiver distance, p is the transmitted power in deci-bels, and β_0 and β_1 are constant parameters that need to be calibrated for a specific environment. Note that this is offsite calibration and no onsite calibration is required [595]. In offsite calibration, one needs to tune the parameters related to the channel environment (e.g., rural, urban). Using the model, the distribution of RSS is characterized as a Gaussian random variable with a mean of μ and a variance of σ^2. In [595], a set of parameters approximating real-world results were used, where $(\beta_0, \beta_1, \sigma) = (-30.00, -10.00, 10.0)$. We used the same set of parameters for our simulation. Figure 15.3(a) shows a snapshot of the RSS in dBm. It can be seen that, because of the large variance of the RSS, the snapshot reveals no obvious RSS peaks (which can be used as approximations for the transmitter locations).

However, if the variance can be reduced to a sufficient level, the snapshot would clearly indicate the RSS peaks as illustrated in Figure 15.3(b). It is therefore reasonable to conjecture that, if one were able to decrease the variance using an appropriate *data smoothing* technique, it may be possible to solve the PST localization problem by using the aforementioned localization approach. For details on data smoothing, we refer the reader to [596].

15.2.4 Simulation Results

The Effects of PUE Attacks

We carried out simulation experiments to showcase the disruptive effects of PUE attacks. In the simulated network, 300 secondary users (which include both legit-imate users and attackers) are randomly located inside a 2000 m $\times$ 2000 m square area, each with a transmission range of 250 m and an interference range of 550 m. These range values are consistent with the protocol interference model used in [597]. Two TV broadcast towers act as primary signal transmitters. Each TV tower has ten 6 MHz channels, and the duty cycle of all the channels is fixed at 0.2. One tower is located 8000 m east of the square area and has a transmission radius of 9000 m; the other tower is located 5000 m south of the square area with a transmission radius of 7000 m.[1] The layout of the simulated network is shown in Figure 15.4(a). Each secondary-user node is randomly placed in the network area and moves according to a random waypoint model by repeatedly executing the following four steps: (1) It randomly chooses a destination in the square area with a uniform distribution; (2) it chooses a velocity v that is uniformly distributed over $[v_{\min}, v_{\max}]$; (3) it moves along a straight line from its current position to the destination with velocity v; and (4) it pauses in the destination for a random period

---

[1]We set the values of 9000 m and 7000 m for the primary users' transmission radiuses based on realistic assumptions. Suppose the following parameters: EIRP (equivalent isotropically radiated power) of the TV towers (transmitters) is 2500 kW, transmitters' effective antenna height is 100 m, receivers' effective antenna height is 1 m, and receivers' energy detection sensitivity is −94 dbm. Under these conditions, one can derive a transmission radius of 8000 m using the rural environment version of the HATA model [593].

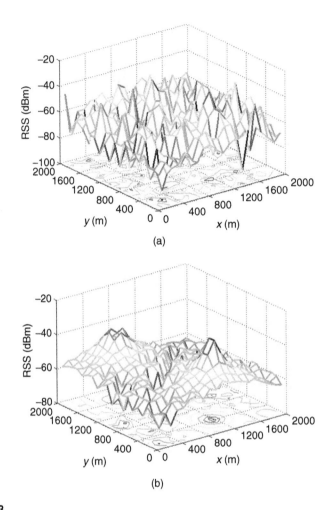

FIGURE 15.3

RSS distributions obtained from the underlying WSN: (a) a snapshot of the RSS raw-data distribution; (b) the RSS distribution in the network when $\sigma = 0$.

that is uniformly distributed over $[0, t_{p-max}]$. We chose the values $v_{min} = 5$ m/s, $v_{max} = 10$ m/s, and $t_{p-max} = 60$ sec. Each simulation instance spans a period of 24 h. Another 1 h before the 24 h was simulated to ensure that the random waypoint model entered a steady state. The number of malicious PUE attackers was varied from 1 to 30 and that of selfish PUE attackers was varied from 1 to 30 pairs. Figures 15.4(b) and 15.4(c) show the simulation results for the selfish PUE attack and the malicious PUE attack, respectively. The y-axis in the figures represents the amount of link bandwidth each secondary user is able to detect. The results show that a selfish PUE attack can effectively steal bandwidth from legitimate secondary users,

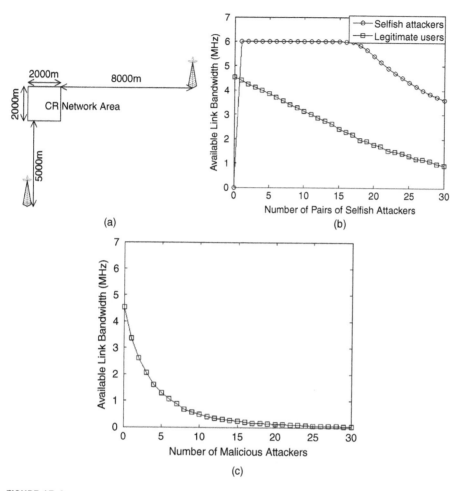

FIGURE 15.4

Simulations to illustrate the effects of PUE attacks: (a) simulation layout; (b) effect of selfish PUE attacks; (c) effect of malicious PUE attacks.

while a malicious PUE attack can drastically decrease the link bandwidth available to legitimate secondary users.

Simulation on the Localization System
Simulation Setting and Objectives

We conducted a set of simulation experiments to evaluate the proposed transmitter localization scheme. Note that verification of signal characteristics

Table 15.1 Simulation Settings for the Localization System

Sensor Density (m^{-2})	Number of Sensors	r (m)	d (m)
2.5×10^{-5}	100	305	1294
5×10^{-5}	200	300	1273
1.25×10^{-4}	500	200	849
2.5×10^{-4}	1000	200	849
5×10^{-3}	2000	100	424
1.25×10^{-4}	5000	100	424
2.5×10^{-3}	10,000	50	212

and measurement of signal energy level are not included in this simulation study.

In the simulation, a 2000 m × 2000 m CR network with an underlying WSN of the same size was assumed and the statistical log-loss propagation model with $(\beta_0, \beta_1, \sigma) = (-30.00, -10.00, 10.0)$ was used. The exact values of these parameters are unknown to the localization system, but we assume that they are estimated using the offsite calibration scheme proposed in [595], where a realistic estimation was given as $(\beta_0, \beta_1, \sigma) = (-32.03, -9.73, 10.0)$. We generated seven simulation settings representing varying degrees of sensor node density in the WSN; these settings are explained in Table 15.1.

We evaluate the system's localization error and computation time. Based on the discussion in Section 15.2.3, the metric of localization error has the following meaning. When a primary signal transmitter is found to be away from any known location of primary users more than the localization error, the transmitter is deemed a PUE attacker. Once a PUE attacker is detected, the localization error defines a range of area for pinpointing the attacker. The computation time is the time to run the localization algorithm but does not include the WSN's network delay for collecting data. The computation time shows the relative computation overhead in different scenarios. It is measured in our specific simulation environment and its absolute value could change as the environment varies.[2]

The Case of a Single Transmitter

We consider three scenarios: a 500 mW primary signal transmitter is in the center, on the border, and on the corner of the WSN; that is, T_1 at (1000 m, 1000 m), T_2 at (1000 m, 50 m), and T_3 at (50 m, 50 m), respectively. The localization errors of the proposed localization system under various settings are shown in Figure 15.5. In the figure, every datum is the average of ten independent simulations. The results

---

[2]In particular, the simulation was run in MATLAB on a P4 2.8 GHz, 512 M RAM PC.

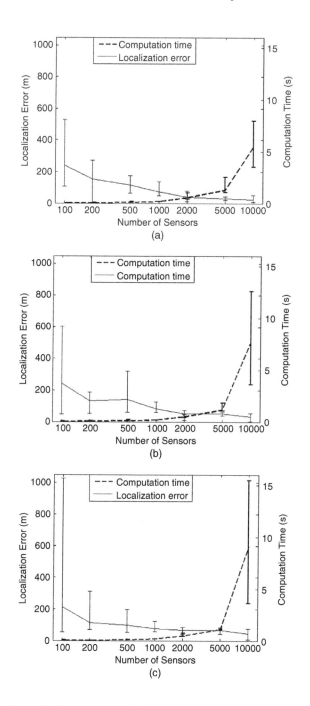

FIGURE 15.5

The localization error of the proposed localization system: (a) $T_1(1000\,m, 1000\,m)$;
(b) $T_2(1000\,m, 50\,m)$; (c) $T_3(50\,m, 50\,m)$.

prove the localization system to be effective. For example, under the 10,000-sensor scenario, the expected space of two adjacent sensors is 20 m, which is close to the localization error of T_1; that is, 21.9 m. T_2 and T_3 have relatively greater localization errors because, on the border or on the corner of the WSN, fewer are sensors are around, resulting in fewer measurements and thus poorer accuracy. Meanwhile, the computation time is shown to be affordable. T_2 and T_3 require relatively greater computation time because fewer measurements means more ambiguity and causes the localization algorithm to sample more regions (i.e., the set R has more elements). Chen and Park [596] also considered other factors that could affect the localization accuracy, such as the use of directional antennas and multiple simultaneous transmissions. For more details, we refer the reader to [596].

15.2.5 Related Research

A large body of work studies localization systems [586–590]. As discussed previously, the existing research is inadequate for solving the PST localization problem. Recently, a number of secure localization schemes have been proposed [598, 599]. These schemes, however, are inappropriate for solving the PST localization problem. The technique proposed in [598] is for receiver localization and cannot be used for transmitter localization. The localization schemes proposed in [598,599] require interaction between the localized object and the localizing devices. In [592], two location verification schemes were proposed for verifying the location of primary users in CR networks. Those schemes, however, have a number of drawbacks. The schemes have no localization capabilities and they are ineffective in a scenario where multiple rogue transmitters mimic incumbent transmitters.

15.3 ROBUST DISTRIBUTED SPECTRUM SENSING

In this section, we study the Byzantine failure problem in the context of the data fusion process of distributed spectrum sensing. The Byzantine failure problem can be caused by spectrum sensing devices that are malfunctioning or carrying out *spectrum sensing data falsification* attacks. A malfunctioning sensing terminal is unable to conduct reliable spectrum sensing and may send incorrect sensing reports to the data collector. In an SSDF attack, a malicious secondary user intentionally sends falsified local spectrum sensing reports to the data collector in an attempt to cause the data collector to make incorrect spectrum sensing decisions. Either case could cause interference to incumbents and result in underutilization of fallow licensed spectrum. We consider the Byzantine failure problem from the perspective of data fusion. We discuss several existing data fusion techniques and propose a new technique, called the *weighted sequential probability ratio test* (WSPRT), to improve robustness against Byzantine failures. Using simulations, the proposed data fusion technique is compared with existing techniques under various conditions.

15.3.1 **Technical Background**

Distributed Spectrum Sensing

Carrying out reliable spectrum sensing is a challenging task for a CR. In a wireless channel, signal fading can result in the "hidden node problem." The hidden node problem in the context of CR networks can be described as an instance in which a secondary user in a CR network is within the protection region[3] of an operating incumbent but fails to detect the existence of the incumbent. Besides the hidden node problem, it is also possible for a secondary user to falsely detect an incumbent because of noise or interference in the wireless environment. Recent research results indicate that these problems can be addressed by requiring multiple secondaries to cooperate with each other in spectrum sensing; that is, DSS. In DSS, each secondary acts as a sensing terminal that conducts local spectrum sensing. The local results are reported to a data collector (or "fusion center") that executes data fusion and determines the final spectrum sensing result. The application of DSS requires that the distance between any two sensing terminals is small relative to their respective distances from an incumbent transmitter.

Existing Data Fusion Techniques

A key component in DSS is the data collector. The data collector needs to employ an appropriate fusion technique to make an accurate spectrum sensing decision. In this subsection, we describe three data fusion techniques recently proposed for DSS in CR networks. To facilitate our discussion, we model the DSS process as a parallel fusion network, as shown in Figure 15.6. In this figure, N_0 is a data collector, N_t $(t = 0, 1, 2, \ldots, m$, where m is the number of N_0's neighboring sensing

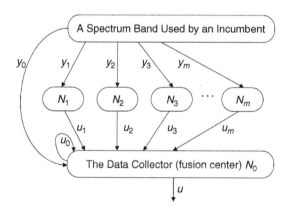

FIGURE 15.6

Modeling DSS into a parallel fusion network.

[3]An incumbent's protection region is defined as the area in which secondaries cannot operate while the incumbent is transmitting so that no interference to the incumbent introduced.

terminals) denotes one of N_0's sensing terminals (N_0 is both a data collector and a sensing terminal), y_i represents the incumbent signal received at N_i, and u_i is the local spectrum sensing report that N_i sends to N_0. The output u is the final sensing decision, which is a binary variable—1 denotes the presence of an incumbent signal, and 0 denotes its absence. The data fusion problem therefore can be regarded as a binary hypothesis testing problem with two hypotheses represented by H_0 and H_1. Correspondingly, each u_i is also binary. To simplify the discussions, the following discussion assumes that spectrum sensing is carried out in a single spectrum band:

- *Decision fusion* [600, 601] requires the data collector to sum up all the values of u_i. A threshold value that is no less than 1 and no greater than $m+1$ needs to be specified. If the sum of u_i values is greater than or equal to the threshold, then the final sensing decision is "occupied"; that is, $u = 1$ and H_1 is accepted; otherwise, the band is determined to be "fallow"; that is, $u = 0$ and H_0 is accepted. Depending on the value of the threshold, decision fusion can have several variants. A threshold value of 1 is an OR fusion rule, a value of $(m+1)$ is an AND fusion rule, and a value of $(m+1)/2$ is a Majority fusion rule.

- *Bayesian detection* [602] requires the knowledge of a priori probabilities of the u_i values when u is 0 or 1; that is, $P(u_i|H_0)$ and $P(u_i|H_1)$. It also requires the knowledge of a priori probabilities of u; that is, $P_0 = P[u = 0]$ and $P_1 = P[u = 1]$, respectively. There are four possible cases. In two cases the sensing decisions are correct, while in the other two cases the decisions are incorrect. The two incorrect decisions are referred to as *misdetection* ($u = 0$ when the band is occupied) and *false alarm* ($u = 1$ when the band is fallow), respectively. The two correct decisions (i.e., $u = 0$ when the band is fallow and $u = 1$ when the band is occupied) are associated with small costs and the incorrect ones are associated with large costs. The case of misdetection of an incumbent may result in interference to the incumbent; hence, this case is the least desirable and accordingly assigned the largest cost. The overall cost is the sum of the four costs weighted by the probabilities of the corresponding cases. Bayesian detection can be represented by the following test, which produces a final spectrum sensing decision that minimizes the overall cost:

$$\prod_{i=0}^{m} \frac{P[u_i|H_1]}{P[u_i|H_0]} \quad \overset{H_1}{\underset{H_0}{\overset{>}{<}}} \quad \frac{P_0(C_{10} - C_{00})}{P_1(C_{01} - C_{11})}, \tag{15.2}$$

where $C_{jk}(j = 0, 1; k = 0, 1)$ is the cost of declaring H_j true when in fact H_k is true.

- *The Neyman-Pearson test* [601, 603] does not rely on the knowledge of the a priori probabilities of u nor on any cost associated with each decision case. However, it requires the knowledge of the a priori probabilities of u_i values when u is 0 or 1. Additionally, either a maximum acceptable probability of false alarm or a maximum acceptable probability of misdetection needs to be defined. The Neyman-Pearson test guarantees that the other probability is minimized while the defined probability is acceptable. The Neyman-Pearson test can be represented as

$$\prod_{i=0}^{m} \frac{P[u_i|H_1]}{P[u_i|H_0]} \begin{array}{c} H_1 \\ > \\ < \\ H_0 \end{array} \lambda, \qquad (15.3)$$

where λ is a threshold calculated from the defined probability of false alarm or misdetection.

As Equations (15.2) and (15.3) show, both Bayesian detection and the Neyman-Pearson test are essentially a fixed-number likelihood ratio test; the only difference is the way that the threshold is chosen.

15.3.2 Weighted Sequential Probability Ratio Test

Byzantine Failure in Data Fusion

The DSS approach is vulnerable to a number of security threats. In particular, Byzantine failure is a major threat to the data fusion process. A Byzantine failure could be caused by either malfunctioning sensing terminals or an SSDF attack. Both cases result in one or more sensing terminals sending false local spectrum sensing reports to a data collector, causing the data collector to make a wrong spectrum sensing decision. Because the two cases have the same impact on data fusion, without loss of generality, we focus only on SSDF attacks in the rest of this section.

To maintain an adequate level of accuracy in the midst of SSDF attacks, the data fusion technique used in DSS needs to be robust against fraudulent local spectrum sensing results reported by malicious secondaries. However, the previously discussed fusion techniques share two common properties that contribute to their vulnerability to SSDF attacks. First, none of these techniques can guarantee both a bounded false alarm probability and a bounded misdetection probability. In an adversarial environment, these probabilities get larger. Second, these techniques treat all sensing terminals indiscriminatingly, regardless of whether a sensing terminal has a history of reporting incorrect data. When an SSDF attacker constantly injects false data, the ideal solution is to filter out the false data and accept inputs only from reliable sensing terminals.

Weighted Sequential Probability Ratio Test

WSPRT is composed of two steps. The first step is a reputation maintenance step, and the second step is the actual hypothesis test. In the reputation maintenance step, a sensing terminal's reputation ratings are allocated based on the accuracy of a sensing terminal's sensing. The reputation value is set to 0 at the beginning; whenever its local spectrum sensing report is consistent with the final sensing decision, its reputation is incremented by 1; otherwise, it is decremented by 1. Under this rule, assuming N_i's reputation value is r_i, the last sensing report N_i sent to N_0 is u_i, and the final decision is u, then r_i is updated according to the following relation: $r_i \leftarrow r_i + (-1)^{u_i+u}$. Obviously, given that the probability of the final decision being true is greater than 0.5, a sensing terminal with a more accurate local sensing report has a higher expected reputation value than a terminal with a less accurate sensing report.

The hypothesis test step of WSPRT is based on a sequential probability ratio test (SPRT), which is a hypothesis test for sequential analysis and supports sampling a variable number of observations [601]. When applied to data fusion for DSS, SPRT has a desirable property—it can guarantee a bounded false alarm probability and a bounded misdetection probability in a nonadversarial environment. Even if each sensing terminal has low spectrum sensing accuracy, SPRT can provide the guarantee by collecting more local spectrum sensing reports. This is an advantage over the techniques discussed previously.

When applying SPRT to data fusion for DSS, one needs to define the following likelihood ratio as the decision variable:

$$S_n = \prod_{i=0}^{n} \frac{P[u_i|H_1]}{P[u_i|H_0]}. \tag{15.4}$$

Note that the number of samples n is a variable and can be different from $m + 1$.

The fusion decision is based on the following criterion:

$$\begin{cases} S_n \geq \eta_1 \Rightarrow \text{accept } H_1 \\ S_n \leq \eta_0 \Rightarrow \text{accept } H_0 \\ \eta_0 < S_n < \eta_1 \Rightarrow \text{take another observation.} \end{cases} \tag{15.5}$$

The values for η_1 and η_0 are

$$\eta_1 = \frac{1 - P_{01}}{P_{10}} \text{ and } \eta_0 = \frac{P_{01}}{1 - P_{10}},$$

where P_{01} and P_{10} are the tolerated false alarm probability and the tolerated misdetection probability, respectively. It can be shown that SPRT minimizes the expected value of n needed to accept either hypothesis H_1 or H_0 [601].

The idea of WSPRT is to modify the likelihood ratio in Equation (15.4) so that the decision variable also takes a sensing terminal's reputation into consideration. The proposed new decision variable is

$$W_n = \prod_{i=0}^{n} \left(\frac{P[u_i|H_1]}{P[u_i|H_0]} \right)^{w_i}, \tag{15.6}$$

where w_i is defined as the weight of N_i and is a function of r_i:

$$w_i = f(r_i). \tag{15.7}$$

For such a decision variable to be robust against SSDF attacks, $f(\cdot)$ should satisfy two requirements:

1. $f(\cdot)$ accepts an r_i with an arbitrary value and outputs $w_i \in [0, 1]$, and $f(\cdot)$ should be a nondecreasing function of r_i; that is, $w_i \geq w_j$ if $r_i \geq r_j (i, j = 0, \ldots, m; i \neq j)$. Also, $f[\max(r_i)] = 1$ and $\lim_{r_i \to -\infty} f(r_i) = 0$.

2. $f(\cdot)$ should ensure that enough weight is allocated to a sensing terminal that has a slightly negative reputation value. This requirement is necessary because, at the beginning of a WSPRT process, a "good" sensing terminal (i.e., a terminal that sends correct sensing reports most of the time and thus eventually obtains a high reputation value) may send incorrect sensing reports due to randomness (e.g., caused by temporary interference) and get a slightly negative reputation value.

Based on these requirements, we use the following function for w_i and $f(\cdot)$:

$$w_i = f(r_i) = \begin{cases} 0 & r_i \leq -g \\ \frac{r_i + g}{\max(r_i) + g} & r_i > -g. \end{cases}$$

The variable $g(>0)$ is used to meet the second requirement. In particular, w_i for a good sensing terminal will not be 0 for the first $(g - 1)$ reputation maintenance steps. For the gth reputation maintenance step, $P[r_i \leq -g] < 2^{-g}$. This probability is very small if g is assigned a relatively small number. (Note that g should be small enough to let the reputation scheme be sufficiently sensitive to incorrect sensing reports.) For example, when $g = 5$, the probability is less than 0.03125.

The following pseudocode describes the WSPRT data fusion algorithm:

```
 1: For all i, r_i = 0.
 2: For each spectrum sensing attempt made by N_0 {
 3:   i = 0, W_n = 1.
 4:   Get a spectrum sensing report u_i from N_i.
 5:   W_n ← W_n · (P[u_i|H_1]/P[u_i|H_0])^{f(r_i)}.
 6:   If η_0 < W_n < η_1, i ← (i + 1) mod (m + 1). Go to step 4.
 7:   If W_n ≥ η_1, accept H_1; that is, output u = 1. Go to step 9.
 8:   If W_n ≤ η_0, accept H_0; that is, output u = 0.
 9:   For each sampled u_i, set r_i ← r_i + (-1)^{u_i + u}.
10: }
```

15.3.3 Simulations

Simulation Environment

We carried out simulations to test and compare all the previously discussed data fusion schemes. In the simulations, N secondaries (secondary users) are randomly

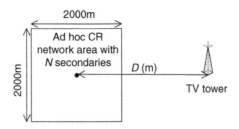

FIGURE 15.7

Simulated network environment.

located in a 2000 m × 2000 m square area, and they form an ad hoc CR network, each node with a transmission range of 250 m. Among the N secondaries are N_a SSDF attackers. We consider two types of SSDF attacks: always false and always free. An always-false attacker always sends sensing reports that are opposite to its local spectrum sensing results, while an always-free attacker always reports spectrum to be fallow. Each secondary moves according to the random waypoint mobility model [604] within the range of the network area. Each node moves with a maximum speed of 10 m/s and a maximum idle time of 120 sec. The incumbent, a TV tower with a duty cycle of 0.2, is located D meters away from the center of the CR network. See Figure 15.7 for the simulation network model.

For the path loss model we selected the HATA model [593], which is the model suggested by the 802.22 working group to model the 802.22 WRAN environment [605]. In our simulation, we assume that the incumbent transmitter's transmission frequency is 617 MHz and the antenna heights of the transmitter and the receiver are 100 m and 1 m, respectively. We assume that the transmitter's effective isotropic radiated power is 100 kW and the spectrum sensing technique used by the secondary users is energy detection. The signal detection threshold is −94 dBm—this is the minimum received power level needed to detect a signal. The mean noise power is −106 dBm. A secondary node acts as both a sensing terminal and a data collector. Each secondary node performs spectrum sensing periodically, at 30-sec intervals. Each simulation experiment lasts for 2 h.

We simulated and compared eight data fusion techniques. For decision fusion techniques, the three variants of AND, OR, and Majority rules are simulated. Since Bayesian detection and the Neyman-Pearson test are both fixed-number likelihood ratio tests (LRTs) with different thresholds, we simulate the two techniques together using the name *LRT* and three thresholds. The first threshold is calculated from the right side of Equation (15.1) by assuming perfect knowledge of P_0 and P_1; that is, $P_0 = 0.8$ and $P_1 = 0.2$. The costs are assigned as $C_{00} = C_{11} = 0$, $C_{10} = 1$, and $C_{01} = 10$, which were also the cost assignments used in [602]. With these values, the first threshold can be calculated as $\lambda_1 = 0.4$. Because determining accurate values for P_0 or P_1 may not be possible in practice, we also ran simulations with two other thresholds: $\lambda_2 = 4\lambda_1$, $\lambda_3 = \lambda_1/4$. We also evaluated SPRT and

WSPRT in the simulations; the associated parameters used in the simulations are $P_{01} = 10^{-5}$, $P_{10} = 10^{-6}$, and $g_i = 5$.

We considered two types of attacks: always false and always free. In always-false attacks, an adversary transmits sensing reports that are false; that is the transmitted reports contain information contrary to the actual local spectrum sensing measurements. In always-free attacks, an adversary always transmits reports that indicate that the spectrum band is fallow.

Simulation Results

Objectives

We are interested in three metrics: misdetection ratio, correct sensing ratio, and number of samples. The first metric has been discussed before. The correct sensing ratio is the number of correct final sensing decisions divided by the number of total sensing decisions. Because these two metrics and the false alarm ratio add up to 1, we can also derive the false alarm ratio from these two metrics. The number of samples refers to the average number of samples a secondary needs to collect from each neighbor to make a final decision, and it measures the overhead of a particular data fusion technique. For decision fusion and the fixed-number likelihood ratio test, the number of samples is always 1. The number of samples changes only for SPRT and WSPRT. Therefore, we study this metric only for SPRT and WSPRT.

Impact of Varying Attack Strength

In this set of simulations, we fix $N = 500$ and $D = 3000$, while changing attack types and varying N_a from 0 to 100 at an interval of 5. Figures 15.8 and 15.9 show the simulation results when we consider always-false and always-free attacks, respectively. In all cases, the decision fusion with an OR rule and an AND rule are not favorable, since they end up with almost always outputting either "occupied" or "fallow."

When always-false SSDF attacks are introduced, Figure 15.8 shows that the correct sensing ratio decreases for all the fusion rules except the OR and AND rules. Among these techniques, SPRT experiences the greatest decrease, which suggests that SPRT is the most vulnerable to always-false SSDF attacks. This vulnerability could be because SPRT may collect multiple reports from malicious secondaries, which amplifies the effect of the attack. In contrast, WSPRT is shown to be the most robust against always-false SSDF attacks. However, the improved correct sensing ratio comes at a cost—the number of required sensing samples has increased by a factor of 5 in the case of WSPRT.

In the set of simulations shown in Figure 15.9, always-free SSDF attacks were considered. The aim of this type of attack is to fool the data collector into thinking that there is no incumbent. Therefore, it can be expected that the attack will cause an increase in the misdetection ratio and a decrease in the false alarm ratio. Our results show that the decision fusion with a Majority rule is most vulnerable to always-free SSDF attacks. Note that a misdetection is considered more harmful

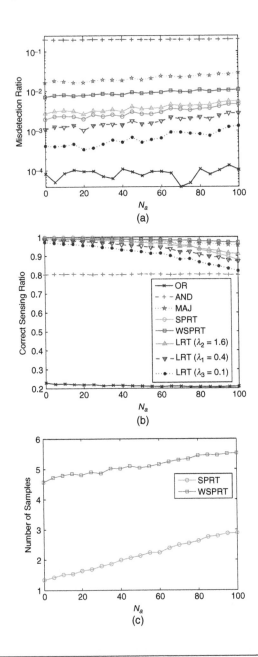

FIGURE 15.8

The performance of eight fusion techniques with different number of always-false SSDF attackers: (a) misdetection ratio, (b) correct sensing ratio, and (c) number of samples.

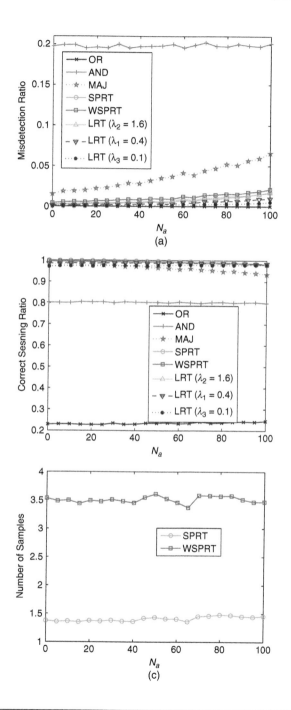

FIGURE 15.9

The performance of eight fusion techniques with different number of always-free SSDF attackers: (a) misdetection ratio, (b) correct sensing ratio, and (c) number of samples.

than a false alarm, since misdetection directly translates into increased interference experienced by incumbents.

15.4 SECURITY VULNERABILITIES IN IEEE 802.22

IEEE 802.22 [582] is the first wireless access standard based on cognitive radio technology. It specifies the air interface for a fixed wireless regional area network that operates in fallow TV broadcast bands. An 802.22 *cell* is a single-hop, point-to-multipoint wireless network composed of a base station and several consumer premise equipment. The BS manages the CPE within its cell and controls medium access via cognitive multiple-access control (CMAC). Throughout this subsection, *incumbent services* refers to TV broadcast services or services involving Part 74 devices.[4] On the other hand, secondary users of the TV bands (i.e., BSs and CPE) are referred to as *802.22 entities*.

The IEEE 802.22 standard mandates that CPE performs distributed spectrum sensing under the control of the BS. In this cooperative spectrum sensing approach, each CPE executes spectrum sensing on its own and sends its "local" spectrum sensing report to the BS, which then makes a final spectrum sensing decision. The presence of Part 74 devices is much more difficult to detect than TV broadcast transmitters, due to their low transmission power. To protect Part 74 communications, 802.22 prescribes two classes of solutions: class A and class B. In the class B solution, class B CPE is deployed to inform collocated 802.22 systems about the presence of Part 74 devices. Information gathered from regular CPE and class B CPE is used by the BS to identify fallow spectrum bands that are free of incumbent signals.

In IEEE 802.22, self-coexistence is an important problem and the standard prescribes several mechanisms for addressing it. There are two main technical challenges in self-coexistence: (1) minimizing the self-interference between over-lapping cells and (2) satisfying the quality of service of the cells' admitted service workloads in a dynamic spectrum access environment. The 802.22 CMAC addresses self-coexistence using the inter-BS dynamic resource sharing mechanisms.

IEEE 802.22 prescribes a *security sublayer* that applies cryptographic transforms to MAC data units. In this section, we describe security threats to 802.22, some of which are not addressed by the security sublayer. In particular, we describe how adversaries can exploit or undermine self-coexistence or incumbent coexistence mechanisms to degrade the performance of 802.22 WRANs and increase the likelihood of those networks interfering with incumbent networks. To provide the readers with sufficient technical background, we briefly discuss the 802.22 air interface and the security sublayer in the next two sections. For a full description of the IEEE 802.22 standard, we refer the reader to Chapter 14.

---

[4]Part 74 devices are wireless microphones, which are licensed to operate in the TV broadcast bands.

The 802.22 Air Interface

One of the most critical design requirements of the 802.22 air interface (i.e., PHY and CMAC layers) is adaptability, which is best embodied in its coexistence mechanisms. In this section, we give a brief overview of the various aspects of 802.22's air interface relevant to coexistence.

PHY Layer Support for Incumbent Coexistence

Spectrum sensing is one of the most important functionalities carried out by 802.22's air interface. The 802.22 entities perform spectrum sensing to identify fallow licensed bands free from incumbent signals. The standard describes a two-stage spectrum sensing approach: *fast sensing* and *fine sensing*. The fast sensing stage is executed before the fine sensing stage, and it typically uses a quick and simple detection technique such as *energy detection*. The measurements from the fast sensing stage are used to determine the need and the duration of the subsequent fine sensing stage. The accuracy of a sensing technique is dependent on various environmental factors, such as the signal-to-interference ratio (SIR).

The 802.22 standard employs a distributed spectrum sensing framework. A CPE is required to report its local spectrum sensing results to its BS (i.e., the BS that controls the CPE) via CMAC layer measurement messages. Using the local spectrum sensing results, the BS determines and adjusts various PHY layer parameters such as channel bandwidth and modulation/encoding rate.

Cognitive Medium-Access Control Layer

The MAC protocol data unit (MPDU) is the smallest unit of transmission/reception in the CMAC. It comprises the MAC header, the MAC payloads, and the CRC (cyclic redundancy checking) field. There are two types of MPDUs, distinguished by their respective MAC headers:

- **General MAC header.** This header is used for intracell general MPDUs. It is used in general MPDUs that contain either higher-layer data traffic or management messages in their payload.
- **Beacon MAC header.** This header is used for intercell beacons. An intercell beacon only carries beacon information elements (IEs) in its payload.

In IEEE 802.22, BSs and CPEs exchange intercell control messages using *intercell beacons*. Intercell beacons play a vital role in incumbent coexistence and self-coexistence mechanisms. Two types of intercell beacons are defined in the standard:

- **BS beacons.** These beacons are used to provide information about the BS's traffic schedule, the current operation channel of the cell, and the like.
- **CPE beacons.** These beacons are used to provide information about a CPE's current cell of attachment as well as information on the traffic flows between the CPE and its BS.

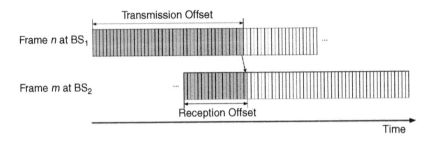

FIGURE 15.10

Synchronization of overlapping BSs.

Intercell Synchronization

To facilitate incumbent signal detection, a BS periodically schedules a quiet period (QP). IEEE 802.22 recommends that neighboring BSs, if possible, synchronize their QPs to improve the reliability of incumbent signal detection. During these QPs, all network traffic is suspended, and 802.22 entities sense the channel for incumbent signals.

Suppose two overlapping cells, with two base stations, BS_1 and BS_2, need to synchronize their transmissions. For every intercell beacon received from BS_1, BS_2 records the frame offset that indicates when it was received. Accuracy of this reception offset[5] is critical for successful synchronization. The transmission offset[6] is indicated in the beacon sent by BS_1. Figure 15.10 depicts the relationship between the transmission offset and the reception offset.

After receiving BS_1's beacon, BS_2 attempts to synchronize with BS_1 by sliding its frames using the following *convergence rule*:

- If $(FDC - O_{tx} + O_{rx} \leq \lceil FDC/2 \rceil)$, slide frames right by $(FDC - O_{tx} + O_{rx})$;
- Otherwise, slide frames left by $(O_{tx} - O_{rx})$,

where O_{tx} is the transmission offset, O_{rx} is the reception offset, and FDC is the frame duration code (i.e., time duration of a frame).

Inter-BS Dynamic Resource Sharing

Every cell requires a certain number of channels to satisfy the quality of service (QoS) of its admitted service workload. When the current channel condition is insufficient to support the required QoS of its workload, a BS in need of spectrum initiates an inter-BS dynamic resource sharing process, so that better channels or more channels can be acquired from neighboring cells. The 802.22 prescribes two

---

[5] The reception offset indicates the offset (in units of slot duration) relative to the start of the first slot of the frame where the beacon was received.

[6] The transmission offset indicates the offset (in units of slots) relative to the start of the first slot of the frame where the beacon is transmitted.

types of inter-BS dynamic resource sharing mechanisms: *nonexclusive* spectrum sharing and *exclusive* spectrum sharing.

After selecting a *target channel*, the BS in need of spectrum has to determine whether nonexclusive sharing of the selected channel is feasible using the following criterion: nonexclusive spectrum sharing is feasible as long as the maximum achievable SIR on the selected channel is higher than the required SIR threshold of the network's supported services. If nonexclusive sharing is feasible, the BS schedules data transmissions on the selected channel with appropriate transmission power control settings. Transmission power control is needed to minimize interference to cochannel neighboring 802.22 systems.

If the maximum achievable SIR on the selected channel is lower than the required SIR threshold, then the BS needs to acquire the spectrum resources through exclusive spectrum sharing. The 802.22 prescribes exclusive spectrum sharing via the *on-demand spectrum contention* protocol [583]. Figure 15.11 illustrates the ODSC algorithm. The BS that initiates the ODSC is called the *contention source*. The contention source randomly selects a *channel contention number* (CCN) that is uniformly distributed in the range $[0, W]$, where W is the contention window size. The CCN is used for determining the "winner" of each pairwise contention. After selecting the target channel, the contention source includes its CCN in a spectrum contention request that it broadcasts to its cochannel neighboring BSs (i.e., *contention destinations*). After receiving a spectrum contention request, a contention destination selects a CCN in the same manner as the contention source. Then the contention destination uses the following contention resolution rule to determine which BS wins this pairwise contention: The BS with a greater CCN value wins the pairwise contention. According to this contention resolution rule, the contention source's probability of winning a pairwise contention is $1/2$. The contention source wins the contended channel only if it wins *all* the pairwise contentions. If the contention source wins the contended channel, all contention destinations perform channel switching to vacate the target channel.

Protection of Part 74 Devices

Part 74 devices are much harder to detect than TV broadcast transmitters due to their significantly lower transmission power. The current 802.22.1 task group is considering options for the protection of Part 74 devices. Two classes of solutions, class A and class B, have been identified. In class A, a separate beacon device is deployed to transmit short wireless microphone beacon (WMB) messages to notify collocated 802.22 systems about the presence of cochannel wireless microphone operations. In class B, the 802.22 system supports a special type of CPE that has specific capabilities to inform collocated 802.22 systems about wireless microphone operations. The 802.22 draft standard states that a single approach is not the best solution.

In the class B solution, a class B CPE transmits WMBs to notify neighboring BSs about the scheduled wireless microphone operation during the QPs of the

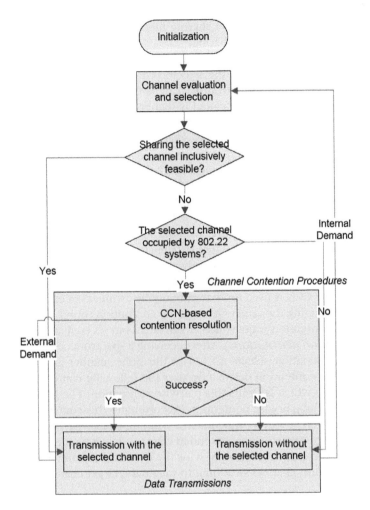

FIGURE 15.11

ODSC algorithm.

BSs. Upon receiving a WMB, the BS acknowledges the reception of the WMB by including a Part 74 acknowledgment in the BS beacons.

15.4.2 An Overview of the IEEE 802.22 Security Sublayer

The security sublayer defined in 802.22 provides confidentiality, authentication, and data integrity services by applying cryptographic transformations to MAC data units carried across connections between CPEs and the BS. The security sublayer has two components: an encapsulation protocol and a privacy key management

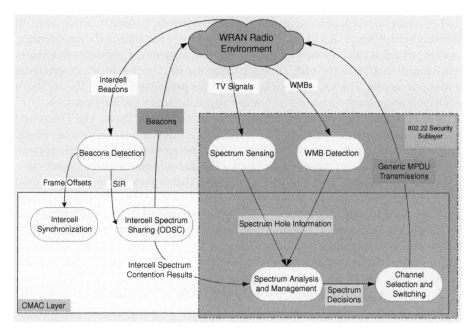

FIGURE 15.12

The 802.22 air interface's functionalities and the ones protected by the security sublayer.

(PKM) protocol. The encapsulation protocol defines a set of supported crypto-graphic suites (i.e., pairings of data encryption and authentication algorithms) and the rules for applying those algorithms to an MPDU payload. The PKM protocol ensures the secure distribution of keying material from the BS to the CPE.

The security sublayer protects network control information by attaching mes-sage authentication codes to CMAC management messages. However, the security sublayer protects only *intracell* CMAC management messages and not *intercell* beacons [606]. Therefore, intercell beacons are vulnerable to unauthorized modi-fication or forgery. Figure 15.12 illustrates the 802.22 air interface's functionalities and the ones protected by the security sublayer.

In an opportunistic spectrum sharing environment, it is necessary to ensure the availability of spectrum for the incumbent users as well as the secondary (WRAN) users. In the context of opportunistic spectrum sharing, a denial-of-service (DoS) attack involves the insertion of forged management messages by rogue terminals to create havoc for the spectrum sensing or spectrum allocation processes. The 802.22 security sublayer provides protection against this type of attack in two ways: (1) PKMv2 is used to provide mutual authentication between a BS and a CPE, thus preventing a rogue terminal from masquerading as a legitimate terminal, and (2) message authentication codes are used to protect the authenticity and integrity of critical management messages exchanged within an 802.22 cell.

The replay of captured messages is a common attack tactic used by adversaries. In an incumbent replay attack [607], an adversary captures and replays the local sensing reports (which is one of many types of intracell management messages defined in 802.22) sent by CPE to their BS. This may cause the BS to make incorrect spectrum sensing decisions. The security sublayer provides protection against the replay of intracell management messages by employing challenge/response protocols.

The IEEE 802.22 thwarts the replay of data packets by using AES (advanced encryption standard) in CCM mode (counterencryption mode with cipher-block-chaining message authentication code). CCM combines countermode encryption (for data confidentiality) and cipher block chaining message authentication code (for data authenticity). The IEEE 802.22 requires that a packet number is inserted into each MPDU. If AES in CCM mode is chosen to encrypt MPDUs, a window has to be used for packet number values to validate the freshness and uniqueness of the packet. A receiver validates the received data packets by verifying that the packets correctly decrypt under AES-CCM and have monotonically increasing packet numbers. The replay countermeasures of the 802.22 security sublayer are inherited from the 802.16e security sublayer.

The IEEE 802.22 prescribes two solutions (class A and class B) to detect the presence of Part 74 devices. In the class B solution, the standard prescribes the use of class B CPE to detect Part 74 signals. If Part 74 signals are detected, a class B CPE sends a WMB to collocated BSs in its vicinity. The 802.22 standard specifies that each class B CPE needs to possess preprogrammed security keys that enable the use of an authentication mechanism to prevent the forgery and modification of WMBs. The security sublayer protects WMBs from replay attacks in the same manner as it protects intracell management messages.

15.4.3 Security Vulnerabilities in Coexistence Mechanisms

One of the most significant security oversights in IEEE 802.22 is the lack of protection provided to intercell beacons. All intercell control messages are vulnerable to unauthorized modification, forgery, or replay. As noted previously, most of the security features of the 802.22 security sublayer are inherited from the 802.16e's security sublayer. Therefore, the 802.22's security sublayer does not take into account the important difference between 802.16e and 802.22; that is, the incumbent and self-coexistence mechanisms. Specifically, 802.22's security sublayer fails to protect the intercell beacons used to carry out coexistence mechanisms.

Next, we describe two attacks—one disrupts the intercell spectrum contention process and the other impedes intercell synchronization. In both attacks, the adversary forges intercell beacons to achieve its attack objective. We coin the term *beacon falsification* attack to refer to such an attack.

Exploiting Intercell Spectrum Contention

A terminal under the control of an adversary first selects the operation channel of a WRAN cell (i.e., "victim cell") as the target channel by eavesdropping on the

BS beacons transmitted by the victim cell's BS. Then the attacker's terminal sends spurious contention requests via forged intercell beacons to the victim cell. This triggers the victim cell to participate in an intercell spectrum contention process via the ODSC protocol. To increase the probability of winning the target channel, the malicious terminal may arbitrarily select *a very large CCN* value. If the victim cell loses the contention, then it vacates the current operation channel (i.e., target channel) and switches to another channel.

If the attacker initiates spectrum contention processes with high frequency and wins most of those contentions, then the victim cell is forced to waste a significant proportion of its network resources in switching channels. This would ultimately lead to significant degradation in network performance. Note that the capture and subsequent replay of contention requests with large CCN values by the adversary produce a similar effect.

Since all traffic activity is suspended during a cell's QP, the attacker's spurious intercell beacons have the best chance of being received by a victim cell during its QPs. This means that the attacker can increase the effectiveness of the attack by synchronizing its transmissions with the victim cell's QPs.

Interfering with Intercell Synchronization

The 802.22 standard states that overlapping cells should synchronize their QPs, when possible, to improve the accuracy of spectrum sensing. As noted previously, no security mechanism is prescribed by 802.22 that protects the intercell beacons against forgery or unauthorized modification. Since neighboring cells coordinate synchronization by exchanging intercell beacons, the intercell synchronization process is just as vulnerable to BF attacks as the intercell spectrum contention process. It is possible for an adversary to insert *false frame offsets* in intercell beacons. Suppose an adversary inserts spurious transmission offset values in intercell beacons and transmits those beacons to two cells that are adjacent to each other. The false information contained in the spurious beacons may cause the two cells to calculate incorrect frame sliding lengths, thus leading to imprecise synchronization of the two cells. In turn, the impreciseness of synchronization leads to increased inaccuracy in spectrum sensing. If the QPs of the two cells are not synchronized, secondary users of those cells need to detect incumbent signals in the midst of secondary signals, which makes spectrum sensing more difficult and may require the use of costly detection techniques, such as cyclostationary feature detection.

15.5 SECURITY THREATS TO THE RADIO SOFTWARE

The programmability of SDR and CR devices is made possible by the signal processing capabilities of the radio software. However, the programmability of such devices raises serious security concerns. Like any desktop software, software for SDR/CR devices are vulnerable to various security threats. Without proper security measures to protect the radio software, the trustworthiness of an entire CR network can be compromised by the insecurity of the software. To clearly grasp the technical challenges in the design and implementation of security measures for radio software, we need to first discuss the unique properties of such software.

Radio software for a CR has unique properties that distinguish it from conventional software. Because of the intrinsic operating characteristics of CRs, software running on them is likely to have the following attributes:

- **Software Portability and Generic Datapath.** The design methodology of existing software-defined radio architectures (e.g., Vanu Radio [608], GNU Radio [609], and Software Communications Architecture (SCA) [234]) suggest that software portability will likely be a primary requirement that drives the hardware and software architectures of most SDR/CR systems. A design approach that aims to maximize software portability leans toward the use of a generic datapath rather than a specialized one customized for a specific application. Many in the SDR industry predict that the importance of portability will lead to the adoption of generic datapaths in most software radios. The importance of portability suggests that security schemes that require support from specific hardware, operating system, or other parts of the computing system may have limited utility.
- **Modular Software Architecture.** Most, if not all, SDR/CR systems adopt a software framework for supporting a distributed, object-oriented, and software-programmable radio architecture. This is true for SCA, GNU Radio, and Vanu Radio. Such an architecture naturally promotes the use of modular software. The modular architecture of radio software needs to be considered when designing a security scheme to protect such software.
- **Real-Time Requirements.** Radio systems have very stringent real-time requirements. One way of satisfying these requirements is to tightly control the timing of software execution. However, this approach limits the radio software to platforms that have very predictable execution rates, and consequently this requirement reduces software portability. To overcome this problem, some existing SDR systems relax the requirement of a predictable execution rate by significantly increasing the buffering carried out at the system input and output. In such an approach, real-time requirements are satisfied as long as the processor is fast enough for the signal processing code's average processing rate to be faster than the input/output rate. Satisfying real-time requirements while maintaining portability and platform flexibility is a challenging task. Mechanisms for protecting radio software (e.g., tamper-resistance schemes) may incur a significant performance penalty in terms of increased runtime overhead, increased memory footprint, and increased executable file size. These penalties complicate the incorporation of security schemes into SDRs and CRs.

Without proper software protection mechanisms in place, CRs are vulnerable to a host of security threats targeting the radio software. These threats include:

- Execution of malicious code: illegal modification of radio software components, resulting in unauthorized changes in signal or protocol parameters.
- Removal of software-based authentication or access control functions.
- Disruption of radio software reconfiguration or obstruction of software management and version control.
- Reverse engineering or pirating of downloaded software components, resulting in intellectual property loss.

Most of these threats can be attributed to one of the following security problems: security vulnerabilities in the software download process, lack of protection against Internet protocol (IP) theft and software pirating, and vulnerability to unauthorized software tampering. Although the first two problems have attracted attention from the research community (e.g., [610, 611]), there is little existing research on the third problem in the context of CR software. The threat posed by the third problem is especially serious because adversaries may attempt to manipulate radio software to gain operational advantages (e.g., transmit at power higher than the authorized limit) or launch attacks against primary-user networks. The prospect of malicious users making unauthorized changes to a CR's operating characteristics is a major concern for regulators and developers.

In recent years, a number of technical approaches have been proposed to protect software IP and thwart exploitation of software vulnerability. These approaches include software watermarking, software diversity, software obfuscation, and tamper resistance. Among these approaches, obfuscation and tamper resistance are appropriate for combating unauthorized software tampering. The technique of obfuscation transforms a program into a functionally equivalent program that is more difficult to understand, thus thwarting reverse engineering. On the other hand, a tamper-resistance scheme detects or prevents integrity violations of the original software. Typically, adversaries carry out reverse engineering as a first step prior to modifying the software since modification requires at least a partial understanding of the target software. Therefore, obfuscation and tamper-resistance techniques are often used together to thwart software tampering. As mentioned previously, when designing and implementing security schemes for protecting radio software, we need to carefully consider the characteristic attributes of the radio software itself.

To ensure the security and reliability of CR software, the implementation of the aforementioned security solutions is not enough. Along with the security mechanisms, potential vulnerabilities within the radio software itself need to be identified and fixed through systematic testing and verification.

15.6 PROBLEMS

1 Based on the discussions in this chapter, describe three security threats unique to CR networks. Also, give an example of a threat common to both conventional wireless networks and CR networks.

2   What two major technical challenges need to be addressed by the noninter-active PST localization scheme to detect PUE attacks? How did the authors of [596] address those challenges?

3   In WSPRT, the value of g should be small enough to let the reputation scheme be sufficiently sensitive to incorrect sensing reports made by malfunctioning sensing terminals, such as $g = 5$. Can we expect a negative impact on the performance of properly functioning sensing terminals if we assign an even smaller value to g, say, $g = 1$?

4   In the BF attack against 802.22 coexistence mechanisms, the objective of the adversary is to maximize its chance of winning the target channel in a spectrum contention process while not arousing the suspicion of the contention desti-nations. One way of achieving this objective is for the adversary to employ the following strategy. The adversary selects a CCN that is uniformly distributed in the range $[W/z, W]$, where $z \geq 1$ is an adjustable parameter and inserts this value in an intercell beacon. Then the adversary emits this forged beacon to neighboring BSs during their QPs. Let α_a denote the attacker's CCN, and α_d denote a contention destination BS's CCN that is uniformly distributed in the range $[0, W]$.
    (a)   Derive the probability that the adversary wins a pairwise contention from one contention destination.
    (b)   Derive the probability that the adversary will win the target channel from k contention destinations.

5   Explain why obfuscation and tamper-resistance techniques should be used together to thwart software tampering.

6   In Section 15.1.1 we discuss the difference between the problem of DSS Byzan-tine failures and the problem of resilient/secure data aggregation. The two problems are different in terms of how *resilience* (or security) is interpreted in the context of each problem. Explain this difference.

Public safety and cognitive radio

16

Marnix Heskamp, Roel Schiphorst, and Kees Slump
University of Twente, The Netherlands

16.1 INTRODUCTION

In public safety applications, robustness and rock-solid technology are paramount. The communication system of rescue workers should always work, even under extreme conditions. So, at a first glance, it may seem that an emergency situation is not something in which one wants to experiment with such a new technology as cognitive radio. There are, however, a number of arguments that justify the application of cognitive radio in this field. First, current-day communication systems used for public safety lack support for multimedia applications that nowadays come standard with low-budget mass-market cell phones. One of the reasons behind this is that it is not sensible from an economic perspective to permanently reserve the large bandwidths required for such applications. Cognitive radio, however, is able to acquire this spectrum on-the-fly only when it is needed. Furthermore, backwards compatibility is very important in the field of professional radio systems because of the large investments in a relatively small market. Therefore, many incompatible standards and new broadband services are main drivers for investigating how cognitive radio can be applied in this field.

One of the most difficult aspects of cognitive radio, in general, is that it is not allowed to cause interference to any primary user. This problem also applies to the public safety field but to a lesser extent. When a large-scale emergency occurs, the importance of the situation makes the rescue workers, at least in principle, primary users of the spectrum. The only primary users that must be carefully sensed for are the legacy systems that may play a role in the rescue operation. Because the number of legacy systems is limited and their standards and frequency bands are well known, most issues can be dealt with by careful engineering of the spectrum sensing system.

This chapter addresses the application of cognitive radio to the field of public safety, disaster relief, and rescue scenarios. First, we give a description of the requirements for public safety communication networks, and explain how cognitive radio technology can be useful. Next we look at the current systems. We focus on the TETRA standard, and its implementation in the C2000 system in The Netherlands. Next we look at a real-world scenario and discuss the main system design aspects, such as spectrum organization, availability of white space, and suitability of certain frequency bands for cognitive radio.

16.1.1 **Requirements**

The next-generation communication system for public safety will have very extensive requirements. These requirements are studied and specified by commissions such as SAFECOM [612] in the United States and Project MESA [613] in Europe. Most of these requirements deal with the higher layers in the network protocol stack. In this chapter we restrict ourselves to the physical layer.

Communication Structure

A public safety wireless network consists of a backbone network, base stations, and handsets. The backbone network is used for inter–base station communication. In case the entire public safety network is down, emergency workers should still have the option for direct communication. In a rescue situation, the network consists of different types of network nodes, such as emergency workers, vehicles, helicopters, airplanes, and robots. Each type of node has different physical layer requirements. For instance, emergency workers carry battery-powered handsets that are energy limited. Helicopters and airplanes on the other hand have very high-speeds, causing Doppler spreading of the signal.

Public safety workers are usually embedded in talk groups. The layout of a talk group, or "fleet map," is determined mainly by the situation at hand and is coordinated by the central control room. Typical talk groups may be a neighborhood police team, a team of parking attendants, or a special group for low-priority conversation. Moreover, the central control room acts like an info/help desk for the rescue workers. If a police officer wants more information about a building, he or she can ask this to the central control room, because they are members of the talk group. If the situation changes, the central control room can dynamically change the members of the talk group.

In a critical situation, rescue workers lack the time to dial a number before they can communicate. They just press the push-to-talk button. A typical system will set up a call within 350 ms. Voice calls are often monitored by support personnel, who are not actively participating in the call but for whom the information exchanged, the identity of the speakers, their emotional state, and other audio indicators are key to their decisions in support of critical operations [614]. For instance, if there is a hostage situation at a school, the police unit that established a perimeter around the school must be able to passively "listen in" on conversations of the SWAT (special weapons and tactics) team.

Reliability

For emergency networks, reliability is an important issue. There are two kinds of reliability: robustness and security.

Robustness is the ability of a system to avoid total failure despite unforeseen conditions or partial damage. A public safety communication system should always be available, especially during large disasters. The network should have guaranteed coverage in the whole service area, including special coverage locations like tunnels. Handsets should have a rugged design and be capable of working under extreme conditions. All base stations have power supply backup batteries, so that the system remains operational for about 4 to 5 h during a power failure. Directly after a power failure is detected, the maintenance crew installs a portable generator. Also if a base station is destroyed, it is very important that the system still has coverage in the service area of the destroyed base station. The reason for this is that most communication of the rescue workers is required in this disaster area. Coverage in this area can be obtained by mobile base stations (on a truck) and by overdesigning the network, so other neighboring base stations can handle the communication needs. In Section 16.2.1 we describe that cellular networks are interference limited. If the signal-to-noise ratio (SNR) requirements are increased at the border of a cell, neighboring base stations also have coverage in the service area of the destroyed base station. A third solution is that handsets should be able to communicate directly with other handsets (without a base station). See also Section 16.2.1. The backbone of the network, i.e., the communication between the base stations, usually consists of optical fiber or microwave radio links. This backbone network should be very robust against failure. Robustness can be obtained by having at least two independent backbone connections to each base station.

Security is the ability of a system to withstand malicious attacks. The communication should be secure against eavesdropping, spoofing, and jamming. It should also be able to block lost or stolen handsets from using the network. In addition, handsets should not contain information that can help unauthorized users access the network.

Broadband

In an emergency situation a picture could say more than a thousand words. Video is even more powerful in providing a clear impression of a complicated situation. So, although voice will always be the primary mode of communication during an emergency, there is a huge demand for multimedia.

From the control room, maps and floor plans of buildings may be uploaded to the device of a rescue worker. Furthermore, the next-generation public safety communication equipment will provide advanced features, like sensors for biomedical and environmental signals.

Paging

In paging communication short, predetermined text messages are sent to mobile devices that are very important for public safety applications. Examples are emergency notifications (e.g., Evacuate Now!), arrival on a predetermined position, and sensor readings. Paging is even more important than voice communication. It is used, for instance, to alarm firefighters when there is a fire alarm. The advantage of such predefined messages is that they convey a lot of meaning in very few data bits.

16.1.2 Commercial Wireless Communication Networks

Commercial wireless communication networks like GSM and 3G networks could potentially provide most of the functionality described in the previous section. However, if a disaster occurs, these networks have two important drawbacks. First, the network gets overloaded, as each person in the vicinity tries to communicate with his or her friends and family. As a result the communication network may collapse. The second reason is that, when a disaster occurs, a part of the infrastructure may be damaged. For instance, the power may be unavailable in the area of the disaster. For economic reasons, commercial networks have no backup for the power supply and hence the affected network is down. In addition, commercial networks lack coverage in rural areas where there are very few customers. Moreover, they also lack coverage in special coverage locations like tunnels and metro stations.

16.1.3 Economic Value of the Spectrum

Spectrum is scarce. However, it is very difficult to estimate the value of the spectrum, because it depends on many parameters, like population and allowed transmit power. Take, for example, the first UMTS auction in The Netherlands in 2000. The area of The Netherlands is about 41.500 km^2 and its population is 16.5 million people. In total there was 155 MHz of spectrum, which was given for 15 years to five operators for 2.7 billion euro; that is, 370 million per MHz [615, 616]. This is only the cost of the license, without the cost for construction of the network and other operational costs.

Spectral Efficiency

With cognitive radio one of the goals is to use the spectrum more efficiently. To get insight in the economics of spectrum usage, a good starting point is the Shannon-Hartley theorem for the capacity of an additive white Gaussian noise (AWGN) channel,

$$C = B \cdot \log_2(1 + \text{SNR}) \quad [\text{b/s}], \tag{16.1}$$

in which C is the channel capacity in bits per second, B the channel bandwidth in hertz,

$$\mathrm{SNR} = \frac{P_{\mathrm{rx}}}{N_0 B} \tag{16.2}$$

is the signal-to-noise ratio, with P_{rx} the received signal power in watts, and N_0 is the power spectral density of the white noise in watts per hertz. The noise in a radio link is caused by various sources. The most fundamental source of noise is thermal noise, which originates from all warm matter with a power of

$$S(f) = N_0 = kT \quad [\mathrm{W/Hz}], \tag{16.3}$$

in which $k \approx 1.3806504 \cdot 10^{-23}$ J/K is the Boltzmann constant. For a temperature of 15.2°C this gives a noise floor of about -174 dBm/Hz.

Based on this Shannon limit, we can identify two "regimes" of operation of a communication link (see Figure 16.1). First, if the SNR is good (≈ 10 dB and more), we have

$$C \approx B \cdot \log_2(\mathrm{SNR}) \approx B \cdot \frac{1}{3} \cdot \mathrm{SNR(dB)}. \tag{16.4}$$

In this regime the capacity depends linearly on the bandwidth but only logarithmically on the transmit power. Therefore, this regime is called *bandwidth limited* or *spectrally efficient*, since most capacity improvement can be gained by increasing the bandwidth. If we choose to divide the bandwidth by a factor N and maintain capacity by increasing the transmit power, we see that the required power grows

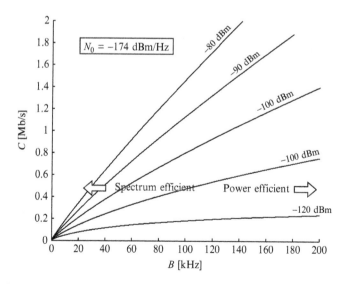

FIGURE 16.1

Channel capacity as function of channel bandwidth for various received signal power levels.

exponentially with N, as

$$B \cdot \log_2(\text{SNR}) = \left(\frac{1}{N}B\right) \cdot \log_2(\text{SNR}^N), \tag{16.5}$$

so the required transmit power for the same capacity is increased by a factor $\text{SNR}^{(N-1)}$. For a low, or even negative, SNR we have

$$C \approx 1.44 \cdot \frac{P_{\text{rx}}}{N_0}. \tag{16.6}$$

In such a "power-limited" regime, increasing the bandwidth has almost no effect, but increasing the received signal power gives a proportionally higher capacity.

16.1.4 Benefits of Cognitive Radio

The general meaning of a cognitive radio is a smart device that does all kinds of useful things for its owner, based on sensory input and machine learning. In a more specific meaning, it is a radio that can opportunistically use white space in licensed bands without causing interference. Due to the special requirements of public safety networks, there are several benefits of cognitive radio technology.

Improved Communication Structure

- **Communication with other networks.** Currently, there exist multiple public safety standards. So, when a large disaster occurs at the border of countries, those countries face a huge challenge if they use different technologies. Also, in emergency situations where military force is required, the same problem occurs. A cognitive radio with support for military standards and other public safety standards would solve this problem.
- **Backwards compatibility.** Because of the large investments and relatively small market, legacy systems are replaced slowly and coexist with new communication networks for a long time. Also, over time, new revisions of standards are released. Cognitive radio allows an upgrade of the existing equipment to this new release without replacing the hardware.
- **Introduction of new services.** New services could be enabled more easily by cognitive radio, as it can adjust its parameters according to the requirements of the new service. It does not have limitations set by existing standards. An example of this is that current-day public safety communication systems lack support for multimedia content that nowadays is normal for mass-market cell phones.

Improved Reliability

A cognitive radio always tries to minimize interference to other networks by changing its frequency if other signals are present. This feature automatically makes a cognitive radio more resilient to jamming.

Enabling Broadband

In case of an emergency, public safety networks are heavily used, and there is demand for more capacity. Implementing this capacity beforehand in the whole network would be very costly. A different approach would be to use cognitive radio to sense for empty frequency bands (white space) and use it as a secondary user to set up an auxiliary communication network. The relatively large bandwidths required for broadband communication could be provided by secondary spectrum usage.

16.2 STANDARDS FOR PUBLIC SAFETY COMMUNICATION

Several communication standards have been developed for public safety applications. The first generation of standards was based on analog modulation (FM/AM) and used an optional speech scrambler to prevent eavesdropping. By the end of the last century, second-generation standards emerged based on digital modulation and trunking. These are now used in most countries. There are three second-generation public safety communication standards: P25 (APCO Project 25), TETRAPOL, and TETRA. APCO Project 25 systems are used by the federal, state/province, and local public safety agencies in North America. An important feature is to be backwards compliant with the old analog APCO Project 16 standard. This means that P25 handsets have support for an analog and digital communication mode. More information about P25 can be found in [617]. TETRAPOL [618] was one of the first digital public safety standards, developed in France. By 1988 it was used by the French Gendarmerie Nationale. The terrestrial trunked mobile radio (TETRA) [619] communication network was developed almost ten years later and is now used in most European and other countries.

16.2.1 TETRA

TETRA was formerly known as *trans-European trunked radio* and standardized by ETSI in 1995. TETRA was specifically designed for use by government agencies, emergency services (police forces, fire departments, ambulance), rail transportation staff, transport services, and the military.

Handhelds can communicate in direct mode operation (DMO), in which they work similar to a walkie-talkie, or in trunked radio mode operation (TMO), in which the TETRA base station infrastructure is used. The DMO allows direct communications in situations where network coverage has been lost. In addition to voice services, the TETRA system supports several types of data communication: status messages, short data services, packet data, and circuit-switched data communication.

TETRA uses time division multiple access (TDMA) with four user channels on one radio carrier and 25 kHz spacing between carriers. Both point-to-point and

Table 16.1 Air Interface Parameters of TETRA

Parameter	Value
Frequency range	150 to 900 MHz
Channel bandwidth	25 kHz
Modulation	$\pi/4$-QPSK
Bits per symbol	2
Transmit filter	Root Nyquist with 0.35 roll-off
Symbol rate	18,000 symbols/s
Raw data rate	36 kb/s
Spectral efficiency	1.44 b/s/Hz
Time slot duration	85/6 = 14.167 ms
TDMA frame	4 time slots = 56.656 ms
Bit rate per channel	36 kb/s per 4 slots = 9 kb/s
Voice codec	ACELP
Handheld rx sensitivity	−104 dBm
Vehicle rx sensitivity	−107 dBm
Mobile tx power classes	1 W, 3 W, 10 W, 30 W

point-to-multipoint transfer can be used. Digital data transmission is also included in the standard though at a low data rate. All voice and data traffic is protected by encryption, so that it is practically impossible to eavesdrop or spoof the communication. Table 16.1 summarizes the air interface parameters, which can be found in [620].

Trunking

For no call to be lost on a communication system even in a worst-case scenario requires an enormous amount of capacity. Moreover, most of this capacity remains unused most of the time. Because such an approach is highly uneconomical, it is almost never used anymore. Instead, practical systems are designed for an average load with some margin for peak loads. The average load, or *traffic intensity*, of a system is defined by

$$A = \frac{\lambda}{\mu} \quad \text{[Erlang]}, \tag{16.7}$$

in which λ calls/sec is the average *arrival rate* of calls, usually measured during a busy period, and μ calls/sec is the average *service rate*. The unit of average load is named after A. K. Erlang, the originator of traffic engineering and queueing theory.

Note that in Equation (16.7) one can recognize Little's theorem, in which μ^{-1} is the average call duration time, or *holding time*. Based on the average load A, the number of channels N has to be chosen. Although in reality λ and μ are random processes that depend on time and other external factors, it is often convenient to regard them as constants. Furthermore, it is convenient to assume that the number of calls per second is Poisson distributed and the call duration is exponentially distributed. In other words, there is an infinite number of potential callers, which have no memory and no interdependencies, which make random calls of random duration.

In a trunking system, all channels are kept in a *pool*, from which they are dynamically assigned to users. A user can request a channel via a *control channel*. After the call is finished the channel is placed back in the pool. On average, enough free channels are available in the pool to handle all incoming requests. However, since the tail of the Poisson distribution extends to infinity, there is always a chance that a large numbers of calls are made simultaneously, so that the pool is exhausted. When that happens there are a number of scenarios. The simplest of them is that the call is just dropped and lost forever. The chance that this happens is given by the Erlang-B equation:

$$\text{Pr\{blocking\}} = B(A, N) = \frac{A^N/N!}{\left(\sum_{n=0}^{N} A^n/n!\right)}. \tag{16.8}$$

Spatial Reuse

Modern mobile radio communication networks are usually based on a cellular architecture. In a cellular system a large geographical area is partitioned into cells. Each cell has its own base station that works with a specific set of frequency channels.

Conceptually it is convenient to model the cell shape as a hexagon because of its nice geometrical properties. If R is the radius of a hexagon then its width is $W = \sqrt{3}R$ and its area is $A = 1.5\sqrt{3}R^2$. Cells are grouped in clusters of size K, and each cell in the cluster is given a different set of frequencies in such a way that neighboring cells have different frequencies. From the geometrical properties of the hexagonal cell shape, it is straightforward to derive that the number of cells in a cluster must be a number $K \in \{1, 3, 4, 7, 9, 12, 13, ...\}$ for which $K = i^2 + j^2 + ij$ holds with i and j positive integers including 0. Furthermore, the distance between two clusters can be shown to be $D = \sqrt{3KR}$.

In a cellular system, the dominant noise contribution in the downlink is not from thermal noise but from interference from neighboring cells that use the same frequency. This interference is called *cochannel interference* (CCI). The system is called *interference limited* rather than *noise limited*. The worst-case CCI occurs if the mobile user is in a corner of a cell, because then it is at the largest distance from its base station. The mobile will receive most of its interference from the six nearest surrounding cells that work on the same frequency. The interference from these *first-tier* cells gives

$$\text{SNR} = \frac{P_{\text{tx}} \cdot (R)^{-\gamma}}{\sum\limits_{n=1}^{6} P_{\text{tx}} \cdot (d_n)^{-\gamma}} = \left[\sum_{n=1}^{6} \left(\frac{R}{d_n}\right)^{\gamma}\right]^{-1}, \qquad (16.9)$$

in which γ is the path loss exponent and d_n the exact distance from the corner of a cell to the nth interfering base station. If we approximate d_n by D this simplifies to

$$\text{SNR} \approx \frac{1}{6}(3K)^{\gamma/2}. \qquad (16.10)$$

From this we get the somewhat counterintuitive result that the SNR, and thus the capacity, is independent of the transmitted power and cell size, and it becomes better with a higher path loss. Furthermore, increasing the number of cells in a cluster increases the SNR but decreases the total system capacity. This is because each cell needs a unique channel, which cannot be used by the other cells in the cluster. However, from Equation (16.5), we saw that, in a bandwidth-limited system, we better give all the cells more bandwidth, at the cost of a lower SNR.

16.2.2 C2000

TETRA is only a standard, and manufacturers make only generic base stations and handsets that implement the standard. To build a complete operational network is still a huge effort. The specification of a communication system for public safety depends on the exact situation in a country, and therefore there are no off-the-shelf solutions. The public safety communication network in The Netherlands is called C2000. It consists of three components:

- **T2000.** A TETRA-based network for voice and low-rate data communication. It uses the frequency band 380–385 MHz for uplink and 390–395 MHz for downlink communication. The network uses around 400 masts and is designed for at least 95% outdoor coverage. It uses both direct mode and trucked radio mode. For special coverage locations, like tunnels and stadiums, the system has additional low-power base stations. To facilitate helicopters and airplanes, a separate overlay network, TETRA AGA (air-ground-air services) was constructed. In this overlay network the distance of the base stations is increased to 83 km.
- **P2000.** Paging is a very important communication application in public safety, where short predetermined texts are transmitted and displayed on pager devices. For instance, they are used to alarm firefighters. C2000 uses a different network for paging. This network, P2000, is based on the FLEX protocol in the 169.650 MHz band. TETRA has also support for paging, but the current outdoor coverage is too low for this application. In fact, no message should

be lost in the system. So, it was more cost effective to construct a separate pager network than increasing the coverage of the T2000 network.

■ **M2000.** M2000 is a software system used in the *public safety answering point* (PSAP). A PSAP is a call center responsible for answering calls to an emergency telephone number for police, firefighting, and ambulance services. M2000 is a software system that helps the employees in the PSAP identify the scale of the emergency and which resources should be allocated to the emergency. Moreover, it facilitates the communication among the different PSAPs. Another important task of the PSAP is to act as an info/help desk in talk groups. So, for instance, a policeman can ask the PSAP for more details about a registration plate. Also, the PSAP can actively monitor the talk groups. The M2000 system facilitates these tasks. Finally, it is also used for network management and network planning. Once a network has been built, it is not finished. New residential areas are developed, new high-rise buildings are constructed that block the radio signal, and so forth. So, every year, several base stations have to be moved or added to the network.

The C2000 network fulfills all important public safety requirements but lacks support for multimedia/broadband Internet communication. In the frequency band in which C2000 operates, the inflexible and fragmented licensing scheme made it difficult to find more static allocated spectrum. On the other hand, a huge amount of this spectrum always is temporarily unused. But traditional radio equipment lacks the technology to find this free spectrum, and there is not yet a legal system that allows ad hoc secondary usage.

16.3 APPLICATION OF COGNITIVE RADIO

In the previous section we looked at public safety communication as it is done today. The requirements for the next-generation system include features that require broadband communication. Cognitive radio may in the future provide a means to find the required bandwidth. In this section we give an example of how a cognitive radio could fit in with existing systems.

16.3.1 The Firework Disaster in The Netherlands

What does a typical disaster look like? This question is difficult to answer, because each disaster is unique and unforeseen. But we can learn from examples from the past. On May 13, 2000, a large disaster took place in the city of Enschede, The Netherlands.[1] On a nice warm Saturday afternoon, a crowd gathered to witness a small fire in their neighborhood, at what they thought was a paper recycling depot. Only the firemen and police knew that actually a fireworks storage was burning.

---

[1] Exact GPS location: 52° 13′ 49″ N, 6° 53′ 41″ E.

What they didn't know was that the safety regulations were violated, and much more and much heavier fireworks were stored than was allowed. After about 20 min the first big explosion occured, followed shortly by a detonation that ruined the whole area. The result was that an area of 1 km^2 was destroyed (400 houses), 23 people were killed, and about 950 injured [621].

Even before the explosions, the central control post and PSAP were flooded with calls. Soon after the final explosion all communication systems collapsed because everyone started calling. Most of the calls over the GSM network involved people notifying their friends and family. Many calls over the public safety network were made by police officers who offered their help after they heard the explosion. However, the first responders that survived the explosion could not reach the control center, because the network was overloaded. Even hours after the explosion, communication remained difficult. For example, much time was wasted because experts could not get reliable information about the risks of explosion of an ammonia cylinder in the refrigeration system of the nearby Grolsch brewery, which also caught fire.

From May 13 until May 25 in total 141 ambulances and 15 helicopters (680 rescue workers), 7980 police officers, and 1675 firefighters were active [621] in the area of the disaster. From these numbers, we estimate that around 2500 rescue workers were active during May 13 in an area of several square kilometers. For normal tasks like house fires or surveillance, only up to 25 rescue workers would be working in this area. So, a large disaster increases the regional/local demand for a public safety communication network by a factor of approximately 100.

16.3.2 Bandwidth Requirements

In this section, we derive the spectral requirements in the case that a cognitive radio system would have been used during this disaster. In the previous section, we estimated that at maximum 2500 rescue workers were present at the disaster location. The primary disaster region was 1 km^2, but rescue workers are of course active in a larger area, so the extended disaster region would be 10 km^2. In this area, we estimate that 25 video streams of 256 kb/s each (MPEG-4 streams) should be uploaded to the central command. Moreover, 25 photos of 1 Mb each are taken per second and should also be uploaded. In total, this results in a gross 31.4 Mb/s stream to the central command.

The cognitive radio network consists of vehicles and rescue workers. We assume that the vehicles can act as relay stations for the rescue workers to communicate with the central command. So, this is similar to normal 2G and 3G networks: The vehicles are base stations and the rescue workers are modeled as mobile terminals. Between the vehicles is a high-bandwidth backbone network for which cognitive radio also may be used.

The noise-limited communication occurs from rescue worker to vehicle and is therefore the most challenging task. For this type of communication, a frequency band has to be selected that has both good propagation conditions and small antennas suited to be mounted on handheld terminals. Frequency bands from 400 MHz to 1 GHz are appropriate bands. On the other hand, communication from the vehicle to central command is not limited by power and therefore higher-frequency bands can be selected, which are less optimal.

So, for rescue worker–to-vehicle communication, we have to derive how much spectrum is required. We assume that the average distance between the rescue worker and the vehicle is 100 m, a spectral efficiency of 1.3 b/s/Hz and an overhead of 50% (encryption, protocol overhead, etc.). Finally, we assume that 70% of the rescue workers can communicate directly with the vehicle and 30% require another rescue worker as relay. The core of the fireworks disaster is the 1 km^2 area, and in the other 9 km^2 surroundings, fewer rescue workers are active. So, the frequency demands are highest in the core of the disaster. If 50% of the 2500 workers are in this area, a total of $31.4 \times 0.5 \times 1.5 \times 1.3 = 30.6$ MHz is needed. If the network can distinguish between primary users and its own network,[2] frequencies can be reused more efficiently. In Section 16.2.1 we introduced a cluster radius D, which determines the number of times a frequency can be used in a square kilometer. In our case the cluster[3] area is 0.165 km^2. So, frequencies can be reused six times in a square kilometer. This makes our spectral requirement in the core of the disaster 5 MHz. So, for this scenario a cognitive radio network has to find 5 MHz of empty space.

16.3.3 Spectrum Organization

In this subsection, we propose to use the cognitive radio as an add-on system to the existing TETRA/C2000 system. The C2000 system uses frequencies around 400 MHz. For easy integration of the cognitive radio add-on, it is beneficial to use frequencies near 400 MHz. In this section we review frequencies from 100 to 800 MHz.

What kind of neighbors and primary users can we expect? Between the FM radio and the UHF television broadcasting band lies a region of the spectrum that is ideal for digital land mobile communication. Below the FM band, it becomes difficult to find a reasonable amount of bandwidth, and above the TV band, indoor coverage becomes difficult unless a lot of base stations are used. The FM band lies worldwide between 88 and 108 MHz, and the television band lies in Europe between 470 and 846 MHz. In the United States, the actual UHF television band starts at 512 MHz, but the numbering of the channels also starts at 470 MHz.

Figure 16.2 shows a simplified overview of the band plan (allocation chart) in The Netherlands between the FM and TV band. Directly above the FM band one finds the aviation band, which in utilization measurements appears much

---

[2]This is possible as each vehicle knows which frequencies are used and the backbone network between vehicles can be used to distribute this information.

[3]We assume in our example that the number of cells in a clusters is seven, which is also used by GSM networks.

FIGURE 16.2

A simplified spectrum band plan of The Netherlands between the FM and TV broadcasting bands.

quieter than the FM band. However, since the signals in this band are used to guide airplanes, the risk of interference is much too large, so that it is unsuitable for cognitive radio and every other type of spectrum sharing.

The band between 137 and 174 MHz is typically used by private mobile radio (PMR) networks used by private security personnel, public transportation, taxies, and the like. Typical equipment in use in this band are analog FM radios that use a 12.5 kHz raster without trunking. Because of this old-fashioned technology, there is a lot of white space in this band, which could be harvested by a cognitive radio.

The band between 174 and 240 MHz was traditionally used for television (VHF band III) but is now used for various purposes and for digital audio broadcasting (DAB). Directly after VHF band III, a large international NATO military band follows. The upper part of this band is shared with the C2000 system.

Between 400 and 470 MHz we find a second band for various applications. Typical usage is for public access mobile radio (PAMR). A PAMR network is a trunked radio system operated by a telecom service provider that licenses capacity to its, usually professional, customers. Digital PAMR systems often use TETRA, and older networks often use the analog MPT1327 standard. Because PAMR is deployed only in the area of its customers, a lot of white space exists in this band, which could be harvested by a cognitive radio.

Above 470 MHz (up to 862 MHz), we find television band IV (UHF). The frequencies in this band are internationally coordinated. This means, in each area, a part of this spectrum is used by terrestrial TV. Neighboring areas use different frequencies to avoid interference. At this moment there is a transition from analog terrestrial TV to digital terrestrial TV. Digital terrestrial TV uses lower transmission power and allows the use of multiple transmitters at the same frequency (single-frequency networks). Basically this means that the spectrum is used more efficiently. In this band a cognitive radio could also harvest a lot of spectrum. The reason for this is that outside the service area is a large surrounding area where this frequency cannot be used by another terrestrial TV transmission due to interference. So, cognitive radio can use these frequencies for local communication, and because this frequency band is used for broadcasting, its transmitters are fixed. So, a cognitive radio would only require its own GPS coordinates and a database of TV transmitters. Spectrum sensing is then only necessary for detecting wireless microphones.

16.3.4 Propagation Conditions

The propagation conditions determine how far a radio wave propagates. It seems beneficial to have good propagation conditions; that is, low path loss. However, for spatial reuse (Section 16.2.1), it is beneficial to have a large path loss. A high path loss allows more spectrum reuse, and this means that the spectrum usage increases. Also, for cognitive radio, a high path loss is better, because it reduces the area in which interference can be caused.

At UHF frequencies, ionospheric reflections play only a small role in the propagation of radio waves. Therefore, we can assume that the range of a signal is in principle limited to the horizon. The distance to the horizon is approximately given by

$$d \approx \sqrt{2krh},$$ (16.11)

in which k is a factor that corrects for the bending of radio frequency waves around the Earth, which is about $4/3$, r is the radius of the Earth, which is about 6371 km, and h the sum of the heights of the transmit and receive antennas. So the signal of a 30 m–high base station cannot be detected beyond about 23 km, and the signal from a handheld cognitive radio cannot cause interference to another handheld radio much farther than 5 km.

A simple path loss model that is often used for land mobile radio is the plane Earth model, which predicts a path loss exponent $\gamma = 40$ dB,

$$P_{rx} = P_{tx} G_{tx} G_{rx} \left(\frac{c}{4\pi f d} \right)^2 \cdot 4 \, \sin^2 \left(\frac{2\pi f h_{tx} h_{rx}}{cd} \right).$$ (16.12)

The first part of Equation (16.12) is just the free-space path loss, which is the dominant factor close to the transmitter. The second factor is caused by the interference between the direct path and the ground reflection. Figure 16.3 compares the path loss predicted by the free-space model with that of the plane Earth model for a frequency of 400 MHz and antenna heights of 1.5 m. As can be seen from this figure, the plane Earth model has two regions, which are separated by the "Fresnel breakpoint distance," d_0, given by

$$d_0 = \frac{4\pi f h_{tx} h_{rx}}{c},$$ (16.13)

which in this example is about 37 m. Before this distance, the path loss varies wildly around the 20 dB/decade line, and after this distance, it falls off smoothly at 40 dB/decade. The last deep-fading dip occurs at $d_0/2\pi$, which in this example is exactly at 6 m.

Note that the equation for the breakpoint distance is somewhat counterintuitive, because it scales inversely with the wavelength, whereas normally dimensions scale proportionally with the wavelength.

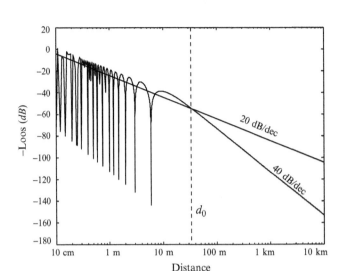

FIGURE 16.3

Negative of the path loss (dB) as function of the distance for a carrier frequency of 400 MHz with antenna height $h_{tx} = 1.5$ and $h_{rx} = 1.5$.

16.3.5 White Space Assessment

To get permission from spectrum regulators to apply cognitive radio in a certain band, one has to convince them the spectrum in this band is structurally under-utilized. This can be done with a spectrum survey. Figure 16.4 shows an overview spectrogram over a full day of a large portion of the UHF spectrum. This spectrogram was recorded by the Dutch radio spectrum authority Agentschap Telecom in October 2005, in the city of Eindhoven in The Netherlands.[4]

At first sight, there appears to be quite some white space in this spectrogram, especially between the TV channels. However, we must be careful not to draw premature conclusions. For example, TV channel 38 (channel 37 in the U.S. channel numbering), just above 600 MHz, is always empty because it is assigned worldwide to radio astronomy. Also the region around 900 MHz seems far less busy than the region around 950 MHz. However, both regions belong to the GSM system, and the uplink frequencies on 900 MHz are in fact paired with the downlink frequencies 45 MHz higher. The uplink channels appear much weaker because a cell phone uses its transmit power much more sparingly than a base station. But because of the symmetry in typical phone conversations, the occupancy in the uplink and downlink must be about equal.

---

[4]Exact GPS location: 51° 27′ 13″ N, 5° 28′ 44.8″ E, at 50 m height.

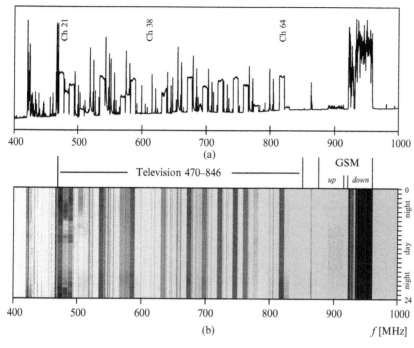

FIGURE 16.4

Overview of the UHF spectrum in The Netherlands: (a) spectrum averaged over 24 h; (b) spectrogram over 24 h.

Figure 16.5(a) shows a close-up of Figure 16.4 in the 400 to 470 MHz band. Again, we see a remarkable amount of white space. Figure 16.5(b) shows the average over a full day, and Figures 16.5(c) and (d) show the maximum and minimum, respectively. If we compare the average and maximum plots, we see that most of the signals have a fairly large peak-to-average ratio, except for the band between 420–425 MHz and 465–470 MHz. These two bands are likely downlinks from cellular networks. The dashed line in Figure 16.5(b) indicates a threshold level, which was set 2 dB above the median power level. Figure 16.5(e) shows the duty cycle obtained by using this threshold. As we can see, a bandwidth of 5 MHz can be found as a contiguous block of spectrum. If our cognitive radio can use noncontiguous spectrum, much more bandwidth can be found.

Why Is the 400 MHz to 1 GHz Band Optimal for Mobile Communication?

Below the 240 MHz, the antenna is too large for mobile communication (i.e., larger than 30 cm). One well-known trick to make an antenna shorter is to roll it up, but this makes it too selective for only one narrow-frequency band. Frequencies from 240 to about 400 MHz are used by military communication and the frequency range

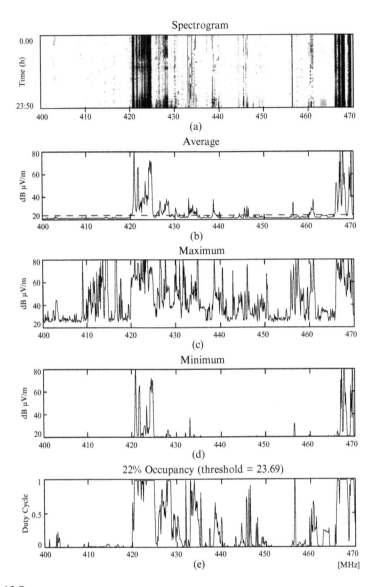

FIGURE 16.5

Overview of the UHF spectrum in The Netherlands: close-up of the 400 to 470 MHz band.

from 1 to 1.4 GHz is in use by aeronautical radio navigation and radio astronomy. Therefore, these bands also are not suitable for cognitive radio. Frequencies below 1 GHz have less indoor penetration loss, less body loss, and bend better around obstacles (less shadowing) compared to frequencies above 1.4 GHz. In [622], the indoor penetration loss and body loss for DAB band III (225 MHz) and the L band

(1465 MHz) are reported. Band III had 3.5 dB less indoor penetration loss and 10 dB less body loss compared to the L band. So, frequencies from 400 MHz to 1 GHz, from a power budget point of view, are the optimal frequencies.

16.3.6 System Spectral Efficiency

A wireless communication system should use the spectrum in an efficient manner. The system spectral efficiency can be defined as

$$\eta \approx \frac{R/B}{K} \quad \text{[b/s/Hz/site]}, \tag{16.14}$$

in which R is the bit rate, B is the bandwidth, and K is the cluster size.

The fraction R/B can be considered as the link spectral efficiency and is linked to the Shannon formula (Equation (16.1)). So, the link capacity increases if the SNR is increased. In a spatial reuse system, as in our example, the SNR is interference limited and Equation (16.10) applies. In this formula the SNR depends on K and the path loss exponent γ. A higher value of both parameters results in a higher SNR and hence a higher link capacity. However, increasing K also decreases the system spectral efficiency. So, it is up to the system designer to choose these parameters in an optimal way. For example, the GSM system has a system efficiency of 0.17, and for the WiMAX (IEEE 802.16) system this value increased to 1.2, which makes the system spectral efficiency seven times higher compared to GSM.

From Equation (16.10) one can derive that it is beneficial for the system spectral efficiency to use a high frequency, as high frequencies have a higher path loss. This is true to get the maximum system spectral efficiency. However, for mobile communication, there is also a power constraint, especially for the uplink to the base station. From this perspective, it is beneficial to use frequencies below the 1 GHz. The latter requirement is more important for our application than higher spectral efficiency.

16.3.7 Antijamming

An important requirement of public safety networks is resistance to jamming. Jamming is the intentional use of a strong radio signal, for instance by terrorists, in an attempt to disrupt communication.

The traditional way of coping with jamming is to use some form of spread spectrum. In a spread-spectrum signal, the signal energy is spread over a much wider bandwidth than the original signal. Since jammers usually have narrowband signals, they disturb only a relatively small part of the signal. There are two well-known spreading techniques: direct-sequence spread spectrum (DSSS) and frequency hopping (FH). In DSSS the signal is multiplied with a pseudorandom spreading code, which is also known at the receiver. To make a signal practically unjammable, the signal bandwidth should be at least several tens of megahertz. This means that, in our example, DSSS would be infeasible due to the limited amount of white space below the 1 GHz. In a frequency-hopping system, the transmitter hops after each packet to another frequency. The hopping sequence is known to both the transmitter and receiver. In a cognitive radio, we have an accurate map of available white space, so we could hop from white space to white space. Usually, a jammer lacks this information, which would make jamming difficult.

Cognitive radio provides inherently some robustness against a simple jammer that works on a fixed frequency. Since a cognitive radio avoids nonempty channels, it simply moves out of the way. On the other hand, a cognitive radio network may be extra vulnerable to a "smart" jammer that follows its target signal. If such a smart jammer is used against a cognitive radio network, the network is required to continuously change its frequency. However, each frequency change must be coordinated with other nodes over a control channel and must be preceded by spectrum sensing. Therefore, the jamming of a single network node affects the whole network.

However, a truly cognitive radio may outsmart even such a smart jammer, because it understands the situation it's in. When a public safety network is clearly under attack, different policies may apply, and the primary-user avoidance scheme can be replaced by a jamming avoidance scheme. In such a jamming avoidance scheme, spread-spectrum techniques are very useful.

16.4 CHAPTER SUMMARY

For the next-generation system of public safety communication, there is a huge demand for broadband applications. One reason for this is that pictures and video are very efficient in describing a situation. Another reason is that modern cell phones have broadband Internet access, and public safety personnel may start to rely on it. But during a large emergency, cell phones are likely to fail. Therefore, broadband Internet access is a requirement for the next-generation public safety communication system.

One problem is that the spectrum in the public safety bands is scarce, and public safety communication has such a high peak-to-average ratio that it is uneconomical to reserve the needed spectrum in the conventional way. Cognitive radio may be a way out of this problem. However, cognitive radio is not allowed to interfere with the primary voice communication in any way.

So before we can add cognitive radio, we first have to acquire a good understanding of how narrowband communication works. In this chapter we therefore review the two foundations of modern public safety communication systems: trunking and cellular frequency reuse. The physical layer of the TETRA standard is reviewed and the implementation in the C2000 system is discussed. Next, we discuss a fireworks explosion as an example of a public safety situation where traditional communication systems failed. First, there was a shortage of voice channels, because hundreds of calls were made simultaneously in a relatively small area. Second, rescue workers could not upload pictures and videos of the situation to the central command post. From this scenario we derived that between 5 and 30 MHz of spectrum must be found by spectrum sensing. From propagation considerations we derived that the preferable spectrum region lies between 400 MHz and 1 GHz. From spectrum surveillance data we see that it is reasonable to expect that such bandwidths can indeed be found in this region of spectrum.

16.5 PROBLEMS

1   In Section 16.3.1 we describe the fireworks disaster in Enschede. In normal day-to-day activities (like house fires), at most 25 public safety workers are active in this area. How much capacity—that is, lines—are required if less than 0.1% of the calls are blocked and, in the busiest hour, there are 0.167 calls per second with an average duration of 30 sec? To calculate this answer, use the Erlang-B model, which can be downloaded from MATLAB-central.

2   If, in the previous question, group communications was used with, on average, five rescue workers in a group, what would be the required capacity in this case?

3   Derive the required capacity on the day of the fireworks disaster. Assume that during the busiest hour there are 100 calls per second with an average duration of 20 sec. How many lines are now required? And what if group communication is used?

4   The TETRA system has four time slots and uses 2×25 kHz of spectrum (TDMA + FDD). If we assume that spectrum per megahertz has the same cost as the UMTS licenses per megahertz (Section 16.1.3), and in addition, we assume that cognitive radio technology could provide 80% of the capacity during a large disaster, what are the economic benefits of using this technology?

5   In an analog cellular network one needs an SNR of at least 18 dB for acceptable audio quality of voice communication. What would be the optimal cluster size K?

6   In a digital cellular network one has a total bandwidth S that must be partitioned over clusters of K cells. Which value of K gives the highest theoretical capacity?

Auction-based spectrum markets in cognitive radio networks

17

Xia Zhou[1], Heather Zheng[1], Maziar Nekovee[2], and Milind M. Buddhikot[3]

[1]University of California Santa Barbara, United States
[2]BT Research and University College London, United Kingdom
[3]Alcatel-Lucent, United States

17.1 INTRODUCTION

Access to the radio spectrum is a key requirement for continuous wireless growth and deployment of new mobile services. Given the fast-growing demand for radio spectrum, regulators around the world are implementing much more flexible and liberal forms of spectrum management, often referred to as *dynamic spectrum management*. This new model dynamically redistributes and reassigns spectrum within and across different wireless systems, adapting spectrum usage to actual demands and achieving much more efficient use of the precious spectrum resource. Within the new model, two prominent approaches are being considered by the regulators: spectrum trading and cognitive spectrum access. In this chapter, we focus on examining the challenges and solutions in the area of spectrum trading.

Spectrum trading is a market-based approach for spectrum redistribution that enables a spectrum license holder (for example, a cellular operator) to sell or lease all or a portion of its spectrum to a third party [623]. The third party can, in principle, change the use of spectrum or the technology to be used, provided certain conditions are satisfied. Note that this is an important departure from the command and control management model, where spectrum licenses are granted by regulators for the provision of a specific service using a predefined technology, and license holders were not allowed to reallocate their spectrum to different technologies or other users. Exposing the radio spectrum to market forces has become

Doi: 10.1016/B978-0-12-374715-0.00017-4

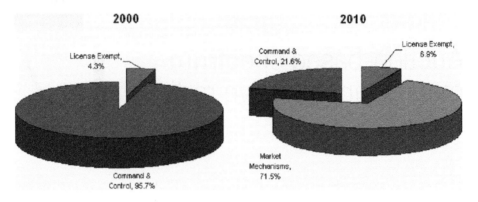

FIGURE 17.1

From command and control to market-driven spectrum allocation in the United Kingdom (courtesy of Ofcom - Spectrum Framework Review, Feb. 2005).

increasingly popular. For example, the U.K. regulator, Ofcom, is aiming that, by 2010, 71.5% of its available spectrum should be operating under market forces [10] (see Figure 17.1). The rationale for the approach is that market mechanisms will allocate spectrum to those that value it most, thereby ensuring that the (economically) most efficient utilization of this resource is achieved. However, at least initially, one expects that such forms of spectrum trading would take place only on a macro-scale (e.g., between two cellular service providers) involving large blocks of spectrum and timescales that are still dictated by complex and cumbersome bureaucratic procedures involved in such wholesale forms of trading.

17.1.1 Dynamic Spectrum Micro-Auctions

Cognitive access to ceratin "publicly owned" licensed bands, such as TV and military bands, are being actively pursued by regulators. However, it is very doubtful that, without any economic incentive, this form of access can be extended to "privately owned" licensed bands, such as 3G spectrum, for which the incumbents have already paid billions of dollars, pounds, or euros to ensure their exclusive use. Therefore, market mechanisms on a micro-scale need to be implemented to create economic incentives for license holders to share their spectrum locally and temporarily with cognitive radios.

For market players (cognitive radios and incumbent systems) to make economically efficient deals, they require a market environment that enables them to negotiate such that mutually acceptable bargains are reached. Auctions are among the best-known market-based allocation mechanisms due to their perceived fairness and allocation efficiency. Indeed, the FCC (Federal Communications Commission) and its counterparts across the world have extensively used auctions for wholesale allocation of spectrum in the last decade and intend to use this mechanism in the future. However, an FCC-style spectrum auction targets long-term national/regional leases, requiring huge upfront investments. In this chapter, on the other hand, our focus is on micro-auction mechanisms that allow for the trading of spectrum rights at the network level. These types of auction mechanisms could be highly attractive to network operators: they provide a flexible and cost-effective means for dynamic expansion of their spectrum resources without the need for costly capital investments in new spectrum. The spectrum obtained through micro-auctions can be used for congestion relief during peak loads in traffic or to enhance existing services and provide new services without the need for acquiring additional spectrum. More generally, users will be able to *dynamically* and *locally* vary their operating frequencies and access the best available spectrum on a just-in-time basis. This may happen either upon instruction from a cognitive base station that acquires spectrum on behalf of users [357] or autonomously by user devices themselves.

17.1.2 The Role of Cognitive Radios

Cognitive functionality is essential in the realization of such types of micro-auctions, because wireless devices can understand the regulatory, technical, and economic context within which they perform the required negotiation and decision-making tasks. The scope of this chapter, however, is not on developing such cognitive functionalities. Instead, we assume that these functionalities will be available in future devices and focus on developing and modeling appropriate auction algorithms to ensure fast and efficient redistribution of the spectrum on the network level. Furthermore, we have no assumptions regarding the underlying network access technologies that a cognitive device uses for its transmissions once it acquires a portion of the spectrum. However, following [624], we envisage that access technologies such as OFDMA will play an important role in enabling our micro-auction mechanisms. These technologies support dynamic bandwidth availability and permit grouping, subdividing, and pooling of pieces of the spectrum into neatly packaged spectrum channels.

17.2 RETHINKING SPECTRUM AUCTIONS

In the past decade, the radio spectrum has been auctioned in terms of prepartitioned bulk licenses that cannot match time-varying market demands. Such mismatch led to several consequences. First, forced to bid in the unit of bulk licenses, buyers face huge upfront costs. As a result, past auctions involved only a very few large (incumbent) players, required significant manual negotiations, and often took months or years to conclude. Second, winning buyers that received the licenses could not efficiently utilize the assigned spectrum because their traffic varies significantly in time and space. Finally, while winning buyers' spectrum sits unused, new entrants and new wireless technologies are either blocked or forced to crowd into highly unreliable unlicensed bands. If not addressed, such inefficiency will soon put a stop to wireless growth and innovation.

Solving such inefficiency requires us to rethink the way the spectrum is distributed, and redesign spectrum auctions to provide networks with spectrum matching their individual demands. Recent works have proposed an eBay-like, open marketplace concept to enable dynamic spectrum trading [625, 626]. In this marketplace, existing spectrum owners (as providers) gain financial returns by leasing their idle spectrum to new spectrum users, and new users (as buyers) obtain the spectrum they desperately need. This marketplace differs significantly from conventional FCC-style spectrum auctions in three aspects:

- **Multiparty trading with spectrum reuse.** Spectrum auctions are fundamentally different from (and much more difficult than) conventional multiunit auctions because of the spectrum's unique property of reusability. Unlike traditional goods (e.g., paintings, bonds, electricity), the spectrum can be spatially reused concurrently. Although two conflicting bidders must not use the same spectrum bands simultaneously, well-separated bidders can. While a conventional auction with n bidders and k bands can have at most k winners, a spectrum auction can have more than k winners. Therefore, unlike FCC-style auctions, which have one provider (i.e., the FCC) and sell one license to only one buyer, the new marketplace supports multiparty trading. Multiple providers can selectively offer their idle spectrum pieces, and each spectrum piece can be sold to multiple "small" buyers. In this way, the new marketplace can exploit spectrum reusability in spatial and temporal domains to improve spectrum usage efficiency.

- **On-demand spectrum trading.** Instead of forcing buyers to purchase predefined spectrum licenses, the new marketplace enables buyers to specify their own demands. Given these demands, the marketplace intelligently selects winners and allocates spectrum to best utilize the spectrum offered by providers and supported by buyers. Such flexibility not only attracts a large number of participants, but also enables the system to effectively multiplex spectrum supply and demand, further improving spectrum utilization.

- **Economic robustness with spectrum reuse.** Without good economic design, spectrum auctions easily can be manipulated by bidders, suffering huge efficiency loss. Auctioneers are forced to apply Bayesian settings, placing strong (and often wrong) assumptions on the distribution of bidder valuations [627]. The heavy overhead and the vulnerability would easily discourage both providers and players from participation. Therefore, only by preventing market manipulation can an auction attract bidders and new entrants and efficiently distribute spectrum to make the best use of this important resource. While conventional auction design has proposed novel solutions to achieve economic robustness, the requirement on spectrum reuse opens up new vulnerabilities in existing solutions [625]. New auction rules are required to achieve economic robustness while enabling spectrum reuse.

With these three requirements in mind, we now describe several ongoing efforts by which to design dynamic spectrum auctions. We start presenting a spectrum allocation algorithm to enable fast auction clearing. We then describe two recent works on adding economic robustness to auction designs, including a single-sided spectrum auction system, where spectrum buyers bid for spectrum from a single auctioneer, and a double spectrum auction system, where spectrum sellers and buyers can trade spectrum by each interacting with an auctioneer.

17.3 ON-DEMAND SPECTRUM AUCTIONS

An on-demand spectrum auction must distribute spectrum on-the-fly to a large number of bidders. Spectrum auctions are multiunit auctions, where the spectrum often is divided into a number of identical channels for sale. Users wish to obtain different amount of spectrum at their desired power levels and may be willing to pay differently depending on the assignment. Toward this goal, we need a *compact bidding language* to allow buyers to conveniently express their desire and do it so compactly, and an *efficient allocation algorithm* to distribute spectrum in real-time subject to the complex interference constraints among bidders.

In this section, we discuss recent ongoing efforts on spectrum allocation algorithms to support dynamic spectrum auctions. We focus on a recent work [628] that proposed a computationally efficient auction framework with simple and effective bidding and fast auction clearing algorithms. Specifically, spectrum buyers (bidders) use a compact and yet expressive bidding format to express their desired spectrum usage and willingness to pay, while an auctioneer executes fast clearing algorithms to derive prices and allocations under different pricing models.

17.3.1 Bidding Format: Piecewise Linear Price-Demand Bids

Assume there are K channels in total, F_i is the set of channels assigned to bidder i, and hence the normalized spectrum assigned to i is $f_i = |F_i|/K$. With the piecewise linear price demand (PLPD), bidder i expresses the desired quantity of spectrum f_i at each per-unit price p_i using a continuous concave piecewise linear demand curve. That is, the bidder would like to pay $p_i \times f_i$ for f_i channels. An PLPD curve can be expressed as a conglomeration of a set of individual linear pieces. A simple example is a linear demand curve:

$$p_i(f_i) = -a_i f_i + b_i, \quad a_i \geq 0, b_i > 0, \tag{17.1}$$

where the negative slope represents *price sensitivity* of buyers—as the per-unit price decreases, demands in general increase.

17.3.2 Pricing Models

Without considering economic robustness, the auction pricing follows directly from each bidder's bid. Bidder i that obtains f_i spectrum is charged $p_i(f_i) \times f_i$ as specified by its bid. In this case, the revenue produced by each bidder is a *piecewise quadratic* function of the price:

$$R_i(p_i) = \frac{b_i p_i - p_i{}^2}{a_i}.\tag{17.2}$$

For linear demand curves, the revenue is a quadratic function of price, with a unique maximum at $p_i = b_i/2$. We can further divide the pricing models into two types: uniform and discriminatory pricing. In *uniform pricing*, the auctioneer chooses a single clearing price p for all the winners. Each bidder obtains a fraction of spectrum $f_i(p) = (b_i - p)/a_i$ and produces a revenue of $R_i(p) = (b_i p - p^2)/a_i$. Any bidder i with $b_i \le p$ gets no assignment. In *discriminatory pricing*, the auctioneer sets nonuniform clearing prices across bidders.

17.3.3 Fast Auction Clearing by Linearizing the Interference Constraints

Given the bids and the pricing model, the auction clearing problem is to maximize the auction revenue $\sum_i R_i(p_i)$ by choosing the winners and their pricing p_i subject to the interference constraints. This optimization problem is in general NP-hard because of the underlining interference constraints grow exponentially with the number of bidders. The authors of [628] propose to reduce the interference constraints into a set of linearized constraints that grow linearly with the number of bidders. Specifically, they propose the *node-L interference (NLI) constraints*. Consider two nodes i and j located at coordinates (x_i, y_i) and (x_j, y_j). Node i is to *the left of* node j if $x_i < x_j$. If $x_i = x_j$, then the node with the smaller index is considered to be to the node to the left. The constraint becomes: Every neighbor of i to the *left* of i, and i itself, should be assigned with different channels:

$$f_i + \sum_{j \in N_L(i)} f_j \le 1, \quad i = 1, 2, \ldots, N,\tag{17.3}$$

where $N_L(i)$ is the set of neighbors of i lying to its left. It has been shown that the new constraints are stricter than the original constraints and lead to a feasible but suboptimal solution within a distance 3 from the optimal solution.

Using the new interference constraints, the auction clearing problem can be solved using linear programming (for the uniform pricing model) or separable programming [629] (for the discriminatory pricing model). Both solutions have polynomial complexity. Readers should refer to [628] for additional details on the algorithms and proofs. In practice, both algorithms run efficiently in real time. Using a standard desktop with a 3.0 GHz processor and 1 GB of RAM and assuming 3500 bidders, the auction clearing finishes in 0.05 sec for the uniform pricing and 80 sec for the discriminatory pricing model.

17.4 ECONOMICALLY ROBUST SPECTRUM AUCTIONS

When it comes to resisting market manipulation, the dominant paradigm is truthful auction design. A truthful auction guarantees that, if a bidder bids the true valuation of the resource, its utility will be no less than that when it lies. Hence, the weakly dominating strategy is for a bidder to bid its true valuation. As we will show, a truthful auction charges a winner independent of its actual bid, which is different from the auction design in the previous section. To bidders, a truthful auction eliminates the expensive overhead of strategizing about other bidders and prevents market manipulation. Thus, it can attract a wide range of network nodes/establishments to engage in the marketplace. To the auctioneer, by encouraging bidders to reveal their true valuations, a truthful auction can help the auctioneer increase its revenue by assigning the spectrum to the bidders who value it the most. For the same reason, many classic auction systems are made truthful, including the sealed-bid secondary-price [630], k-position [631,632], and VCG auctions [633,634].

While prior works have enforced truthfulness in conventional auctions, existing truthful designs either fail or become computationally prohibitive when applied to spectrum auctions. The fundamental reason is that, unlike goods (e.g., paintings, bonds, electricity) in conventional auctions, the spectrum is reusable among bidders subject to the spatial interference constraints. Because interference is only a local effect, bidders in close proximity cannot use the same spectrum frequency simultaneously, but well-separated bidders can. These heterogeneous interdependencies among bidders make secondary-price and k-position auctions no longer truthful. Furthermore, these constraints make the problem of finding the optimal spectrum allocation NP-complete, and hence a real-time spectrum auction with many bidders must resort to greedy allocations that are computationally efficient. Unfortunately, it has been shown that the VCG auction loses its truthfulness under greedy allocations.

In the following, we describe VERITAS [625], a truthful dynamic spectrum auction framework. VERITAS achieves truthfulness with computationally efficient spectrum allocation and pricing mechanisms, making it feasible for online short-term auctions. In addition, VERITAS provides the auctioneer with the capability and flexibility of maximizing its customized objective and allows bidders to request spectrum by the exact number of channels they want to obtain or by a range defined by the minimal and maximal number of channels.

Consider the typical sealed-bid auction in Figure 17.2. The auctioneer sells k channels by running an online auction periodically. Each bidder requests spectrum by the number of channels and the per-channel price it would like to pay. After receiving the bids, the auctioneer determines the winners, their spectrum allocations, and prices, based on the bids and the interference condition among bidders. As shown in Figure 17.2, the interference condition is represented by a conflict graph [635] $G = (V, E)$, where V is the collection of the bidders and E is the collection of edges where two bidders share an edge if they conflict. Table 17.1 summarizes the notations used to define an auction problem.

Using these notations, we now define a truthful auction, and a truthful and efficient spectrum auction.

Definition 1. *A truthful auction is one in which no bidder* i *can obtain higher utility* u_i *by setting* $b_i \neq v_i$.

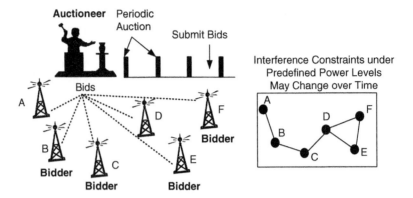

FIGURE 17.2

A dynamic spectrum auction scenario: (left) an auctioneer performs periodic auctions of spectrum to the bidders; (right) a conflict graph illustrates the interference constraints among bidders.

Table 17.1 Summary of Auction Notations	
Notations	**Description**
Channel request d_i	The number of channels requested by bidder i
$D = \{d_1, d_2, ..., d_n\}$	The set of demands across all the n bidders
Per-channel bid b_i	The per-channel bid submitted by bidder i or the maximum price bidder i is willing to pay for a channel
$B = \{b_1, b_2, ..., b_n\}$	The set of bids submitted by all the n bidders
Per-channel valuation v_i	The true valuation a bidder i has for a channel. In most cases, v_i is private and known only to bidder i
Channel allocation d_i^a	The number of channels an auction winner i receives
Clearing price p_i	The price charged to an auction winner i; in a truthful auction, $p_i \leq d_i^a \times v_i$
Bidder utility u_i	The utility of bidder i, or the residual worth of the channels: $u_i = v_i \cdot d_i^a - p_i$ if i is an auction winner and 0 otherwise

In the context of spectrum auctions, the design must ensure truthfulness and enable spectrum reuse across auction winners to improve spectrum utilization.

Definition 2. *An efficient and truthful spectrum auction is one that is truthful and maximizes the efficiency of spectrum usage subject to the interference constraints.*

In building a truthful and efficient spectrum auction, VERITAS integrates a greedy spectrum allocation with a carefully designed pricing mechanism. Let us start from a simple scenario, where bidders' channel requests are strict: a bidder i requests d_i channels and accepts allocations of only either 0 or d_i channels.

17.4.1 Spectrum Allocation

In determining the auction winners, VERITAS applies a greedy solution. It first sorts the bid set B by a descending order of b_i, then allocates bidders sequentially from the highest one to the lowest one. From each bidder i, the algorithm first checks whether there are enough channels to satisfy i's request d_i. If so, it assigns i with d_i's lowest indexed channels that have not been assigned to i's conflicting peers. Such monotonic allocation is critical to achieve auction truthfulness.

17.4.2 Winner Pricing

VERITAS charges each winner i with the bid of its critical neighbor multiplied by the number of channels allocated to i. The price reflects the minimum value of i's bid to win the auction. It is independent of i's actual bid, and is always no more than i's actual bid multiplied by the number of channels allocated to i. This property is also referred to as the *individual rationality*. A critical neighbor is defined as follows:

Definition 3. *Given* $\{B \setminus b_i\}$, *a critical neighbor* $\mathbb{C}(i)$ *of bidder* i *is one of* i*'s neighbors where if* i *bids lower than* $\mathbb{C}(i)$, i *will not be allocated, and if* i *bids higher than* $\mathbb{C}(i)$, i *will be allocated.*

At first sight, finding the critical neighbor for each bidder i seems computationally expensive. It requires inserting i's bid immediately after each of its neighbors and running the allocation algorithm repeatedly. VERITAS overcomes this problem, using an intelligent pricing algorithm that identifies the critical neighbor for each bidder by running the allocation algorithm once. For each bidder i, the algorithm first removes i from the sorted bid set and runs the allocation. When assigning channels to i's conflicting peers, the algorithm removes the assigned channels from i's available channel set. The first winning conflicting peer that makes i's available channels less than its demand d_i is i's critical neighbor. The detailed algorithm description can be found in [625].

17.4.3 Supporting Other Bidding Formats

VERITAS enables bidders to use diverse demand formats. A bidder can request spectrum by the exact number of channels it would like to obtain (strict requests) or by a range defined by the minimum and maximum number of channels (range requests). Using the range request, a bidder i can request d_i channels but accept any number of channels between 0 and d_i. To ensure truthfulness under this request, VERITAS applies an advanced allocation and pricing mechanism. When allocating channels, if the number of available channels is less than t's demand d_t, the algorithm allocates whatever is possible. When determining prices, the algorithm needs to find multiple (rather than one) critical neighbors for each winner because bidding below each critical neighbor results into being allocated with different numbers of channels. For each set of additional channels obtained by bidding higher than the last critical neighbor, the algorithm charges the winner with the bid of its last critical neighbor. The clearing price is the sum of prices charged for all of the bidder's assigned channels.

17.4.4 Supporting Different Auction Objectives

VERITAS provides the auctioneer with the capability and flexibility of maximizing its customized objective. By sorting the bid set differently, the auctioneer can configure the order of allocation to maximize the auction revenue or the social welfare. For example, it has been shown that, to maximize the sum of winning bids, known as the social welfare [632], the best-known greedy algorithm is to assign channels following the descending order of $b_t/|N(t) + 1|$ [636], where $N(t)$ is the number of conflict peers of bidder t. VERITAS enables this flexibility by allowing different sorting metrics as long as they are an increasing function of the bid b_t and are not affected by the bids of other bidders. Example sorting metrics are: b_t, $b_t/|N(t)| + 1$, or $b_t \times |N(t)|$.

17.4.5 VERITAS Performance and Complexity

It has been shown that the VERITAS auction design is truthful by combining the monotonic spectrum allocation and the critical-neighbor-based pricing algorithm [625]. The computational complexity of VERITAS is on the order of $O(N^3K)$, where N is the number of bidders and K is the number of channels auctioned. Among them, $O(N \log N + K|E|)$ is from the allocation algorithm and $O(NK|E|)$ is from the pricing algorithm, where $|E|$ is the number of edges in the bidder conflict graph. Such polynomial complexity makes VERITAS suitable for dynamic, on-demand spectrum auctions.

Figure 17.3(a) compares VERITAS's spectrum utilization to that of the best-known greedy allocation algorithm [636], where VERITAS performs similarly to the greedy solution. Figure 17.3(b) examines its auction revenue as a function of the number of channels auctioned. VERITAS exhibits an interesting trend: As the number of channels auctioned increases, the revenue first increases then decreases. This is because VERITAS charges winners by their critical-neighbors' bids. Increasing the number of channels reduces the level of bidder competition. As the number of winners increases to include all the bidders, the price charged each winner also decreases to 0. To maximize its revenue, the auctioneer can choose to control the number of channels to be auctioned. To prevent bidder manipulation, the auctioneer must make decision *prior to* the auction execution. Determining the optimal number of channels is a challenging question given the complex interference constraints. A simple heuristic was proposed in [625].

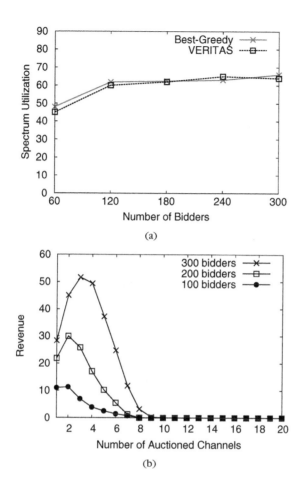

FIGURE 17.3

VERITAS performance: (a) Spectrum allocation efficiency versus the number of bidders. VERITAS performs similarly to the best-known greedy algorithm [636]. (b) VERITAS auction revenue versus the number of channels auctioned. The auction revenue depends heavily on the level of bidder competition. As the number of channels auctioned increases, the level of competition decreases and the winners' prices reduce.

17.5 DOUBLE SPECTRUM AUCTIONS FOR MULTIPARTY TRADING

We have described an auction design where an auctioneer sells its spectrum channels to buyers. In this section, we describe a double spectrum auction design where multiple spectrum sellers and buyers can trade spectrum flexibly by interacting with an auctioneer. As shown in Figure 17.4, the auctioneer is a matchmaker

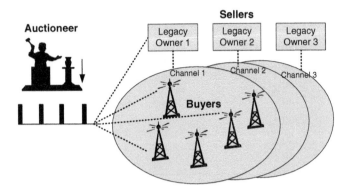

FIGURE 17.4

Multiparty spectrum trading based on double auctions. The auctioneer performs an auction among both sellers and buyers. Sellers provide idle spectrum pieces dynamically with regional coverage, while buyers request spectrum channels in local areas based on their demands. Each channel contributed by a seller can be reused by multiple, nonconflicting buyers.

between sellers and buyers. It buys spectrum pieces from sellers and sells them to buyers. In this way, existing spectrum owners (as sellers) can obtain financial gains by leasing their selected idle spectrum to new spectrum users; new users (as buyers) can access the spectrum they desperately need and in the format they truly desire. By multiplexing spectrum supply and demand in time and space, dynamic auctions can significantly improve spectrum utilization.

To model a double spectrum auction, we define the bid, true valuation, price, and utility of both sellers and buyers. The notations for buyers, B_n^b, V_n^b, P_n^b, and U_n^b, follow those in Table 17.1, and the notations for sellers are defined in Table 17.2. In addition to truthfulness and spectrum reuse, a double spectrum auction must also achieve two additional properties: *individual rationality* and *budget balance*.

Table 17.2 Summary of Double Auction Notations Related to Sellers

Notations	Description
Seller's per-channel bid B_m^s	The minimum payment required by seller m to sell one channel
Seller's per-channel valuation V_m^s	The true valuation a seller m has for a channel
Seller's price P_m^s	The payment a winning seller m receives by selling a channel
Seller's utility U_m^s	The utility of seller m $U_m^s = P_m^s - V_m^s$ if m wins the auction and 0 otherwise; this is different from the buyer case

Definition 4. *A double auction is* individual rational *if no winning buyer pays more than its bid (i.e., $P_n^b \leq B_n^b$), and no winning seller gets paid less than its bid (i.e., $P_m^s \geq B_m^s$).*

This property guarantees nonnegative utilities for bidders who bid truthfully, providing them incentives to participate in the auction.

Definition 5. *A double auction is* ex-post budget balanced *if the auctioneer's profit is $\Phi \geq 0$. The profit is defined as the difference between the revenue collected from buyers and the expense paid to sellers: $\Phi = \sum_{n=1}^{N} P_n^b - \sum_{m=1}^{M} P_m^s \geq 0$.*

This property ensures that the auctioneer has incentive to set up the auction.

In the following, we describe TRUST [626], a double spectrum auction framework that achieves the four required properties: spectrum reuse, truthfulness, individual rationality, and budget balance. Table 17.3 compares TRUST to existing double auction designs. Conventional double auction designs (VCG [637] and McAfee [638]) achieve truthfulness but do not consider spectrum reusability. VERITAS [625], on the other hand, addresses only single-sided buyer-only auctions and loses the truthfulness when directly extended to double auctions [626].

TRUST [626] breaks the barrier between spectrum reuse and truthfulness in double spectrum auctions. In essence, it enables spectrum reuse by applying a spectrum allocation algorithm to form buyer groups. It achieves the three economic properties via the bid-independent group formation and a reusability-aware pricing mechanism. TRUST consists of three components: grouping buyers, determining winners, and pricing.

17.5.1 Grouping Buyers

TRUST groups multiple nonconflicting buyers into groups so that buyers in each group do not conflict and can reuse the same channel. This is done privately by the auctioneer performing a spectrum allocation algorithm and organizing buyers

Table 17.3 Comparison of Various Double Auction Designs

Existing Double Auction Designs	Spectrum Reuse	Truthfulness	Ex-Post Budget Balance	Individual Rationality
VCG	X	✓	X	✓
McAfee	X	✓	✓	✓
VERITAS extension	✓	X	✓	✓
RUST	✓	✓	✓	✓

assigned to the same channel to a group. Unlike VERITAS, the group formation is independent of the buyer bids to prevent bidders from rigging their bids to manipulate the group size and its members.

The group formation can cope with various interference models by using different spectrum allocation algorithms. If the buyer interference condition is modeled by a conflict graph, the group formation is equivalent to finding the independent sets of the conflict graph [639,640]. If the interference condition is modeled by the physical signal-to-interference and noise ratio (SINR) [641], TRUST finds multiple sets of buyers who can transmit simultaneously and maintain an adequate received SINR [642]. TRUST performs this allocation only to form buyer groups, not to assign specific channels to buyers.

17.5.2 Determining Winners

Next, TRUST treats each buyer group as a superbuyer and runs the conventional double spectrum auction algorithm to determine the winning sellers and superbuyers. Let $G_1, G_2, ..., G_L$ represent the L groups formed. For any group G_l with $n_l = |G_l|$ buyers, the group bid π_l is

$$\pi_l = \min \left\{ B_n^b | n \in G_l \right\} \cdot n_l. \tag{17.4}$$

TRUST sorts the seller bids in nondecreasing order and the buyer group bids in nonincreasing order: $\mathbb{B}' : B_1^s \leq B_2^s \leq ... \leq B_M^s$, and $\mathbb{B}'' : \pi_1 \geq \pi_2 \geq ... \geq \pi_L$. Define k as the last profitable trade:

$$k = \text{argmax}_{l \leq \min\{L,M\}} \pi_l \geq B_l^s. \tag{17.5}$$

Then the auction winners are the first $(k - 1)$ sellers and the first $(k - 1)$ buyer groups.

17.5.3 Pricing

To ensure truthfulness, TRUST pays each winning seller m by the kth seller's bid B_k^s and charges each winning buyer group l by the kth buyer group's bid π_k. This group price is evenly shared among the buyers in the group l:

$$P_n^b = \pi_k/n_l, \quad \text{for all } n \in G_l. \tag{17.6}$$

No charges or payments are made to losing buyers and sellers. The uniform pricing is fair because buyers in a winning group obtain the same channel and thus are charged equally. In addition, to ensure individual rationality, a group bid must not exceed the product of the lowest buyer bid in the group and the number of buyers in the group, which is used in the process of determining winning groups. With such pricing mechanism, the auctioneer's profit becomes $\Phi = (k - 1) \cdot (\pi_k - B_k^s)$ and it is easy to show that $\Phi \geq 0$.

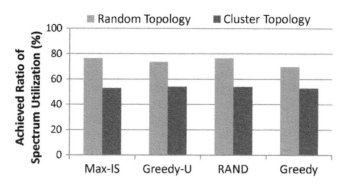

FIGURE 17.5

The percentage of spectrum utilization achieved by TRUST compared to that of pure allocations without economic factors. Four allocation algorithms are considered, including Max-IS [640], two greedy allocation algorithms [639], Greedy-U and Greedy, and random allocation RAND.

17.5.4 **TRUST Performance and Complexity**

As shown in [626], by integrating the monotonic winner determination and bid-independent pricing, TRUST achieves truthfulness, ex-post budget balance, and individual rationality while enabling spectrum reuse to improve spectrum utilization. One key advantage of TRUST is that it can use any spectrum allocation algorithm in forming buyer groups. Thus, its complexity depends heavily on the allocation algorithm used.

On the other hand, ensuring these economic properties comes at a cost in spectrum utilization. This is because TRUST selects winning buyer groups by the minimum bid in the group multiplied by the group size, so that groups of different sizes have equal opportunity in being chosen. On the other hand, the convectional spectrum allocation algorithms always choose large groups, leading to an advantage in spectrum utilization. Figure 17.5 illustrates the ratio of TRUST's spectrum utilization to that of conventional spectrum allocations without economic consideration [639, 640]. It examines the performance using random and clustered topologies. In random network topologies, TRUST achieves 70–80% spectrum utilization of the conventional spectrum allocation, while in clustered network topologies, TRUST sacrifices roughly 50% of spectrum utilization in exchange for economic robustness. This is because, in clustered topologies, the group sizes become much more diverse and TRUST could select a set of small buyer groups, which degrades the overall spectrum utilization.

17.6 **CHAPTER SUMMARY AND FURTHER READING**

In this chapter we examine the challenges and solutions in the area of spectrum trading. Unlike the conventional command and control management model, spectrum trading is an open, market-based approach for redistributing the spectrum. New users can gain access to the spectrum they desperately need and existing owners can gain financial incentives to "lease" their idle spectrum. We focus mainly on dynamic spectrum auctions, because auctions are among the best-known market-based allocation mechanisms. Dynamic spectrum auctions are fundamentally different from (and much more difficult than) conventional multiunit auctions because of their unique requirement of spectrum reusability. With this in mind, we introduce three recent works on on-demand spectrum auctions, truthful spectrum auctions, and truthful double spectrum auctions. Together, they provide the basic building blocks for constructing an efficient, economically robust, real-time dynamic spectrum marketplace.

It is important to note that there have been numerous contributions and ongoing efforts on dynamic spectrum allocation, pricing, trading, and auctions. A small set of them are summarized next. Building on these extensive contributions, the use of spectrum trading is moving from pure research to several commercial deployments and ideally will expand to the general public in the near future.

Exploitation of market mechanisms for dynamic allocation and redistribution of spectrum has been the topic of several other recent research investigations, and the literature on this topic is growing. Important also, the use of such mechanisms is starting to move from the realm of pure research into that of development and commercial exploitation. For example, Spectrum Bridge Inc. [643], a U.S.-based company, developed a real-time online marketplace that enables spectrum owners and users to buy, sell, and lease FCC-licensed spectrum. According to the company's web site, its online marketplace, SpecEx, provides access to over 200 billion units of spectrum that the FCC made eligible for secondary market transactions. For the benefit of the reader we summarize in the remaining of this section some of the most recent research on the use of market mechanisms for dynamic spectrum access.

A framework for coordinating dynamic spectrum access among service providers was proposed in [357]. The scheme proposed in this work relies on a spectrum broker that controls the allocation of spectrum among the requesting operators. This work was later extended to cases where the interference among bidders is modeled by pairwise and physical interference models and the bidders can bid for heterogeneous channels of different widths using generic bidding functions [644].

The price dynamics of a dynamic spectrum market was explored in [645]. The authors considered a marketplace consisting of spectrum agile network service providers and users. Competition among multiple primary users to sell their spectrum are modeled in this work as a noncooperative game. An interesting feature of this work is that the analysis takes into account differences in evaluation of the quality of the offered spectrum by buyers. For example, radio waves at lower frequencies, such as UHF, travel longer distances and penetrate more readily through walls. Therefore, buyers may value spectrum in such bands highly for applications that require good penetration properties. Also, depending on their operating wireless technology, some spectrum buyers may value contiguous segments of spectrum higher than noncontiguous ones.

The dynamics of multiple-seller and multiple-buyer spectrum trading in dynamic spectrum access networks is considered in [646]. In this work it is assumed that the secondary users can adapt their spectrum buying behavior to the variations in price and quality of spectrum offered by different primary users. At the same time the primary users can adjust their behavior in selling their spectrum to achieve the highest utility. Similar to [645] the competition among primary users in selling spectrum is modeled using a noncooperative game formulation. At the same time, evolution in the spectrum buying behavior of secondary users are analyzed using the deterministic and stochastic models of evolutionary games.

One of the early papers that explores the use of auctions for dynamic allocation of spectrum is [647]. The authors consider a scenario where multiple code division, multiple-access operators bid for the spectrum to a spectrum manager. They present an optimal bidding and pricing mechanism with the objective of maximizing the revenue of the operators based on the willingness of users to pay. Auction-based mechanism for dynamic spectrum access are also explored in [648], where an optimization problem is formulated to maximize the revenue of spectrum owners through pricing and spectrum assignment. In [624] the authors describe a combinatorial clock auction mechanism for trading of spectrum in the context of an OFDMA-based cognitive radio network. Combinatorial clock auctions [649] are used when a range of items are on sale that may be logically grouped together into many packages to suit either the buyer, the seller, or both. In these auctions, bids for such packages are made throughout a number of sequential open rounds and a final sealed-bid round. During the sequential rounds buyers have an opportunity to explore the bid space as their bids are either accepted or rejected until there is no change in the winners or no new bids are submitted. The authors of [624] present a modified version of the combinatorial clock auctions to reduce the complexity of the mechanism for cognitive radios that attempt to buy and sell spectrum on behalf of users.

17.7 PROBLEMS

1   What is spectrum trading? How does it differ from the command and control model?

2   This chapter deals mainly with auction-based mechanisms to enable efficient deals between spectrum buyers and sellers. Name and explain at least one other market mechanism that can be used for trading spectrum.

3   How does spectrum differ from other natural resources, such as gas and electricity? How do these differences affect the use of auctions in trading spectrum?

4   Explain what a conflict graph is. Use MATLAB to construct and visualize conflict graphs for a collection of 100 nodes uniformly distributed in a 1 km^2 rectangular area. Assume that all nodes have a 100 m transmission radius.

5   What are the limitations of FCC-style auctions?

6   What are the differences between single-sided auctions and double auctions?

7   Consider the first-price auction where the winner is charged by its bid. What revenue trend would you expect as the number of channels auctioned increases? Explain your conjecture by comparing it to Figure 17.3.

8   *Critical neighbor* is defined in VERITAS for determining each winner's price. Consider a winner i in VERITAS, and among i's unallocated neighbors (i.e., i's neighbors who did not win any channel in the auction), let j be the one with the highest bid. Is j always i's critique neighbor? If it is, explain the reason, and if it is not, give a counterexample.

9   Buyer group formation is an important step for TRUST to enable spectrum reuse. Given the conflict graph of buyers and the set of bids of sellers, do you think it is a good idea to make the buyer group size more balanced? Explain your conclusion.

GNU radio for cognitive radio experimentation

18

**Michael J. Leferman, Di Pu, and
Alexander M. Wyglinski**
Worcester Polytechnic Institute, United States

Rapid prototyping platforms enable the implementation of new radio system designs. An accessible prototyping platform is a combination of GNU Radio (GR) for software development and the universal software-defined radio peripheral (USRP) for the hardware that transmits the data. This chapter introduces the reader to these two development tools, which will enable a software-defined radio (SDR) developer to implement cognitive radios.

18.1 INTRODUCTION

A *cognitive radio* (CR) is a *software-defined radio* that is also capable of sensing its environment, track changes, and reacts upon its findings. A CR is an autonomous unit in a communications environment that frequently exchanges information with the networks it is able to access as well as with other CRs. From our point of view, a CR is a refined SDR [650]. Presently, GNU Radio is widely used to implement cognitive radio designs, enabling both experimentation and research. Therefore, we focus mostly on GNU Radio in this chapter.

18.1.1 Introduction to GNU Radio

An SDR system is a radio communication system where components that have typically been implemented in hardware (e.g., mixers, filters, amplifiers, modulators/demodulators, detectors) are instead implemented using software on a personal computer or other embedded computing devices [651]. As a result, it transforms radio hardware problems into software problems. The fundamental characteristic of software radio is that software defines the transmitted waveforms and software demodulates the received waveforms. This is in contrast to most radios, in which the processing is done with either analog circuitry or analog circuitry combined with digital chips.

Software radio is a revolution in radio design due to its ability to create radios that change on-the-fly, creating new choices for users. In theory, software radios can theoretically perform the same tasks as traditional radio systems. However, software radio systems also possess a substantial amount of flexibility, which provides the users and even the radio designers the options to create and employ advanced features deemed difficult by conventional radio systems. Controlling a computer, with necessary hardware supports, to play with radios should be easier, interesting, and attractive.

GNU Radio is a free software toolkit for learning about, building, and deploying software radios [652]. GNU Radio provides a library of signal processing blocks and the connection to tie them all together. It is open source, which provides complete source code such that everyone can see how the system is built. To use GNU Radio, the communications systems engineer should be familiar with the following: First, some degree of competence with objected-oriented programming (OOP) is necessary, since GNU Radio employs both Python and C++ programming languages. Second, familiarity with the Linux operating system is required, since GNU Radio is particularly well-suited for this operating system. Finally, expertise in wireless communication systems, digital signal processing, basic hardware, and circuit design are important. All of these will help you understand how GNU Radio works and implement your own system.

18.1.2 The Software

GNU Radio's software is organized using a two-tier structure, as shown in Figure 18.1. All the performance-critical signal processing blocks are implemented

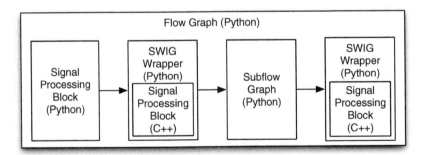

FIGURE 18.1

GNU Radio software architecture.

in C++, while the higher-level organizing, connecting, and gluing are done using Python. Many frequently used signal processing blocks have been implemented and provided to us as parts of the GNU Radio software.

This structure has some similarity with the OSI seven-layer data network model. The lower layer provides service to the higher layer, while the higher layer does not care about the implementation details carried on in lower layers, except for necessary interfaces and function calls. In GNU Radio, this layer transparency exists in a similar way. From the Python point of view, what it does is just selecting necessary signal sources, sinks, and processing blocks; setting correct parameters; then connecting them together to form a complete application. In fact, all these sources, sinks, and blocks are implemented as classes in C++. The parameter setting, connecting operations correspond to some sophisticated functions or class methods in C++. However, Python cannot see how C++ is working. A piece of lengthy, complicated, and powerful C++ code is nothing but an interface to Python.

As a result, no matter how complicated the application is, the Python code is almost always short and neat. The real heavy load is thrown to C++. One rule of thumb should be kept in mind: *For any application, what we need to do at the Python level is to draw a diagram showing the signal flow from the source to the sink, sometimes with the support of graphical user interfaces (GUIs).*

Knowledge of Python is crucial in learning GNU Radio. Python is a powerful and flexible programming language. However, if one has a sufficient C/C++ background, it is not that difficult to learn. Considering that Python has some special characteristics when applied to GNU Radio and some of its useful features may not be necessary in GNU Radio, in Section 18.2 we talk more about combining the Python programming techniques and software radio concepts.

18.1.3 **The Hardware**

The *universal software radio peripheral* is designed to allow general-purpose computers to function as high-bandwidth software radios. In essence, it serves as a digital baseband and IF section of a radio communication system.

The basic design philosophy behind the USRP has been to do all the waveform-specific processing, like modulation and demodulation, on the host CPU. All the high-speed general-purpose operations, like digital up- and down-conversion and decimation and interpolation, are done on the field-programmable gate array (FPGA).

Figure 18.2 shows the picture of a USRP board equipped with four daughterboards (two for the receiver (rx) and two for the transmitter (tx)). A detailed introduction about USRP is presented in Appendix C.

18.1.4 **GNU Radio Resources**

Started in 1998, GNU Radio is now an official GNU project [609]. All the essential material about GNU Radio and USRP can be found on their wiki page

Receive Channel B   Altera FPGA   Transmit Channel A

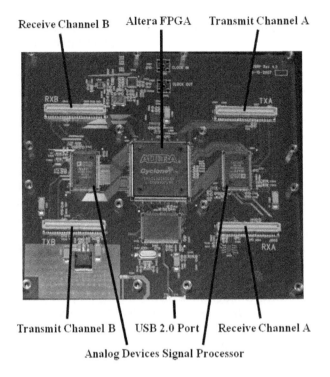

Transmit Channel B   USB 2.0 Port   Receive Channel A

Analog Devices Signal Processor

FIGURE 18.2

The USRP board: one motherboard, two rx/two tx daughterboards.

(www.gnuradio.org/trac). One can download the latest version of GNU Radio, acquire the most updated progress, and make individual contributions. An emailing list for GNU Radio or USRP is also available for subscription.[1]

18.2 ANALOG RECEIVER

As mentioned in the previous section, the software structure of GNU Radio contains two levels. All the signal processing blocks are written in C++ and Python is used to create a network or graph and connect these blocks together. Many frequently used blocks have been provided by the GNU Radio project, such that in many cases you do not need to use the C++ programming language to make your own modules, but rather just Python to connect all the blocks.

In this section, we use an FM receiver example to illustrate how to write our own Python code. By running this code (please refer to Appendix E), you should be able to listen to the FM radio station or other FM signals if an audio device is available.

---

[1]www.gnu.org/software/gnuradio/mailinglists.html.

Almost all the GNU radio Python programs can be divided into three basic parts from the beginning to the end: the *radio class*, the *main function*, and the *program start*. We generalize such a framework in Appendix D. The details of each part can be added according to the real applications. Although we define these three parts in such a sequence, their actual execution order is reversed. When we run a program from the command line, the program's start part first checks to see if this module is being run by itself or imported by another module. If it is being run by itself, the main function is called and the radio defined in radio class works.

Next, the code of the FM receiver is analyzed line by line, and the Python programming, signal processing techniques, software radio concepts, and some hardware configuration are talked about along the way.

18.2.1 **The First Line**

After reading the code of some other GNU Radio Python programs, you will find that the first line of these programs is almost always:

```
1 #!/usr/bin/env python
```

The Python scripts can be made directly executable if we put this line at the beginning of the script and give the file an executable mode. The #! must be the first two characters of the file. You can run this program in the command line using: ./FMReceiver.py arguments.

Consequently, the Python interpreter is invoked and the code in this script executed line by line orderly. Python is an interpreted language, similar to MATLAB. As a result, no compilation and linking is necessary.

18.2.2 **Importing Necessary Modules**

The next step is to import several items into the program environment:

```
1 from gnuradio import gr, gru, modulation_utils, optfir
2 from gnuradio import usrp, blks2, audio
3 from gnuradio.eng_option import eng_option
```

Understanding these statements requires the knowledge of "module" and "package" concepts in Python. If we quit the Python interpreter and enter it again, all the functions and variables we had defined are lost. Consequently, we want to write a somewhat longer program and save it as a script, containing functions and variable definitions and maybe some executive statements. This script can be used as the input of the Python interpreter. We may also want to use a certain function we have written in several programs without copying its definitions to each program. To support this, Python provides a module/package organization system. A *module* is a file containing Python definitions and statements, with the suffix .py. Within a module, the module's name (as a string) is available as the value of the global variable _ _name_ _. Definitions in a module can be imported into other modules or into the top-level module. A *package* is a collection of modules that have similar functions, which are often put in the same directory. The _ _init_ _.py files are required to make Python treat the directories as packages. A package could contain both modules and subpackages (which can contain sub-subpackages).

Modules can import other modules using the Import command. It is customary to place all "import" statements at the beginning of a module. Note that the import operation is quite flexible. We can import a package, a module, or just a definition within a module. When we try to import a module from a package, we can either use import packageA.moduleB, or from package A import module B. When using from package import item, the "item" can be either a module/subpackage of the package or some other names defined in the package, like functions, classes, or variables.

It is worth taking a while to introduce the modules used in this example since these modules or packages are frequently encountered in GNU Radio. The top-level package of GNU Radio is gnuradio, which includes all GNU Radio–related modules. It is located at: /usr/local/lib/python2.4/site-packages.

The gr is an important subpackage of gnuradio, which is the core of the GNU Radio software. The type of *flow graph* classes is defined in *gr* and it plays a key role in scheduling the signal flow. The eng_notation is a module designed for engineers' notation convenience, in which many words and characters are endowed with new constant values according to the engineering convention. The module usrp provides the interfaces to control the USRP board. The usrp is often used as the signal source or sink. We see the details about them later in this chapter.

Looking at the next line, we see the following command:

```
1  from gnuradio.eng_option import eng_option
```

We can either import a complete module/subpackage or just a function, class, or variable definition from this module. In this case, eng_option is a class defined in the module gnuradio.eng_option. We do not need the whole module to be imported, but just a single class definition. The gnuradio.eng_option module does nothing but add support for engineering notation to optparse.OptionParser.

At this point, it is necessary to emphasize again that these imported modules may contain executable statements as well as the function or class definitions. The statements are executed immediately after the modules are imported. After importing the modules and packages, many of the variables, classes, and modules defined in them have been initialized.

18.2.3 The Initialization Function

We now enter the first part of the design process: the radio class part. In this program, we define a specific radio class called digital_radio, which is derived from the parent class gr.top_block. In GNU Radio, many other classes are also derived from gr.top_block. This large "BLOCK family" makes GNU Radio programming neat and simple; it also makes the scheduling of the signal processing clear and straightforward.

Then we implement the method (or function) __init__ of the class gr.top_block to initialize the parent class. The syntax for defining a new method is def funcname(arg1 arg2 ...).

The __init__ is an important method for any class. After defining the class, we may use the class to instantiate an instance. This special method __init__ is used to create an object in a known initial state. Class instantiation automatically invokes __init__ for the newly created class instance.

One important feature of Python worth mentioning before we talk about the details in the function __init__ is that we notice this piece of code possesses no explicit signs of class definition or the start or end of a function. Usually in other programming languages such as C++, we use a "begin" and "end" pair, or a pair of "{" and "}" explicitly to denote the two ends of a group of statements. However, in Python, this is not the case. Rather, statement grouping is performed by indentation instead of beginning and ending brackets. Consequently, it is advisable to be careful about your editing and layout of the code when you write programs using Python. Now let us continue with the function __init__.

The following line of code,

```
1  def __init__(self):
```

declares the initialization method __init__ with one argument. Conventionally, the first argument of all methods are often called self. This is nothing more than a convention: the name self has absolutely no special meaning to Python. However, methods may call other methods by using method attributes of the self argument, such as self.connect(), which we discuss later.

18.2.4 Constructing the Graph

Constructing the graph is the main task of the radio class part, as well as the whole program. As mentioned at the beginning of this section, all the signal processing blocks are written in C++ and Python is used to create a graph and connect these blocks together. Therefore, it is time to define the source, sink, and all the signal processing blocks between them to construct a proper radio.

Listing 18.1

Construct an FM receiver graph

```
1  self.usrp = usrp.source_c()
2  adc_rate = self.usrp.adc_rate()
3  usrp_decim = 200
4  self.usrp.set_decim_rate(usrp_decim)
5  usrp_rate = adc_rate / usrp_decim
6
7  chanfilt_decim = 1
8  demod_rate = usrp_rate / chanfilt_decim
9
```

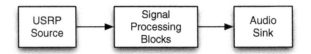

FIGURE 18.3

FM receiver block diagram.

```
10 audio_decimation = 10
11 audio_rate = demod_rate / audio_decimation
12 print "audio rate: %d" % audio_rate
13
14 mux_value = usrp.determine_rx_mux_value(self.usrp, daughterboard_port)
15 print "mux value: %d" % mux_value
16 self.usrp.set_mux(mux_value)
17
18 self.daughterboard = usrp.selected_subdev(self.usrp, daughterboard_port)
19 print "Using Rx daughterboard %s" % (self.daughterboard.side_and_name(),)
20
21 chan_filt_coeffs = optfir.low_pass (1,
22                                     usrp_rate,
23                                     80e3,
24                                     115e3,
25                                     0.1,
26                                     60)
27
28 self.channel_filter = gr.fir_filter_ccf (chanfilt_decim, chan_filt_coeffs)
29
30 self.receiver_block = blks2.wfm_rcv (demod_rate, audio_decimation)
31
32 self.sink = audio.sink(int(audio_rate), audio_device)
33
34 self.connect(self.usrp, self.channel_filter, self.receiver_block, self.sink)
```

In our program, we are building an FM receiver such that the source is USRP, which can receive the incoming FM signals, and the sink is an audio device, which allows us to hear the signal. Since the output data rate of USRP is 64 megasamples/second (MS/s), which is much higher than the data rate of an audio device (32 kHz), we need some intermediate processing blocks to transfer between them. The whole graph is shown in Figure 18.3.

In this example, we are using the usrp block to be the signal source for the FM receiver.

Listing 18.2

Set USRP function

```
1 self.usrp = usrp.source_c ()
```

Here, usrp is the module we imported at the beginning. The usrp module is located at /usr/local/lib/python2.4/site-packages/gnuradio/usrp.py.

It tells us the module usrp is a wrapper for the USRP sources (receiving) and sinks (transmitting). When a source or a sink is instantiated, the usrp module first probes the USB port to locate the requested board number, then it uses the appropriate version-specific source or sink. The `source_c` is a function defined in the `usrp` module. It returns a class object that represents the data source. The suffix `_c` means the data type of the signal is complex,[2] because the signal coming into the USRP is complex (actually a real/imaginary pair).

From the next line, we start seeing the real signals coming in and being processed. Note that, in Python, a comment starts after the symbol #. All statements after # in each line are ignored by Python interpreter. Besides, we save the details of signal processing for the next part.

The last line in this part finally completes our signal flow graph.

Listing 18.3

Connect flow graph function

```
1  self.connect (self.usrp, self.channel_filter, self.receiver_block, self.sink)
```

This method is designed for the flow graph to bind all the blocks together. The signal flow graph is done at this point.

When we use MATLAB to do simulation, it is believed that, to write the code cleanly and efficiently, we need to memorize a number of MATLAB built-in functions and tool boxes well and use them skillfully. The same applies to GNU Radio. Approximately 100 frequently used blocks come with GNU Radio. For a certain number of applications, we can complete the designing using these existing blocks, programming only on the Python level, without the need to write our own blocks. Therefore, being familiar with some frequently used blocks makes it very convenient to write a particular Python code. Furthermore, some useful documents are generated using `Doxygen` that you can also use. After installing the `gnuradio-core` and `usrp` modules, you can find two html packages located at `/usr/local/share/doc/gnuradio-core-x.xcvs/html/index.html` and `/usr/local/share/doc/usrp-x.xcvs/html/index.html`.

You can refer to them concerning almost all the built-in blocks, their usage, and relations.

18.3 DIGITAL TRANSMITTER

Modern telecommunications systems use source and channel encoding to make the digital data being sent more robust. This means the most common type of communication is digital, even voice cell phone and VoIP calls. This example develops a basic digital radio that transmits a random bit stream it generates. Figure 18.4 shows the top-level blocks implemented in our radio.

The data are generated by the Python module running our flow graph, where they are then made into a packet by the `mod_pkt` block. The `mod_pkt` block

---

[2]The data type can be c (complex), f (float), i (4 byte integer), or s (2 byte short integer).

FIGURE 18.4

Top-level blocks implemented in a digital transmitter.

is provided by the GR module blks2 and is itself a flow graph. First, the code explaining how to build the radio is introduced in Section 18.3.2, which explains the rest of the code necessary to run the radio in the GR framework. The reader is encouraged to quickly review the full program in Appendix F to get an idea of where the following code listings fit into the overall program.

18.3.1 Building the Radio

In this example module, a class for the radio is created called digital_radio. The class inherits from gr.top_block, which provides the GR framework the functions and variables needed to treat this object like a flow graph object. Listing 18.4 implements the described digital radio transmitter. Whenever an instance of an object is created, Python automatically calls the __init__ function, if one is implemented.

Listing 18.4

Digital radio class

```
 1  class digital_radio(gr.top_block):
 2      def __init__(self):
 3          gr.top_block.__init__(self)
 4
 5          mods = modulation_utils.type_1_mods()
 6          modulator = mods['dbpsk']
 7
 8          self.packet_modulator = \
 9          blks2.mod_pkts(modulator(**modulator_args),
10                          access_code = None,
11                          msgq_limit = 4,
12                          pad_for_usrp = True,
13                          use_whitener_offset = False)
14
15          self.amp = gr.multiply_const_cc(Tx_amplifier)
16          self.setup_USRP()
17
18          self.connect(self.packet_modulator, self.amp, self.usrp)
```

Since this system consists of only three blocks, the flow graph is very easy to construct. The first line creates the digital_radio class. Putting gr.top_block in parenthesis tells Python to make this class inherit from the gr.top_block class. Inheriting from a parent class provides all of the default functions of the parent class that have not been overwritten in the implementation of the subclass class. Line 2 defines the `__init__` for this subclass, overwriting the default init function. To avoid duplicating the code written in the parent default init class, line 3 calls the default init function. The list of modulators implemented by GR is read from the `modulation_utils` module on line 5.

Line 6 creates an instance of the differential binary phase shift keying (DBPSK) modulator. Lines 8 through 13 create the packet modulation block. The modulator object needs modulation-specific arguments passed into it; in this case, they are defined as global variables and passed into the modulator on line 9. The list of possible arguments can be checked by looking at the modulator code in the file at the path `GRROOT/gnuradio-core/src/python/gnuradio/blks2impl/dbpsk.py`. Additional details on the arguments passed into the `mod_pkts` block can be found in the `pkt.py` file in the same folder. To figure out in what files these modulators were implemented, the text search tool grep was used. This command, explained in the Appendix A, allows the user to search for text inside files, allowing the user to search for the name of the class and returning the line with the matching string and the file containing that line.

Line 15 creates a block that multiplies the input samples by a constant, here set by the global variable `Tx_amplifier`. The `setup_USRP` function is called on line 16, creating the USRP and daughterboard objects, then setting them up. Since those objects are created with the prefix `self.OBJECT`, they are attached to this object and can be accessed from anywhere that has access to this object. The last line uses a GR function to connect these three blocks, creating the flow graph. Next, the `packet_modulator` block is further discussed.

Modulating the Data

The `mod_pkt` block provided by the `blks2` module is a general-purpose block for modulating data. As seen on line 9 of Listing 18.4, this block is simple to create, needing a modulator and whatever arguments are needed for the specific radio that are not the default. The block contains a Python flow graph, as seen in Figure 18.5, which can use any modulation scheme implemented by GR. Figure 18.5 shows the flow graph needed to implement DBPSK.

The data source can be a bit stream from a previous block, or a packet of binary data can be entered into the flow graph using the provided function `send_pkt`.

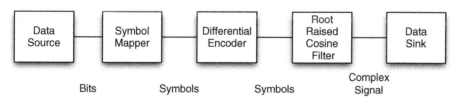

FIGURE 18.5

DBPSK encoder.

Listing 18.5 shows the function, which calls the function provided by this block, simply passing the payload and eof arguments to the `packet_modulator` object. The function returns a Boolean value that indicates the successful transmission of the packet, so this function then returns that information also using the return command. The main function uses this function to put the generated data into the flow graph to be transmitted.

Listing 18.5

Packet creation function

```
1  def send_pkt(self, payload='', eof=False):
2      return self.packet_modulator.send_pkt(payload, eof)
```

Digital modulation schemes are used to transmit bits over the air. This process includes converting bits to symbols, which are then converted into analog signals and up-converted to be broadcast. There are a variety of digital modulation schemes, which use different amplitudes, phases, frequencies, or combinations of the three. This radio uses phase shift keying (PSK) as covered in Section 3.3.5, but quadrature amplitude modulation is covered in Section 3.3.6. To reduce the error rate, a process called *differential encoding* is used, resulting in the name differential phase shift keying (DPSK).

PSK assigns a waveform to a bit, a 1 is represented by a signal with one phase, and a 0 is represented by a signal with a different phase. Differential encoding assigns bit values to a transition: To transmit a 0, the signal stays the same and to transmit a 1 the signal is changed to the other phase. This encoding requires that both the transmitter and receiver know which signal is broadcast initially. The waveforms of different phases are represented by discrete symbols, which are converted to analog signals by the next block.

A filter is used to transform these signals into waveforms. Specifically, a *raised cosine filter* is a low-pass filter frequently used for pulse shaping in digital modulation, as introduced in Section 3.8. A raised cosine filter also reduces *intersymbol interference* (ISI). To prevent ISI the filter is divided into two parts, each being a *root raised cosine* (RRC). The frequency response of a perfect raised cosine filter is symmetrical, about 0 Hz, and is divided into three parts: It is flat (constant) in the passband, sinks in a graceful cosine curve to 0 through the transition region, and is 0 outside the passband. ISI is reduced due to the gradual transition of the response to 0, instead of an abrupt change, which would cause more noise. The result of this filter is a series of sink pulses, which can be easily transformed from baseband to passband using a properly configured USRP.

Setting up the USRP

To make code more readable, a complex set of commands can be moved to a separate function. Here, a dedicated set of commands are moved to a separate function, even though this function is not called more than once. Listing 18.6 is the function used to set up the USRP and its daughterboard. This function is called by the __init__ function of the digital_radio class, the class in which this function is defined.

Listing 18.6

USRP setup function

```
 1  def setup_USRP(self):
 2      self.usrp = usrp.sink_c();
 3      dac_rate = self.usrp.dac_rate();
 4
 5      self.usrp.set_interp_rate(interpolation_rate)
 6
 7      self.usrp.set_mux(usrp.determine_tx_mux_value(self.usrp,
 8                                              daughterboard_port))
 9      self.daughterboard = usrp.selected_subdev(self.usrp,
10                                              daughterboard_port)
11      self.daughterboard_gain = self.daughterboard.gain_range()[2]
12      self.daughterboard.set_gain(self.daughterboard_gain)
13
14      result = self.usrp.tune(self.daughterboard._which,
15                              self.daughterboard,
16                              center_frequency)
17      if result == False:
18          print "Failed to set daughterboard frequency"
19
20      self.daughterboard.set_enable(True)
21
22      print self.daughterboard.side_and_name()
23      print self.daughterboard_gain
24      print self.daughterboard.gain_range()
```

Line 1 of Listing 18.6 sets up the function, accepting an argument as to which object is calling the function, in this case the digital_radio object. The USRP block is created on line 2 using usrp.sink_c; the "c" denotes the samples the USRP expects should be of type complex. Line 3 reads the sample rate of the DAC off the USRP. For more advanced systems, this rate is used to determine the interpolation rate. For this example, the interpolation rate is declared as a global variable and simply set on the USRP on line 5. Line 6 sets the multiplexer value for the transmission of data over the USB connection. This value configures the FPGA to send the received data to the correct daughterboard port.

Lines 9 and 10 poll the daughterboard to determine which daughterboard model is plugged into the specified port; the port is specified by the global variable daughterboard_port. The range of the built-in gain of this board is read on line 11, which returns an array of values. The element of the array indicating the maximum gain is used with the bracketed numeral 2, which returns the third element of the array. Line 12 sets the maximum gain on the daughterboard. Lines 14 through 16 tune the daughterboard to the desired center frequency specified by the global variable center_frequency. The function setting the center frequency returns a Boolean value reflecting whether or not that frequency has been successfully set. That value can then be checked and errors thrown or the frequency adjusted and attempted to be set again; lines 17 and 18 just print the error to the command window.

Line 20 is the most important, and easiest to overlook, line of the configuration. Line 20 turns the transmitter on, actually enabling the broadcast of the data. Lines 22 through 24 print out information about the setup to the command window for debugging purposes. The constructed and configured transmitter is ready to be run, as covered in the next section.

18.3.2 Running the Transmitter

The Main Function

With the class object completed, a function is developed to set up and utilize that class. This is the function where adjustments can be made to make the radio adaptive. For this simple program, only one function is needed to set up and run the radio. The code in Listing 18.7 instantiates the radio, starts the flow graph, generates random data, and sends the data to the packet modulator.

Listing 18.7

The main function

```
1  def main():
2      radio = digital_radio()
3      radio.start()
4
5      while repeat_packets:
6          pkt_num = 0
7          while pkt_num < max_number_packets:
8              data = (packet_size - 2) * chr(pktno & 0xff)
9                  payload = struct.pack('!H', pkt_num & 0xffff) + data
10                 radio.send_pkt(payload)
11             pkt_num += 1
12             sys.stderr.write('.')
13
14      send_pkt(eof=True)
15      radio.wait()
```

Line 2 creates an instance of the digital radio class, and the __init__ function of that class is called automatically. Line 3 starts the flow graph. Line 5 starts the outer while loop to send packets, while line 6 resets the packet counter used in the inner loop started on line 7. Notice the use of white space to indicate where the code to be run in the loops starts and ends. The data for the packet are created on line 8, filling with a series of ones but leaving space for the packet header. Line 9 attaches the packet header to the data, putting the packet into the flow graph in line 10. The packet counter is incremented on line 11. Line 12 uses a standard error output to write a period to the command line, indicating the program is running. An error is used here to print the character immediately; if a standard output is used, the characters are collected internally and written to the command line in one large chunk.

If the transmitter is set to send only one set of packets, these loops are exited and the code on line 14 tells the data source that no more data will be sent to it. Line 15 has the program wait for all the data in the flow graph to be processed before it exits the radio. As this code is written and with the repeat_packets variable set to true, the set of packets are continually run and the radio has to be exited manually. The code in the next section allows classes in this module to be imported without this radio being run automatically.

Starting the Program

The Python language is parsed using spaces. When a module is run, lines with no spaces are run as a Python script. Declared classes, such as the `DigitalRadio` class introduced in the previous section, and functions, such as the main function that is about to be introduced, are not run until they are called by this script. Listing 18.8 is the code that calls the main function.

Listing 18.8

Calling the main function

```
1  if __name__ == '__main__':
2      try:
3          main()
4      except KeyboardInterrupt:
5          pass
```

Line 1 of Listing 18.8 checks to see if the module is being run as a standalone program or is being called from another module. The `__name__` attribute exists for all Python modules and is set to `__main__` when the module is the top-level module. If the module is being run as a standalone program, lines 2 and 3 run the main function until it finishes naturally. Line 4 sets up an exception, listening to the keyboard interrupt to exit from the main function abruptly. Line 5 is needed because of Python's dependency on white space. After certain commands, like the `except` on line 4, Python requires a command. In this case, the program is to quietly exit; the pass command is used as a dummy command to prevent a parsing error and let the script natuall finish. Line 5 could be replaced with the command `sys.exit()` to force the script to exit. To ensure the Python interpreter has included all the lines of the module, the code in Listing 18.8 is the last to be included in any module.

For this basic transmitter, the main function is where the data are randomly generated and sent to be transmitted. If the source of the data is also a block, this function would only create and start the radio and possibly make adjustments in adaptive applications.

The Beginning of the Module

Listing 18.9 shows the code needed to set up the Python environment correctly. In addition to code needed to make the program work, global variables are defined in this section of the program. Global variables should be used sparingly; the variables should be defined on the correct scope, and this level is very rarely appropriate. Here global variables are used so the radio can be configured easily in one location. The correct way to configure radios is using an options parser and command line arguments, but this method was used to simplify this example. See example code in GR to learn about using the options parser.

Listing 18.9

The beginning of the module

```
 1 #! /usr/bin/env python
 2
 3 # This program transmits randomly generated digital data, using
 4 #   Differential Binary Phase Shift Keying (DBPSK)
 5
 6 from gnuradio import gr, gru, modulation_utils
 7 from gnuradio import blks2, usrp
 8 from gnuradio import eng_notation
 9
10 center_frequency = '100M' interpolation_rate = 64 Tx_amplifier =
11 12000 daughterboard_port = (0,0) daughterboard_gain = 0
12 max_number_packets = 10000 packet_size = 1000 repeat_packets = True
13 roll_off = 0.35 modulator_args = {'samples_per_symbol': 2,
14                   'excess_bw': roll_off,
15                   'verbose': False,
16                   'gray_code': True,
17                   'log': False}
```

Line 1 of Listing 18.9 sets the terminal to run this script in the Python environment. Line 2 offers a brief description of the program, enabling other people to use this program and making debugging easier in the long term. Lines 5 through 7 tell the Python interpreter from which modules to retrieve classes, functions, and values. The rest of Listing 18.9 are global variable declarations. Line 13 specifies which port and antenna to use. Following the format (a, b), *a* specifies the port on the USRP motherboard—a 0 is Port A and 1 is Port B. The *b* in this format is to specify which antenna to use if the daughterboard in the given port supports multiple antennas. This code assumes the Basic tx board daughterboard, the frequency set on line 10 has to be adjusted for a different daughterboard.

This concludes the DBPSK transmitter, this code can be run using the ./DBPSKTx.py command with the terminal set to the path containing that file. The corresponding receiver is covered in the next section.

18.4 DIGITAL RECEIVER

The receiver in a communications system is typically more complex than the transmitter. GR modulations schemes are implemented in modulator/demodulator pairs making this high level seem very similar despite the subsystems being drastically different. This layer of abstraction enables the simple implementation of the receiver flow graph, as seen in Figure 18.6. Synchronization and demodulation are handled by the *packet_demodulator block*, as will be reviewed later in this section. A visualization of the result of a fast Fourier transform and controls for the center frequency enable the receiver to be tuned to the frequency of the transmitter. First, the creation of the radio is covered, followed by creating GUI and finally the code needed to run these modules.

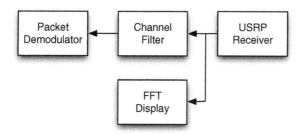

FIGURE 18.6

Flow graph of DBPSK receiver.

18.4.1 Building the Radio

The receiver radio is extremely similar to the transmitter. Listing 18.10 shows the digital radio class that implements the receiver. The amplifier block is replaced with an FIR filter that removes the noise in the signal coming from anything outside the transmission channel. In addition to the filter, a *fast Fourier transform* (FFT) block is added to view the raw received signal. This display allows the user to make sure the receiver is tuned to the correct frequency.

Listing 18.10

Digital receiver radio

```
 1 class digital_radio(gr.top_block):
 2     packets_received = 0
 3     packets_correct = 0
 4
 5     def __init__(self, FFT_usrp):
 6         gr.top_block.__init__(self)
 7
 8         self.setup_USRP()
 9
10         sw_decim = 1
11         chan_coeffs = gr.firdes.low_pass(1.0,
12                                          sw_decim * self._samples_per_symbol,
13                                          1.0,
14                                          0.5,
15                                          gr.firdes.WIN_HANN)
16
17         self.channel_filter = gr.fft_filter_ccc(sw_decim, chan_coeffs)
18
19         demods = modulation_utils.type_1_demods()
20         demodulator = demods['dbpsk']
21
22         self.packet_demodulator = \
23         blks2.demod_pkts(demodulator(**demodulator_args),
24                          access_code = None,
25                          callback = self.receive_pkt_callback,
26                          threshold = -1)
27
28         self.connect(self.usrp, self.amp, self.packet_demodulator)
29
30         self.connect(self.usrp, FFT_usrp)
```

Listing 18.10 shows the digital receiver class, with the first line indicating that the class inherits from `gr.top_block`. Lines 2 and 3 set up two variables to store the total number of packets received and the number of packets correctly received. This information can be used in adaptive communications to automatically change aspects of the radio or, in this case, simply be displayed for the user. Line 5 defines the function called when a new instance of this class is created. Unlike earlier init functions, this takes a parameter, a handle to the signal processing block associated with the *fast Fourier transform* display created for the GUI. This block has to be connected to this flow graph. Line 7 calls the function that sets up the USRP and the appropriate sample rates. Lines 9 through 14 set up a channel filter to remove excess noise. Line 10 starts the multiline description of the filter coefficients. Line 16 creates the filter, applying both the software decimation factor and the generated filter coefficients. Lines 18 through 25 set up the packet demodulator block the same way the transmitter set up the packet modulator block. The difference here is on line 25, where a callback function is passed to the block. This callback function is called each time a full packet has been received. Line 28 connects each of these blocks together. Line 30 connects the FFT block, created by the GUI class, directly to the USRP.

Demodulating the Data

While demodulation is provided by GR, it is important to understand what this block is doing. The receiver has to perform synchronization, demodulation, and error detection. Since DBPSK is being used to transmit the data, DBPSK obviously is used for the receiver. The method for synchronization changes depending on the type of modulation, so GR puts them together in one module. Figure 18.7 shows the flow graph for the DBPSK demodulator.

The data source for the digital receiver comes from the USRP board and is in I/Q format. The prescaler is used to lower the data rate to make processing the signal

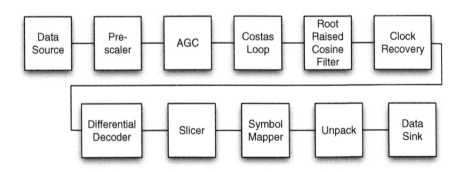

FIGURE 18.7

Flow graph of DBPSK decoder.

easier to do with slower components. A prescaler is an electronic counting circuit used to reduce a high-frequency electrical signal to a lower frequency by integer division. Prescaler circuits are useful in many applications, such as clock generation in digital circuits and *phase-locked loop* (PLL) circuits. It is usually desired to divide a clock signal by an integer N. Although programmable divider circuits can serve the same purpose, they may have frequency limitations, so a prescaler circuit is more commonly used.

Then, the amplitude of incoming signal is adjusted using *automatic gain control* (AGC). AGC is a signal processing technique where the amount of increase is adjusted automatically based on the strength of the incoming signal. Weaker signals receive more gain; stronger signals receive less gain or none at all. The average output signal level is fed back to adjust the gain to an appropriate level for a range of input signal levels.

Next, a costas loop is used for carrier phase recovery, which is the first step for synchronization. The costas loop is actually a PLL circuit. In the usual implementation, a local voltage-controlled oscillator provides quadrature outputs, one to each of two phase detectors; that is, product detectors. The same phase of the input signal is also applied to both phase detectors and the output of each phase detector is passed through a low-pass filter. The outputs of these low-pass filters are inputs to another phase detector, the output of which passes through a loop filter before being used to control the voltage-controlled oscillator.

After the carrier phase recovery, a root raised cosine filter is used to do the pulse shaping. This completes the first half of the filter implemented in the transmitter to prevent intersymbol interference. Some digital data streams, especially high-speed serial data streams, are sent without an accompanying clock. It is the case in our digital receiver, so the receiver generates a clock from an approximate frequency reference, then phase aligns to the transitions in the data stream with a PLL. This process is commonly known as *clock recovery*, and it is our second measurement to assure synchronization. For this scheme to work, a data stream must transition frequently enough to correct for any drift in the PLL's oscillator.

A differential encoder is needed in the receiver path to inverse the differential encoding process at the transmitter. The differential decoder computes the phase changes between two filtered constellation points and decodes the bits. The differential decoder requires the current and previous detector decisions.

The last special block used here is the slicer. A slicer is similar to a modem. For example, it takes audio signals from a radio and converts them to a digital format and feeds the data stream to a computer through the serial port. There are many digital signals on the airwaves that your scanner can receive but cannot decode. A data slicer can convert these signals to a format your computer can understand. With the right software, you can view the messages and system activity contained in these digital signals. Decoding these digital signals opens up a whole new world of radio monitoring. Software programs are available for download on the Internet that can decode the majority of these data signals.

Consequently, the digital signals go through the symbol mapper and are unpacked before they arrive at the data sink, which is actually our computer. These two steps are conventional demodulation operations in digital communication.

Setting Up the USRP

The USRP setup function is, again, very similar to the transmitter. Instead of using information about the transmit chain, like the DAC rate, the information for the receive chain is used, the ADC rate. The rest of the code in Listing 18.11 corresponds to the code for the transmitter in Listing 18.6; please refer to Section 18.3.1 for more details.

Listing 18.11

USRP receiver setup

```
1  def setup_USRP(self):
2      self.usrp = usrp.source_c();
3
4      adc_rate = self.usrp.adc_rate();
5
6      self.usrp.set_decim_rate(decimation_rate)
7      self.USRP_rate = adc_rate / decimation_rate
8      self.FFT_usrp.sample_rate = self.USRP_rate
9
10     self.usrp.set_mux(usrp.determine_rx_mux_value(self.usrp,
11                         daughterboard_port))
12     self.daughterboard = usrp.selected_subdev(self.usrp,
13                         daughterboard_port)
14
15     self.daughterboard_gain = \
16     self.daughterboard.gain_range()[1]
17     self.daughterboard.set_gain(self.daughterboard_gain)
18
19     result = self.usrp.tune(self.daughterboard.which(),
20                  self.daughterboard,
21                  center_frequency)
22     if result == False:
23         print "Failed to set daughterboard frequency"
```

Packet Callback Function

Python allows certain functions to be called on some sort of action. The functions, called *callback functions*, can be called with data when something changes on a GUI or, in this case, when a full packet has been received by the `receive_packet` block. The block performs a cyclic redundancy check (CRC) to verify the data and calls the specified callback function with two parameters, a Boolean indicating the result of the CRC check and the data that have been decoded. Listing 18.12 shows the callback function being used for this radio. This simply prints the total number of packets and the number of packets correctly decoded to the command line.

Listing 18.12

Packet received callback function

```
1      def receive_pkt_callback(ok, payload):
2          (pktno,) = struct.unpack('!H', payload[0:2])
```

```
3 │        packets_received += 1
4 │        if ok:
5 │            packets_correct += 1
6 │
7 │    print "ok = %5s  pktno = %4d  n_rcvd = %4d  n_right = %4d" % (
8 │            ok, pktno, n_rcvd, n_right)
```

The first line defines the function and defines the who parameters. The second line takes the first 3 bytes of binary data from the received packet and decodes that data as the packet number. Line 3 increments the total number of packets counter, and if the CRC check passed, line 5 increments the counter for the number of packets received correctly. Lines 7 and 8 display the status of the receiver in the terminal window from which the program is being run.

18.4.2 Creating the User Interface

GR uses a GUI framework called wxWidgets, a cross-platform multilanguage toolkit. In addition to the basic functionality of wxWidgets, GR provides some useful parent classes from which to inherit, such as the stdgui2. These modules can be found in the gr-wxgui folder in GR. In this example, the GUI is created separately from the radio for clarity, but nothing prevents the two from being combined into the same function. In this section, wxWidgets is reviewed, followed by an explanation of the class creating the GUI and finally an explanation of the callback function that is used when a GUI action is detected.

Panels, Frames, and Sizers

To lay out a GUI in wxWidgets, a few containers are used to set the order of the elements. The GUI is set up by overriding the OnInit function of the parent app class. The top-level container is the app, which has a top-level frame, which can contain any number of frames or panels. Menus, status bars, and toolbars can be added to this frame. Sizer boxes are used to lay out GUI elements, or widgets, to this panel.[3] GR often uses two types of sizers: the horizontal box sizer and the vertical box sizer. These boxes add widgets incrementally in either a column or row. These sizers can be nested to create a very specific GUI layout, as depicted in Figure 18.8 .

Figure 18.8(a) places three elements onto a panel, numbered 1 through 3. It first adds elements 1 and 2 to a *horizontal sizer box* (hbox) then adds that sizer box to a *vertical sizer box* (vbox). Finally, element 3 is added to the vbox to provide this layout. Figure 18.8(b) shows a different layout, with elements 1 and 2 added to the vbox. The vbox is then added to a hbox followed by element 3. Sizers can be nested and combined with widgets to lay out the desired displays and controls. The terms hbox and vbox are simply variable names commonly used in GR for these objects.

---

[3] http://docs.wxwidgets.org/stable/wx_sizeroverview.html.

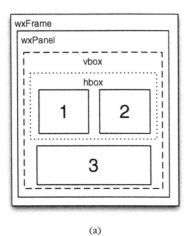
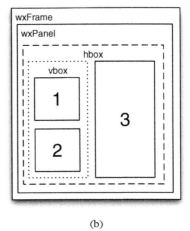

(a)   (b)

FIGURE 18.8

GUI layouts.

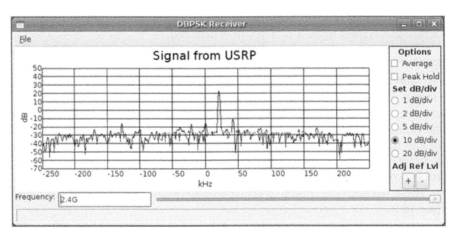

FIGURE 18.9

Receiver GUI.

Creating Displays and Controls

Using wxWidgets, a GUI is created for the receiver. This GUI consists of two parts, a display of the FFT of the received signal and controls that allow the user to change the center frequency of the receiver. The code in Listing 18.13 shows the class that creates the GUI shown in Figure 18.9.

Listing 18.13

User interface class

```
1  class receiver_display(stdgui2.std_top_block):
2      def __init__(self, frame, panel, vbox, argv):
3          stdgui2.std_top_block.__init__(self, frame, panel, vbox, argv)
4
5          self.frame = frame
6          self.panel = panel
7
8          self.radio = digital_radio()
9
10         self.FFT_usrp = fftsink2.fft_sink_c(self.panel,
11                             title = "Signal from USRP",
12                             fft_size = 512,
13                             average = False,
14                             sample_rate = self.radio.USRP_rate,
15                             ref_scale = 32768.0,
16                             ref_level = 50,
17                             y_divs = 12)
18
19         self.radio.add_fft(self.FFT_usrp)
20
21         vbox.Add(self.FFT_usrp.win,4, wx.EXPAND)
22
23         self.controls = controls = form.form()
24         freq_box = wx.BoxSizer(wx.HORIZONTAL)
25
26         def update_frequency_slider(key_values):
27             return self.set_frequency(key_values['freq'])
28
29         controls['freq'] = form.float_field(
30             parent = self.panel, sizer = freq_box,
31             label = "Frequency", weight = 1,
32             callback = \
33             controls.check_input_and_call(update_frequency_slider))
34
35         freq_box.Add((5,0), 0)
36
37         controls['freq_slider'] = \
38             form.quantized_slider_field(parent = self.panel,
39             sizer = freq_box, weight = 3,
40             range = (87.9e6, 108.1e6, 0.1e6),
41             callback = self.set_frequency)
42
43         vbox.Add(freq_box, 0, wx.EXPAND)
44
45         self.connect(self.radio)
```

The first line creates a separate class to create the GUI for the program. It is not necessary to put the GUI components into a separate class, but it is done here for clarity. When a new instance of the GUI is created, the __init__ function is automatically called. Lines 5 and 6 create the frame and the panel on which the GUI elements will be placed. An instance of the radio class previously introduced is created on line 8. Lines 10 through 17 create and configure an FFT block that will display the signal being received. The radio class includes a function to add this FFT block to the receiver, which is called on line 19. Line 21 puts the display of the FFT into a vbox, making this display the top element of the GUI. GR provides an easy way to connect action listeners (or callback functions) to elements of the GUI. It does this with its form framework. The form module, in `gr-wxgui`, allows the setup of these GUI elements in one command. An edit box and slider are created to adjust the center frequency. The form is created on line 23 and a horizontal sizer box on line 24. Lines 26 and 27 provide a callback function for the frequency text box, which gets the frequency from the `key_values` variable and calls the `set_frequency` function. Lines 29 through 33 create a special text box provided by the "form" module for inputting floating point numbers. The framework provides a way to access the value set in the edit box, storing all values from the form in `key_values` and line 27 shows how to read the frequency value from this. The code automatically adds the floating point edit box to the horizontal slider. In Python, a back slash tells the interpreter to concatenate two lines that would otherwise not be concatenated, as seen on line 32. Line 35 inserts a space between this edit box and the next element put into this horizontal box. Lines 37 through 41 create a slider control for the frequency and adds it to the horizontal box as well. The range of frequencies represented by the slide bar is set on line 40—here the FM bands are assumed. The range consists of three values: minimum, maximum, and step size. Line 43 adds the frequency controls to the main sizing box, vbox. GR treats this class as the top-level flow graph, so the radio needs to be added to its flow graph, as done on line 45. With the GUI set up, the function to set the frequency is developed.

The Callback Function

The callback function used by the GUI can be found in Listing 18.14. This function attempts to set the frequency and displays the result.

Listing 18.14

Frequency set callback function

```
 1    def set_frequency(self, target_frequency):
 2        r = usrp.tune(self.u, 0, self.subdev, target_frequency)
 3
 4        if r:
 5            self.freq = target_freq
 6            self.myform['freq'].set_value(target_freq)
 7            self.myform['freq_slider'].set_value(target_freq)
 8            self.update_status_bar()
 9            self._set_status_msg("OK", 0)
10            return True
11
12        self._set_status_msg("Failed", 0)
13        return False
```

The function is defined and the new target frequency is set on the first line. The second line attempts to set the USRP to the new frequency and returns the result to the variable "r." If the tuning is successful, lines 5 through 10 update the elements of the GUI to the new frequency, update the status bar, and make this function return true. Lines 12 and 13 are never run if the tuning is successful, but if it is not, they report the failure.

18.4.3 **Running the Receiver**

Similar to the analog receiver and digital transmitter, the digital receiver needs a number of Python and GR modules, which are set up using the code in Listing 18.15. In addition to importing modules, a number of global variables are used to set up different aspects of the receiver. As discussed in the transmitter section, this is bad programming practice but used in the example to make it easier to follow. See the examples built into GR for using the options parser to configure these aspects of the radio.

Listing 18.15

The beginning of the module

```
 1  #! /usr/bin/env python
 2
 3  from gnuradio import gr, gru, modulation_utils
 4  from gnuradio import blks2, usrp
 5  from gnuradio import eng_notation
 6  from gnuradio.wxgui import slider
 7  from gnuradio.wxgui import stdgui2, fftsink2, form
 8
 9  import wx
10
11  center_frequency = '100M'
12  decimation_rate = 64
13  daughterboard_port = (0,0)
14  daughterboard_gain = 0
15  # Raised Cosine rolloff factor
16  roll_off = 0.35
17  gray_code = True
18  demodulator_args = {'samples_per_symbol': 2, 'excess_bw': roll_off,
19  'verbose': False, 'gray_code': True, 'log': False}
20  alpha = 0.001
```

Line 1 of Listing 18.15 tells the computer to use the Python interpreter to run this code. Lines 3 through 7 import the necessary GR modules and line 9 imports wxWidgets. The rest of this listing controls the knobs this radio provides the user. These parameters must be set correctly to the parameters in the corresponding transmitter.

With these classes and modules, the receiver is ready to be run. Listing 18.16 shows the script at the bottom of the receiver module. After the Python interpreter has gone through the rest of the code in this module, this code is run, creating a new instance of the program and running its main loop. Since Python is running the GUI, putting a wait command here (radio.wait( )) would be inappropriate; waiting here would prevent the GUI from exiting its init function and the GUI would not start. The program instead runs MainLoop( ) and waits for action listers attached to elements in the GUI.

Listing 18.16

Instantiating the receiver

```
1  if __name__ == '__main__':
2      program = stdgui2.stdapp (receiver_display, "DBPSK Receiver")
3      program.MainLoop ()
```

This receiver is run just like the transmitter. After the window opens, the center frequency can be set to the transmitter, and the signal is displayed. Attaching probes and other displays aid in the debugging process. Accessing the signal at the correct parts of the built-in subsystem blocks is somewhat complicated, as changes to commonly used blocks affect other radios using those blocks.

18.5 COGNITIVE TRANSMITTER

A cognitive radio is defined here as a transmitter that is aware of its environment and can combine this awareness with knowledge of its user priorities, needs, operational procedures, and governing regulatory rules. It adapts to its environment and configures itself in an appropriate fashion. The radio learns through experience and is capable of generating solutions for communications problems unforeseen by its designers.

One powerful application of SDR is developing a cognitive engine, an intelligent software package that "reads the meters" and "turns the knobs" of any attached SDR platform. Using an electric combination of artificial intelligence techniques, including case-based decision theory, multiobjective genetic algorithms, and neural networks, it implements a system of nested cognition loops. Applied to public safety communications, this technology is the basis of a working prototype public safety cognitive radio that can scan the public safety spectrum (multiple bands and multiple waveforms, all incompatible) and configure itself to interoperate with any public safety waveform that it finds within 0.1 sec of determining that a signal is present.

IEEE 802.22 is the first standard that uses cognitive radio techniques for wireless regional area networks (WRAN) utilizing white spaces in the TV frequency spectrum on a noninterference basis. The intent of this standard is to provide broadband access in rural and remote areas with performance comparable to DSL and cable modems. The television spectrum was selected for this application due to its propagation characteristics and underutilization. In this example, the transmitter uses time division to send data, then stops sending data to check noise in the channel, similar to the scheme required for the 802.22 standard. The structure of the transmitter is shown in Figure 18.10. The idea of time division is illustrated in Figure 18.11. In the example, data are sent for 9 sec and the noise is calculated in 1 sec. These times can be adjusted to change the balance between transmission and noise calculation and, in the 802.22 standard, is on the order of milliseconds.

18.5.1 Building the Radio

The transmitter flow graph is exactly the same as that for the digital transmitter discussed earlier in this chapter. In a separate flow graph, a receiver is created to calculate the power of the noise in the channel. This information is used to make adjustments to the transmit power, saving power in low-noise environments. Code Listing 18.17 shows the class that creates the noise calculation flow graph. The

USRP receiver is set up as discussed earlier in this chapter and connected to a power calculation block.

Listing 18.17

Channel noise flow graph

```
 1 class noise_calculator(gr.top_block):
 2     def __init__(self):
 3         gr.top_block.__init__(self)
 4         self.setup_USRP()
 5
 6         sw_decim = 1
 7         samples_per_symbol = 2
 8         chan_coeffs = gr.firdes.low_pass (1.0,
 9             sw_decim * samples_per_symbol,
10             1.0,
11             0.5,
12             gr.firdes.WIN_HANN)
13         self.channel_filter = gr.fft_filter_ccc(sw_decim,
14                                         chan_coeffs)
15
16         self.power_calculator = gr.probe_avg_mag_sqrd_c(-60)
17
18         self.connect(self.usrp, self.channel_filter,
19                     self.power_calculator)
```

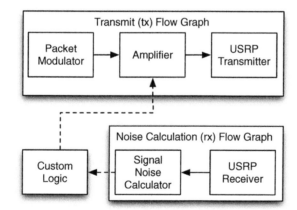

FIGURE 18.10

Flow graph of cognitive transmitter.

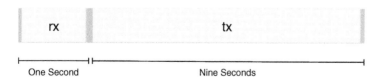

FIGURE 18.11

Time division of cognitive transmitter.

This radio is very similar to the digital receiver introduced in a previous section. The USRP is set up in exactly the same way and a channel filter is used to remove noise in irrelevant channels in the same way. Instead of demodulating transmitted data, a power calculation block is used to detect other signals in the channel:

$$y[k] = x[k]^2 + \alpha \times y[k-1].$$

The block implements the equation in C++ to calculate the average magnitude squared. This is an infinite impulse response (IIR) filter that multiplies past outputs by a factor of α then adds that to the input squared.

18.5.2 Running the Transmitter

Timer Control of Flow Graphs

These two radios are controlled using timer objects to get the desired functionality. The timer objects are built into Python and they take two arguments, a time in seconds and a function to be called when the timer runs out. This is one of the ways to apply a cognitive engine to GNU Radio. In a class similar to this one, any number of "dials" can be read and decisions to adjust "knobs" can be made. It is important to balance the efficiency of the radio configuration and time it takes to calculate that configuration in real-time systems.

Listing 18.18

Cognitive transmitter class

```
 1  class adaptive_transmitter():
 2      def __init__(self):
 3          self.tx_radio = transmitter()
 4          self.noise_radio = noise_calculator()
 5
 6          self.Tx_Callback()
 7
 8      def transmitter_on(self):
 9          return self.Transmitting
10
11      def Tx_Callback(self):
12          self.Transmitting = False
13          self.tx_radio.stop()
14          through
15          self.tx_radio.wait()
16          print "\n Stopped Tx Radio, starting noise calculation"
17          self.noise_timer = threading.Timer(noise_time, self.Noise_Callback)
18          self.noise_radio.start()
19          self.noise_timer.start()
20
21      def Noise_Callback(self):
22          print self.noise_radio.power_calculator.level()
23          self.noise_radio.stop()
24          self.noise_radio.wait()
25          print "Stopped noise calcualtion, starting transmission"
26          self.tx_timer = threading.Timer(tx_time, self.Tx_Callback)
27          self.tx_radio.start()
28          self.Transmitting = True
29          self.tx_timer.start()
30
31      def send_pkt(self, payload='', eof=False):
32              return self.tx_radio.send_pkt(payload, eof)
```

A transmitting flag has been created to control the flow of data into the transmitter. The flag is just a Boolean value that is read by the function declared in Listing 18.18 on lines 8 and 9. The main function uses this function to know when to pass data into the flow graph.

Lines 21 through 29 cover the function called when the noise calculation finishes. This function changes the transmitting flag to true, reads the noise level from the probe block, and adjusts the transmit power of the block. Line 23 stops the noise calculation and line 24 waits for the information currently in the flow graph to flow through it. A new timer object is created on line 26, including passing the function handle to be called when the timer runs out. Next, the transmit flow graph is started along with the timer. A very similar timer callback function is created on lines 11 through 19, which is called when the transmission is finished. The function mirrors the noise calculation callback function.

The Main Function

The main function of the adaptive radio is also very similar to the digital transmitter. Listing 18.19 shows the function with the extra control logic needed to handle the transmitter being off for part of the time. The changes include gating new data into the transmitter to prevent buffers from filling while the noise calculation is taking place. The transmitting flag is checked on line 7 of this code listing. If the transmitter is not on, this function will wait on line 14 for a percentage of the noise calculation time.

Listing 18.19

Main function

```
 1  def main():
 2      radio = adaptive_transmitter()
 3
 4      while repeat_packets:
 5          pkt_num = 0
 6          while pkt_num < max_number_packets:
 7              if radio.transmitter_on():
 8                  data = (packet_size - 2) * chr(pkt_num & 0xff)
 9                  payload = struct.pack('!H', pkt_num & 0xffff) + data
10                  radio.send_pkt(payload)
11                  pkt_num += 1
12                  sys.stderr.write('.')
13              else:
14                  time.sleep(noise_time / wait_factor)
15
16          radio.send_pkt(eof=True)
```

Listing 18.20 shows the script that calls the main function. One important change to this function from previous examples is on line 5. In past examples, a simple pass is placed here, but the timers in this example are created in separate threads. These threads need to be stopped explicitly, so the sys.exit() command is used to destroy all the threads and objects associated with this program. This change allows the program to exit cleanly.

Listing 18.20

Calling the main function

```
1  if __name__ == '__main__':
2      try:
3          main()
4      except KeyboardInterrupt:
5          sys.exit()
```

The way to run this program is identical to the other programs presented in this chapter. This transmitter can be used with the digital receiver introduced in this chapter or with the adaptive receiver created in the Problems section.

18.6 CHAPTER SUMMARY

This chapter introduces the reader to GNU Radio for cognitive radio experimentation. The Introduction section provides an overview of software-defined radio and the GNU Radio toolkit. The hardware commonly used with this software, the universal software radio peripheral, is also introduced.

The analog communication section focuses on the Python level, introducing some Python basics and how Python is used in GNU Radio to connect signal processing blocks and control the flow of the digital data. An FM receiver is introduced as a common example of simple radio development. As the radio is reviewed line by line, the syntax of Python and the simple software radio concepts are introduced in parallel.

The most common type of modern communication is digital. Building upon the analog communication section, a digital radio is developed to transmit a randomly generated bit stream modulated using DBPSK. The corresponding receiver is then developed with a user interface to aid in tuning the center frequency.

Software-defined radio is an enabling technology for cognitive radio development. In the last section of the chapter, a time division protocol is implemented to enable channel noise calculations by the transmitter. This time division is similar to the one specified by the 802.22 standard and outlines how cognitive engines can be added to GR. The reader is then asked to implement a cognitive engine in a receiver based on the packet error rate in the Problems section.

An introduction to basic Linux commands and the installation procedure are in Appendices A and B for users new to Linux. Full code listings for each of the radios can be found in Appendices F, G, and H.

18.7 PROBLEMS

The first two problems require access to a USRP with GSDR installed and make assumptions that imply the daughterboard in port A is used and that daughterboard is capable of transmitting at 2.4 GHz.

1 **Sine wave generator.** In this problem, the built-in Python program usrp_siggen.py is located, run, and used to generate a sine wave and transmit it at a certain frequency.

(a) What Linux command shows the path to the Python file usrp_siggen.py? From the top-level GSDR folder, what is the resulting path?

(b) What Linux command changes the working directory of the terminal to the folder containing the usrp_siggen.py file?

(c) What command runs the usrp_siggen.py program? What command line option displays the available options? What command line options set the center frequency to 2.4 GHz?

(d) What is displayed in the terminal after the program is run?

(e) How can the amplitude and frequency of this sine wave be changed?

2 **Graphical user interface.** This problem lets you observe the sine wave you generated in the previous problem, using the built-in usrp_oscope.py program.

(a) What Linux command shows the path to the Python file usrp_oscope.py? From the top-level GSDR folder, what is the resulting path? What command changes the working directory to this folder?

(b) What command runs this program? What command line options configure the oscilloscope program to receive the sine wave at the same center frequency as it was transmitted?

(c) Run both the signal generator set-up in Problem 1 and the oscilloscope in this program. How does the figure change as the signal generator frequency and amplitude change?

(d) Add some noise to this sine wave at the transmitter. How does the signal change with and without noise?

(e) What settings on the X and Y scales display just over one period of the waveform? What settings property configure the trigger?

(f) Repeat steps 2 through 4 using the usrp_fft.py program.

3 **FM transmitter.** Read Section 18.2 first, then go through the following questions. This problem lets you come to your own Python code to implement an FM transmitter.

(a) What is the purpose of the first line of the program?

(b) What modules or packages are necessary for the transmitter?

(c) Draw a block diagram of the transmitter.

(d) Write a Python class implementing this transmitter calling it AnalogFMTx.

(i) How are the source and sink for the transmitter configured?

(ii) How are the signal processing blocks between the source and sink configured?

(iii) What Python command connects the blocks together?

(e) Run your FM transmitter program, then run the FM receiver program in Section 18.2. Can you hear anything?

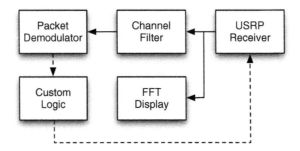

FIGURE 18.12

Flow graph of adaptive receiver.

4 **Adaptive receiver.** Read Section 18.5 first, then go through the following questions. This problem lets you come to your own Python code to implement an adaptive receiver that changes amplification based on packet error rate. Hint: The structure of the receiver is shown in Figure 18.12.
   (a) Write the callback function containing the logic needed to check the packet error rate, adjust the gain of the receiver, and reset the packet count counters.
   (b) Starting with the digital reciter in Section 18.4, write the receiver with a timer object and the callback function.
   (c) What Linux command runs this program?

Cognitive radio platforms and testbeds

19

Danijela Cabric[1], David Taubenheim[2], Gio Cafaro[2], and Ronan Farrell[3]

[1] University of California Los Angeles, United States
[2] Motorola, United States
[3] National University of Ireland Maynooth, Ireland

In this chapter we examine three experimental platforms, each of which represents a different approach to address various complexities of cognitive radio. (Here, we take *platform* to mean a physical implementation of a cognitive radio architecture, supporting a number of applications and experiments.) The motivations to create a platform are many: There is a need to understand how a cognitive radio system, subsystem, or algorithm behaves in realistic scenarios that are not readily or accurately contrived in simulation; analytical studies of cognitive radio behavior often cannot account for the interaction between a device, its wireless neighbors, and its radio frequency (RF) environment as a system, but a physical implementation can bridge the gap between theory and reality; reducing algorithms to practice in a real-time system can bring to light its strengths and weaknesses. For regulators, a successful demonstration of a cognitive radio platform is important to instill the confidence to move forward with the legal framework to open the spectrum to shared use.

19.1 INTRODUCTION

The idea of cognitive radios created great interest in academic and industrial research. As a result, there is a large number of proposals for their physical and network layer functionalities. However, most of research studies and results that evaluate cognitive radios rely on a theoretical analysis or computer simulations. Given the novelty of the functionalities and the lack of proper models of shared spectrum environments, these analytical approaches cannot fully characterize cognitive radio

539

performance. Furthermore, these studies often neglect practical system limitations and dependencies among different functionalities. It is evident that, to enable this technology and fully understand system design issues in its implementation, proposed cognitive radio systems should be verified and demonstrated in realistic scenarios through physical implementation and experimental studies.

From the system design perspective, it is far from clear what mechanisms are best suited to implement cognitive system architecture. Conventional design methodologies rely on layering and clearly defined interfaces among different functionalities, so that they can be independently developed. While a plethora of techniques in the literature propose individual functions, none of them has been demonstrated as a part of the system in a real-world scenario. In addition to demonstration, it is desired to characterize their performance through extensive experimentation in a predefined set of test cases and establish metrics for comparison of different implementations. The need for experiments is also stressed by the inability to realistically model all random sources encountered in a wireless channel, receiver circuitry, and interference environment.

One approach to cognitive radio system design exploration is to use a test bed platform that provides flexibility and rapid reconfiguration for implementing different approaches and allows experiments under controlled but realistic environments. A use of real-time test bed operation enables a large set of experiments for various primary-user types, receiver settings, and network configurations. This is particularly important for a comprehensive evaluation of a statistical behavior that computer simulation addresses with extensive Monte Carlo simulations. Besides addressing their performance limitations, a physical implementation of complex signal processing algorithms and protocols provides estimates of their hardware complexity.

In communications system research, testbeds have been used predominantly to test single point-to-point links and measure established metrics such as bit error rate versus signal-to-noise ratio and effective throughput under different wireless channel propagation environments. Certainly, cognitive radio research requires new test bed capabilities, not only because of new functionalities, but also due to complex interaction between physical and network layers that have to be addressed jointly. A set of requirements that a cognitive radio test bed should satisfy includes:

- Real-time baseband processing for spectrum sensing and agile transmission with high computational throughput and low latency.
- Integration of physical and network layers through real-time protocol implementation on embedded processors.
- Sufficiently wide bandwidth radio front end with agility over multiple subchannels and a scalable number of antennas for spatial processing.
- Central processing of information exchange between multiple radios for controlled physical and network layer development and analysis (e.g., control channel implementation).
- Ability to perform controlled experiments in different environments (e.g., shadowing and multipath, indoor and outdoor environments).

- Support for multiple radios that can be emulated as either primary users or cognitive radios.
- Reconfigurability and fast prototyping through a software design flow for algorithm and protocol description.

19.2 COGNITIVE RADIO PLATFORM BASED ON BERKELEY EMMULATION ENGINE

19.2.1 Test Bed Architecture

The BWRC cognitive radio test bed hardware architecture consists of a Berkeley emulation engine (BEE2) [653], reconfigurable 2.4 GHz radio modems, and fiber link interface for connection between BEE2 and radios. The software architecture consists of the Simulink-based design flow and BEE2-specific operating system that provide an integrated environment for implementation and simple data acquisition during experiments.

BEE2 Emulation Engine

Real-time baseband processing implemented using BEE2 (Figure 19.1) consists of five Xilinx Vertex-2 Pro 70 field programmable gate arrays (FPGAs) [654] in a single compute module with integrated 500 giga-operations per second (GOPS). These five FPGAs provide plenty of parallelism, which can be used to implement computationally intense signal processing algorithms including multiple radios. In addition, FPGAs offer rapid reconfigurability, which can speed up architectural exploration for the final application-specific integrated circuit (ASIC) implementation. Furthermore, the design complexity of FPGA designs (in terms of multipliers, logic slices, and memory) could be used for area estimation of an ASIC with the same functionality. Due to limited on-chip memory resources, each FPGA is connected to a 4 GB DDR memory. These memory capabilities are very useful for logging experimental statistics over a long time period for trace processing.

In addition to dedicated logic resources, each FPGA embeds a PowerPC 405 core for minimized latency and maximized data throughput between the microprocessor and reconfigurable logic. To support protocol development and interact with other networked devices, the PowerPC runs a modified version of Linux and a full IP protocol stack. Since FPGAs run at clock rates similar to those of the processor cores, system memory, and communication subsystems, all data transfers within the system have tightly bounded latency and are well suited for real-time applications.

For this real-time processing engine to interact with radios and other high-throughput devices, multigigabit transceivers (MGTs) on each FPGA are utilized together with a physical XAUI 4X (IB4X) electrical connection to form 10 Gbps full-duplex links. A total of 18 such interfaces per BEE2 are on board allowing independent connections to 18 radios. Each MGT channel is software configurable to communicate and exchange data at any rate below 10 Gbps.

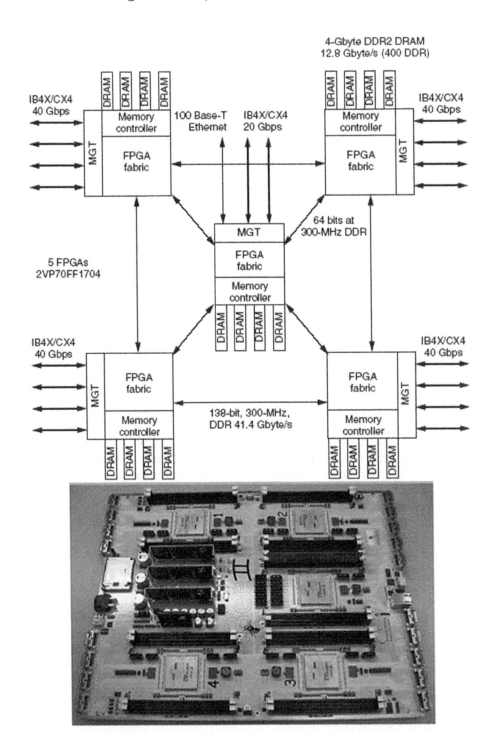

FIGURE 19.1

BEE2 system architecture (top) and implementation (bottom).

Reconfigurable Radio Frequency Front End

Support for wireless networking and flexibility to implement various sensing algorithms and adaptive transmission techniques require highly reconfigurable radio modems. Such radio modems are designed to operate in an unlicensed 2.4 GHz ISM (industrial, scientific, and medical) band, where it can be adaptively tuned to over 85 MHz of bandwidth with programmable center frequency and several gain control stages. A top-level block diagram and implementation are presented in Figure 19.2. Both received signal strength (RSSI) and automatic gain control (AGC)

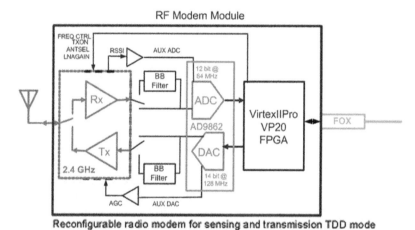

Reconfigurable radio modem for sensing and transmission TDD mode

FIGURE 19.2

Reconfigurable 2.4 GHz radio modem: architecture (top) and implementation (bottom).

are measured in real time to support optimal signal conditioning on the receiver end. This is particularly important in spectrum sensing, where any nonlinearity or noise amplification greatly degrades receiver sensitivity. It also features dual-antenna configuration for switched antenna diversity. This feature allows calibration of receiver noise and gain through signal processing rather than physical connection of a noise meter.

The analog/baseband processing is implemented with 14-bit 128 MHz digital-to-analog converters (DACs), 12-bit 64 MHz analog-to-digital converters (ADCs), and 32 MHz–wide baseband filters. The high resolution of signal converters is chosen so that there is insignificant degradation of receiver sensitivity when digital signal processing algorithms are evaluated. Digital signal processing algorithms such as fast Fourier transforms (FFTs), matched filters, and others can be implemented with on-board Xilinx Virtex-IIPro20. In addition, this FPGA is used to implement radio control functions, calibration of analog impairments, and real-time access to programmable radio registers. To make ADC and DAC samples available for further processing on the BEE2 side, XAUI Infiniband connectors are also integrated on the radio board.

Software Design Flow

To make effective use of the developed hardware, a software design flow for rapid implementation and experimentation of algorithms and protocols is needed. Commonly used algorithm and protocol description languages are standard MATLAB and C. Therefore, the software tool chain is built around MATLAB/Simulink from MathWorks [655] coupled with the Xilinx System Generator [654] for mapping high-level block diagrams and state machine specifications to FPGA configurations. This environment supports simultaneous development of signal processing algorithms and digital design description for their hardware realization. Therefore, no translation is required so that signal processing researchers can realize hardware implementation of developed algorithms.

The original Xilinx System Generator library is enhanced by a set of parameterizable blocks to support interfaces with hardware components such as ADCs, radio configuration registers, and DDR memory. Simulink design is translated directly to FPGA configuration through BEE2-enhanced Xilinx Platform Studio (BAPS) [656]. Furthermore, the tool provides the developer with hardware estimates of the design in terms of number of multiplications, logic slices, and memory. This is extremely important in the design optimization process.

One of the key features in the design flow is the ability to communicate and control hardware registers, block RAMS, DRAMS, and software running on control FPGA in real time. This feature allows rapid postprocessing of acquired signals in MATLAB or access to radio configuration registers during the experiment via automated scripts. Furthermore, it allows an implementation of protocols in C programming language and its direct integration with underlying hardware. This was enabled by enhancing the Linux operating system through the abstraction of hardware registers and memory on the user FPGA using file mapping [657]. BEE2

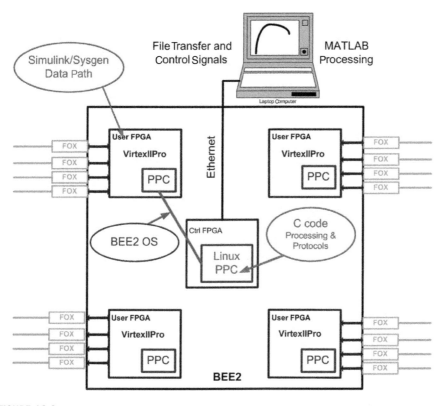

FIGURE 19.3

Software design flow for mapping of algorithms and protocols on BEE2 and experimental control.

can be connected to the local area network so that registers and memory can be accessed and transferred to laptops or PCs via Ethernet. Figure 19.3 illustrates the mapping process of algorithms and protocols on BEE2 as well as experiment control via Ethernet.

With defined cognitive system architecture and a developed test bed platform for its implementation, we proceed with the development of spectrum sensing functionality.

19.2.2 Supported Configurations

Networking Capabilities

While the XAUI interface can provide a transfer of raw ADC bits and DAC samples to the BEE2 signal processing engine, it can extend up to only a couple of meters distance due to losses and noise issues associated with electrical connection. To overcome this limitation, a fiber link between the BEE2 and radio modems is deployed by using optical transceivers compatible with XAUI Infiniband connectors. An optical cable can connect radio modems at distances up to one-third of a mile from the BEE2. In addition, the optical link provides good analog signal isolation on the front-end side from the digital noise sources created by BEE2. This capability allows radios to be spatially distributed while the information is centrally processed on BEE2, which is particularly suitable for implementation of network cooperation or relaying.

An Ethernet interface to BEE2 is provided to support common connection of other networked devices, such as laptop computers with 802.11 and Bluetooth radios or arbitrary signal generators. This capability enables a controlled way of experimenting with cognitive radios, implemented on reconfigurable radio modems, in the presence of legacy users or emulated primary users. A central FPGA processor with a Linux IP stack controls the experiment setup through a Linux networking socket layer interface. For example, an experiment control can include generation of traffic patterns of an arbitrary signal generator or 802.11/Bluetooth radios, including rate, modulation, and power, as well as setup and time recording of on/off transmission intervals. Figure 19.4 illustrates an example experiment

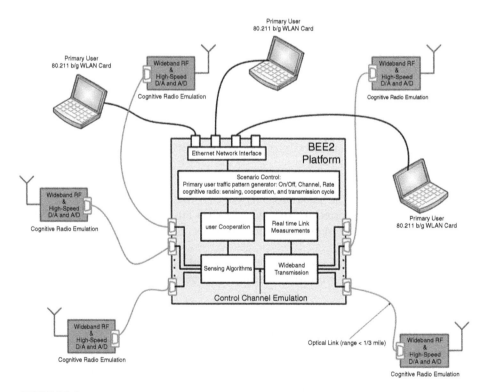

FIGURE 19.4

Emulation of cognitive radios and primary users using a BWRC test bed.

scenario where cognitive radio network uses cooperative spectrum sensing to minimize interference to 802.11 wireless local area network (WLAN) users.

Multiple-Antenna Processing

Advanced physical layer capabilities such as multiple antennas are highly desirable for exploration of techniques like antenna diversity, spatial filtering, or space/time coding. Given the large number of parallel interfaces, a multiple-antenna front end can be created using existing hardware. However, a simple parallel connection of multiple radios does not create a multiple-antenna transceiver. To ensure synchronous operation, every radio must have the center frequency derived from the common local oscillator (LO) and the same digital clock (CLK). Otherwise, uncorrelated noise and phase noise sources degrade the performance of a multiple-antenna system. The system for LO, CLK, and power distribution is designed for scalability up to 16 antennas. Figure 19.5 presents a four-antenna subsystem integrated with a BEE2 platform.

19.2.3 **Case Study: Spectrum Sensing**

Next, we focus on spectrum sensing algorithm implementation and experiments using the BEE2 platform. Spectrum sensing has been identified as a key enabling functionality to ensure that cognitive radios do not interfere with primary users, by reliably detecting primary-user signals. In general, primary-user systems require protection even in the worst-case scenarios when the received signal at a cognitive radio sensor radio could be far below the noise floor. As a result, the new problem that cognitive radios face is the reliable detection in negative SNR (signal-to-noise ratio) regimes.

A plethora of techniques are proposed in the literature for cognitive radio spectrum sensing. In principle, those techniques can be divided into two categories based on the number of sensors used for spectrum sensing. The first category uses a single sensor and relies on physical layer signal processing to meet the required probability of detection under a minimum specified SNR. Here, we consider approaches based on energy and feature detection. The second category of spectrum sensing approaches uses multiple sensors to exploit diversity in their received SNRs and overcome the worst-case channel conditions.

Energy Detector: Implementation and Experimental Results

In terms of implementation, there are a number of choices for energy detection–based sensors. The main design goals are to optimally filter the signal bandwidth, and minimize the contribution of the out-of-band noise and interfering signals for increased sensitivity. Analog implementations (as seen in Figure 19.6(a)) require an analog prefilter with a fixed bandwidth, which becomes quite inflexible for simultaneous sensing of narrowband and wideband signals. Digital implementations offer more flexibility by using FFT-based spectral estimates. FFT-based architecture inherently supports various bandwidth types and allows sensing of multiple signals

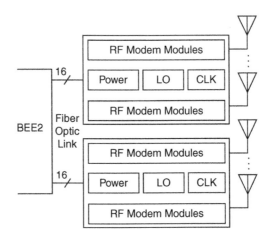

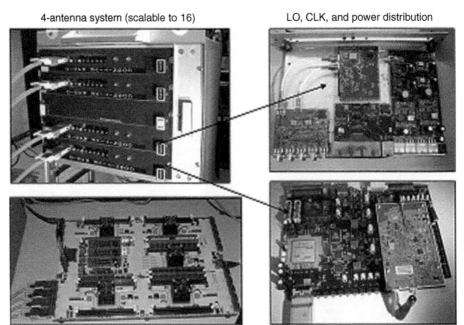

FIGURE 19.5

Multiple-antenna radio front end: architecture (top) and implementation (bottom).

simultaneously. The size of the FFT is the critical parameter, because larger FFT size improves the bandwidth resolution and detection of narrowband signals, but at the same time increases the sensing time. Figure 19.6(b) presents the architecture for a wideband energy detector spectrum sensor.

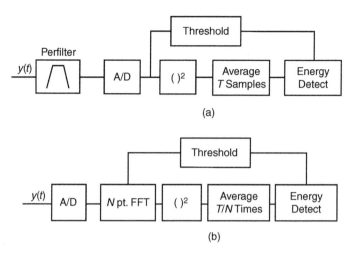

FIGURE 19.6

Energy detector implementations: (a) narrowband architecture and (b) wideband architecture.

In practice, it is common to choose a fixed FFT size to meet the desired resolution with moderate complexity and low latency. Then, the number of spectral averages becomes the parameter used to meet the detector performance goal. We consider this approach in our experiments. The energy detector is implemented on the FPGA of the reconfigurable wireless modem. It is designed using a 1024-point FFT with a fully parallel pipelined architecture for the fastest speed. Therefore, the measured sensing times do not include any computational latency. Due to ADC sampling at 64 MHz, this implementation has 62.5 kHz FFT bin resolution. Each block of FFT outputs is averaged using an accumulator with a programmable number of averages.

The goal of our experimental study was to evaluate and verify the theoretical results of the performance and possible limitations of the wideband energy detector. The measurements were performed for a 4 MHz–wide quadrature phase shift keying signal at 2.485 GHz carrier, in the SNR regime from -7 to -23 dB. For each signal level, we collect two sets of energy detector outputs: one in the presence and the other in the absence of the quadrature phase shift keying (QPSK) signal. From "no input signal" data, we estimate the detection threshold to meet the specified probability of false alarm, $P_{fa} = 5\%$. Then, we apply the threshold to the data where the signal is present and compute the P_d. To accurately estimate the P_d and P_{fa}, each detection measurement is repeated 1000 times. From these, measurements sensing the time versus SNR relationship are derived for $P_{fa} = 5\%$ and $P_d = 90\%$. Figure 19.7 presents the experimental results.

The results show that the theoretically predicted performance holds for SNRs above -20 dB. However, below a -20 dB SNR, the detection becomes progressively harder, and at -23 dB the signal could not be detected regardless of the sensing

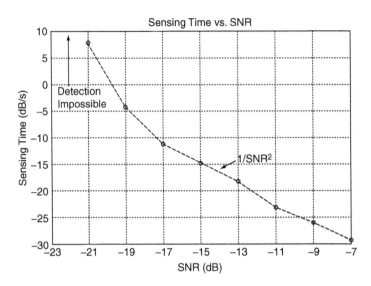

FIGURE 19.7

Measured sensing time versus SNR for energy detector.

time duration. Unfortunately, this deviation has two major consequences. First, energy detection is not a robust method for sensing of very weak signals. Second, this deviation from theoretical results shows that modeling highly negative SNR regimes is not adequate.

Cyclostationary Feature Detector: Implementation and Experimental Results

Cyclostationary feature detectors were introduced as a complex two-dimensional signal processing technique for recognition of modulated signals in the presence of noise and interference. Recently, they were proposed for the detection of weak primary-user signals in the context of spectrum sensing for cognitive radios.

Implementation of the cyclostationary feature detectors requires computation of the spectral correlation function (SCF). The frequency domain estimation methods require computation of an N_{FFT} point FFT plus cross-correlation of all bins and averaging over a period of detection time. However, for the specific signal of interest only the deterministic region of the SCF needs to be computed. Therefore, the multiplication-intensive cross-correlation can be limited to the frequency bins occupied by a signal of interest. We developed a parameterizable architecture for computation of any segment of $N_{FFT} \times N_{FFT}$ matrix representing the SCF.

Figure 19.8 shows a detailed block diagram. To match the input rate of N_{FFT} samples and the output rate of a maximum N_{FFT}^2, K frames ($K < N_{FFT}$) are buffered and arranged so that cross-correlation can be performed through a simple scalable data path of multiplexers, delay lines, and multipliers. We implemented a 256-point FFT and computed SCF of size 256×16 ($N = 256, K = 16$). The implementation

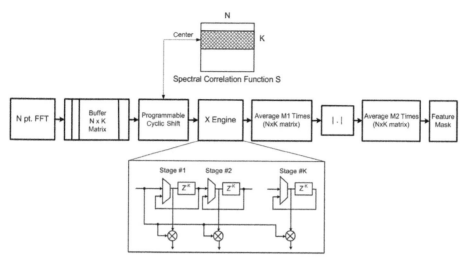

FIGURE 19.8

Implementation of a cyclostationary feature detector for robustness under sampling clock offsets.

Table 19.1 Implementation Complexity Comparison between Energy and Feature Detectors			
Parameter	**Energy**	**Feature (K = 16)**	**Feature (K = 32)**
Multipliers	18	82	162
16 kb–block RAM	4	83	165
4-input LUTs	10,353	21,200	24,389
Flip-flops	12,155	20,559	23,609

complexity is compared with a 256-point FFT energy detector and $K = 32$ cyclostationary detector. Design summaries are reported in Table 19.1. The number of multiplications increases by an order of magnitude with respect to the energy detector. If computation of the entire SCF (256×256) is required, the number of multipliers increases to 1044. The proposed implementation is scalable to any K being a power-of-2 number in the range from 1 to 256.

Through experimentation with the cyclostationary feature detector related to the symbol rate, we find that the sampling of an incoming signal is critical to the feature detector performance. In the theoretical analysis, it is often assumed that the sampling clock is an integer multiple of the symbol rate. However, in the presence of a sampling clock offset, there is a drift in sampling point instances within a symbol time. Intuitively, a random sampling of a periodic signal does not resemble its periodicity. Figure 19.9 presents experimental measurements of

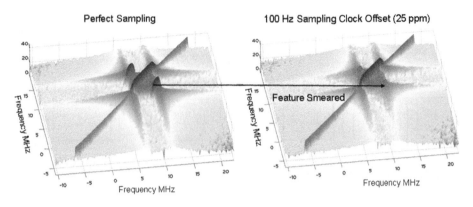

FIGURE 19.9

Spectral correlation function of 4 MHz QPSK signal with perfect sampling (left) and with 100 Hz sampling offset (right).

spectral correlation functions under perfect sampling clock synchronization and under typical sampling clock offset (25 ppm). As can be observed from the figure, in the spectral correlation domain, a sampling clock offset causes smearing of signal features.

To test the robustness of feature detectors under varying noise and adjacent interference where the energy detector could not operate, we experiment with the adjacent channel interference coming from the commercial 802.11 g WLAN with continuous traffic generated by video camera data transfer between two laptops. Figure 19.10 shows the performance of both energy and feature detectors in the presence of this strong adjacent band interference. On the one hand, the sensing time of the feature detector is comparable to that of the energy detector within a constant factor. However, the main advantage of feature detection is that there is no limit on the achievable sensitivity.

19.2.4 Lessons Learned

To efficiently develop and use the large-scale cognitive radio test bed with hardware and software components, a design flow that simplifies both hardware and software development is necessary. Furthermore, for easy adaptation, this design flow should be easy to learn by researchers with a broad spectrum of hardware and software codesign experience.

- **Computing devices in the loop.** The Linux environment on the control FPGA was necessary to enable BEE2 connection to the local area network via Ethernet interface. This further allowed for the full remote access to the test bed through the secure shell (SSH) protocol. In fact, many of the experiments described in this chapter were controlled remotely. The software

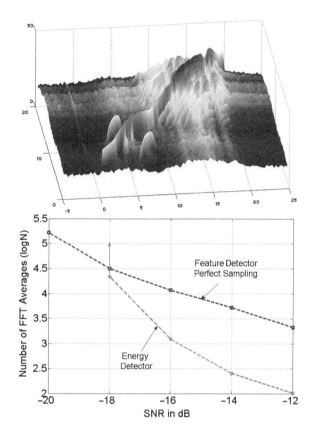

FIGURE 19.10

Features of a QPSK signal and the adjacent 802.11 g signal (top) and detector performance in nonstationary noise due to adjacent band interference (bottom).

environment allowed for remote programming of the user FPGAs, configuration of sensing parameters, and data transfer from the BEE2 to a remote workstation for further processing in MATLAB and verification.

- **Lab equipment in the loop.** To further expedite the experiments, the lab equipment was connected to the Ethernet in the laboratory. In particular, the ESG E4438C signal generator was connected to the Ethernet to allow for remote configuration of generated test vectors. The ESG supports the standard command for a programmable instruments (SCPI) interface. The SCPI provides a uniform and consistent language for the control of test and measurement instruments. The same commands and responses control corresponding instrument functions in SCPI equipment, regardless of the supplier or the type of instrument.

19.3 MOTOROLA 10 MHZ–4 GHZ CMOS-BASED, EXPERIMENTAL COGNITIVE RADIO PLATFORM

19.3.1 Introduction

The development of a fundamentally flexible, low-power transceiver integrated circuit (IC) enabled our team to realize a small form-factor, experimental cognitive radio (CR) platform, shown abstractly in Figure 19.11. Through the flexible programming of this RFIC (radio frequency IC), the platform can receive and transmit signals of many wireless protocols—standard or experimental—at carrier frequencies from 10 MHz to 4 GHz with channel bandwidths from 8 kHz to 20 MHz. This section describes some of the considerations and challenges in designing the RFIC. Then the section examines a CR platform the RFIC enables. The section closes with an example subsystem that takes advantage of the platform's resources to perform cyclostationary analysis, a type of spectrum sensing that can be used in CR systems to assess on-air activity.

19.3.2 Integrated Radio Front End: The RFIC

The RFIC is the cornerstone of the platform's flexibility. Indeed, the RFIC allows the platform to operate on many classes of signals, but a designer will be met with many challenges in implementing the RF sections of CR systems, stemming from their (often desirable) broadband nature. For example, receiver architectures must be chosen to minimize the need for filtering, since low-loss, broadband, tunable filters are not practical in today's technology [658–660], especially in size-constrained applications. A direct conversion approach is preferred for this reason, although it comes with its own disadvantages. Quadrature LO generation over a wide-frequency range is another challenge and is typically mitigated with multiple high-frequency voltage controlled oscillators (VCOs) and complex divider schemes [661]. Unfortunately, this approach often leads to spotty coverage that targets a subset of predefined frequencies. Direct digital synthesis offers some attractive features, like fast switching and continuous coverage, but again, it has disadvantages that require particular attention. Novel circuit architectures have

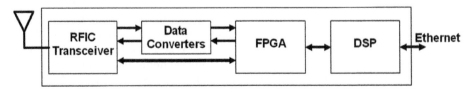

FIGURE 19.11

High-level block diagram of the CR platform, putting RFIC in context.

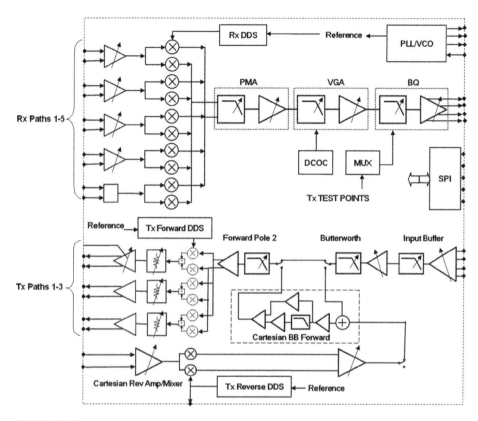

FIGURE 19.12

RFIC architecture.

been applied in the 90 nm CMOS RFIC to overcome problems that encumbered wideband transceivers in the past. Solutions to these and other challenges are now discussed, beginning with an architectural overview.

Overview of RFIC Architecture

Referring to Figure 19.12, three independent direct digital synthesizers (DDSs) use a single 1 GHz phase-locked loop reference to provide differential quadrature LO signals to the receiver, transmitter, and transmitter linearization Cartesian feedback mixers. Direct conversion is used in the receiver, and similarly a direct launch quadrature modulator is used in the transmitter. In the receiver one of five paths is selected to drive a common analog baseband low-pass filter section with programmable corner from 4 kHz to 10 MHz. Provisions are made for a receiver AGC, DC offset correction, and in-band and out-of-band receiver signal strength indicator. Dynamic matching is applied in the direct conversion mixers to improve the second-order intermodulation intercept point (IP2), reduce flicker noise, and minimize DC offset. Differential baseband analog in-phase and quadrature receiver signal outputs are provided for external connection to an analog-to-digital converter and digital signal processing.

In the transmitter, the direct launch quadrature modulator supports both linear and constant envelope modulation formats to cover baseband bandwidths from 4 kHz to 10 MHz (channel bandwidths of 8 kHz to 20 MHz) and RF carrier frequencies from 10 MHz to 4 GHz. There are two fully differential modes of operation: classic in-phase/quadrature (I/Q) fed by external digital-to-analog converters and polar (with an external power amplifier (PA)). In either case, the baseband bandwidth and output power are programmable to meet a given spectral mask requirement.

The RFIC contains five differential receiver inputs that drive differential, passive, quadrature (I/Q) mixers. Two of the five inputs are dynamically matched using chopping. The user selects one of these inputs via a serial interface register to continue on to the postmixer amplifier (PMA) and subsequent processing; the multiple receiver inputs give the RFIC the ability to be used in applications that, for example, switch between bands needing different antennae. Four of the receiver inputs include an inductorless low-noise amplifier (LNA) of which the gain may be set by applying an external voltage to a pin on the RFIC. (Having inductors in the LNAs would have reduced the operational bandwidth—not what we want in a CR front end.) In our CR platform, the voltage is produced by a DAC programmed by an AGC algorithm running in an FPGA. Using this partitioning, the AGC algorithm is flexible by application, not locked into the silicon of the RFIC. Unlike the four other inputs, the fifth input has no internal LNA; instead it couples directly to the down-conversion mixers. This difference allows an external very low-noise LNA to be used in some cognitive radio applications, such as spectral sensing.

The quadrature mixers on rx inputs 1, 3, and 5 are nonchopped mixers built with a quad ring of CMOS transmission gates. Because the mixer design is passive (with active CMOS devices acting as switches), excellent power drain, linearity, and noise figure are achieved. The current drain from the 1.2 V supply of the LO buffers for the nonchopped I/Q mixers is 3.7 mA at 1 GHz, while the third-order intermodulation product of the mixers is +17 dBm. The noise figure of the mixers is 5 dB, essentially equal to the conversion loss. All the mixer current drain comes from the LO buffers and multiplexers, since there is no DC current drain in the actual switching mixer CMOS devices.

The quadrature mixers on rx inputs 2 and 4 use dynamic matching (chopper stabilization [662]) to improve IP2, flicker noise, and DC offset, to mitigate the limitations and device mismatches of an RF system implemented in a CMOS process. (Flicker noise from CMOS can degrade sensitivity and noise figure, while DC offset can affect the error vector magnitude in narrowband and OFDM applications.) The chopping is implemented with three mixers in series, where each mixer is built with a quad ring of CMOS transmission gates. The measured improvement of chopper stabilization is shown in Figure 19.13.

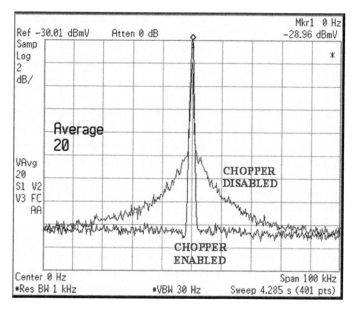

FIGURE 19.13

RFICs chopper stabilization.

The baseband filter architecture has four poles of filtering with two real poles and one complex pole pair in the Sallen-Key BiQuad. The filter bandwidth is programmable via the SPI from 4 kHz to 10 MHz in 6.25% steps or less. The user has independent control of the pole locations of the PMA, voltage gain amplifier (VGA), and BiQuad, as well as control of the BiQuad filter Q. This gives the user the flexibility to trade off filter shape and attenuation for passband amplitude and phase distortion to suit the CR application. Baseband filter gain control is accomplished at three points: at the PMA input, in the VGA, and in the output buffer. The entire baseband filter lineup has a maximum gain of 64 dB and a minimum gain of 4 dB.

Transmitter

Once again referring to Figure 19.12, the transmitter's baseband section provides filtering, programmable attenuation, level shifting, and buffering for the DAC inputs; it drives the forward RF section and the Cartesian baseband forward path. The input buffers provide stepped attenuation for the incoming baseband signals. Programmable active RC reconstruction filters limit the amount of far-out quantization noise and alias images. A resistive-capacitive tracking network provides automatic filter pole adjustment for the reconstruction filters. Closed-loop correction of baseband DC offsets and I/Q phase gain imbalance is performed during a pretransmit warmup period.

For narrowband modulations, the transmitted signal can optionally be power corrected by the Cartesian feedback path, which is essentially a highly linear direct conversion receiver with low sensitivity. The correction path works as follows: First, the output of the external power amplifier is coupled into the RF inputs of this block, where the signal is either amplified or attenuated and down-converted to baseband using the down-mixer and feedback LO. Then, two baseband amplifiers provide programmable gain for the feedback signal before being fed into the Cartesian forward path. With its programmable control, this block provides the feedback gain in the Cartesian system that ultimately controls the output power of the transmitter. Finally, two 6-bit DACs tune out any DC offset errors with a granularity of 2.5 mV per step to correct for any nonlinearity-induced errors in the forward transmission path.

The transmitter's direct launch quadrature modulator (which is not shown in the figure) supports both linear and constant envelope modulation to handle signals with baseband bandwidths of 4 kHz to 10 MHz (channel bandwidths of 8 kHz to 20 MHz) at RF carrier frequencies from 10 MHz to 4 GHz. There are two fully differential modes of operation: classic I/Q and polar (with an external PA). In either case, baseband bandwidths and output power are programmable to meet the spectral mask requirements of a given protocol.

Finally, the forward RF chain contains three separate RF mixer/driver paths along with the associated biasing and gain control. Each path is independently programmable to trade off bandwidth, power control range, and linearity according to the physical layer specification being processed.

Frequency Synthesizer

Direct digital synthesis is used to supply LO signals for the receiver and transmitter (both the forward and the reverse paths) and serves as the clock for the chopping mixers. DDS has a feature set that makes it attractive for CR applications: a very wide tuning range can be achieved with a single VCO. In the RFIC, a single integrated VCO provides the 1 GHz clock to the digital processing blocks of multiple independent DDSs, each of which can be tuned independently and with phase-coherent properties. This arrangement is immune to VCO pulling (transmit remodulation) [663], since the VCO is not operating at the DDS output frequency. Each DDS generates square wave outputs with fast rise and fall times, which are ideal for driving the switching mixers in the receiver and transmitter. Finally, the cycle-to-cycle frequency switching enables unique transceiver capabilities not possible with traditional phase-locked loops with their associated lock times. For example, there is no downtime between switching frequencies, as there is with PLLs, so a CR's channel scanning activity is expedited.

Traditional direct digital synthesizers have two disadvantages compared to PLL-based synthesizers: higher power consumption and greater spurious frequency content. Additional effort during the design of the DDS architecture [664–666] resulted in typical power consumption below 120 mW, using a ROM-less architecture, and spurious frequency components below -35 dBc, using a nonzero-mean dither technique [667].

The differential I and Q outputs can switch frequency on a glitch-free, cycle-to-cycle basis anywhere from 10 MHz to 4 GHz by using multiples of the DDS frequency. With 15 Hz resolution and measured phase noise of -123 dBc/Hz at 25 kHz offset and -150 dBc/Hz at 20 MHz offset, the DDS has the flexibility, noise performance, and switching time performance needed for CR applications.

19.3.3 Experimental Cognitive Radio Platform

The RFIC just described enables a small form-factor CR platform, allowing the team to transition cognitive radio ideas from concept to in situ testing, either at the bench or in the field, while maintaining a high degree of physical portability and RF flexibility. Prior to developing this experimental platform, the design team often had observed that, because of their size, physically large cognitive radio systems—while perhaps possessing enormous processing power and RF filtering capability—were impractical to evaluate in important radio environments. On the other hand, one can design a radio to an arbitrarily small size at the expense of flexibility, which also would work against the goal. Another concern was that any forthcoming innovation could suffer, since large systems discourage simply picking them up and trying out an idea with five or six units dispersed at various locations.

These realizations became motivation to consider carefully the scope of the experiments that would need to be conducted. For example, the platform did not need the ability to run at 4×4 multiple input, multiple output with a 50 MHz wide flexible orthogonal frequency division multiplexing physical layer and Turbo coding, just to verify empirically the validity of a cognitive behavior algorithm. In today's technology, those features necessitate at least one, and probably more than one, top-tier FPGA, resulting in a larger-size and higher-cost platform. Another consideration was whether a given spectral sharing algorithm is robust everywhere, based on successful trials in only one or two radio bands. Even though reducing the flexibility of the radio reduces its complexity and size, spectral agility is important for the evaluation of robust communication. Additionally, it is recognized that the platform's design, as an experimental system and not a product, does not necessarily need to run on battery power, so plugging it into mains allows it to consume as much power as could be dissipated as heat. Ultimately, in order of importance, the design's priorities were ranked with flexibility at the top, followed by size as a close second, then power consumption.

Overview of Platform

This section explores the architecture in more detail, beginning with a brief description of its mechanical architecture. Then the focus shifts to its hardware architecture, first describing the system's hardware components, then moving on to its logical structure.

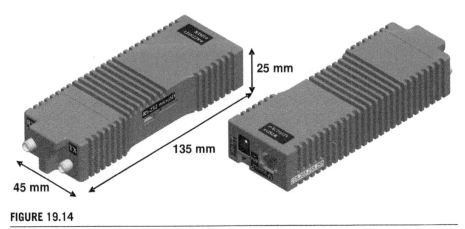

FIGURE 19.14

Exterior view of CR platform.

Mechanical Architecture

The physical manifestation of the design goals is shown in the isometric view of Figure 19.14. The dimensions of the unit are 45 mm × 135 mm × 25 mm, giving a volume of 152 cm^3. Since the platform is intended to operate both inside and outside the laboratory, its housing must absorb shock, shield the interior RF circuitry, and dissipate heat. The result is a robust machined aluminum clamshell of which the partition meets at the carrier board. The surface area of the housing is increased by including fins over which surrounding air can pass to cool the circuits inside, which are coupled to the housing with a formable carbon heat spreader. Exterior ports include the DC power jack (AC to DC conversion for mains power is handled by an off-the-shelf power pack), an on/off switch, an Ethernet port, two LED windows, a micro-SD card slot, two antenna connectors, and a small RS232 port.

Hardware Components

The hardware architecture is implemented on three printed circuit boards (PCBs) (Figure 19.15): the processor board, the transceiver board, and the carrier board. The PCBs are stacked to allow for a low-footprint external housing.

Processor Board

The processor board contains an Analog Devices Blackfin 561 dual-core processor, 64 MB of RAM, and 8 MB of nonvolatile (NV) flash memory. Clocked at 500 MHz, the Blackfin runs uClinux (Linux for microcontrollers) as its operating system (OS) and can perform many digital signal processor (DSP) functions efficiently. Applications for the system are developed on a Linux workstation and cross-compiled to execute under uClinux on the processor. The flash memory is presented as a mounted directory in the OS; this is where we store the OS, configuration bitfiles for the FPGA, small applications, and various data files. Compression is enabled for the

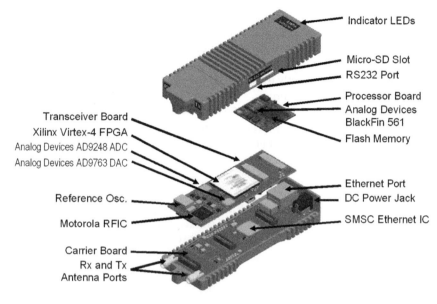

FIGURE 19.15

Exploded view of CR platform and its components.

flash, to increase the usable capacity beyond 8 MB. For larger applications and data files, the platform provides a micro-SD slot, which is also mounted as a directory under the OS.

To access the platform on the Internet or an intranet, the developer may telnet to the OS on the 561 and also may FTP files via its Ethernet connection. Besides using the standard telnet and FTP interfaces, access to the TCP/IP stack is available through an abstraction layer in the OS, so one may customize the way data are consumed or produced for the benefit of applications running on a workstation.

Transceiver Board

The transceiver board is where the RFIC, data converters, and FPGA are located. The data converters are two separate ICs: the dual-input Analog Devices AD9248 ADC and the dual-output Analog Devices AD9763 DAC. Dual-channel converters are used since the RFIC's baseband inputs and outputs are complex signals; that is, one path is the in-phase (I) component, and the other path is the quadrature (Q) component. The AD9248 ADC provides 14 bits of resolution for the received I and Q signals at sampling frequencies up to 65 MHz. On the transmitting side, the AD9763 DAC accepts 10-bit I and Q signals at sampling frequencies up to 125 MHz. A DAC capable of a sample rate higher than the ADC's is selected for the sake of presenting an oversampled, filtered signal to the RFIC to transmit to reduce sampling images on the air.

The data converters are connected to a Xilinx XC4VSX35 FPGA, which performs the digital signal processing for cognitive radio operations, such as spectrum sensing, signal detection, modulation, and demodulation. Of course, depending on the complexity of the (de)modulation, the FPGA can be configured to pass the data converters' signals directly to the Blackfin to be processed by software, rather than in the FPGA's fabric resources.

The FPGA is interfaced to the Blackfin processor via two 16-bit buses: a control bus and a direct memory access (DMA) bus. Working with 32-bit data words is often preferable, so logic on the FPGA and driver software on the processor handle concatenating two 16-bit bus transactions into a single 32-bit word and vice versa.

Carrier Board

While the other two boards contain RF and processing resources, the carrier board contains support components, such as a voltage regulator, an Ethernet port, an Ethernet physical/medium-access layer (PHY/MAC) IC, and antenna ports. In many ways, it is the base board of the system, as the transceiver board plugs into it. This modular design supports the possibility of changing the transceiver board in favor of one with a different FPGA or processor, for example.

Logical Structure

Figure 19.16 shows a logical structure of the CR platform. There are certainly other ways to partition processing between the FPGA and the processor, but

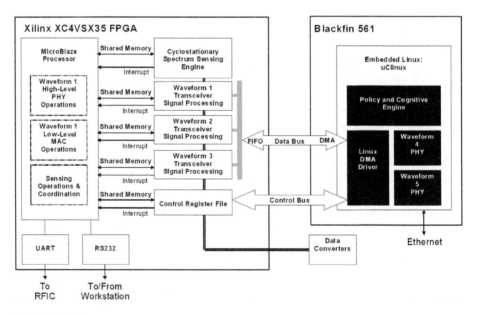

FIGURE 19.16

Logical structure of the CR platform.

this arrangement is suitable for current CR projects. In the logical fabric of the Xilinx XC4VSX35 FPGA, an embedded MicroBlaze processor is instantiated to perform several tasks: high-level PHY operations such as packet parsing and error detection, low-level MAC layer operations such as packet address filtering, and sensing operations such as initiating spectrum sensing and preprocessing sensing data. These tasks were coded in the C language and compiled into an executable form that occupies block memories in the FPGA as program RAM. MicroBlaze also controls the RFIC through an instantiated UART peripheral, and similarly an RS232 peripheral allows for standard input and output calls for debugging the embedded C code on a workstation.

Digital signal processing for high-complexity or high-rate operations—perhaps too intensive to implement in DSP software for real-time use—is performed in the FPGA. For the platform described here, such operations include cyclostationary analysis for spectrum sensing and direct sequence spread spectrum (DSSS) or OFDM waveforms with channel bandwidths greater than 10 MHz.

Depending on the complexity of the waveforms, it might be possible to accommodate more than one waveform's PHY in a single configuration bit file. (The bit file contains the definition of the behavior of the FPGA's logic fabric.) In this system's architecture, the signal processing elements are actually peripherals of the MicroBlaze processor, linked through shared memory and interrupts. Shared memory is a technique that allows a single memory buffer to be accessed by more than one subsystem in the FPGA's fabric. For example, a transmit packet buffer implemented as shared memory can be filled by the MicroBlaze's embedded C code and read out by a waveform's hardware for modulation. To software, the shared memory looks like an array, while to hardware, it looks like a RAM.

Cognitive radio subsystems that are not so processing-intensive can be coded for the Blackfin processor. Examples include waveforms similar in complexity to 2G cellular telephony and some public safety modulation schemes. Algorithmic processing that is more convenient to use within the framework of an OS is better suited to DSP software as well. In the case of the platform described here, a CR's policy engine and cognitive engine are good candidates for the Blackfin.

Linux driver software in the Blackfin allows for high-speed data transfers to and from the FPGA's signal processing subsystems via the data bus and direct memory addressing. On the FPGA end of the bus, first-in, first-out (FIFO) buffers control the flow of data to and from the Blackfin processor. Linux drivers also send commands or queries to the FPGA's control register file; the processor is the bus master, and the FPGA is the bus slave. The data converters can be logically connected to any of the signal processing subsystems in the FPGA fabric or piped to the processor via the DMA data bus.

It is important to point out that one may take advantage of the reconfigurability of the FPGA by storing more than one bit file in the processor board's NV memory or on the micro-SD card, then instructing an application on the Blackfin to load a new FPGA configuration as appropriate. Further, any FPGA configuration may contain a MicroBlaze soft-IP processor or only signal processing elements to be controlled by the processor.

19.3.4 Case Study: Cyclostationary Analysis

This section reviews a short case study that checks the validity of a signal processing subsystem of which the function is to perform cyclostationary spectrum sensing (CSS) on a channel received by the radio. Examples of signal processes that exhibit cyclostationarity include modulation of the amplitude, phase, or frequency of a sine wave and modulation of the amplitude, position, or width of a pulse train [117]. For cognitive radio applications, cyclostationary signal analysis can be applied during spectrum sensing to determine, for instance, whether the energy on a channel is noise or a valid signal. This kind of signal analysis can also be applied to classify modulation types without demodulating the signal [668].

For this RFIC-enabled platform, the team chooses to implement the CSS block in FPGA logic, the signal-processing resources of which are well suited for such an operation with a high degree of parallelism. While there are numerous ways to perform the cyclostationary calculation, the technique illustrated in the block diagram in Figure 19.17 was selected. First, a frame's worth of the complex time-domain input signal is frequency shifted by $\pm\alpha$, where α is an array of frequencies. Then, the positively shifted signal is cross-correlated by the complex conjugate of the negatively shifted signal. The result is stored in a matrix in which each row is a frame of correlation results produced by a different α value. This matrix is averaged a desired number of times with the cross-correlation of a new frame of samples. (Averaging helps mitigate the effect of noise, to reveal cyclostationary features more clearly.) Finally, an FFT of the averaged correlation matrix along the α rows is performed. Note that, for severely band-limited signals, it might be necessary to square the signal before running its analysis [669], so the implementation is designed with such an option.

Like in Berkeley's BEE2 cognitive radio system, many of the signal processing subsystems in the platform, including the CSS, were built using the Xilinx system

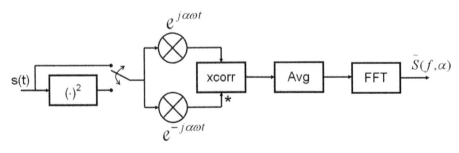

FIGURE 19.17

Signal processing operations for cyclostationary spectral sensing.

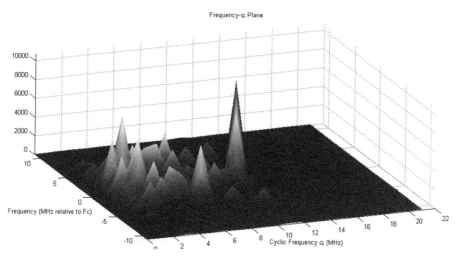

FIGURE 19.18

MATLAB plot of on-channel f-α plane calculated by CSS block in FPGA.

generator. This tool allows designers to build and simulate systems in Simulink, using a library of synthesizable blocks, then generate HDL automatically for a desired Xilinx FPGA. The resulting FPGA implementation is bit and cycle accurate with the Simulink simulation.

In a benchtop test, the team seeks to verify whether the system could determine the chip rate of a DSSS signal received at a carrier frequency of 800 MHz. In actuality, the chip rate of the signal is known a priori to be 11 million per second, so one would expect a correlation peak in the f-α plot in the $\alpha = 11$ MHz row at or near a 0 MHz carrier frequency. (Since a transmit filter is used to limit the DSSS signal's bandwidth, the received signal is squared to recover its chip rate, as described previously.) The peak is evident in the plot of Figure 19.18.

19.3.5 Lessons Learned

The design phase of the platform and the implementation of signal processing algorithms for cognitive radio bring many learning experiences for those involved. Here, we share a few lessons learned.

- **Heat dissipation.** Keeping the hardware cool enough to function reliably and avoid damage can preclude upgrades to digital components. For instance, it would have been possible to use a higher-capacity FPGA or a faster processor in the platform without changing its size, but the housing is unable to dissipate the extra heat produced by the more powerful components. Lesson: Heat dissipation will become an even bigger challenge in future hardware with more capability.

- **Sampling frequency generation.** In this platform's design, a single reference oscillator supplies the FPGA with its clock signal. While a digital clock manager (DCM) in the FPGA can create a new clock zone by multiplying up and dividing down the input clock frequency, the permitted ratios are limited and might not hit a waveform's natural sampling frequency. Therefore, a significant quantity of the FPGA's logic must be used to resample the captured signal to the desired rate using digital signal processing techniques, such as a polyphase filter. In turn, since the resampled data are to be consumed at a rate different than their original sample rate, they must be processed in batch mode, rather than in real time, leading to additional logic. Lesson: For cognitive radio systems in which many unrelated sample rates are likely, consider a PLL for frequency generation. External PLL ICs are available off the shelf, but some FPGAs contain PLLs for flexible clock generation.

- **FPGA memory.** While this platform's FPGA has 432 kB of memory in the form of block RAMs, some operations important to cognitive radio experiments require more. For example, in the cyclostationary spectrum sensing block described in this text, it might be necessary to average results over hundreds or thousands of epochs to pull features out of the noise. However, since the calculations require phase coherence, epochs must be contiguous. Contiguity requires a buffer capable of holding all desired epochs-worth of samples. (It is presently not practical to design the cyclostationary sensing subsystem to run at real time due to the high sample rate and large number of correlations that must be performed, so buffering is necessary.) Lesson: When designing a platform or test bed for cognitive radio concepts, external RAM for the FPGA should be considered if buffer-intensive signal processing is within the realm of possibility.

19.4 THE MAYNOOTH ADAPTABLE RADIO SYSTEM

19.4.1 Introduction

Software-defined radio (SDR) technologies are rapidly maturing, and it is becoming feasible to consider their use in commercial systems. Much of the emphasis in software and cognitive radio research has focused on the challenges faced by the mobile user; however, the fixed base station infrastructure can benefit greatly from the application of SDR/CR techniques. Many immediate cost benefits drive the increased use of SDR techniques in base stations, for example, improved time to market in the face of evolving standards or the benefits of a single platform for multiple standards. There are also operator benefits. Mobile phones and other client devices are consumer devices with minimal cost implementations. Base stations that follow a software radio paradigm can support multiple devices within a single wireless access network, allowing network operators to offer a wider range of services. Again from the network perspective, the move toward femtocells to offer increased broadband services to existing mobile clients requires femtocells to detect, configure, and self-manage in an independent distributed fashion. This will drive further deployment of software-defined and cognitive radio technologies into the network access infrastructure.

Wireless base stations offer certain advantages that encourage deployment of SDR technologies: They are typically wire powered and expensive. This means that they can more reasonably afford the additional processing power and costs required by SDR systems. On the other hand, base stations present a different set of challenges to the SDR community. To ease the performance requirements of a mobile device, base stations have a significantly higher performance requirement, in terms of sensitivity to weak received signals, transmit power, linearity, and bandwidth. Recently there has been a convergence of some aspects of base stations and clients: femtocells base stations mean that power and computational resources are more limited; wideband communication schemes mean that clients must now match the base stations in spectral bandwidth. However, the linked issues of linearity and sensitivity remain. A decrease in linearity or sensitivity would have unacceptable impacts on overall network performance or mobile client power consumption.

In 2004, the National University of Ireland Maynooth joined a research collaboration, the Centre for Telecommunications Value Chain Research. This was a consortium of five universities and Bell Labs, now part of Alcatel-Lucent. As part of this collaboration we began an investigation into software-defined radio platforms for use in base stations. The objective of this program was to investigate the challenges in developing a practical system from a base station perspective and implement a fully integrated software and hardware SDR/CR platform in partnership with the IRiS software radio framework of Trinity College Dublin [670]. In the following subsections, the key issues faced in designing the Maynooth adaptable radio system (MARS) are discussed.

Platform Objectives

When designing any radio platform, a number of design choices need to be made to meet the objectives and constraints of the radio, for example, operating frequency, resilience to interference, signal processing requirements, and available technologies. For a reconfigurable radio platform these issues become more complex. When a standard is developed the performance objectives and constraints are balanced in such a way as to achieve the optimal feasible performance. In a reconfigurable radio platform that is targeted at supporting multiple standards, typically the radio design must take into account the worst-case constraint for each standard and cannot rely on any leniency an individual standard may have provided. As a result, with some notable exceptions [671,672], most experimental cognitive radio platforms do not attempt to match standards but to provide frequency and waveform flexibility.

The MARS platform had the original objectives of being a personal computer–connected radio front end, where all the signal processing is implemented on the computer's general-purpose processor (Figure 19.19). The platform endeavors to deliver a performance equivalent to that of a future base station and wireless

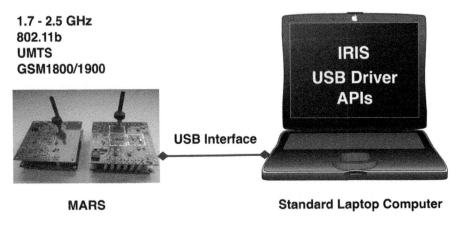

FIGURE 19.19

The Maynooth adaptable radio system platform.

communication standards in the frequency 1700 to 2450 MHz. These simple high-level objectives drive a range of technical design choices:

- **Future base stations.** In 2005, when this project started, most base stations supported a frequency band no greater than 5 MHz. However, there is strong interest in a base station that could support distinct and separated bands of frequencies, enabling base station sharing between operators or where operators may own different bands of frequency. This drove a specification that full-band support should be explored, 70 MHz over an approximate 700 MHz range. Since the start of the project, wideband schemes such as wCDMA and WiMAX have become increasingly popular and bandwidths of at least 25 MHz need to be supported.
- **General-purpose computer connected.** Much of the work on software and defined and cognitive radio is undertaken by researchers more familiar with general-purpose processors than FPGA or DPS devices. The vision of this project is to provide an interface with a general-purpose computer in which modulated baseband data are passed between the computer and the radio platform. This can be easily identified as a performance bottleneck, as one must choose a standardized interface. At the start of this project, high-performance interfaces were limited and, on the assumption that it must be widely supported and not require alteration of the computer, the USB2 standard was selected. This choice has had a significant impact on system performance, which is detailed later.
- **Communication modes between 1700 and 2450 MHz.** This range of frequencies is comparatively narrow but is the most congested frequency range for personal communications. As a project specification we identified the following communication modes that are to be supported: GSM1800, PCS1900, IEEE 802.11b/g, UMTS (TDD and FDD).

In addition, the Irish communications and spectrum regulator (ComREG) licensed to our university two 25 MHz bands of spectrum at 2.1 and 2.35 GHz.

19.4.2 **Design Motivation**

The motivation for the development of this platform is to examine whether it is possible to support multiple communication modes within a single platform. In a single or multiband radio, aggressive filters mitigate the impact of adjacent communications. In a reconfigurable radio, such filters are not available as they inherently restrict frequency flexibility. In the absence of such filters, it is necessary to design a radio system, and in particular a radio receiver, that can satisfy the worst-case requirements. If this is not possible, then the receiver should endeavor to ease the filtering requirements so as to ease the challenges in designing a reconfigurable filter.

In this project the intention is to support a number of wireless communication modes such as IEEE 802.11b, GSM1800, GSM1900, and UMTS/CDMA. To determine the RF system specifications it is necessary to analyze the individual parameters and spectral masks for each standard and integrate them to produce a single worst-case specification. The primary parameters of interest for the design of the platform are: receiver sensitivity, receiver third-order intermodulation product (IP3), receiver noise figure (NF), transmitter power levels, and transmitter phase noise. These parameters determine the blocking performance of the receiver, the spectral and spurious masks of the transmitter, and the expected receiver bit error rate.

One of the most challenging requirements is that of capturing the minimum allowable signal in the presence of blockers. Under the assumption that strong filtering does not exist (as the system is frequency flexible), the radio system must have sufficient dynamic range for digital signal processing to extract the desired signal in the presence of blockers and interferers. Figures 19.20 and 19.21 show

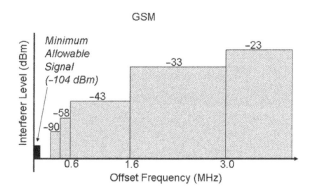

FIGURE 19.20

GSM900 receiver interference profile.

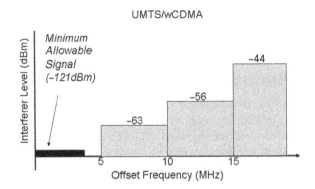

FIGURE 19.21

The wCDMA receiver interference profile.

typical interference profiles for the GSM and wCDMA standards. The GSM standard (at all frequencies) presents the most challenging requirement, as it requires successful reception of a −104 dBm signal (in the base station, −102 dBm for a GSM/GPRS handset) in the presence of a 0 dBm blocker. As our signal capture band is targeted at a complete communication band (for example, the complete GSM band), to capture the smallest required signal in the presence of such a blocker would suggest an analog-to-digital converter with dynamic range of over 106 dB, assuming an ideal receive signal chain. As a first-order approximation this is acceptable, though a more detailed analysis shows that, for a given bit error rate, a lower dynamic range can be used [673, 674].

Table 19.2 displays the receiver requirements for each of the communication standards. A composite specification for the receiver can be calculated by taking the most stringent requirements.

Radio Platform Architecture

Several architectures may be used for a radio, for example, super-heterodyne or direct conversion. In modern digital radio systems the main questions are whether

Table 19.2 RF Specifications for Various Standards

	GSM	UMTS	802.11b
Noise figure (dB)	9	9.6[2]	9[3]
IP2 (dBm)[1]	43	8.0	10
IP3 (dBm)	−18	−21.0	−18

[1] IP2 is required for zero-IF or low-IF architectures.
[2] Assuming a processing gain of 25 dB.
[3] Assuming a processing gain of 10.4 dB.

to use an analog intermediate frequency and at what point in the system to place the digital/analog interface. As the receiver and transmitter face different constraints, it is possible, and common, to make a different choice for each part. The two most common options are to use a low intermediate frequency with subsequent translation to a baseband implemented digitally or to translate the signal band of interest directly to a baseband for digital processing.

With improvements in DAC and mixers, it is increasingly popular to implement a single step transmitter using a DAC to generate a modulated output and a single mixer stage. This is then fed directly to a power amplifier. The input is commonly two baseband signals (I and Q); however, it is increasingly feasible to digitally generate a combined signal, modulated onto a low intermediate frequency. This is simpler than for receivers as there are fewer noise-related issues.

For receivers, traditionally a two-step process is selected for easier implementation. Such an architecture has particular benefits in a frequency-agile radio receiver where there is a fixed bandwidth. In this scenario more aggressive filtering may be applied at the intermediate frequency to allow flexibility in the RF stage. However, in a bandwidth-flexible radio system selection of an intermediate frequency becomes problematic. When selecting an intermediate frequency one must undertake a frequency planning exercise, the objective of which is to pick an intermediate frequency such that harmonics and intermodulation products do not coincide with a signal band of interest. These unwanted frequency components are typically generated from the local oscillator, nonlinearities in the mixers, or through multiplication products. Though these frequency components are generally, though not always, weaker than the desired signal, if they coincide with the signal band, they introduce signal-dependent distortion, which has a significant impact on overall performance. For an application with a wide bandwidth and a highly variable carrier frequency, it is very difficult to identify an intermediate signal frequency that is suitable for the radio of communication bands. On the other hand, direct-to-DC conversion places additional constraints on the receiver, with signal chain performance dependent on linearity and to a large degree on IP2 performance. In addition, there have historically been issues with local oscillator leakage resulting in DC distortion in the receiver. As most communication schemes have content at and near DC, this has been a reason to avoid direct conversion architectures in favor of low intermediate frequency or heterodyne solutions. Recently, low-noise amplifiers and mixers have shown significant improvements and direct-conversion solutions are increasingly viable. Given these considerations, it was decided that the MARS platform would be implemented using a direct conversion architecture for both receiver and transmitter (Figure 19.22).

Given our direct-conversion architecture, the performance of the data converters is critical. A very important aspect of reconfigurable digital radio design is that frequency flexibility and ADC performance are fundamentally linked. Flexibility requires relaxation of the band-select filters near the antenna. If these filters are relaxed then there is an increase in noise, unwanted signals, and blockers.

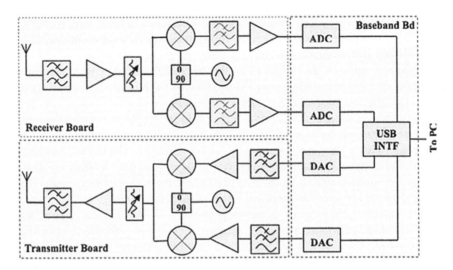

FIGURE 19.22

MARS platform architecture.

To satisfy the sensitivity requirements in this more hostile environment requires a higher-performance analog-to-digital converter so that subsequent digital signal processing can recover the signal. If higher-performance ADCs are available at an acceptable financial and power cost, then the filters can be relaxed. Recently, commercial products started appearing that offer this high-speed, high-resolution performance, but they are expensive and power hungry.

In recent years significant advances have been made in the development of reconfigurable or tunable filters [675, 676]. These new filter designs can be categorized into two groups: switched filters and tunable filter components. Tunable filter components utilize varactors and other variable components to adjust the characteristics of the filter but with limited control and flexibility. Switched filters are more common and switch between banks of components or entire filters as needed to adjust to the communication mode. Base stations use fundamentally different filter technologies. Handset devices use a maximum of 3 W for very small periods of time and more commonly transmit milliwatts of power. Newer base stations transmit at a 60 W peak but sustain an average of between 10 W and 20 W. Older 2G base stations typically transmit at five times, if not more, this power level. In both cases, heat dissipation of the power losses in the filter is a significant problem and traditionally milled metal cavity structures have been used. These blocks of metal are not amenable to electrical tuning, though some mechanical tuning solutions have been developed. Given this, the preference in base station design is to employ an expensive ADC and gain an overall power and cost benefit through a more relaxed filter.

Platform-Host Communications Interface

The purpose of this platform is to interface with an external processing unit, specifically a general-purpose computer. There is a trend in wireless infrastructure design toward increased modularity. The interface between the radio front end and digital signal processing units is becoming increasingly well defined. Two developing and similar interface standards are the common public radio interface (CPRI, supported by Siemens, Ericcson, Huawei) and the open base station architecture initiative (OBSAI, supported by Nokia, Samsung, Hyundai). Given this partition, the performance of the interface is critical to overall system performance. When considering the impact of the performance of this data link, a number of factors need to be taken into consideration:

- Maximum sustained data rate (Mbps), after link management overhead.
- Latency or transport delay.
- Data sample resolution on receive and transmit channels.
- Clock recovery encoding.
- Control channel for reconfigurable radio platforms.
- Number of independent transceivers (multiple-input, multiple-output systems).

Several of these are defined by the link technology chosen or the target communication mode; however, forward error correction and data sample resolution are influenced by design choices. Earlier it was noted that, for the receiver, a high-resolution ADC could substantially ease the design of the band-select filters; however, an implication of this is that these higher-resolution data samples must be transferred to the signal processing unit across this data link or else truncation results in an effective loss of sensitivity. The transmitter on which the signal is being generated and any quantization noise due to limited numerical resolution experience the same attenuation as the transmitted signal. However, without strong filters at the antenna, it is important to minimize any potential out-of-band noise. A second design choice is whether to embed a high-quality clock into the data. When dealing with any RF front end, it is important to ensure that the system is precisely frequency controlled; delivering a high-quality frequency reference through the data to all base stations is a convenient choice and one proposed in the CPRI/OBSAI standards. Such a clock can be embedded in several ways, but one of the most popular is to use symbol mapping. One such scheme is 8b/10b encoding, where 8 bits are maps to a 10-bit symbol, with a 25% overhead. These design choices have a significant impact on the required performance of the data link, as is shown by the examples given in Table 19.3.

For the application we use 16 bits in each direction to provide the necessary receive sensitivity and minimum out-of-band noise. We choose not to utilize clock embedding. For our target of a 70 MHz bandwidth, a data rate of approximately 10 Gbps is beyond the scope of any standard PC interface. In 2006, the best choice we had available was USB2, which had a maximum sustained throughput of about

Table 19.3 Platform Interface Communications Requirements

	Minimum	Typical
RF bandwidth (MHz)	25	25
Sampling frequency (MSps)	50	60
IQ channels	Yes	Yes
rx resolution (bits)	14	16
tx resolution (bits)	10	16
8b/10b clock embedding	No	Yes
Total bit rate (Mbps)	2400	4800

380 Mbps, allowing us a bandwidth of about 3 MHz (simplex) or 1.5 MHz (duplex). Modifying the sample resolutions allows us to double the throughput. This was a fundamental performance bottleneck for the platform with only two solutions: place a processor or FPGA on the board or use a higher-performance link. For the initial development, these options were not pursued and the RF performance was throttled to match the USB interface.

Software Radio Engine

This adaptable radio platform was developed in partnership with Prof. Linda Doyle of Trinity College Dublin, who developed a software radio framework called IRiS, which has being extended to include cognitive radio features. IRiS has been under development at Trinity College Dublin since 1999. It is a highly flexible and highly reconfigurable software radio platform for a general-purpose processor running either Windows or Linux. Designing an integrated platform requires collaboration among the different technical disciplines, and the difficulties that can arise range from the different technical languages (for example, a radio means different things for an RF engineer than for a cognitive radio engineer) to the hardware/software physical interface. From a hardware perspective, a software radio engine must interact with the hardware platform in a sustained, high-speed continuous manner. With a non-real-time operating system such as a general-purpose processor, this cannot be guaranteed; and it is necessary to develop a customized interface module to buffer the data and optimize the flow of data across the USB interface. As a software issue, this may appear to be isolated from the radio hardware, but any sufficiently long gaps appear as interruptions in the modulated radio signal. In addition, most communication standards have tight latency requirements, and excessive buffering exacerbates the latency issues that all development platforms experience.

The IRiS architecture is illustrated in Figure 19.23 and comprises DSP components configurable through an extensible markup language (XML) file. Examples for

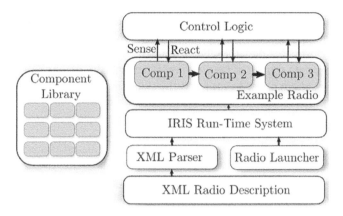

FIGURE 19.23

IRiS architecture.

such components are modulators, framers, and filters. Each of the components has a set of parameters and an interface to the control logic, which allow for reuse in different radio configurations. The control logic is a software component designed for a specific radio configuration; that is, it is aware of the full radio chain while the processing components are not. This control logic can subscribe to events triggered by radio components and change radio parameters or reconfigure the radios structure. This enables the IRiS framework to support cognition through this control mechanism.

To design a radio with IRiS, an XML configuration file is written that specifies the radio components and their parameters and connections. Optionally the radio designer can implement a control logic manager for dynamic radio reconfiguration. On startup, the XML file is parsed and the run-time engine creates the radio by instantiating and connecting the specified components. The run-time engine then loads the control logic and attaches it to the components. Finally, the radio is started and blocks of data generated by the source component are processed by each of the components in the radio chain. The control logic can react to events triggered by components with anything from diagnostic output to a full reconfiguration of the radio.

To integrate the two platforms, two additional components were written: an optimized Linux USB driver and an API to manage all interactions. The role of the API module is to ensure smooth guaranteed transport of data between the two systems. To avoid data loss in the receiver chain, it is required to read continuously from the RF transceiver and deliver the data to the IRiS receiver components. This is accomplished by collecting the data in a thread running independently of the signal processing component. This thread continuously reads data from the hardware and inserts such data into a buffer shared with the IRiS component. When this component processes the data, it reads the data from this buffer and outputs them to the next component in the radio chain. One of the major issues in software radio frameworks is the movement of data between components, which can incur significant performance penalties. For sustained performance care should be taken when planning data transport between modules and the planned use of memory. The transmitter chain proved simpler as the IRiS component is generating the data and the only requirement is that the buffers remain filled.

Radio Receiver

In a direct-conversion receiver architecture, a direct trade-off is made between RF band-select filtering and the performance requirements of the ADC. In the absence of strong filters, the ADC must have sufficient resolution to support the dynamic range required to separate interference from weak signals. An ADC with a signal bandwidth of 70 MHz and 106 dB (in excess of 17 bits) resolution is highly challenging, but devices available at the time of development were capable of delivering 16-bit performance at high speed though with high power consumption. We selected a family of pin-compatible ADCs from Linear Technologies that can deliver up to 105 MSps (LTC220* family). This enables lower-performance ADCs to be used seamlessly where the baseband signal processing cannot support higher speeds.

The RF low-noise amplifier selected is the the Freescale MBC13720. This part is a low-noise amplifier with a bypass switch. It generates a gain of 12 dB and noise figure of 1.55 dB at a frequency of 2.4 GHz. The LNA is able to operate in a frequency range from 400 MHz to 2.4 GHz. It features two enable pins to control the amplification stage, which is software controlled. The gain at this stage has limited programmability. For noise mitigation, maximizing early-stage gain is the preferred option, with greater gain control available at the baseband stage.

The performance of the demodulator is important in a direct-conversion architecture. The AD8347 device from Analog Devices was chosen. It is a direct quadrature demodulator with RF and baseband AGC amplifiers. Its noise figure is 11 dB at maximum gain, and it provides excellent quadrature phase accuracy of one degree and I/Q amplitude balance of 0.3 dB. This high accuracy is achieved by the polyphase filters employed by the local oscillator quadrature phase splitter. The DC offset problem is minimized by an internal feedback loop. Any remaining DC offset effects can be corrected by digital correction, but this was not implemented in the current prototypes.

In a frequency flexible system an agile local oscillator is required. Often a clock/data recovery circuit is used to lock onto the transmission frequency; however, in an SDR architecture a band of frequencies is captured and clock recovery is undertaken digitally. The primary criteria for the local oscillator, in an SDR RF front end, is agility and low phase noise. We selected a low-power delta-sigma fractional-N PLL from National Semiconductor (LMX2470) with a mini circuit VCO ROS-2500. The sigma-delta modulated fractional-N divider was designed to drive close-in spur and phase noise energy to higher frequencies. The modulator order is programmable up to the fourth order, permitting us to alter the phase noise characteristics at different frequency offsets. The device can operate in the range

500–2600 MHz with a phase noise of −200 dBc/Hz. It is optimally operated in a smaller range, but this can be adjusted by changing the local oscillator frequency.

Radio Transmitter

The three main components in a direct conversion transmitter are the power amplifier, the modulator, and the DAC.

The modulator chosen is the Analog Devices AD8349. It is a quadrature modulator that is able to operate with an output frequency range of 700–2700 MHz. It features a modulation bandwidth from DC to 160 MHz and a noise floor of −156 dBm/Hz. Dual different I/Q inputs are provided from the DAC and, to improve the noise performance, the local oscillator drive. The output power generated by the modulator is within the range of −2 to +5.1 dBm.

The power amplifier is constructed as a two-stage element: a fixed gain power amplifier and a digitally controlled variable gain amplifier. The power amplifier used is the MGA-83563 from Avago, which is a broadband high-linearity amplifier. It works in the frequency range of 40–3600 MHz and achieves a small signal gain of 20 dB with a noise figure of 4.1 dB. This variable gain amplifier is the Analog Devices ADL5330, which operates from 10 MHz to 3 GHz frequencies, with a gain control range of 60 dB. The combined system can deliver 22 dBm of power in 256 programmable steps.

Digital-to-analog converters are more capable than ADCs for any given technology. For this application it was possible to get a dual-path 16-bit DAC from Maxim (MAX5875) that can support output rates of up to 200 MSps. It features an integrated +1.2 V bandgap reference and control amplifier to ensure high-accuracy and low-noise performance. The output rate is adjustable based on the provided clock frequency.

Final Design

The objective of the platform is to determine the feasibility of implementing a base station–orientated front end using off-the-shelf components. The final architecture utilizes high-performance ADCs and DACs and direct conversion architecture for both the transmitter and the receiver. To provide reconfigurability, a control plane is added to allow for control of frequency, receiver gain, transmit power, and data conversion rates.

The implementation of the MARS platform is as two separate simplex elements: a receive-only and a transmit-only board (shown in Figure 19.24). Duplex operation is avoided due to the limitations of the USB throughput. To overcome this issue, an FPGA-enhanced version of the baseband board exists that allows for duplex operation with some on-board processing.

19.4.3 Experiments and Use Cases

The development of the platform was an exploration of the challenges in implementing a base station–orientated reconfigurable platform. As such it provided us

FIGURE 19.24

The modular MARS transmit and receive platforms.

with many insights on how the technical issues are subtly different than those experienced in handheld designs. The completed platform is only the first stage in a continuing program of improvements in this platform. The current system is now being used at NUI Maynooth to support research projects in the area of novel wireless communication modes and as a means for experimenting with new radio component designs. As an integrated software (IRiS) and hardware (MARS) platform with an ongoing research program, it allows us to explore hardware components (for example, novel digital power amplifiers) that offer new capabilities without having to wait for a commercial system with these elements.

One of the more interesting experiments that we have undertaken is to explore the interoperability of the MARS platform and the universal software radio platform (USRP). Due to the scarcity of radio platforms, it can become difficult to separate overall system performance from the characteristics of the hardware platform. To explore this issue, we used the IRiS framework with the USRP to transmit images to an equivalent MARS receiver. The high-level experimental setup is depicted in Figure 19.19 and the overall SDR platform was tested under a number of use cases, for example,

- Spectrum sensing.
- Still image and video transmission.
- Novel communication schemes.
- Interoperability testing with the USRP.

One test of the integrated SDR platform, comprising MARS and IRiS, was successful transmission of streamed images. To isolate platform artifacts, a USRP and a MARS platform were used interchangeably as transmitter and receiver. The IRiS software engine has appropriate software interfaces for the two platforms. The IRiS software engine reads a bitmap image, frames the data using a simple structure, with appropriate data whitening and error correction encoding. Differential quadrature phase shift keying (DQPSK) is used to modulate the data into four symbols. To limit the spectral footprint of the signal, the data are up-sampled and filtered with a root raised cosine pulse shaper. The resulting I/Q samples were delivered over USB to the radio front end. At the receiver, the MARS platform demodulates the data and delivers unprocessed I/Q samples over USB to the software engine. IRiS then undertakes filtering, clock data recovery, and demodulation. The data are then deframed and reconstructed into the image. In this experiment we used a 1 MSps transfer rate. In this mode of operation we can operate over six times faster, limited primarily by the processing performance of the PC or laptop used. The results of this experiment are shown in Figure 19.25, where the resulting image and constellation diagram are presented. The constellation diagram provides an indication that the error vector magnitude is acceptably small and good communication is possible.

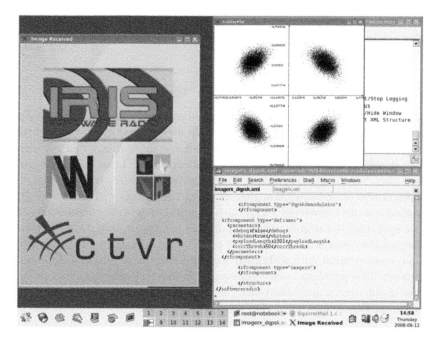

FIGURE 19.25

Transmitted image with constellation diagram.

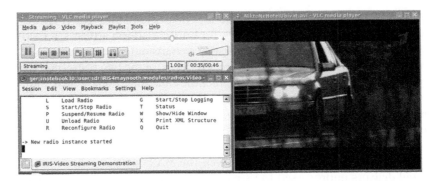

FIGURE 19.26

Example of video received using the SDR platform with performance statistics.

In another experiment, a video sample was transmitted and received using two MARS platforms (Figure 19.26). A DBPSK modulation scheme was used. The transmitted signal bandwidth was approximately 300 kHz with an I/Q sample rate of 2 MSps. This proved acceptable for video transmission but higher throughput could be obtained with higher-order modulation schemes. The error vector magnitude suggests that a more dense constellation diagram could be implemented without significant impairment of performance. The limitation on using a higher-modulation scheme lies in the software engine, and this is likely to improve with time and processing power.

19.4.4 Lessons Learned

The experience of developing this radio platform highlights a number of design choices that dominate the design process. A number of these are discussed in detail but a quick review of their titles is a sufficient argument to say that the greatest issue in designing a reconfigurable radio platform is the interdisciplinary nature of the team that must be involved. The original design team that commenced this project was biased toward analog/RF design engineers with limited digital or embedded systems knowledge. With the presumption of many engineers that their area of speciality is the most challenging and surely the other areas are easier, insufficient attention was initially paid to the digital design issues. The project required collaboration with additional domain expertise to successfully complete the project. This problem occurs repeatedly in industry and academia, and for a successful outcome, a multidisciplinary team needs to be involved from the project specification stage.

Partitioning of Processing Capability

During the project specification stage, there was an argument whether an FPGA or onboard processor should be added to the design, as in the universal software radio platform. It was decided that an 8051 microcontroller would be sufficient for managing the control plane and all signal processing would occur elsewhere. The immediate impact of this decision is that raw data need to be transferred over the USB link. Even minimal processing onboard would greatly reduce the data rate and allow increased signal bandwidth to be supported. A second important issue is whether a general-purpose processor can support the 70 MHz bandwidths we were attempting to achieve. All software radio frameworks to which we have had access are currently incapable of achieving these speeds without some form of hardware acceleration (such as with an FPGA or dedicated graphics chip [677]). With the improvement in software engines and processing power, this should be resolved in time, but currently most cannot deliver the necessary performance. Finally, in our design it became unwieldy to manage and route the control plane and data channels between the receiver and transmitter and the USB device. An FPGA with its ability to internally route data among its many pins provides a powerful implementation tool, even if no signal processing is undertaken. Our conclusion after this process is that, though the CPRI/OBSAI standards state that no signal processing will occur at the RF elements, a very powerful argument can be made that a dedicated processor, and most likely a small FPGA (for example, a Spartan3 or Virtex5 device from Xilinx), is needed to manage the communication link, the digital control plane, and to ease the routing challenges.

Host/Platform Connectivity

For any platform where a significant portion of the physical and data link layers are not implemented on the board, a substantial data transport mechanism is required. In our platform, like other platforms developed at the same time, we chose to use USB2. This was relatively new at the time, widely available on computers, and offered reasonably high-speed connections. As shown earlier, USB2 does not begin to approach the speeds required to support any reasonable-sized bandwidth. There will always be a need for greater bandwidth and one should be ambitious in the connectivity included in the design. One option is to consider using fiber optic links. The past few years have seen significant improvement in their performance and maturity, and many FPGAs now natively support gigabit communication links, such as PCI-express, Infiniband, or gigabit Ethernet. Using an optical technology provides a convenient upgrade path to higher performance with little change in architecture.

Embedded Software Design

When visualizing a software radio platform, often the first thoughts to mind are policies, software, and the radio hardware. We discovered that a critical aspect of a successful design is the appropriate use of embedded or device-level software. On the radio platform, the microcontroller needs to interface with the PC over a USB device and manage the digital control plane that configures all the hardware elements in such a way as to be implemented while receiving data. As in obtaining maximum performance from the USB link, a highly optimized Linux USB driver has to be written that buffers then organizes the data in optimal bundles for transmission. When functioning correctly, this code is transparent and easily overlooked. However, the complexity and difficulty of the implementation require specialist skills commonly outside the scope of a typical software engineer or computer scientist.

19.4.5 Future Plans

The Maynooth adaptive radio system is a continuing program of research into software-defined and cognitive radio. The current platform has been very useful in exploring the base station–related issues for reconfigurability. As part of our ongoing program, students at NUI Maynooth are continuing development in a two-step process. We are currently finalizing designs for a Xilinx Spartan3 FPGA to be placed onboard to manage routing and provide some processing capability. This allows increased flexibility of the board and some limited standalone functionality.

The second step is more aggressive and focuses on developing a system that is scalable to an arbitrarily large number of radio transceivers connected through optical links. This design uses a baseband processing board with significant FPGA processing capability, not unlike that of the Berkeley emulation engine shown earlier. This board supports multiple wideband RF transceivers through CPRI-compatible optical interfaces supporting at least 1.5 Gbps. The board communicates with a PC through a PCI-express adaptor card, which can deliver at least 2 Gbps in each direction per lane, with multiple lanes available on most modern computers. This provides sufficient connectivity to support wideband communication schemes such as IEEE 802.16e (mobile WiMAX). The objective is to develop a platform that can support a variety of applications where individual RF transceivers can be driven from a single point:

- Multiple-input, multiple-output applications.
- Remote RF base stations.
- Advanced beamforming with large arrays.
- Simultaneous multimode cognitive radios.
- Wideband communication schemes.

This platform is currently at early stages of development. Fortunately, a substantial number of the components are now available commercially, and this should assist in what we have learned through experience to be a highly challenging and complex problem. This project is being developed as part of an open collaboration with local industry and other universities.

19.5 CHAPTER SUMMARY

The design and implementation of a cognitive radio testbed depends highly on its intended application. In this chapter three experimental testbed platforms were presented, which focused on the exploration of cognitive radio systems from the varying perspectives of spectrum sensing, the mobile unit, and the base station element. Each design faces unique challenges and in this chapter we present the motivations, architectures, challenges, and learned experiences from each testbed.

The first platform is developed using the Berkeley emulation engine, reconfigurable 2.4 GHz radio modems, and fiber link interface for connection between BEE2 and radios. The software architecture consists of Simulink-based design flow and the BEE2-specific operating system that provide an integrated environment for implementation and simple data acquisition during experiments. The testbed is particularly suited for the development of physical and network layer functionalities and their experimental characterization in realistic wireless scenarios, with advanced capabilities, such as real-time high-speed signal processing and protocol implementation, support for multiple network interactions, and multiple antenna operation. This testbed is used for the design, implementation, and experimental studies of a spectrum sensing functionality.

Motorola describes a fundamentally flexible, low-power transceiver RFIC and how it enables a small form-factor, experimental CR platform. The platform can receive and transmit signals of many wireless protocols at carrier frequencies from 10 MHz to 4 GHz with channel bandwidths from 8 kHz to 20 MHz. Selected design considerations and challenges in designing the RFIC are first examined, followed by a description of a CR platform based on the RFIC. The design of the platform ranks flexibility at the top, followed by size, then power consumption. At a volume of $152\,\text{cm}^2$, the platform contains the RFIC, a Xilinx XC4VSX35 FPGA, an Analog Devices Blackfin 561 processor running uClinux, nonvolatile memory, high-frequency data converters, and Ethernet connectivity. After discussing the platform's hardware and logical architecture, a brief case study is examined. Last, a selection of lessons learned during the design of the platform are recounted.

The third system presented is the Maynooth adaptable radio system, which was developed to explore software-defined and cognitive radio principles from the base station perspective. The RF front end was designed with the perspective of supporting an adjustable bandwidth of up to 70 MHz in the range of 1.8–2.5 GHz with all signal processing undertaken off board using a general-purpose processor. The design ranks bandwidth and signal performance over power consumption, which is less of a concern in a fixed installation such as a base station. After discussing the motivations and resulting design, the challenges encountered, the lessons learned, and the technology constraints are discussed and possible routes for improvement in future designs are presented.

19.6 PROBLEMS

1 How would you improve the wideband energy detector implementation with respect to robustness to the noise uncertainty?

2   How would you improve the wideband energy detector implementation with respect to sensitivity to adjacent band interferers?

3   Would an oversampling help improve the performance of (a) an energy detector or (b) a cyclostationary feature detector? Explain your answer.

4   How would you minimize memory requirements for a cyclostationary feature detector?

5   What is the difference between multiple-antenna front ends that have a common LO and separate LOs for each radio? What applications require a common LO?

6   What are the advantages of having a separate sensing radio front end?

7   What other approaches to heat dissipation might have been considered for the design of Motorola's experimental platform?

8   Discuss the trade-offs that would need to be considered in Motorola's platform, if battery operation were a requirement.

9   What is an advantage of designing signal processing subsystems in the FPGA to be peripherals of the processor embedded in the fabric?

10   In Motorola's RFIC, why is the off-channel energy detector feature important to cognitive radio?

11   What are some of the most memory-intensive operations in a cognitive radio platform?

12   Given the usefulness of FPGAs, what advantages and disadvantages are to be incurred through the use of a digital down-converter as opposed to doing this in the analog domain?

13   Calculate the required ADC performance for a given hostile environment with and without a filter, assuming the blocker and interferer specifications for GSM900.

14   Identify within a transceiver front-end architecture those specific aspects most challenging to implement within the context of a reconfigurable radio platform.

15   Take a communication scheme of your choice and analyze the characteristics of that scheme in the context of assessing its performance for use in a partitioned SDR platform (for example, such issues as sustained performance, peak versus burst performance, and latency).

16   Given the four communication schemes GSM900, GSM1800, UMTS, and 802.11b, and assuming that you wish to design a base station that could capture the full signal bands in each case (say 50–70 MHz), identify and sketch those frequencies where an intermediate frequency may be placed so

unwanted spurious elements (second- and third-order products) do not interfere with the receive signal or other sensitive radio frequencies (for example, GPS).

17 In any reconfigurable radio implementation, power consumption is a major concern. In the three presented examples, identify which elements consume the most power and identify potential alternative solutions that could reduce the power levels.

Cognitive radio evolution

20

Joseph Mitola III

Stevens Institute of Technology, United States

20.1 INTRODUCTION

Introduced in 1998–1999, cognitive radio has become an all-encompassing term for a wide variety of technologies from dynamic spectrum access to reconfigurable networks. As originally published [6,678], the concept emphasized enhanced quality of information (QoI) for the user, with spectrum agility framed as a means to an end and not as an end in itself. The first research prototype cognitive radio (CR1), for example, learned to initiate a simulated Bluetooth link to exchange business cards in a wireless manner based on the detection of the user's a prototypical setting of "introductions," in a simulated verbal exchange among users. This intelligent agent embedded in the programmable digital assistant (PDA) had not been specifically programmed for this use case but rather learned this behavior from observing a prior user-initiated business card exchange. The embedded machine learning technology of case-based reasoning (CBR) in this architecture "observed everything" and therefore was able to associate the user's prior manual use of its own Bluetooth chipset to exchange business cards, associating contemporaneous cues in the speech domain such as phrases like "May I introduce" and "Very pleased to meet you." CR1 synthesized a CBR template to autonomously perform the actions Power-up Bluetooth, Exchange Business cards, Power-down Bluetooth when the situation Introductions is detected. Because of CBR, CR1 also could "learn by being told," which moves the expense of creating business logic rules from the service provider (where programming business logic is an overhead cost) to the device and the user, where the costs have to do with user acceptance of the embedded CBR technology. In addition, embedded CBR enables cognitive radio to learn etiquettes for sharing radio spectrum with legacy radios.

During the past five years, the world's radio research and engineering communities have been developing software-defined radio (SDR) and cognitive radio (CR) for dynamic radio spectrum sensing, access, and sharing [679–681], revealing many regulatory, business, market, and open architecture needs implicit in the broad potential that cognitive radio architecture (CRA) introduces [17, 575, 682]. Radio architectures from wearable nodes and radio access points to the larger converged networks have evolved from the niche market of single-band, single-mode car phones of the 1970s to today's ubiquitous multiband, multimode fashion statements. This chapter characterizes architecture evolution, including the near-term multimedia heterogeneous networks that converge traditional cellular architectures with Internet hot spots. This chapter also looks ahead toward the evolution of policy languages [683] and to sentient spaces [365, 684], integrated wireless environments that merge wireless technologies with increasing interplay of radio engineering with related information services of computer vision [685] and human language technologies (HLT) [686].

Contributions to CRA evolution span leading institutions across the globe, including (alphabetically) CVTR, Trinity College, Ireland; General Dynamics; Harris Corporation; Harvard University; IMEC, Belgium; Intel, KTH, the Royal Institute of Technology, Stockholm; Microsoft; Motorola; NICT Japan; Philips; RWTH Aachen, Germany; Samsung; Stanford University; Teknische Universitat Karlsruhe, Karlsruhe, Germany; TU Delft and Twente, The Netherlands; University of California at Berkeley; Virginia Tech; Virtual Center of Excellence (VCE) Wireless, United Kingdom; Zhejiang University, Hangzhou, China; to all of whom the author is particularly indebted.

Although the concept of cognitive radio stimulated research, development, and regulatory policy evolution, contemporary CRA emphasizes dynamic spectrum access [182]. Questions remain regarding the cost of implementing dynamic spectrum access regulatory and business policies in evolvable hardware, software, and network services. This chapter therefore considers this challenging problem from the perspective of cognitive linguistics, a branch of the humanities concerned with neonatal development of human language and its use in social settings. The current phase of CRA evolution includes greater integration of the user as the eighth layer of the protocol stack, more efficient network and node reconfigurability, and enhanced regulatory, business, security, and network operations policy languages.

20.1.1 Organization

This chapter first reviews the concept of radio architecture, including prototypical architectures for dynamic spectrum and embedded agents. It then describes the apparent lack of a comprehensive meta-level architecture for distributed heterogeneous networks and their related meta-level superstructures, including regulatory rule making and spectrum auctions. Changes in use case drive wireless architecture, so Section 20.3 shows how the historically significant striving for ubiquity and high data rate is beginning to give way to evolved value propositions where high quality of service (QoS) is the starting point for higher QoI. Section 20.4 therefore develops the potential for greater integration of cross-discipline information sources like video surveillance and human language technology in future cognitive radio architectures. To help guide this evolution, QoI is characterized along its several dimensions in Section 20.5, while Section 20.6 introduces the contributions of cognitive linguistics to policy language evolution and Section 20.7 offers a review of challenges and opportunities before the conclusion in Section 20.8.

20.2 COGNITIVE RADIO ARCHITECTURES

Radio architecture is a framework by which evolving families of components may be integrated into an evolving sequence of designs that synthesize specified functions within specified constraints (design rules) [688]. A powerful architecture facilitates rapid, cost-effective product and service evolution. An open architecture is available to the public, while a proprietary architecture is the private intellectual property of an organization, government entity, or nonpublic consortium. Figure 20.1 illustrates functional components integrated to create an SDR device, which may be wearable, mobile, or a radio access point (RAP) in a larger network. The set of information sources includes speech, text, Internet access, and multimedia content. Today's commercial radio frequency (RF) channel sets typically have four chipsets (GSM 900, GSM1800, CDMA, and Bluetooth, for example), evolving in the near term to a dozen band-mode combinations with smart antennas and multiple-input, multiple-output (MIMO) emerging [689]. In addition, a channel set may include a cable interface to the public switched telephone network (intermodulation intercept point, IP, or synchronous digital hierarchy, SDH) as well as a radio access point. Any functions may be null in any realization, eliminating the related components and interfaces from a given product for product tailoring and incremental evolution.

With the continued progress of Moore's law, increasingly large fractions of such functionality are synthesized in chipsets with software-definable parameters; in the field-modifiable firmware of field programmable gate arrays (FPGAs), in software for niche instruction set architectures (e.g., digital signal processor chips), and

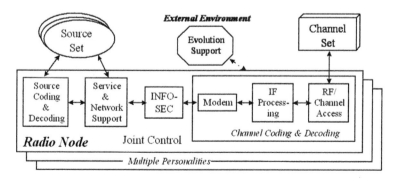

FIGURE 20.1

Set theoretic model of SDR architecture [687].

increasingly on blade server(s) and single-chip arrays of general-purpose processors (GPPs) like IMEC Belgium's SIMD4 [690].

Software-defined radios may be implemented from reusable code bases of millions of lines of code [691], the deployment, management, and maintenance of which pose configuration challenges. SDR software typically is organized as radio applications objects layered upon standard infrastructure software objects for distributed processing such as the SDR Forum's SCA (software communications architecture [682,692]), which originally was based primarily on CORBA [19,693]. The Object Management Group's evolved SCA has a platform-independent model (PIM) with platform-specific models (PSMs) for software-based communications (SBC). Infrastructure layers of such architectures are illustrated in the larger context of Figure 20.2. Prior to circa 2005, such architecture was overkill for handsets, but radio access networks have grown to millions of lines of code, consisting of the kinds of software objects with the types of layering illustrated in the figure, and now are applicable to handsets and systems on chip (SoC). As Petri Mähönen (RWTH, Aachen, Germany) was among the first to clearly differentiate [324], software radio and cognitive radio are "interlinked and are family members, but they also have distinctive roadmaps" as the evolution of cognitive radio architecture illustrates. Again from Mahonen, "There are still formidable hardware and algorithm development problems (such as AD/DA converters ...) before full (ideal) all-in-one software radio

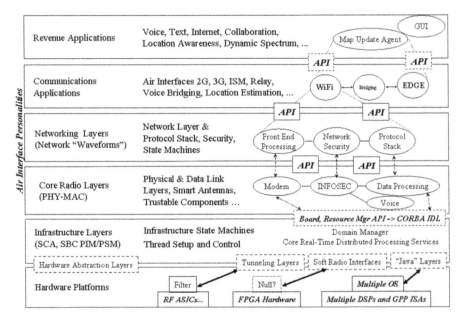

FIGURE 20.2

SDR object and API layering [688].

can be built." However, "the basic paradigm in the cognitive radio is to provide technologies, which enable radio to reason about its resources, constraints, and be aware of users/operators' requirements and context." What are the resources and constraints?

Arguably since the early 1900s, conventional radio architecture has been constrained by government regulatory frameworks accurately characterized as lanes in the road: bands allocated to specific uses, in the public interest. That regulatory regime addressed the public interest within the constraining economics of radio devices and related infrastructure (such as large, expensive television broadcast towers). This was economically efficient (e.g., Pareto efficient) from the "transistor radio" and television era to the deployment of first- and second-generation cellular radio. However, today's low-cost multiband, multimode wearable wireless fashion statements, the proliferation of cellular infrastructure, the gigabit-per-second core intermodulation intercept point (IP) networks, and wireless local area network consumer products have proliferated wireless access points of all sorts in the home, workplace, and seemingly just about every place else in developed economies. The new wireless ubiquity and heterogeneity offers rapidly emerging alternatives to the lanes in the road that include dynamic spectrum access.

20.2.1 Dynamic Spectrum Access

Dynamic spectrum access is the process of increasing spectrum efficiency via the real-time adjustment of radio resources, such as via a process of local spectrum sensing, probing, and the autonomous establishment of local wireless connections among cognitive nodes and networks. As originally proposed, cognitive radio envisioned real-time spectrum auctions among diverse constituencies, using for one purpose, such as cellular radio, spectrum allocated and in use for another purpose, such as public safety, and conversely, to multiply both the number of radio access points for public safety and for more efficiency use public safety spectrum commercially during peak periods [678]. Although that initial example has yet to be fully realized, the U.S. Federal Communications Commission (FCC) encouraged the application of that technology to the secondary use of underutilized television spectrum, such as in ad hoc, short-range wireless local area network (WLAN) in the spectrum allocated to another primary purpose, such as broadcast television. In addition, the principles of cognitive radio for dynamic spectrum access also apply to enhance the efficiency of use within and across each "lane in the road," such as via the intelligent selection among multiple alternative PHY-MAC layers (alternative lanes in the spectrum road) by cognition across network, transport, and application layers of the protocol stack [694]. Researchers characterize the advantages of short-term localized dynamic spectrum auctions [695, 696], including rigorous and comprehensive treatments in the European Community's precompetitive end-to-end reconfigurability (E2R) program [697]. In spite of commercial proposals [698], only long-term large-capacity anonymous leasing appears to be established in the marketplace [699].

FCC endorsement of cognitive radio in secondary markets in the United States offered opportunities for improved spectrum utilization [60]. In addition, the National Institute of Information and Communications Technology (NICT), Yokosuka, Japan, characterized SDR and cognitive radio from technical [700, 701] and regulatory [702] perspectives. Ofcom, the regulatory body of the United Kingdom, remains appropriately skeptical of the economics of dynamic spectrum [703]. On the other hand, the Commission for Communications Regulation (COMREG), Ireland, imposes constraints [704] but also encourages innovation, such as by allocating over 100 MHz of spectrum for experiments and demonstrations during IEEE DySPAN 2007 in Dublin. Guatemala [705] employs Titulos de Usurfrucato de Frecuencias (TUF), specifying spectrum-use parameters in great detail, which establishes a strong reference point for the liberalization of spectrum allocation toward dynamics [706]. In Europe, countries including Austria, Sweden, and the United Kingdom apparently have sanctioned de facto transfers of spectrum rights among spectrum licensees, while a recent EU Framework Directive empowers all EU countries to introduce secondary trading of spectrum usage rights [695].

The SDR Forum's CR working group (CRWG) and the inclusion of CR in its annual academic challenge promotes the global interchange among academic research and industry development of cognitive radio in SDR [707]. DARPA's XG program [708, 709] put substantial emphasis on the near-term potential of smart radios to share the radio spectrum dynamically, leading among other things to the success of the IEEE Dynamic Spectrum (DySPAN) conferences in 2005 and 2007, where XG research results were demonstrated [710]. XG simplified the ideal cognitive radio architecture (the iCRA [7]) to a simple rule-engine that controls the radio's air interfaces to conform to spectrum-use policies expressed in a rule-based policy language. This yields a simple, flexible, near-term, dynamic spectrum access (DSA) architecture clearly articulated by Simon Haykin [11].

20.2.2 The Haykin Dynamic Spectrum Architecture

Wireless ubiquity is more of a fact than a goal even in many developing economies. In most developed economies, evolved GSM and code division multiple access networks are both competing and cooperating with an Internet gone wireless for the revenue from voice and Internet traffic, with voiceover Internet protocol (VoIP) growing. Integral to this evolution is the potential for some adjustment to the protocol stack. As illustrated in Figure 20.3(a), Haykin's DSA emphasizes the need for cognitive radio to be aware of the many occupants of a radio environment by analyzing the radio scene to avoid interference, to operate in spectrum holes, and to provide channel state information that enhances the transmission. One implication for the protocol stack is the integration of cognitive nodes into cognitive networks via the universal control channel (UCC) of some sort, as shown in Figure 20.3(b), supplemented by group control channels. Spectrum sensing emerges in this architecture as the key enabler for greater agility in the use of spectrum for best possible QoS. Such a cognitive radio can quantify channel occupancy

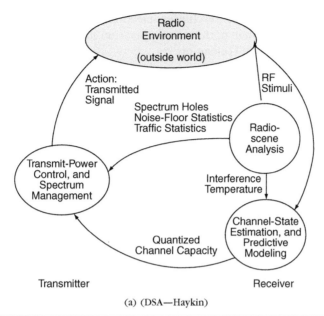

(a) (DSA—Haykin)

Contextual Control Layer: The User

Applications Layer

Session/ Presentation Layers

Transport Layer

Network Layer

Link Layer

| Group Management | Link Management | MAC |

Universal Control Channel (UCC)

Group Control Channel (GCC)

Physical Layer

| Spectrum Sensing | Channel Estimation | Data Transmisson |

(b) (Implications Haykin—Mitola)

FIGURE 20.3

Haykin dynamic spectrum access architecture.

and identify opportunities for RF chipset selection, signal in space transmission control [711], and other high-performance spectrum management features in the physical layer, such as MIMO operation.

20.2.3 The Ideal CRA

The ideal cognitive radio (iCR) architecture [182] differs from the DSA in its degree of integration of self-awareness, user awareness, and machine learning. The cognition cycle developed for the research prototype cognitive radio, CR1 [182], is shown in Figure 20.4. This cycle implements the full embedded agent and sensory perception capabilities required of an iCR, differentiating proactive planning from reactive behaviors and learning. Sensory stimuli enter the cycle via sensory perception (e.g., location, motion, temperature, vision, speech). Object-level change detection initiates the cognition cycle.

The iCR continually observes (senses and perceives) the environment, orients itself, creates plans, makes decisions on its own and in conjunction with the user and external networks, then acts. Actions may be physical, such as transmitting a signal, or virtual, such as associating a user's action with the current situation. The iCR may observe user actions (e.g., via keystroke capture) to form a macro-sequence to be applied in similar situations, such as searching for a wireless business card when introductions are detected via voice in some future setting. Actions of intelligent agents include movement in the environment to improve the likelihood of achieving a goal. Early planning systems used rule bases to solve simple

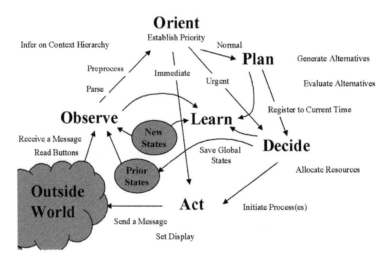

FIGURE 20.4

Ideal cognition cycle: The observe, orient, plan, decide, act, learn (OOPDAL) cycle of the iCRA embeds a self-aware, user-aware, RF-aware agent into SDR to improve QoS and QoI.

planning problems like the monkey and the bananas [712], stimulating the development of an entire subculture of planning technologies now integrated into a broad range of applications from factory automation to autonomous vehicles [713] and RoboCup Soccer [714], integrating learning with planning [715]. These planning technologies apply to radio domains for further architecture evolution.

During the wake epoch illustrated in the figure, the reasoning components react to changes in the environment. For example, the iCR might analyze speech or text of radio broadcasts, such as checking the weather channel or stock ticker tapes for changes of interest to the user. Any RF-LAN or other short-range wireless broadcasts that provide environment awareness information may be also analyzed for relevance to the user's needs inferred by the iCR via machine learning, say, assisted via a library of behavior stereotypes. In the observation phase, a CR also reads location, temperature, light level sensors, and the like to infer the user's communications context. Since the iCRA was based on agent technology, it leverages the continuing advances in agent communications language (ACL), the Java agent development environment (JADE), and multiagent systems (MAS) [716]. During the wake epoch, the detection of a significant change, such as the presence of a new radio network in the RF domain, of a new physical object in a visual scene, or of a topic in the speech domain, initiates a new cognition cycle. For example, IMEC Belgium developed low-power burst signal detection and presynchronization techniques to characterize changes in the RF scene [717].

Sleep epochs allow for computationally intensive pattern analysis, self-organization, and autonomous learning. A prayer epoch during sleep provides autonomous interaction with higher authorities, such as cognitive networks and regulatory authorities, for constraints, advice, and solutions to problems unavailable locally.

20.2.4 Networking and CRA Evolution

Neither the DSA nor the iCRA provides a comprehensive architecture for cognitive wireless networks (CWNs) [718]. At the Dagstuhl workshop in 2003, a consensus emerged that CWN significantly expands the research framework and architecture evolution possibilities to a mix of ad hoc and fixed wireless networks with self-awareness and greater spectrum efficiency, to mobile device awareness, to substantial memory in the network [719], and to distributed machine learning [718]. The DSA and iCRA per se do not provide such a rich research and development framework for legacy and cognitive devices and heterogeneous networks with regulatory policy constraints. In addition, without a supportive distributed network architecture, policy language, and methods of payment, short-term real-time spectrum auctions over small geographic areas seem unlikely to emerge, not withstanding much research and position papers before regulatory bodies [698].

Ideas for CWN CRA are characterized by cooperation among intelligent entities [679]. Cooperation results from game theoretic considerations [679], from considerations of power optimization [720, 721]; relays [722], ad-hoc discovery, and routing [723]; diversity [724]; cross-layer optimization [725]; stability and security [726, 727]; and spectrum efficiency considerations [720]. Distributed antennas [728, 729], cooperative header compression [730] and coding [731], and distributed spatial channel control [732], among others, result from such a focus on cooperation in cognitive wireless networks.

Qusay Mahmoud (Canada), similarly, brings together ideas for CRA migration toward self-awareness [733]. In this context, researchers at Osaka University, Japan, point out that biological systems from molecular processes and immune systems and to social insects to predator–prey relationships exhibit robustness in the face of catastrophe, a property desired in communications networks. Architectural properties of such biological systems include (group) membership perception, network awareness, buffer management (pheromone decay), and message filtering [734]. Also in this context, John Strassner (Motorola) and his group in Waterford, Ireland, report a refined autonomic network architecture, in some sense a distributed networked version of the iCRA [735]. Strassner makes a strong case for cognitive network architecture to address network semantics more completely, using the border gateway protocol [736] as a motivating example. There is no lingua franca for networking that bridges vendor-specific syntax and semantics, modalities, functions, and side effects, as Strassner dramatically illustrates. He shows the potential contributions of autonomic networking in future cognitive networks via a framework of (user) experience and (wireless and wire line) connectivity architectures with FOCALE [737]. Despite many research investigations into the role of semantic web technologies in CRA evolution [738, 739], none as yet appears to address Strassner's key issues in a sufficiently intuitive, computationally feasible, compact, and efficient way as to have become widely adopted. In part because of the shortfalls of the XML and semantic web technologies alone, the IEEE 802.21 and P1900.5 standards committees are pursuing behavior modeling of cognitive radio nodes and networks in their policy language deliberations [740].

In addition, Manoj, Zorzi, and Rao's [694] cognition plane organizes cross-layer reasoning of a joint layer optimization module by placing cognition and control modules in the PHY, MAC, network, transport, and applications layers. The cross-layer cognition bus applies the OODA loop of the iCRA with get/set access to the networks to bypass intervening layers, such as direct linkage from the application to the MAC to, for example, analyze the MAC layer to avoid default wireless channels during congestion. This CogNet AP can, for example, identify a preferred channel based on expected traffic during any hour of the weekday. They analyze Neel's et al. game-theoretic treatment of cognitive networks [741] to characterize challenges of sustaining Nash equilibriums in myopic, s-modular games and potential games, showing how potential games may be realized in practice. In spite of such promising work and positive rule making, apparently as yet no widely accepted comprehensive architecture (functions, components, and design rules) within which the potential for fair, stable dynamic spectrum (potentially including micro-scale space/time/RF auctions) is being realized in the marketplace.

A deeper understanding of the related technical, social, and economic factors seems to be an important open meta-level issue for CRA evolution.

Clearly, the Haykin DSA, cooperative and self-aware networks research, and cross-layer cognition provide crucial foundations for cognitive radio architecture evolution. Where power management remains important, simplification of the DSA brings cognitive behavior in the RF domain closer to practice. The pace with which systems evolve from the practical focus of the current architectures to address larger issues may be driven by evolved use cases.

20.3 ARCHITECTURE EVOLUTION AND USE CASE EVOLUTION

Commercial wireless use cases continue to evolve. The use cases that captured market share and propelled radio engineering to its current levels of success have been based on the proliferation of cellular wireless networks, on the one hand, and on the affordability of fiber-optic core networks and short-range WLAN of the Internet. Ubiquity has brought with it a shift of use case from mere ubiquity toward affordable differentiated multimedia services in purely commercial markets as well as greater integration of historically distinct market segments such as commercial and public safety wireless, such as Block D in the U.S. 700 MHz wireless auctions. Block D challenges have been characterized by the SDR Forum as "meeting the divergent needs of commercial and public safety users, coverage, shared operational control, robustness, adaptability, and spectrum use in the absence of network buildout." Such public commentary reflects an evolution of use case that drives wireless architectures from the relatively monolithic cellular radio networks with gateways to the public switched telephone network toward greater integration with the Internet as characterized in Table 20.1. The lines of the table without italics have been well established during the past few years, and thus need little elaboration, but set the stage for the more speculative use case projections in italics.

20.3.1 Product Differentiation

With ubiquity in developed economies, wireless service providers have suffered profit erosion and are beginning to compete for multimedia services integration across the broad domains of personal information (voice, text, personal games, Internet access, and email) and entertainment (digital radio and TV broadcast, network games, and Internet broadcast modalities like YouTube and MySpace) with many forms of infotainment taking shape. Wireless has been both a perpetrator and a beneficiary of the infotainment mega-trends, and to remain relevant, cognitive radio architectures must make it easier (and more affordable) for the service provider to deliver highly user-specific (differentiated) services, whether at home, at work, traveling, or on holiday. This need along with other factors tend to drive the CRA from today's DSAs toward greater use of technical parameter profiles that

Table 20.1 Wireless Use Cases Drive Radio Architecture Evolution

Use Case Parameters	Foundation (1990–2005)	Evolution (2005–2020)
Core wireless use cases	Toward ubiquitous access	Toward integrated services
Profit margins	High (handsets infrastructure) then handset profits declining	Low (handsets infrastructure) to high for differentiated services
Value proposition	QoS (Connectivity, data rate)	QoI (User is the eighth OSI layer)
PSTN integration	SS7 [742], SDH [743]	IP-SIP [744], Mobile IP, or IPv6 [745]
Reconfigurable hardware	Not worth the cost vs. chipset	Transitioning to mainstream?
Location awareness	Niche applications	Ubiquitous
Multimedia	Infeasible to feasible	Strong differentiator
Spectrum awareness	Within allocated band	Across multiple bands
Spectrum auctions	Large blocks for long term	Small space/time holes short term
Public safety	Distinct markets	Integration with agility
Data rate framework	Stationary, walking, vehicle	Hot spot, traveling, emergency
Sentient spaces	Video surveillance markets	Elder care and home robotics

are tailored to each particular user's infotainment practices, such as learned by a recommender system embedded in the radio [746]. In fact, semantic web technologies are making it increasingly easy to represent computationally and reason with ever more subtle and sophisticated aspects of the needs, habits, and preferences of individual users. This leads one to postulate an emerging value proposition founded on QoS but expanded to reflect ubiquitously high QoS across service providers with QoI as an increasingly central driver of wireless architecture evolution.

20.3.2 Protocol Stacks

Although asynchronous transfer mode (ATM) [747] established transport efficiency with predictable QoS for its high value in core networks of the foundational era, IP seems to be the ultimate beneficiary of efficiency and QoS in the era ofnetwork convergence. This applies both in the current transitional patchwork of IP with network address translation (NAT), VoIP via session initiation protocol (SIP), and in the longer term toward IPv6 perhaps. This expectation sets some characteristics of the networking layers of the protocol stacks of handsets, vehicular radios, and radio access network infrastructure and thus of CRA evolution as well.

20.3.3 OA&M

Self-awareness is not evident in the DSA, but the costs of operation, administration, and maintenance are moving self-awareness for autonomous configuration management toward center stage in the European Community's end-to-end efficiency (E3) wireless initiative [748,749]. Self-awareness and self-examination properties of agent-based evolved CRAs may help address the challenges of configuring software stacks for mobile device and infrastructure releases, as well as offering additional protection from inadvertent misconfiguration. The iCRA incorporates the necessary self-awareness and its mathematical theory [329] draws on Godel theory to establish the basis in computability for the self-referential but computationally stable self-examination. Thus, the DSA provides an appropriately simplified transitional architecture that is now beginning to evolve toward the iCRA's promise of autonomous operations, administration and management (or maintenance) (OA&M) [733].

20.3.4 Location Awareness

A microcosm of evolution from QoS to QoI has occurred in location-based services. During the early foundational era, location seemed to be potentially useful but did not rise to the status of a killer app on its own. Government mandates for 150 m accurate location information for the delivery of emergency services to cell phone users helped to transition location awareness from a niche to the mainstream. But at the same time, inventions for routing as a function of location [750] and services like GEOPRIV [751,752] made it possible to customize access to knowledge about personal location, so the role of location information in wireless architectures continued to grow. Today, wireless location-based services are differentiated, based on ease of use (e.g., Google Maps versus MapQuest), QoI parameters [182], rather than mere availability (a QoS parameter), or time delay of the results (another QoS parameter). Multimedia services may also undergo such a transition from QoS to QoI as wireless multimedia coverage continues to expand, soon becoming expected even in developing economies. Thus, the ability of cognitive radio architectures to enhance multimedia in terms of both cost of availability (cost of a QoS service level agreement) via spectrum efficiency with QoI as a mobile user value proposition may propel cognition for either or both purposes into a more central role in architecture evolution.

20.3.5 Spectrum Awareness

Spectrum awareness now too is beginning to move to center stage, as suggested in Table 20.1. Historically, a commercial wireless device had to be awareof the network but not much else. The network told the radio what to do and that was that. Today's handsets, PDAs, and even some laptops are armed with multiple chipsets capable of accessing GSM (ideal for global roaming) and code division multiple access (CDMA) (e.g., for medium data rates in a larger coverage area) as well as for accessing 54 Mbps WiFi hot spots. Today's user puts up with the tedium of picking the WiFi network, with whatever security risks that entails, while host cellular networks deal with most of the other choices of radio band and mode for the user. However, regulatory rulings on dynamic spectrum and the commercial success of Internet alternatives to cellular wireless and the PSTN (e.g., Skype over WiFi) render the autonomous mediation of radio bands and modes into an opportunity for CRA evolution. As something like the U.S. 700 MHz Block D rules emerge, it will become useful for wireless devices of the future to autonomously recognize prototypical emergency situations without being told to by a network that is inoperative because of the emergency. This raises the performance bar for DSA and places the autonomous determination of the user's situation at center stage: Is the user a victim who has priority for assistance, a first responder authorized to assist in the emergency, or a bystander who should yield spectrum to those who need it most? Integration of diverse sensor modalities may be needed to effectively address such situations [182].

20.3.6 Spectrum Auctions

In the foundation era, the radio spectrum was allocated in relatively large blocks, the lanes in the road, that raised substantial government revenue. Within the past few years, private enterprise has offered Web-based tools and services for incumbent spectrum licensees to anonymously cross-license small slices of spectrum for relatively short periods of time—notionally, a few megahertz for a few months at a time. As Frederick Jondral's group at TU Karlsruhe characterized in some detail [753], the sale of chunks of primary spectrum as small as 5 sec in duration for prices as low as a few cents per chunk for enhanced email services and web browsing could provide an increase in spectrum utilization of between 15% and 25%. Thus, CRA could improve secondary markets from today's megahertz-months toward the more efficient kbps-seconds, although just where the revenue break points might be is yet to be determined. Linda Doyle's group at Trinity College, Ireland, showed that what is feasible still appears to be orders of magnitude from theory [754]. Indeed, the government of Ireland dedicated over 100 MHz of spectrum to the experimental characterization of the potential for IEEE DySPAN 2007. To characterize performance in spectrum overcrowding, DySPAN 2008 has intentionally smaller spectrum allocations [755].

20.3.7 User Expectations

The way users value QoS parameters like coverage (probability of a mobile dial tone) and data rate may also change with the agility of spectrum access. Users readily recognize that megabit-per-second data rates are available in WiFi hot spots

FIGURE 20.5

Entrance to the SAS Royal Viking hotel, facing the street.

but not in remote areas, so they adjust their expectations and plans for the use of a flexible PDA accordingly. Today, mobile data rates are expressed in a framework that reflects the mobility of the user within fully built-out and relatively monolithic cellular infrastructure: Stationary users have a higher data rate than those walking, and they have higher data rates than those in moving vehicles. The 3G technology recognizes data rate differentiation indoors, reflecting additional nonhomogeneity of the networks. During the next few years, most homes and businesses could become multi-megabits-per-second wireless hot spots, potentially via B3G femtocells or Internet WLANs or both, accelerating CRA evolution.

The identification of a specific hot spot may be based in part on the global positioning system (GPS), but in complicated urban settings other sensor modalities like computer vision, speech, and other human language technologies may play a role [182]. For example, Figure 20.5 shows the entrance to the Royal Viking Hotel in Stockholm, facing from a revolving door toward the street. The glass foyer provides great GPS and GSM coverage, but the GPS does not establish whether the radio is inside or outside the hotel. In addition, GSM fades deeply when one traverses the entrance and most cell phones lose a call in progress here. If instead of merely reacting to a deep fade, the cell phone were aware of the user's precise location and direction, then a more aggressive adaptive equalizer could be invoked for the transit through the tunnel so that the call is not lost.

Generally, it is impractical to operate a GSM network-handset pair with the high-performance equalizer and network compensation required to transit such a tunnel. That is, you cannot operate a network profitably if constantly configured for such worst-case situations. However, if the cell phone autonomously detects the lobby and reliably predicts the tunnel transit, it may employ expensive measures autonomously (its high-performance equalizer) coordinating with the network to maintain affordable connectivity. The resulting perception of never dropping a call (exceptional QoS via multisensory CRA) could be a market differentiator. Although consumers are not likely to wear a cell phone camera to gain such a minor advantage, the information prosthetic value of such a camera might create market value. A first responder might wear such a camera phone to transmit images to the incident commander. Location fine structure includes altitude, trajectory, indoors-outdoors, weather, and other characteristics of location. Thus, a CRA that enables higher radio performance by the opportunistic exploitation of larger situation information (such as the visual cues to the transit of the tunnel) may facilitate smoother transition to higher QoS for emerging applications like the commingling of public safety and commercial services in the recent 700 MHz spectrum auctions in the United States.

20.3.8 First Responder Situation Awareness

Public safety and military users refer to the detailed knowledge of physical state as *situation awareness* [756, 757]. With DSA, a cognitive radio bases its actions on little more than GPS and instantaneous fade data. However, by taking advantage of the video surveillance streams that public safety incident commanders employ in increasing numbers to manage major incidents, the cognitive radio of the future may be able to optimize its use of radio resources to reflect finer grain aspects of the user's situation: specific location, surroundings (e.g., in dense smoke), movement (e.g., trapped), and potentially intent (trying to rescue a victim versus trying to escape a cleared area). The value to the radio of greater awareness of the user's physical setting in space and time may reduce uncertainty and promote better situation-based radio resource management, such as which of the first responders gets MIMO resources for video to assist rescue versus perhaps location-only low-power, low-data-rate radio resources, such as when on standby in an assembly area. Historically, such needs have not been met with corresponding financial resources, but governments around the world may be more inclined to invest for the evolution era of Table 20.1 than they were during the foundation era.

20.3.9 Commercial Sentient Spaces

Spatial spectrum *confluence domains* may be defined as the radio environments created (usually unintentionally) where a wide diversity of commercial and other wireless products, services, and networks are brought together in a single location, such as a home, business, apartment building, factory, or other social space, becoming the venue for a variety of broadcast (e.g., HDTV, DBS), WLAN (e.g., 802.11a/b/g, WiFi, HomeRF/Zigbee, Bluetooth, UWB), cellular (e.g., 3G femtocells), and broadband radio resources (e.g., WiMAX). Confluence domains create a combinatorially explosive set of opportunities for interference as is often true today or for cognitive load balancing and cooperative power management [758] as architectures evolve.

Elder care may be an emerging market where such tight integration of space, time, and RF makes economic sense during the evolution era of the table. The sentient home of the future may include video cameras and voice interaction to assist elderly residents in remembering whether they took their prescription drugs, removing a shoe from the stairs, turning off the stove, and performing other tasks that promote health and avoid accidents [759]. Other sentient space applications include child care [182], infotainment, and interactive games, where the wireless devices situated in the sentient space enhance the interactive experience. Usually in the United States there is a wireless point of sale terminal to check in your rental car at the airport. Although the parking lot with only such devices falls short of the sentient space vision, the application shows how wireless technologies move sensors and data entry from a desk, where it is convenient for the administrator, to the parking lot, where it is convenient for the user. The iCR architecture's emphasis on the user enables more aggressive redeployment of sensing, such as cognitive replacement of data entry and fiduciary with machine vision to reduce costs and enhance the commercial customer experience. Wireless networks meet several needs in commercial spaces, including low-cost deployment and removal, high mobility for the users, and modular evolution of the sentient space as medical and wireless technologies continue to advance.

In part in response to industry interest in sentient spaces for elder care and agricultural robotics [760], the Object Management Group (OMG) developed specifications that facilitate the deployment of embedded intelligence. So far OMG released specifications for smart transducers and super-distributed objects, logical representations of hardware/software components that perform well-known functionality and services [761]. The architecture appears to rely on autonomy and cooperation among a massive number of such objects, where the very number of interacting objects reduces the effectiveness of conventional plug-and-play technologies. Super-distributed objects may include wireless devices, software modules for radios, transducers, video cameras, servers, smart light bulbs and switches, electric motors, and other massively distributed components. In such applications, radio is a means for distributed control among hardware-software artifacts, and the need for trustable wireless connectivity is substantial. Although it is far too early to tell, there may be a transition during the evolution period of Table 20.1 where the majority user of wireless becomes IPv6-enabled devices instead of people. The potential disruption could be significant in an evolution to autonomous networked devices as the primary wireless user. People may interact with a house full of smart devices via ubiquitous computer vision and HLT, both spoken and written. These technologies may proliferate for the commercialization of sentient spaces apart from radio applications, but the information about the user's situation that results could be used by cognitive radios to autonomously adjust radio resource priorities to changing needs.

20.4 SENSORY PERCEPTION IN THE EVOLVING CRA

To characterize the potential for evolution toward commercially viable sentient spaces requires a brief review of computer vision and human language technologies. Each of these technologies is so broad that there is no hope of providing even a survey of the state of the art in this brief treatment. The intent instead is to identify important aspects of these technologies for future cognitive radios to interact with users, peer radios, and external sensory-perception networks (such as a cognitive video surveillance network), including speaker identification via voice biometrics, context-aware voice commands, and keyword extraction from email to detect stereotypical situations for proactive wireless services synthesis.

20.4.1 Machine Vision

Machine vision includes video analysis [762] that reliably identifies people and objects in complex scenes and detects events of interest, such as an illegal turn via a traffic camera. Cognitive vision [685] systems continuously observe a video scene to perform such event detection tasks. Video analytic applications programmers' interfaces (APIs) have evolved for video surveillance [763, 764] and Internet retrieval [765] among others.

Video analytic products offer few open architecture standards suited for CRA evolution, but as video object APIs and interface standards emerge, CRs may interact with cognitive video systems, such as a surveillance network in a snowy, deserted parking lot where the user has a serious fall after hours, is injured, and unable to call for aid. In this use case, autonomous collaboration between the users' CR and the wireless video surveillance network could yield an accurate diagnosis of the user's state and timely emergency response for enhanced QoI as the cognitive PDA acts as an agent for the injured user [766]. An important cognitive vision, cognitive radio API research issue is the rapid characterization of visual scenes with fidelity (granularity and accuracy) appropriate to radio use cases. The video scene API should assist the CR in connecting with a data rate and setting the priority appropriate to the user's situation and larger context of others contending for spectrum access. For example, radio access may be expedited for a small number of users in and near a traffic accident until first responders arrive, but such simple priority schemes may be counterproductive in a tsunami-class event. Thus, the exchange of situation information with cognitive wireless networks seems to be important in the evolution toward interdisciplinary integration with cognitive vision, toward the sentient-networked CRA.

20.4.2 Human Language and Machine Translation

Computer processing of human language includes both real-time speech recognition and high-performance text processing as well as machine translation. During the evolution period of Table 20.1, CRs may perceive spoken and written human language (HL) with sufficient reliability to detect, characterize, and respond appropriately to stereotypical situations, unburdening the user from the counterproductive tedium of identifying the situation for the radio. Machine translation in the cell phone may assist global travelers with greetings, hailing a taxi, understanding directions to the restaurant, and the like. Such information prosthetics may augment today's native language facilities. With ubiquity of coverage behind it, CRA evolves toward more accurately characterizing the user's information needs, such as via speech recognition and synthesis to interact with wearable wireless medical instruments, opening new dimensions of QoI.

Computer Speech

Computer speech technology offers opportunities for machine perception of content in well-structured audio channels such as 800 directory assistance. Although deployed with all Windows XPTM laptops, speech recognition does not appear to be in wide use for interaction with personal electronics or machine dictation. However, the technology now is mature enough to transcribe carefully spoken speech in benign acoustic environments such as a (quiet) home office, with 3–10% raw word error rates, reduced in structured utterances such as dictation to less than 2%. In situations where the speech is emotional, diffluent, heavily accented, or focused on a rare topic, the word error rates increase to about 25%, but even with these high word error rates topic spotting for geographical topics can yield 14.7% precision, improved by an order of magnitude during the past five years [767].

Speaker identification technology [768] has equal error rates (equal probability of false dismissal and false alarm) of <10% for relatively small collections of speakers (<100). Such algorithms are influenced (usually corrupted) by acoustic backgrounds that distort the speaker models. Speaker recognition may be termed a soft biometric since it could be used to estimate a degree of belief that the current speaker is the usual user of the CR to align user profiles. Such speaker modeling could contribute to multifactor biometrics to deter the theft of personal information from wireless devices.

Text Understanding

Business intelligence markets are deploying text understanding technology that typically focuses on the quantitative assessment of the metrics of text documents, such as to assess and predict the performance of a business organization. Quantitative analysis of databases and spreadsheets often do not clearly establish the causal relationships needed for related business decision making. Causal cues typically are evident in the unstructured text of business documents like customer contact reports, but the extraction of those relationships historically has been excessively labor intensive. Therefore, businesses, law enforcement, and government organizations are employing text analysis to enhance their use of unstructured text for business intelligence [769], with rapidly growing markets. These products mix word sense disambiguation, named entity detection, and sentence structure analysis with business rules for more accurate business metrics than is practicable using purely statistical text mining approaches on relatively small text corpora.

Google depends on the laws of very large numbers, but medium to large businesses may generate only hundreds to thousands of customer contact reports in the time interval of interest. For example, IBM's unstructured information management architecture (UIMA) [770] text analysis analyzes small samples of text in ways potentially relevant to cognitive radio architecture evolution such as product defect detection.

In addition, Google's recent release of the Android open handset alliance software [771] suggests a mix of statistical machine learning with at least shallow analysis of user input (e.g., for intent in the Android tool box). Google has become a popular benchmark for text processing. For example, a query tool based on ALICE (artificial linguistic Internet computer entity) is reported to improve on Google by 22% (increasing proportion of finding answers from 46% to 68%) in interactive question answering with a small sample of 21 users. Of the half of users expressing a preference, 40% preferred the ALICE-based tool (FAQchat) [772]. The commercial successful text retrieval market led by Google is stimulating research increasingly relevant to cognitive radio architecture evolution. For example, unstructured comments in wireless network service and maintenance records can yield early insight into product and service issues before they become widespread. This can lead to quicker issue resolution and lower aftermarket service and recall costs. Cognitive wireless networks (CWNs) might analyze their own maintenance records to discover operations, administration, and maintenance issues with less human oversight [773], enhancing the cost-effectiveness of cognitive devices and infrastructure compared to conventionally maintained wireless products and networks. This might entail a combination of text processing and functional description languages (FDL) such as the E3 FDL [749, 774].

The global mobility of the foundation era of Table 20.1 spurred the creation of world-phones such as GSM mobile phones that work nearly everywhere. During the evolution era, text analysis, real-time speech translation [775], text translation [776], image translation [777], and automatic identification of objects in images [778] will propel CRA evolution toward the higher layers of the protocol stack, ultimately with the user (from preferences to personal data) as the eighth layer. For example, if a future cognitive PDA knows from its most recent GPS coordinates that it is near the location of a new user's home address and recognizes light levels consistent indoors as well as a desk and chair, then it could enable Bluetooth to check for a home computer, printer, and other peripherals. Detecting such peripherals, it could ask the user to approve sharing of data with these peripherals so that the PDA could automatically acquire information needed to assist the user, such as the user's address book. The user could either provide permission or explain, for example, "You are my daughter's new PDA, so you don't need my address book. But it is OK to get hers. Her name is Barbara." Objects prototypical of work and leisure events such as golf clubs could stimulate proactive behavior to tailor the PDA to the user, such as wallpaper of the great Scottish golf courses.

20.4.3 **Situation Perception Architectures**

With a set of APIs for cognitive vision, video analysis, speech recognition and synthesis, and text analysis, individual cognitive radios and CWNs may evolve toward the organization of machine perception, illustrated in Figure 20.6.

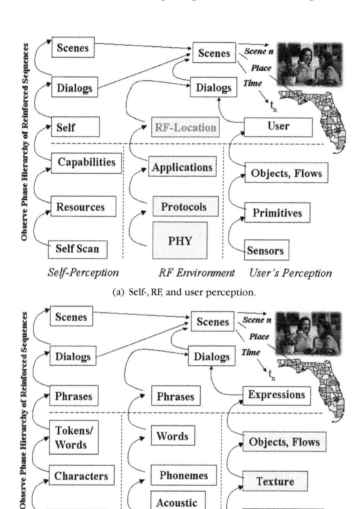

(a) Self-, RF, and user perception.

(b) Media stream contributions.

FIGURE 20.6

Comprehensive situation perception architecture.

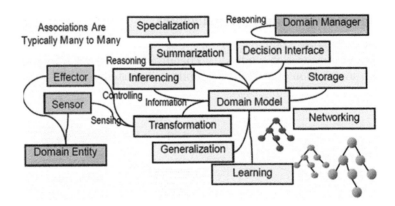

FIGURE 20.7

The complexity of situation assessment [779].

Perception of the self and the user in a specific situation may be aided by the recognition of stereotypical interchanges between the user and the wireless networks in which the user employs the radio more or less directly, as is the case today, while the cognitive agent observes the dialog, as well as text, speech, or visual cues to situation changes which may be reinforced in one of the other domains. Different participants perceive such situations differently as a function of role and information need [779], as illustrated in the figure.

The complexity of such situations (see Figure 20.7) must be fully addressed with the user as the eighth layer of the protocol stack, fully characterized in machine-readable form to reach the next level of QoI.

20.5 QUALITY OF INFORMATION

QoS concerns the availability, data rate, and timing of bit streams, addressing user needs at the physical through network layers of the protocol stack. QoI concerns the information that meets a specific user's need at a specific time, place, physical location, and social setting, climbing the protocol stack through the applications layer to a postulated user layer, the eighth layer of the protocol stack. QoI may augment QoS to help guide the evolution of radio architecture from SDR and dynamic spectrum access toward the iCRA. One expression for QoI [182] is as follows:

$$\text{QoI} = \text{Quantity} \times \text{Precision} \times \text{Recall} \times \text{Accuracy} \times \text{Detail} \times \text{Timeliness} \times \text{Validity}. \quad (20.1)$$

Each of the parameters of this equation has many possible valid forms. The requirements for the purposes of characterizing QoI are:

1. Each parameter is real, occurring in the range [1.0, 0].
2. Each parameter reflects the best QoI at its maximum of 1.0.
3. Parameters monotonically approach 0 in proportion to degradation of QoI.

The parameters need not be linear. Given the product form, any term that reaches 0 reduces net QoI to 0, so the forms most helpful in decision support would retain nonzero value in all terms where potentially useful information is provided. Each term is briefly discussed.

20.5.1 **Quantity**

The quantity term represents the amount of information displayed or provided by the wireless device or network via the device to the user. If the device has no information for a given situation, then quantity is 0. Information provided to a user in response to a need may not be provided in real time, but may be made available from cache or a priori knowledge. Thus, while QoS measures the quality of a connection, QoI measures the service provided to the user in terms that matter to the user. QoS metrics do not ascribe value to cached data, but QoI ascribes value to autonomous caching of data that the CR infers might be helpful at some point in time (and for which space is available). Users know that, if the most current information is not available, then older information may be better than nothing and may be as good as real-time information if it has not changed. Maps, for example, may be safely cached for relatively long periods of time since roadways change relatively infrequently. As CRs learn user preference, the ability to cache relevant data may increase, and if that is the case, then QoI characterizes the value to the user of such proactive caching.

Proactive management of quantity may shape network traffic as well. A cognitive radio that displays a cached street map advises the network of the date of its cached map, and thus avoids downloading a megabyte of data, has provided the user with accurate information nearly instantaneously, and offloaded a megabyte of traffic from a potentially busy network. In the foundation era, when revenue was based to a first-order approximation on increasing the use of the network, this might not be such a smart thing for a profit-making network to do. However, in the evolution era, when revenue is based on market share through product differentiation, instantaneously accurate information valued by the consumer might capture market share without burdening the network. Thus, the quantity term in the QoI metric expresses value not just in terms of connectivity and data rate, but in terms of information made available without using the network unnecessarily. On the other hand, suppose sufficient information is present, so that quantity = 1.0. Precision, recall, and accuracy further characterize the quality of the quantity provided.

20.5.2 **Quality: Precision and Recall**

Precision and recall reflect the degree to which the information corresponds to the user's need. In information retrieval, recall is the fraction of relevant documents retrieved from a corpus by a query. Precision is the fraction of documents retrieved that are relevant. Recall of 1.0 indicates that all relevant documents are retrieved, while precision of 1.0 indicates that no irrelevant documents have been retrieved. Adapting this well-known metric to QoI, one may apply precision and recall to items provided to a user by a wireless device, with or without the assistance of one or more networks and with or without prior caching. Users may provide feedback by obtaining more items, rejecting, or not using the items retrieved.

20.5.3 Quality: Accuracy

Accuracy characterizes the quantitative aspects of the information provided. Accuracy is reduced by errors such as lack of factual correctness (e.g., spell the President's name wrong). Numerical accuracy in QoI reflects any numerical errors, whether from the original source, via transmission, or via misformatting the results. Numerical precision may limit accuracy. If the accuracy required by the user is met, then accuracy = 1.0. The rate of degradation of the accuracy metric from 1.0 may depend on the situation and may take any form (linear, quadratic, exponential, fractal, defined by table lookup, etc.).

20.5.4 Timeliness

Timeliness is defined in terms of the user's timeline along which the information is to be employed. If the information is needed immediately, then the quality may be characterized as inversely proportional to excessive time delay. To avoid division by 0, one may consider timeliness to be 1.0 if the information is available before a minimum delivery time:

$$T_{min}(\text{time, place, social setting, topic}). \tag{20.2}$$

Situations include time, place, and social setting, such as shopping or needing medical attention. Suppose the shortest time delay in such a setting is ε so the maximum contribution of timeliness to QoI would be $1/\varepsilon$. If timeliness is normalized by ε, then maximum timeliness is 1.0. In medical situations, there typically is a window after which the information is of marginal value if any, so the timeliness parameter may fall off sharply after such a window. Similarly, in some situations, timeliness may be decremented from 1.0 if the delivery time is less than ε. There is value in meeting users' exact timelines in the same sense that a wakeup call should be delivered when requested, and a user is not happy if the wakeup call is 15 minutes early.

20.5.5 Quality: Validity

Validity is 1.0 if the information provided is true and approaches 0 if false with fuzzy set membership, for validity values in [1.0, 0.0].

20.5.6 **Quality: Level of Detail**

Finally, if sufficient detail is provided that the user regards the information provided as complete, then detail = 1.0, gradually dropping toward 0 if insufficient elaborating detail is provided.

The representation of the user as the eighth layer of the protocol stack becomes a computational reality in what may be called the *value plane* [780] to the degree that the user's preferences in the QoI dimensions are made computationally accessible via semantic web, computational ontology, user modeling, and other languages for expressing user needs. QoI has cross-layer implications. Given intermittent connectivity of wireless devices, QoI can guide a cognitive radio in its choice among a collection of candidate wireless access points. User preferences modeling guides the CR in proactive caching when connectivity and data rate are available, particularly in a future where there is no additional cost to such caching (such as from WiFi and the Internet service provider at home). To the degree that a cognitive radio can independently estimate user state or can infer that state from a CWN or a collaborating network, such as a wireless video surveillance network, the embedded cognitive agent may be able to manage radio resources so as to maximize the QoI for its own user, whose needs and preferences it has learned.

20.6 COGNITIVE RADIO POLICY LANGUAGES

This section characterizes the influence of policy languages on CRA evolution. IEEE standards activity P1900.5 is developing a cognitive radio policy language architecture (CRPLA) to express policies for dynamic spectrum sharing. Regulators and service providers require that the related spectrum use and service policies reflect social contracts and business logic. There currently is a lack of consensus on the scope and needs of policy languages, with roots in differences among regulatory, financial, business, and technical domains. Cognitive linguistics assist in formulating the evolution of CRA to include cognitive radio policy language (CRPL) that is more fully responsive to regulatory and business needs than is available today.

20.6.1 **What Is a Policy Language?**

Business logic establishes mechanisms for achieving value propositions, codified in computationally viable schemas, rules, procedures, and protocols. This enables enterprises to realize value via wireless networks, including unlicensed radio bands, where radio access is based on best effort, and licensed bands in which QoS differentiates networks, while overall quality of experience (QoE) driven by QoI differentiates service providers. Multimedia wireless services combine voice, data, and video clips and streams at home, at work, on the road, and to the handset.

Business concepts like "gold service" imply supplier-specific network configuration and use constructs with a myriad of details that capture the nuances and business intent of the provider. Cognitive radio is employed by such enterprises, including conventional cellular service providers, multinational corporations on large campuses, and small businesses using commercial products. CRPLs have the potential to orchestrate network behaviors from PHY-MAC to applications layers, shaping the QoE for the consumer within regulatory constraints.

Strassner [683] defines *policy language* as an overloaded term for the business goals of an organization, regulatory rules that need to be enforced (e.g., spectrum masks), the goals the system must try to achieve (e.g., "gold service"), configuration changes (e.g., backup modes), and system behavior (e.g., priority for first responders). The U.S. Institute of Electrical and Electronic Engineers (IEEE) established task group P1900.5 within Standards Coordinating Committee (SCC) 41 to develop a CRPL that supports as many of these needs as practicable.

20.6.2 Policy Language Needs

Policy languages relevant to CRPLA evolution occur in the following broad information technology (IT) endeavors:

1. Information assurance (IA) and information security (INFOSEC).
2. Spectrum management (e.g., DARPA's XG [781, 782], CoRaL [783], and Maude [784] radio policy languages).
3. Network management [785].
4. Configuration management [697, 786].

CPRL scope includes the following domains of the related entities [787]:

Domain	Entities
Regulatory	Spectrum manager, regulator entities
Trusted third party	Certification and security entities
Privileged third party	Reconfiguration manager, software providers, equipment manufacturers
Service providers	Content providers, service
Network operators	Fixed network operators, mobile network operators, virtual network operators, pilot channel providers
User domain	User, subscriber, equipment

Within this collection of participants, regulatory, business, and legal domains for policy conformance come together. However, these entities come to only limited consensus on the attributes of their languages in these CRA domains. There is apparent agreement that CPRL needs to be accreditable, unambiguous, and extensible and that there must be interoperability among dialects [788]. There is less consensus on expressiveness and computational modeling.

Expressiveness includes the question of whether to incorporate established security primitives into CPRL. Should one address privacy as a distinct language (e.g., GeoPriv), as an overlay (e.g., in REIN), or embedded into enterprise governance language(s) [789, 790]? Mobility management [791], business logic, and efficient end-to-end reconfiguration ([786]) each has unique language primitives.

Computational modeling includes the modeling of functions and computational structure. Radio regulatory domain functions require spectrum allocations and power control masks, space/time knowledge to detect and avoid legacy users [329], and etiquette for distributed spectrum sharing [679]. Radio network layer domain models include the generic link layer (GLL) for mobility management and GeoPriv for privacy policy modeling [792]. Although UML supports these needs (use case models, data abstractions) and RDF to OWL, OWL-Lite, and so forth, augmented with rules support ontology needs, there are many ways to address the functional issues. REIN, for example, provides a dialect of RDF-OWL with extensions for interoperability based on question answering with APPEL privacy overlay.

Computational structure differs among logic, rules, and games. Approaches based on logic lack consensus on questions of representation and proof. CoRaL limits policy language expressions to types, terms, and formulas, while CLMM admits fuzzy logic. Maude, referenced previously, extends CoRaL using the semantic web rule language SWRL [793], which combines OWL and RuleML, to return answers to XG policy-compliance questions that include explanations of what is needed to become policy compliant. Extensions for rules lack consensus on domain objects and relationships. XG rules use selectors, opportunities, and constraints, while business logic admits a wide array of forms, formats, and tools. Each constituency has a different approach to orderings (lack of consensus on relationships to be ordered), Turing computability, composability (within versus across languages/dialects), and comparability (within versus across languages/dialects, e.g., XACML). At present OWL, rule languages, and object-oriented modeling each contributes something, but no single language has the domain-specific expressive power, computational viability, and theoretical properties for a single CRPL.

20.6.3 What Is Language?

The diversity of views regarding needs for policy languages in part can be traced to a fundamental difference of opinion regarding the nature of language. The computer science community of Turing, Post, Chomsky, et al., formulates language as a set of symbols $s \in S$ along with rules of syntax that identify subsets of s^* that constitute the language. The semantic web community [794] augment syntax with ontology via the resource definition framework (RDF) evolved to OWL and its multiple variants. One path toward CRPL lies in UML actors, OWL ontologies, rules, and richer behavior modeling such as radio propagation modeling for regulatory conformance.

However, policy makers tend not to be computer scientists but lawyers, economists, financiers, and business people. They focus on revenue generation via sustaining value propositions, balancing social contracts between the public interest and open markets. CPRL for policy makers must address the synthesis of IT systems that support the value propositions with rapid, cost-effective reconfiguration. This set of needs draws one to consider cognitive linguistics for greater insight. One cannot do justice to this discipline here, but the following synopsis of relevant tenets provides a starting point:

1. **Experiential embodiment.** Language is not just a set of symbols with rules, but structured references to prior sensorimotor experience. The term *tiger evokes* curiosity at the zoo and fear in the jungle, not based on rules for t_i and so on over s*, but on experience with an actual tiger in the field, at the zoo, or in effigy [795, 796].
2. **Thing centric.** Children learn about things first, and things are in some sense the most fundamental constructs of human language. Places are defined via sensorimotor interaction with referential things (e.g., "under the table"; in radio, propagation radius from a handset). Paths are sequences of places, actions are things moving along (abstract and concrete) paths, and causes are things that set other things in motion [797].
3. **Dialectical.** Languages are not top down (ontological), but middle out, with the most common things being most broadly understood and specialization following social structure [798].

20.6.4 Cognitive Linguistics for CRPLA

The cognitive linguistics (CL) principles just summarized more fully characterize the scope of the CRPL domain via the following illustrative use case for "gold service via TV white space." In this use case, the actors are consumers, first responders, service providers, network operators, equipment manufacturers, government regulators, and lawmakers. The actors and equipment may be modeled as CL things:

Actors as things	Equipment, things proximate to actors
Consumers	Cell phone, PDA, laptop
First responders	Public safety radios, towers; command centers
Service providers	Sales force (as things), air time, bills and billing
Network operators	Cellular infrastructure, technicians, shareholders
Equipment manufacture	Manufacturing lines, R&D Labs, sales force, shareholders
Regulators	Legal documents, rule and order, publicity
Lawmakers	Laws, news reports, public opinion

The CRPL technical community needs to satisfy these actors whose needs come into conflict in situations like the 700 MHz D-block of the U.S. spectrum auctions and in TV white space during a serious emergency. The social contract that regulators and lawmakers find important concerns public safety in a serious emergency like a Hurricane Katrina or Indonesian tsunami. Networks are not provisioned nor configured by network operators to deal with such rare contingencies because it is not cost-effective for the shareholders. One unrealized potential of cognitive radio is the expansion of network capacity using TV white space as needed to achieve commercial "gold service" and to go further during such emergencies, for example, by opportunistically leveraging white space to deliver "gold service" to first responders.

The related things, places, paths, actions, and causal relationships of cognitive linguistics for radio policy domains are as follows. Cognitive linguistics includes abstract and concrete things that induce places that combine to form paths over which actions occur because of causes.

Table 20.2 begins to characterize the cognitive radio policy domain in terms of shared experience via competing for and reconfiguring resources that include the radio spectrum, capital, facilities, equipment, and people. The necessity of representing radio interference via two- and three-dimensional radio propagation measurements and models strongly differentiates CPRL from most other policy domains. Interference often is transient and nonlinear; and often interference is an aggregate effect of multiple sources of radiating energy. In addition, enforcement requires causal reasoning over relationships among stakeholders, artifacts, and infrastructure. These two fundamental needs differentiate the CPRL needs from other policy enforcement domains.

20.6.5 CRPLA Evolution

This brief treatment has provided an overview of some ongoing considerations of the cognitive radio policy language community. For human beings, language is a pointer to prior experience. Cognitive linguistics modeling of the cognitive radio domains reflect the richness of the behaviors experienced by the radios themselves, radio propagation, interference, and reconfiguration for emergency situations. To the degree that the languages enable the policy makers, service providers, network operators, and radios to express how the radios help or hinder the people in these situations, CPRL will help accrue greater the potential value to society.

Although it is helpful to view CPRL in terms of ontology and rules, the richness of interaction among legacy and cognitive radio handsets, vehicular radios, and cognitive wireless infrastructure requires the incorporation of full three-dimensional spatial representations of behavior as well as changes in these behaviors over time and in aggregates, particularly in emergency response situations. In addition, the accurate diagnosis of policy compliance faults and policy failures seems to call for broad-scope causal reasoning across the rich sets of relationships that become evident when modeling the cognitive radio use cases via the principles of cognitive linguistics. By orthogonalizing domains in terms of common object, sensorimotor interaction among those objects (places), paths, actions, and causes, the cognitive linguistics–based approach should enhance our ability to enhance network

Table 20.2 Cognitive Linguistics: Abstract and Concrete Things Involve Places that Include Paths over Which Actions Occur Because of Causes

Things	Places	Paths	Actions	Causes
Actors				
Consumers	Checkbook	Spending choices	Buy, use*, return to manufacturing	Satisfaction, dissatisfaction
First responders	Emergency venue; speech	Roads, sea, air; commu- nications*	Render aid; give orders	Humanitarian; coordination
Sales force	Sales offices	Argument	Sell	Better product*
Service providers	Corp HQ, congress	Management, finance	Hire, invest, promote	Mission, RoI
Network operators	Networks, handsets*	Via RF and cyberspace*	Control, configure*, repair	Innovation*, fault
Manufacturers	Manufacturing centers, R&D labs	Research*, sales	Design, produce, distribute	Engineering*, RoI
Radio regulators	HQ, labs, test facilities	Policy making*	Promulgate rules*, monitor*	Social contract
Lawmakers	Congress, courts	Argument, campaigns	Lawmaking, lawsuits	Public interest, concerns*
Equipment				
Cell phone	Consumer, responders hand	Spectrum*, RF, networks	Talk, listen, use data and text*	Need to communicate
PDA, laptop	Consumer, responders lap	Spectrum*, RF, networks	Talk, listen, use data and text*	Need to use large data sets
Public safety radios	First responders hand, belt	Spectrum*, RF, networks	Command, coordinate*	Maintain order; social contract
Safety infrastructure	Station, vehicles	Broadcast*, RF, networks	Transmit, receive; talk, listen	Events, policies
Sales force	First responders*	Social networking	Understand new needs	Competition, policies
Air time	RF*, networks	Order service*; pay bills	Make available; generate bills	RoI
Cellular infrastructure	Towers, backhaul, offices	RF*, financial, repair	Provide service	RoI (CAPEX vs. OPEX)
Manufacturing	Leadership	Innovation*	Experiment*, improve	RoI, pride of workmanship
RD Labs	Leadership	Innovation*	Hypothesize, experiment	Science & engineering*
Other Things				
Legal recourse	Satisfaction	Litigation	Bring lawsuit	Dissatisfaction*
Public opinion	Popular, unpopular	Advertising, news	Advertise	Person, topic, organization
Shareholders	Rewarded	RoI	Invest	Better RoI
Consumer opinion	Popular, unpopular*	Publicity, word of mouth	Advertise, promote, deprecate	Experience*

* Indicates key role of three-dimensional space/time radio spectrum occupancy and interference via compli- cated interactions among infrastructure, mobile nodes (vehicular and handheld), and other electromagnetic systems like cable TV networks, power lines, and electrical equipment.
RoI = return on investment.

and node autonomy, empower constructive behavior, and minimize disruption, achieving the intent of the policy makers.

20.7 CHALLENGES AND OPPORTUNITIES

Challenges and opportunities abound in cognitive radio architecture evolution. One challenge is the accurate characterization of dynamic spectrum opportunities. For example, spectrum measurements reported in the literature do not fully caveat the feasibility of spectrum sharing. Measurements that show 5% occupied spectrum often do not account for GPS and other navigation aids that cannot be detected via spectrum scanning but require cross-correlation receivers. Other measurements do not reflect the duty cycles of the radar bands where a pulsed radar listens for most of its duty cycle, contributing 0.1% to spectrum occupancy but 100% to airport surveillance. Pulsed radar spectrum therefore may not be shareable in a meaningful way, but often radar bands are included in the spectrum scan statistics without clear caveats. Space communications typically require high-gain receiving antennas, some 60 ft across, and the signals from the spacecraft are not detected in the spectrum scans either. Ad hoc networking in the apparently unoccupied downlink band would raise havoc with space systems. Important counterexamples to the unintentional oversimplification of spectrum occupancy characterization include the rigorous spectrum sensing campaigns of Walter Tuttlebee's Virtual Center of Excellence (VCE) conducted by Mark Beach's group at the University of Bristol in the UK [799, 800] and the data sets of the Crawdad site [801]. Finally, even in the evolution era, emergency channels simply must be kept clear for the relatively infrequent emergency communications to experience maximum SNR, so they should in fact show no occupancy most of the time. Analog AM voice is audible 6 dB below the 0 dB tangential noise floor because of the sinusoidal nature of voice, but if that emergency broadcast channel is full of ad hoc network traffic, then the safety officer may not hear "Mayday."

There are business implications for overstating the case for spectrum sharing, ultimately causing problems for the dynamic spectrum community. The opportunity, then, is to include in the evolving CRA sufficient computational intelligence about navigation aids, radar, and emergency communications that cognitive radios know how to listen to legacy communications as a function of band type, know how to avoid jamming location services, creating hot clutter in radar tracks, and generating other spectrum artifacts that detract from the trust of the evolving cognitive radio architectures.

Radio propagation is notoriously ragged in its spatial extent, even in the sweet spot between 300 MHz and 3 GHz, because of multipath, knife-edge diffraction, Fresnel zones, and other well-known phenomena. Yet the potential contributions of communities of cognitive radios to space/time fine structure of the radio spectrum may not be fully realized until the CRAs come to include high-performance spatial knowledge. A technical paper on the compact representation of such spatial knowledge was deemed best paper at DySPAN 2007, for example [802]. Several proposals have been made for general spectrum auction frameworks [803, 804] and many technical papers written on this topic [805], with at least one commercial web site that offers anonymous rental of underused spectrum today [699]. Yet at present, no technical architecture has been deployed for real-time small space/time/RF spectrum auctions.

20.8 CHAPTER SUMMARY

The transformation of radio engineering by SDR and cognitive radio is exciting. SDR places demands on analog devices to access spectrum in increasingly larger chunks, to generate increasingly pure transmissions with increasingly efficient power conversion. The technologies, products, networks, and related systems are increasing in complexity to meet rising expectations. Cognitive radio architecture provides an evolving series of frameworks for research, development, and product deployment. As users transition from the question of "Can you hear me now?" to "What have you done for me lately?" the radio engineering community may transition from QoS as the coin of the realm toward metrics more like QoI and QoS*QoI. Architectures will evolve greater QoI for lower cost via interdisciplinary information integration, where radio engineers are better able to leverage information about the user's location, preferences, and current situation to deliver multimedia infotainment in one instant and emergency response in the next. High value added and profitability are needed to transition smoothly to continue to balance market needs and public interest. During this evolution, the radio research community will continue to lead the way with mathematical foundations, focused use cases, sensory-perception integration, and radio engineering evolution, toward ultimate realizations of the cognitive radio value proposition: better use of radio spectrum for the user. In addition, bridging the gaps among policy makers, business imperatives, network complexity, and device technologies seems to call for renewed emphasis on the integration of the human perspective reflected in policies and business perspectives with the varied hardware, firmware, software, and services technologies of the technical perspective. Cognitive linguistics–oriented policy language architecture offers a rich new perspective for more cost-effective approaches to these challenging problems.

GNU radio experimentation

Essential linux commands

GNU Radio (GR) currently works best on a Linux-based platform. The following are some common commands that are helpful while developing on a Linux platform. Commands in Linux are implemented as separate programs and anything with the permissions set to executable, including text files, can be run.

man COMMAND Displays the manual for the specified COMMAND. The vast majority of Linux commands use man pages for documentation. The page provides explanations of the function of the command, arguments and flags you can pass to the command and other useful information.

cd PATH Changes the working directory of the terminal. If the PATH starts with a '/' then the path will be in relation to the root directory; that is, '/home'. If the path just starts with a folder name, then it will be relative to the current directory; for example, if the current directory is /usr, then 'cd share' would go into the /usr/share folder. To go up a level use '..', for example if the current directory is /usr/share and the desired directory is /etc, then the command would be 'cd ../../etc', which goes up two directory levels then back down into the /etc folder. If no path is specified, 'cd' sets the path to the home directory of the current user.

ls PATH ARGUMENTS Lists the contents of the directory specified by PATH; if no PATH is specified then it uses the current directory. Arguments like -a show all files, specifically hidden files denoted by a period (.) in front of the name. Command arguments can be combined and are case sensitive, so -lR would use the long listing format to list all the files including subdirectories. Using the -l displays the permissions of the files in the folder. Using this you can verify that a file has permission to be executed.

sudo COMMAND Runs the command COMMAND as root. Root is the administrative account in Linux and has permission to run any command in the system. The user account being used needs to be in the root group to use this command. This command enables changing permissions, writing to restricted folders, and whatever else a budding Linux user may need to do with administrative access. This command gives you all the rope you need to hang yourself, so use it wisely.

chmod PERMISSIONS FILENAME Sets the permissions on a file or folder. Use this command with 'sudo' to change the permissions of a file of which you are

Doi: 10.1016/B978-0-12-374715-0.00021-6

621

not the owner. There are three sets of permissions to be set: the owner's permissions, the group's permissions, and other users' permissions. These groups are indicated by user (u), group (g), and others (o), respectively. There are three levels of permission: read (r), write (w), and execute (x). These abbreviations are put together to set the permissions for each group in the format WHO = PERMISSION. The command 'sudo chmod go=rx FILENAME' means both the group and others can read the file and execute it. This command comes up commonly in GSDR development when setting a Python file to have execute permissions, such as 'sudo chmod a = rwx FILENAME', which sets the permissions for all three levels of users to have full access to the file. Permissions can also be set graphically by right-clicking on the file in a Nautilus file browser window. Select properties then the permissions tab. Run 'sudo nautilus' to change permissions of files you do not own using the graphical method.

locate FILENAME Searches for the FILENAME in any folder on the system. If the installer is missing a file, 'locate' will find the path to the file, assuming the file is on the system. The command can search for fragments of the file name and wild cards will work; for example, the command 'locate *.py' will list all of the Python files on the system. The 'locate' command uses a database of the files on the system; to find newly added files the 'updatedb' command has to be run before searching.

updatedb Updates the database that 'locate' uses to find files. After installing new software, this command has to be run or the new files will not be found. It is a good idea to know when the last time this command has been run to ensure the results of a 'locate' search are accurate. The database is written to a restricted folder on the computer, so you *must* run this command as root ('sudo updatedb'). The command asks for your password, runs for a second, and exits quietly with no output. If you are not a part of the root group on your system you will get errors.

grep ARGUMENTS EXPRESSION QUALIFIERS Searches contents of files that meet the criteria specified by QUALIFIERS for the given EXPRESSION. The search starts in the current directory, but the -R argument makes the command also search subdirectories. The command returns the file containing the line with the EXPRESSION and the contents of that line.

./FILENAME Executes the file specified by FILENAME. If the file named is a text file, it will try to run the file as a script, such as the bootstrap and configure scripts. The first line of the Python programs developed in this appendix tells the computer to run the script in the Python environment. When you are running commands that are on the path, you need not use this command and can just type the name of the command.

GNU radio installation guide

B

This is the install build guide for GNU radio (GR). There may be other options in this installation process, but this is the method suggested by the authors of this appendix. This guide reviews how to install GR from the latest build, called *trunk*. Synaptic packages also are available for GR, but it takes much longer for these packages to be updated. GR is set up under version control (using Smart Subversion (SVN)) and both software packages should be installed in a common area for use by everyone working on the system. This guide is meant to be a supplement to the GR build guide found at http://gnuradio.org/trac/wiki/BuildGuideBuildGuide. This appendix assumes you already installed Ubuntu and chose not to use the built-in synaptic packages. Those new to Linux should read Appendix A to review commonly used Linux commands. This appendix first helps with the installation of the needed dependencies, an SVN server to get the most up-to-date code from GR, then the installation of GR.

B.1 INSTALL DEPENDENCIES

The easiest way to install the dependencies needed by GR is to use the scripts found at http://gnuradio.org/trac/wiki/UbuntuInstall to get the dependencies needed for GR. If you would like to install the packages manually, that page includes a list of the needed software packages as well. You can start this installation while we set up Smart SVN since you need not pay close attention to it.

If you are using Ubuntu 8.10, this command installs all the dependencies you need. The backslashes at the end of each line (\) tell the terminal to concatenate the successive lines together; when typing this command into a terminal they are not necessary.

```
sudo apt-get -y install swig g++ automake1.9 libtool python-dev
                                              fftw3-dev \
libcppunit-dev libboost1.35-dev sdcc-nf libusb-dev \
libsdl1.2-dev python-wxgtk2.8 subversion guile-1.8-dev \
libqt4-dev python-numpy ccache python-opengl libgsl0-dev \
python-cheetah python-lxml doxygen qt4-dev-tools \
libqwt5-qt4-dev libqwtplot3d-qt4-dev python-qwt5-qt4
```

If you are running Unbuntu 8.04, then this is the command you have to run:

```
sudo apt-get -y install swig g++ automake1.9 libtool python-dev
                                                      fftw3-dev \
libcppunit-dev sdcc libusb-dev libasound2-dev libsdl1.2-dev \
python-wxgtk2.8 subversion guile-1.8-dev libqt4-dev python-numpy-ext \
ccache python-opengl libgsl0-dev python-cheetah python-lxml doxygen \
libqwt5-qt4-dev libqwtplot3d-qt4-dev qt4-dev-tools python-qwt5-qt4
```

You may have to install libboost separately, however, since version 1.35 is required and the version in synaptic for Ubuntu 8.04 is 1.34. Boot can be downloaded from the www.boost.org/doc/libs/1_37_0/more/getting_started/unix-variants.html boost web site. As versions continue to be updated, other dependencies may be needed at other versions. While the specifics change, this appendix covers the basic steps needed to manually install a software package. After installing the dependencies, run 'updatedb' to make sure the files added to the systems can be found using the 'locate' command.

If you get an error that says

```
E: Could not get lock /var/lib/dpkg/lock - open (11 Resource
                                        temporarily ...
E: Unable to lock the administration directory (/var/lib/dpkg/), is
                                                      another...
```

you have another instance of synaptic or apt—get running, close the other instances, and run the command again.

B.2 INSTALL AN SVN CLIENT

SVN servers enable version control, allowing the user to use older versions of code if the current version is broken. Any SVN client can be used with an SVN server, but the authors use Smart SVN because it works cross-platform, so the users need to learn one tool and can use it on whatever platform they are currently developing on. Using a SVN client to download GR makes it easy to use an older version if necessary or download the latest code. If the user does download the latest code, GR has to be rebuilt for many of the changes to take effect.

Start by downloading Smart SVN from the web site www.syntevo.com/smartsvn/download.html. Unzip the program and move the folder to /usr/share. The easiest way to do this is to open a file browser window as root using 'sudo nautilus', browsing to /usr/share, and clicking and dragging the file into the window, moving files with the root user's permissions. Unlike other operating systems, running as root is dangerous and should be done as infrequently as possible; system files can be overwritten and break the system. This simply copies the folder to /usr/share, and you can then delete the original.

While in the root window, create a folder called 'gnuradioTrunk' on the /usr/share folder. Right-click on the folder and go to the permissions tab. The owner is root, change it to your user name.

Make sure you have the current Java runtime environment (JRE) installed. At the time of this writing, we use the synaptic package 'sun-java6-jre' and its dependencies. If Java is not installed, you will see an error that looks like this when you try to run Smart SVN:

```
user@computer:/usr/share/SmartSVN 5.0.1/bin ./smartsvn.sh
Warning: /bin/java does not exist
Error: No java environment found
```

Enter the directory using 'cd /usr/share/SmartSVN 5.0.1/bin/' and run the Smart SVN using './smartsvn.sh'. The first time this is run, you will enter a setup utility. After the setup is run, that command simply opens Smart SVN. Accept the license agreement, Tell it to use the Free Professional edition; after the 31 days it reverts to the limited free edition. The limited edition provides all the functionality (as of now) that is needed. Select 'My repositories are already set up'. If you want to use an SVN server for your own work, you have to set one up, but that is outside of the scope of this guide.

The program window now opens and asks you to start a project. Select 'Check out project from repository' and click 'OK'. On the new screen is a 'Manage' button, opening a window that allows you to add repositories. Click 'Manage', then the 'Add' button, and finally the 'Enter SVN URL' button. The URL for GR is http://gnuradio.org/svn/gnuradio/trunk. Smart SVN then fills in the required information from the URL: click 'Next' twice; the repository can be renamed, but the default is OK; then click 'Finish'. Click 'OK' in the Repository Profiles screen. Make sure the correct repository is selected in the 'Check Out Project' screen, then click 'Next'. After the directory tree loads, click 'Next' again; make sure the top level of the repository is selected. Enter '/usr/share/gnuradioTrunk' as the Local Directory and click 'Next' again. One more 'Next' and then a 'Finish', Smart SVN then downloads the most up-to-date version of GR to the specified local directory (/usr/share/gnuradioTrunk). We are now ready to start installing GR (make sure the dependencies have finished). After getting the repository, run 'updatedb' to make sure the new files can be found using 'locate'.

B.3 INSTALL GR

In a terminal, go to the GR root directory. As set up earlier in this appendix, this will be '/usr/share/gnuradioTrunk'. The command to change your directory is 'cd /usr/share/gnuradioTrunk'. This guide is in parallel to the one found at GR's web site http://gnuradio.org/trac/wiki/BuildGuide but helps with some pointers and debugging suggestions. Each section here discusses the commands you run in the terminal. You do not put the ' ' characters in the command (so 'COMMAND' is put into the terminal as just COMMAND).

Run './bootstrap'

This creates the install files and should be used only when installing from 'trunk', which gets the source code from the SVN. If you are installing using the tarball (zip file) of the latest stable release, this is not necessary. If you get an error that looks like this (or some subset of this):

```
./bootstrap: 25: aclocal: not found
./bootstrap: 26: autoconf: not found
./bootstrap: 27: autoheader: not found
./bootstrap: 28: libtoolize: not found
./bootstrap: 29: automake: not found
./bootstrap: 25: aclocal: not found
./bootstrap: 26: autoconf: not found
./bootstrap: 27: autoheader: not found
./bootstrap: 29: automake: not found
```

then you have not successfully installed all the dependencies. Go back to find which program was not installed successfully or check if a program is still being installed.

Run './configure'

This command runs a script that checks each of the modules of GR to see which ones should be installed. Errors may occur if the dependencies were not installed correctly or if they were not properly added to the path. The path is the list of places where programs, including the terminal, look for other programs. To access the Help for this script, use the command './configure -h'. This lists all of the options. Here you can specify the location of a library if it is not detected automatically using '- -with-XXXX'. For example, the flag '- -with-boost-libdir=LIB_DIR' allows you to specify the path of the Boost libraries, a common set of libraries to go missing. When a library or program is missing but installed on the system, the 'locate' command can find the correct path. Here, LIB_DIR would be replaced with the result of the 'locate' command.

A module can also be forced to be included; if there is an error configuring that module, the configuration stops and you can see that error. The command '- -enable-usrp2' stops the configuration if there are problems with the usrp2 module. The command with these two examples looks like './configure - -with-boost-libdir=LIB_DIR - -enable-usrp2'.

Baring any errors, the output looks like this:

```
********************************************************************
The following components were skipped either because you asked not
to build them or they didn't pass configuration checks:

usrp2-firmware

These components will not be built.


********************************************************************
```

```
The following GNU Radio components have been successfully configured:

config
gruel
omnithread
gnuradio-core
pmt
mblock
usrp
usrp2
gr-usrp
gr-usrp2
gr-msdd6000
gr-audio-alsa
gr-audio-oss
gr-atsc
gr-cvsd-vocoder
gr-gpio
gr-gsm-fr-vocoder
gr-pager
gr-radar-mono
gr-radio-astronomy
gr-trellis
gr-video-sdl
gr-wxgui
gr-qtgui
gr-sounder
gr-utils
gnuradio-examples
grc

You my now run the make command to build these components.

**********************************************************************
The following components were skipped either because you asked not
to build them or they didn't pass configuration checks:

gcell
gr-gcell
gr-audio-jack
gr-audio-osx
gr-audio-portaudio
gr-audio-windows
gr-comedi

These components will not be built.
```

The output should be checked to make sure the modules you need are being installed. Some of the modules overlap and are platform specific, so some of the modules are expected *not* to be installed. Here, additional dependencies are needed for firmware development, so 'usrp2-firmware' was not installed. Different modules for audio have been developed for each operating system to make sure one of them is installed; here 'gr-audio-alsa' and 'gr-audio-oss' are installed, so the other gr-audio-XXXXX options are not necessary.

If there are errors, use the 'locate' command with '- -with' flags to get the configure script to find everything it needs.

Run 'make'

In this step the computer actually compiles the source code for your system, probably using the GNU compiler, gcc. This takes awhile, probably just less than an hour, so grab a coffee or read slashdot (er, study for that big exam). Errors here are probably from errors missed in the ./configure script; check to see what is missing, try to find it using 'locate', and if the program is not installed, install it and then re-run './configure'.

You can tell the terminal to run a second program after the first exits without errors, so to run 'make check' right after 'make'—use the command 'make && make check'.

Run 'make check'

This checks the compile for any errors by running tests built into and written for the source code. If there are errors, a dependency is either not installed or the path to it could not be found. Review the output of './configure' and make for errors and warnings as to what is the missing file. The command may have to be rerun as the output may have filled up the command line buffer.

If last part of the output includes a bunch of lines of 'Nothing to be done for "XXXX"', then GR was complied properly. The complied libraries can now be installed.

Run 'sudo make install'

This actually installs GR onto the system by copying the compiled libraries into the common path locations of the system. Assuming it exits without errors, we are ready to set up access to the universal software radio peripherals (USRPs). The only error the authors have seen here are related to permissions. This command must be run as root, using the sudo command.

B.4 SET UP USRPS

Now with GR installed, a USRP group has to be set up so multiple users can access the USRP and the computer knows what to do when a USRP is plugged in. This is necessary only the first time GR is being installed on a system; if you are just recompiling a newer version of GR, then this is unnecessary.

The commands in this section can be copied directly into the command line, with the exception of actually entering a user name where it says '⟨YOUR_USERNAME⟩'. After these commands have been run, you need not rerun them, with the exception of the 'sudo addgroup ⟨YOUR_USERNAME⟩ usrp' command. This command has to be run for each user that needs to access the USRP. Ubuntu does offer graphical ways of adding users to a group and is better suited to adding multiple users.

```
sudo addgroup usrp
sudo addgroup <YOUR\_USERNAME> usrp
echo 'ACTION=="add", BUS=="usb", SYSFS{idVendor}=="fffe", \
  SYSFS{idProduct}=="0002", GROUP:="usrp", MODE:="0660"' > tmpfile
sudo chown root.root tmpfile
sudo mv tmpfile /etc/udev/rules.d/10-usrp.rules
```

The first line creates a group 'usrp', and the second line adds the specified user name to be a member of that group. The third line creates a udev rule and writes it to a tmpfile. The details of the third line are outside the scope of this guide. The fourth line makes 'root' the owner of the tmp file, so it can be moved into a restricted folder that is checked for udev rules when the udev process is started or restarted.

At this point, Ubuntu is configured to know what to do if it detects the USRP on the USB, except that 'udev' needs to reload the rules to include the newly created one. Use the following commands to reload the rules. If neither of these methods work, you may have to reboot the system.

```
sudo /etc/init.d/udev stop
sudo /etc/init.d/udev start
```

or

```
sudo killall -HUP udevd
```

To make sure the USRP is being correctly recognized, use the command 'ls -lR /dev/bus/usb | grep usrp'. The result for each USRP should look like this:

```
leferman@Katahdin:/etc/udev/rules.d ls -lR /dev/bus/usb/ | grep usrp
crw-rw- - - - 1 root usrp 189, 897 2008-12-12 14:24 002
```

Each device file is listed with group 'usrp' and mode 'crw-rw - - - -'. If this is not the result of the command, the first thing to try is rebooting the computer. Occasionally this check works, but the USRP still does not get recognized; again first try restarting the computer.

Now that the USRP group exists, the group can be assigned access to the gnuradioTrunk folder. In the terminal, cd to /usr/share and, after replacing USERNAME with your user name, enter the command 'sudo chown USERNAME.usrp -R gnuradioTrunk/'. This sets the group of all the GR files to 'usrp' and makes sure members of the group can read, write, and execute the files in this folder. Running this command allows anyone using the USRP (who is in the USRP group) access to the GR files.

B.5   TEST USRP

From the root directory of GR (it should be /usr/share/gnuradioTrunk), the built-in example can be run to see if the host computer can send data to the USRP. First, change the digital examples directory using 'cd gnuradio-examples/python/usrp'. Make sure you have a USRP1 plugged into a USB port on the computer and turned on. Run './usrp_benchmark_usb.py' and look for this output:

```
Testing 2MB/sec... usb\_throughput = 2M
ntotal     = 1000000
nright     = 999418
runlength = 999418
delta      = 582
OK
Testing 4MB/sec... usb\_throughput = 4M
ntotal     = 2000000
nright     = 1997079
runlength = 1997079
delta      = 2921
OK
Testing 8MB/sec... ./usrp\_benchmark\_usb.pyusb\_throughput = 8M
(and so on.....)
```

The USRP1 should be able to achieve the full 32 MB/s throughput.

Possible Errors

Keep in mind that some lines, specifically the connect() function, can be broken up into multiple lines, allowing you to find which part of the command is throwing the error.

- **RuntimeError: Unable to find USRP #0**: This means the USRP is not plugged in. Make sure it has power and the USB cord is plugged into a working USB port. You can run the command 'ls -lR /dev/bus/usb | grep usrp' to list all of the devices plugged into the USB ports, then search for the USRP. If you get the following output, then you know the USRP is plugged in and set up correctly. Despite this appearing to work correctly, you may need to restart your system for changes in the rules (changes in the "Setup USRPs" section) to take effect. If this does not show up, go through the steps in the "Setup USRPs" section to create the permissions correctly.

  ```
  crw-rw- - - - 1 root usrp 189, 658 2008-12-11 10:40 019
  ```

- **ImportError: libusrp.so.0: cannot open shared object file: No such file or directory**: If this error is seen, the path to the file to the cache of shared libraries needs to be set. Use 'locate' to find the path to the missing library, and use these commands to add that path to the shared library cache. The file containing the paths to be included in the cache is '/etc/ld.so.conf'. The command 'sudo nano /etc/ld.so.conf' opens a command line text editor to add the path to the file. To update the cache use the command 'sudo ldconfig'; this make the changes take effect.

B.6   GENERAL INSTALLATION NOTES

1. Older versions of software on the computer can confuse installers and get the wrong versions of software linked together. Make sure to remove old versions (especially the GNU Radio packages from Synaptic) before installing.
2. Machines are used by more than one user. Installing programs like Boost into your home directory will cause headaches later. These programs should be installed in /usr/share unless suggested otherwise by the program. Because you are installing in common areas, root access (sudo) is needed, and you have to set the permissions as discussed in this guide.
3. Always check the GNU Radio and Ubuntu web sites for the most updated tips, tricks, and instructions.

Universal software radio peripheral

C.1 THE MAIN ELEMENTS ON THE USRP BOARD

A typical setup of the universal software radio peripheral (USRP) board consists of one motherboard and up to four daughterboards, as shown in Figure C.1.

The USRP has four high-speed analog-to-digital converters (ADCs) and four high-speed digital-to-analog converters (DACs). These four input and four output channels are connected to an Altera Cyclone EP1C12 FPGA [806]. The FPGA, in turn, connects to a USB2 interface chip, the Cypress FX2, and on to the computer. The USRP connects to the computer via a high-speed USB2 interface only and does not work with USB1.1.

AD/DA Converters

The board has four high-speed 12-bit AD converters. The sampling rate is 64 M samples per second. In principle, it could digitize a band as wide as 32 MHz. The AD converters can bandpass sample signals of up to about 150 MHz, though. If we sample a signal with the intermediate frequency larger than 32 MHz, we introduce aliasing. Actually the band of the signal of interest is mapped to some places between −32 MHz and 32 MHz. The higher the frequency of the sampled signal, the more the SNR is degraded by jitter. The recommended upper limit is 100 MHz. Note that we can use other sampling rates if desired. All the available rates are submultiples of 128 MHz, such as 64 mega-samples per second (MS/s), 42.66 MS/s, 32 MS/s, 25.6 MS/s, and 21.33 MS/s.

At the transmitting path, there are also four high-speed 14-bit DA converters. The DAC clock frequency is 128 MS/s, so Nyquist frequency is 64 MHz. However, we probably want to stay below about 50 MHz or so to make filtering easier. So a useful output frequency range is DC to about 50 MHz.

So, in principle, we have four input and four output channels if we use real samplings. However, we have more flexibility (and bandwidth) if we use complex (I/Q) sampling. Then we have to pair them up, so we get two complex inputs and two complex outputs.

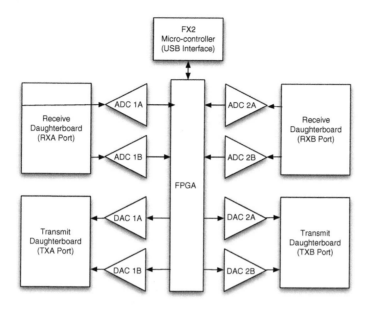

FIGURE C.1

The USRP1 board.

Daughterboards

The daughterboards are used to hold the radio frequency (RF) receiver interface or tuner and the RF transmitter. On the motherboard, there are four slots, where you can plug in up to two tx daughterboards, labeled TXA and TXB, and two corresponding rx daughterboards, RXA and RXB. Each daughterboard slot has access to two of the four high-speed AD/DA converters (DAC outputs for tx, ADC inputs for rx).

This allows each daughterboard that uses real (not I/Q) sampling to have two independent RF sections, and two antennas (four total for the system). If complex I/Q sampling is used, each board can support a single RF section, for a total of two for the whole system. Normally, we can see that there are two SMA connectors on each daughterboard. We usually use them to connect the input or output signals.

Several kinds of daughterboards are available now, which allows for maximum flexibility in frequency planning for the daughterboards.

FPGA

Probably understanding what goes on the USRP field programmable gate arrays (FPGAs) is the most important part for the GNU Radio users. Upon understanding FPGA, we also get to know how data flow on the USRP. As shown in Figure C.1, all the ADCs and DACs are connected to the FPGA. This piece of FPGA plays a key role in the USRP system. Basically, what it does is to perform high-bandwidth math and reduce the data rates to something you can squirt over USB2. The FPGA connects to a USB2 interface chip, the Cypress FX2. Everything (FPGA circuitry and USB microcontroller) is programmable over the USB2 bus.

C.2 DATA FLOW ON THE USRP

In this section we will concentrate on the hardware side, introducing how data flow on the USRP board. After reading this section, you should have a clear understanding of the signal processing blocks in the FM receiver example and better insight on the USRP board.

To make things clearer, the data flow is introduced in two kinds, rx path and tx path, as shown in Figure C.2.

RX Path

At the rx path, we have four ADCs and four DDCs. Generally speaking, on each daughterboard, the two analog input signals are sent to two separate ADCs. The digitized samples are then sent to the FPGA for processing. Upon entering the FPGA, the digitized signals are routed by a multiplexer, or MUX, to the appropriate digital down-converter (DDC).

There are four DDCs. Each DDC has two inputs, I and Q. The MUX is like a router or a circuit switcher. It determines which ADC (or constant 0) is connected to each DDC input. We can control the MUX using the usrp.set_mux() method in Python. This allows having multiple channels selected out of the same ADC sample stream.

Now let us see the digital down-converter. What does it do? First, it down-converts the signal from the intermediate frequency (IF) band to the baseband. Second, it decimates the signal so that the data rate can be adapted by the USB2 and is reasonable for the computer's computing capability. Figure C.3 shows the block diagram of the DDC. The complex input signal (IF) is multiplied by the constant frequency (usually also IF) exponential signal. The resulting signal is also complex and centered at 0. Then we decimate the signal with a factor M.

Our standard FPGA configuration includes digital down-converters implemented with four stages cascaded integrator-comb (CIC) filters. CIC filters are very high-performance filters using only adds and delays. For spectral shaping and out-of-band signal rejection, 31 tap halfband filters also are cascaded with the CIC filters to form a complete DDC stage. The standard FPGA configuration implements two complete digital down-converters. Also there is an image with four DDCs but without halfband filters. This allows one, two, or four separate rx channels. It is possible to specify the firmware and FPGA files that are to be used by loading the corresponding rbf files. By default, an rbf is loaded from /usr/local/share/usrp/rev{2,4}. The one used unless you specify the *fpga_filename* constructor argument when instantiating a USRP source or sink is std_2rxhb_2tx.rbf. Table C.1 lists the current three rbf files and their usage.

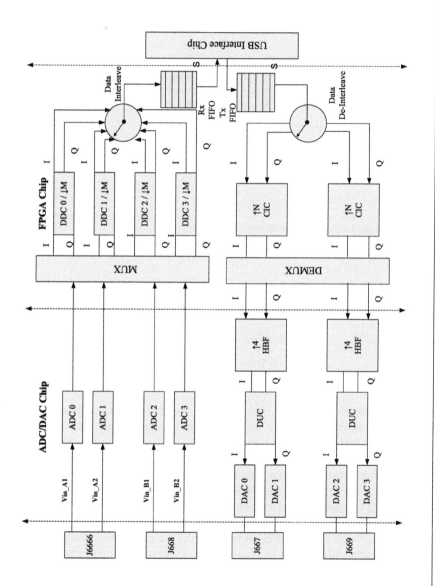

FIGURE C.2

USRP receive and transmit paths [807].

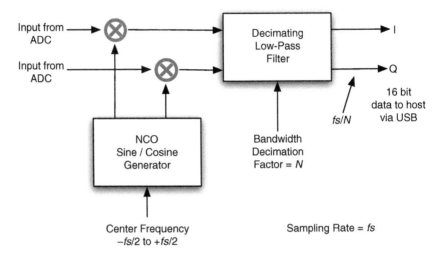

FIGURE C.3

The digital down-converter and decimation stage in rx path.

Table C.1 rbf Files and Their Usage

rbf File Name	FP-GA Description
multi_2rxhb_2tx.rbf	There are two or more USRPs
std_2rxhb_2tx.rbf	Contains two rx paths with halfband filters and two tx paths
std_4rx_0tx.rbf	Contains four rx paths without halfbands and zero tx paths

Note that, when there are multiple channels (up to four), the channels are interleaved. For example, with four channels, the sequence sent over the USB would be 'I0 Q0 I1 Q1 I2 Q2 I3 Q3 I0 Q0 I1 Q1', and so forth. The USRP can operate in full duplex mode. When in this mode, the transmit and receive sides are completely independent of one another. The only consideration is that the combined data rate over the bus must be 32 MB/s or less.

Finally, the I/Q complex signal enters the computer via the USB. That is the software world.

To illustrate the effect of a halfband filter in the rx path, two situations are discussed. The bandwidth of each rx channel is f. If we set the center frequency at 0 Hz, then, at frequency domain, it is $[-f/2, f/2]$.

Situation 1

Suppose each of four users (u1, u2, u3, and u4) uses one of the rx channels, as shown in Figure C.4. In the frequency domain, the bandwidths they occupy are all from $-f/2$ to $f/2$. In this situation, the FPGA rbf file should be std_4rx_0tx.rbf,

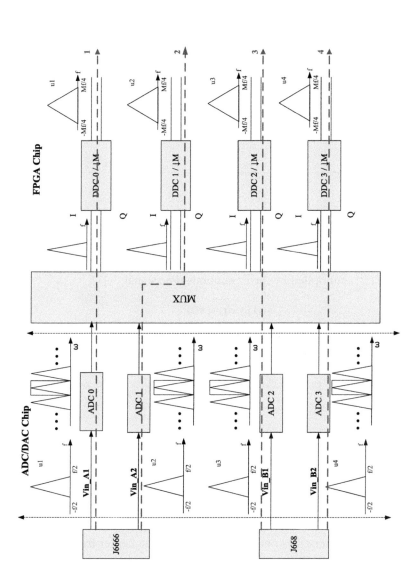

FIGURE C.4

Four rx channels (labeled 1, 2, 3, and 4), starting from the inputs A1, A2, B1, and B2, passing through one of the ADCs and one of the DDCs, then reaching the USB in the end.

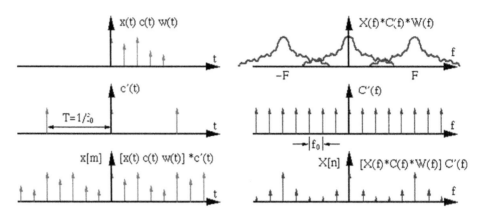

FIGURE C.5

Impulse train sampling.

which means the FPGA firmware contains four rx paths without halfbands and zero tx paths.

In an analog-to-digital converter, continuous signals are converted to discrete digital numbers. A common and useful way to do this is through a periodic impulse train signal [90], $c(t)$, multiplied by the continuous time signal:

$$x_c(t) = x(t) \times c(t), \tag{C.1}$$

$$c(t) = \sum_{n=-\infty}^{\infty} \delta(t - nT), \tag{C.2}$$

where T is the sampling period. This is known as *impulse train sampling*, as shown in Figure C.5.

By definition,

$$x_c(t) = \sum_{n=-\infty}^{\infty} x(t)\delta(t - nT) = \sum_{n=-\infty}^{\infty} x(nT)\delta(t - nT) . \tag{C.3}$$

One property of the Fourier transform is the multiplicative property: $X_c(j\omega)$ is a periodic function of ω, consisting of a superposition of shifted replicas of $X(j\omega)$, scaled by $1/T$.

Then, the low-pass filter takes out the signal around the center frequency and suppresses the rest as replicas.

If the decimation value of the DDC is $M/2$, after entering the DDC, the bandwidth of each signal will be expanded to $[-Mf/4, Mf/4]$. The sum of all the bandwidths is $2Mf$.

Situation 2

Suppose each of two users (u1 and u2) uses one of the rx channels, as shown in Figure C.6. In a frequency domain, the bandwidths they occupy are both from

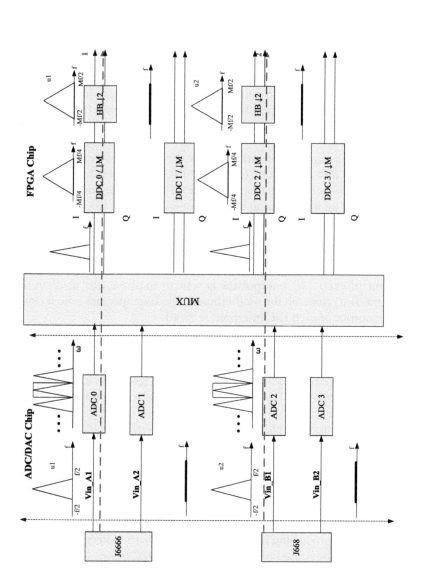

FIGURE C.6

Two rx channels (labeled 1 and 2), starting from the inputs A1 and B1, passing through one of the ADCs and one of the DDCs with a halfband filter, then reaching the USB in the end.

$-f/2$ to $f/2$. In this situation, the FPGA rbf file should be std_2rxhb_2tx.rbf, which means FPGA firmware contains two rx paths with halfbands and two tx paths. If the decimation value of DDC is $M/2$, after entering the DDC, the bandwidth of each signal will be expanded to $[-Mf/4, Mf/4]$. Then, after entering the halfband filter, the bandwidth of each signal is further expanded to $[-Mf/2, Mf/2]$. The sum of all the bandwidths is still $2Mf$.

TX Path

At the tx path, the story is pretty much the same, except that it happens in reverse. We need to send a baseband I/Q complex signal to the USRP board. The digital up-converter interpolates the signal, up-converts it to the IF band, and finally sends it through the DAC.

More specifically, first, interleaved data sent from the host computer are pushed into the transmit first in, first out (FIFO) on the USRP. Each complex sample is 32 bits long (16 bits for in-phase, and 16 bits for quadrature). These data are deinterleaved and sent to the input of an interpolation stage. Assuming the user has specified an interpolation factor of L, this interpolation stage interpolates the input data by a factor of $L/4$ using CIC filters.

The output of the interpolation stage is sent to the demultiplexer, or DEMUX. The DEMUX is less complicated than the receiver MUX. Here, the in-phase and quadrature output of each CIC interpolator is sent to in-phase and quadrature inputs of one of the DAC chips on the motherboard. The user specifies which DAC chip receives the output of each CIC interpolator [808].

Inside of the AD9862 CODEC chip, the complex-valued signal is interpolated by a factor of 4 using halfband filter interpolators. This completes the requested factor of L interpolation. After the halfband interpolators, the complex-valued signal is sent to a digital up-converter. Note, at this point, the signal is not necessarily modulated to a carrier frequency. The daughterboard might further up-convert the signal.

The in-phase and quadrature output of the DUC are sent as 14-bit samples to individual digital-to-analog converters, which convert them to analog signals at a rate of 128 mega-samples per second. These analog signals are then sent from the AD9862 to either daughterboard interface J667 or J669, which represent slots TXA and TXB, respectively.

Different from the digital down-converters, the digital up-converters on the transmit side are actually contained in the AD9862 CODEC chips, not in the FPGA. The only transmit signal processing blocks in the FPGA are the interpolators. The interpolator outputs can be routed to any of the four CODEC inputs [807].

As mentioned before, all the multiple rx channels (one, two, or four) must be the same data rate (i.e., same decimation ratio). The same applies to the one, two, or four tx channels, each of which must be at the same data rate (which may be different from the rx rate) [809].

GNU radio python program structure

```
#! /usr/bin/env python
# The previous line allows this program to be run from the linux
# command line using: ./PythonShell.py

# Import nessary objects from other modules, the nessary objects
# depend on the real application
from gnuradio import gr, gru
from gnuradio import usrp
from gnuradio import eng_notation


# This pseudo program shows how a GNU Radio Python program can be
# divided into three basic parts


####### Radio Class #######
# Here, we inherit the parent class, 'gr.top_block', for our class
# 'digital_radio'
class digital_radio(gr.top_block):

        # Initialization function, sets up the radio
        def __init__(self):
                # Initilizes parent class
                gr.top_block.__init__(self)

                # Setup blocks for the radio
                self.source = setup_source()

                self.sink = setup_USRP()

                # Connect the blocks in a flow graph
                self.connect(self.source, self.sink)


####### Main Function #######
# This function is called when we run this program from the
# command line
def main():
```

```
        # Create the radio we created, the __init__ function is called
        # automatically
        # Here I could pass parameters to the class
        # but none are used for simple example
        radio = digital_radio()

        # Start the flow graph
        radio.start()

        # After the radio is started, code adapting the radio can be
        # added here

        # Wait for source to finish
        radio.wait()


####### Program Start #######
# Checks to see if this module is being run by itself, or imported by
# another module
if __name__ == '__main__':
    # If it is being run by itself, try to run the main function
    try:
        main()
    # If there is a keyboard interrupt, exit mail function before
    # it finishes
    except KeyboardInterrupt:
        # 'pass' is a placeholder, used when nothing is to be done,
        # but a command is needed syntatically
        pass
```

Analog receiver code

```python
#! /usr/bin/env python
# The previous line allows this program to be run from the linux
# command line using: ./FMReceiver.py

# Import nessary objects from other modules
from gnuradio import gr, gru, modulation_utils, optfir
from gnuradio import usrp, blks2, audio
from gnuradio import eng_notation

# These are global variables
# Making these parameters global variables is bad programming pratice.
# These should be set using command line options, for sake of
# simplicity they have been set this way.
# See other GNU Radio examples to see how to use the options parser.

# Set audio device to be used by the program
audio_device = ""
frequency = 100.7e6
# Set which daughterboard port to use, (0,0) for RXA and (1,0) for RXB
daughterboard_port = (0,0)


####### Radio Class #######
# Here, we inherit the parent class, 'gr.top_block', for our class
# 'digital_radio'
class digital_radio(gr.top_block):

    # Initialization function, sets up the radio
    def __init__(self):
        # Initilizes parent class
        gr.top_block.__init__(self)

        # Set up blocks for the radio
        # Create USRP, usrp is data source
        self.usrp = usrp.source_c()
        # Read ADC Rate from usrp, 64 MS/s
        adc_rate = self.usrp.adc_rate()
        # Create a decimation rate
        usrp_decim = 200
        # Set decimation rate on USRP
        self.usrp.set_decim_rate(usrp_decim)
                # usrp rate= (64MS/s)/200= 320 kS/s
```

```python
usrp_rate = adc_rate / usrp_decim

        # channel filter decimation rate= 1
        chanfilt_decim = 1
        # demodulate rate= 320 kS/s
demod_rate = usrp_rate / chanfilt_decim

        # audio decimation rate= 10
        audio_decimation = 10
        # audio rate= (320kS/s)/10= 32 kHz
audio_rate = demod_rate / audio_decimation
        # print the audio rate for the audio device
        print "audio rate: %d" % audio_rate

        # set mux value between ADC and DDC
        mux_value = usrp.determine_rx_mux_value(self.usrp,
        daughterboard_port)
print "mux value: %d" % mux_value
self.usrp.set_mux(mux_value)

        # select a daughterboard to receive the FM signals
self.daughterboard = usrp.selected_subdev(self.usrp,
daughterboard_port)
print "Using Rx daughterboard %s" % (self.daughterboard.
side_and_name(),)

# taps for channel filter
chan_filt_coeffs = optfir.low_pass (1,      # gain
                        usrp_rate,      # sampling  rate
                        80e3,           # passband cutoff
                        115e3,          # stopband cutoff
                        0.1,            # passband ripple
                        60)             # stopband attenuation

        # channel filter on the receiver path
        self.channel_filter = gr.fir_filter_ccf
            (chanfilt_decim, chan_filt_coeffs)

        # demodulate the FM signal
self.receiver_block = blks2.wfm_rcv (demod_rate,
                            audio_decimation)

        # select a specific audio device as final sink
self.sink = audio.sink(int(audio_rate), audio_device)

# connect the blocks in a flow graph
self.connect(self.usrp, self.channel_filter,
            self.receiver_block, self.sink)


####### Main Function #######
# This function is called when we run this program from the
# command line
```

```
def main():

    # Create the radio we created, the __init__ function is called
    # automatically Here I could pass parameters to the class, but none
    # are used for this simple example
    radio = digital_radio()

    # Start the flow graph
    radio.start()

    # After the radio is started, code adapting the radio can be
    # added here

    # Wait for source to finish
    radio.wait()


####### Start of program #######
# This checks to see if this module is being run by itself, or imported
# by another module
if __name__ == '__main__':
    # If it is being run by itself, try to run the main function
    try:
        main()
    # If there is a keyboard interrupt, exit mail function before
    # it finishes
    except KeyboardInterrupt:
        # 'pass' is a placeholder, used when nothing is to be done,
        # but a
        #  command is needed syntatically
        pass
```

Digital transmitter code

```
#! /usr/bin/env python
# The previous line allows this program to be run from the
# linux command line, letting the shell know to run this script
# in the python enviroment
# The command would be:  ./DBPSKTx.py


# This program transmits digital data using Differencial Binary Phase
#    Shift Keying (DBPSK)


# Import nessary objects from other modules
from gnuradio import gr, gru, modulation_utils
from gnuradio import blks2, usrp
from gnuradio import eng_notation

# Needed for struct
import struct
# Needed for System error
import sys

# These are global variables
# Making these parameters global variables is bad programming pratice.
#    These should be set using command line options, for sake of
#    simplicity they have been hard coded. See GSDR examples
#    to see how to use the options parcer.
center_frequency = 2400000000
interpolation_rate = 64
Tx_amplifier = 32768
daughterboard_port = (1,0)
daughterboard_gain = 0
max_number_packets = 100000
# Specify the size of the packets to be sent, in bytes
packet_size = 1000
repeat_packets = True
# Raised Cosine rolloff factor
roll_off = 0.35
# Gray code modulator
gray_code = True
# Create modulator arguments
modulator_args = {'samples_per_symbol': 2,
                  'excess_bw': roll_off,
```

Doi: 10.1016/B978-0-12-374715-0.00026-5

```
                    'verbose': False,
                    'gray_code': True,
                    'log': False}


######## Radio Class ########
# Here, we inherit the parent class, 'gr.top_block', for our class
#    'digital_radio'
class digital_radio(gr.top_block):

    # Initialization function, sets up the radio
    def __init__(self):
        # Initilizes parent class
        gr.top_block.__init__(self)

        # Set up blocks for the radio

        # Create Transmitter
        # Random data packets will be created in the main function and
        #   inserted into this block to be modulated and transmitted

        # Get a list of the possible modulations
        mods = modulation_utils.type_1_mods()
        # Create a dbpsk modulator object
        modulator = mods['dbpsk']

        print getattr (modulator, "__name__")

        self.packet_modulator = \
        blks2.mod_pkts(modulator(**modulator_args),
                       access_code=None,
                       msgq_limit=4,
                       pad_for_usrp=True,
                       use_whitener_offset=False)


        # Create amplifier
        self.amp = gr.multiply_const_cc(Tx_amplifier)

        # Setup USRP, this function will add self.usrp and
        # self.daughterboard
        #   to this object
        self.setup_USRP()

        # Connect the blocks in a flow graph
        self.connect(self.packet_modulator, self.amp, self.usrp)

    # Sets up USRP
    def setup_USRP(self):
        # Create USRP Block
        self.usrp = usrp.sink_c();

        # Read the DAC rate from the USRP
        dac_rate = self.usrp.dac_rate();
```

```python
        self.usrp.set_interp_rate(interpolation_rate)

        # Set up chosen daughterboard
        self.usrp.set_mux(usrp.determine_tx_mux_value(self.usrp,
                                          daughterboard_port))
        self.daughterboard = usrp.selected_subdev(self.usrp,
                                          daughterboard_port)
        # Set gain of daughterboard to max
        self.daughterboard_gain = self.daughterboard.gain_range()[2]
        self.daughterboard.set_gain(self.daughterboard_gain)

        # Set center frequency of daughterboard
        result = self.usrp.tune(self.daughterboard.which(),
                            self.daughterboard,
                            center_frequency)
        # If the center frequency could not be set, print an error
        if result == False:
            print "Failed to set daughterboard frequency"

        # Turn on Transmitter
        self.daughterboard.set_enable(True)

        # Dispay different aspects of the duaghterboard
        print self.daughterboard.side_and_name()
        print self.daughterboard_gain
        print self.daughterboard.gain_range()

    # Calls the function on the packet modulator to send packets
    def send_pkt(self, payload='', eof=False):
        return self.packet_modulator.send_pkt(payload, eof)


####### Main Function #######
# This function is called when we run this program from the
# command line
def main():

    # Create an instance of the radio we created, the __init__ function
    #   is called automatically. Here parameters can be passed to the
    #   class, but none are used for this simple example.
    radio = digital_radio()

    # Start the flow graph
    radio.start()

    # After the radio is started, code adapting the radio can be added
    #   here.  In this case, we create random data to be transmitted.
    while repeat_packets:
        # Reset packet counter
        pkt_num = 0
        # Craete and send packets
        while pkt_num < max_number_packets:
```

```
                    # Create data in the size of a packet
                    data = (packet_size - 2) * chr(pkt_num & 0xff)
                    # Put together data with a packet header
                    #### TO DO ##### explain this line more
                    payload = struct.pack('!H', pkt_num & 0xffff) + data
                    # Send packet to radio to be transmitted
                    radio.send_pkt(payload)
                    # Increment packet counter
                    pkt_num += 1
                    # Let the user know the radio is running by printing
                    #   something to the command line.  An error is used
                    #   so it appears immediately, instead of waiting
                    #   and written in a large block.
                    sys.stderr.write('.')

        # GSDR will exit if all the blocks report that they have
        #   emptied their queues.  This command tells the source block
        #   it is done transmitting.
        radio.send_pkt(eof=True)

        # Wait for radio to finish, if set to repeat, this radio
        #   will never finish.
        radio.wait()

####### Start of program #######
# This checks to see if this module is being run by itself, or
#   imported by another module
if __name__ == '__main__':
    # If it is being run by itself, try to run the main function
    try:
        main()
    # If there is a keyboard interrupt, exit mail function before
    # it finishes
    except KeyboardInterrupt:
        # 'pass' is a placeholder, used when nothing is to be done,
        #   but a command is needed syntatically.  In other programs
        #   may have to be sys.exit()
        pass
```

Digital receiver code

```python
#! /usr/bin/env python
# The previous line allows this program to be run from the linux
#    command line using: ./DBPSKTx.py

# This program transmits digital data using Differencial Binary
#    Phase Shift Keying (DBPSK)

# Import nessary objects from other modules
from gnuradio import gr, gru, modulation_utils
from gnuradio import blks2, usrp
from gnuradio import eng_notation
from gnuradio.wxgui import slider
from gnuradio.wxgui import stdgui2, fftsink2, form

import wx

import pdb

# These are global variables
# Making these parameters global variables is bad programming
#    pratice.  These should be set using command line options, for
#    sake of simplicity they have been hard coded. See GSDR examples
#    to see how to use the options parcer.
center_frequency = 2400000000
decimation_rate = 128
daughterboard_port = (1,0)
daughterboard_gain = 0
max_number_packets = 10000
# Specify the size of the packets to be sent, in bytes
packet_size = 1000
repeat_packets = True
# Raised Cosine rolloff factor
roll_off = 0.35
# Gray code modulator
gray_code = True
# Create modulator arguments
demodulator_args = {'log': False, 'samples_per_symbol': 5,
            'omega_relative_limit': 0.0050000000000000001,
            'mu': 0.5, 'verbose': False}
#
alpha = 0.001
# Something threshold, in dB
```

```
_____threshold = 30

####### Radio Class #######
# Here, we inherit the parent class, 'gr.top_block', for our class
#    'digital_radio'
class digital_radio(gr.hier_block2):
    # Create class variables to keep track of the number of packets
    #    recevied and the number correct.
    packets_received = 0
    packets_correct = 0

    # Initialization function, sets up the radio
    def __init__(self):
        # Initilizes parent class
        #gr.top_block.__init__(self)
        gr.hier_block2.__init__(self,
                    "digital_radio",
                    gr.io_signature(0, 0, 0), # Input signature
                    gr.io_signature(0, 0, 0)) # Output signature


        # Set up blocks for the radio

        # Set up USRP receiver, this function will add self.usrp
        #   and self.daughterboard to this object
        self.setup_USRP()

        # Design filter to get actual channel we want
        sw_decim = 1
        samples_per_symbol = 2
        chan_coeffs = gr.firdes.low_pass (1.0,  # gain
            sw_decim * samples_per_symbol, # sampling rate
            1.0,                     # midpoint of trans. band
            0.5,                     # width of trans. band
            gr.firdes.WIN_HANN)      # filter type

        # Receive channel filter, reduces out of channel noise and
        # complex in and out, float taps
        self.channel_filter = gr.fft_filter_ccc(sw_decim,
                                            chan_coeffs)


        # Get a list of the possible modulations
        demods = modulation_utils.type_1_demods()
        # Create a dbpsk modulator object
        demodulator = demods['dbpsk']

        print getattr (demodulator, "__name__")

        # blks2.mod_pkts(self._modulator_class(**mod_kwargs),
        self.packet_demodulator = \
        blks2.demod_pkts(demodulator(**demodulator_args),        .
                            access_code = None,
```

```
                        callback = self.receive_pkt_callback,
                        threshold = -1)


        # Connect the blocks in a flow graph
        self.connect(self.usrp, self.channel_filter,
                    self.packet_demodulator)
        #self.connect(self.usrp, self.packet_demodulator)



    # Set up USRP
    def setup_USRP(self):
        # Create USRP Block
        self.usrp = usrp.source_c();

        # Read the DAC rate from the USRP
        adc_rate = self.usrp.adc_rate();

        self.usrp.set_decim_rate(decimation_rate)

        # Calcualte the sample rate from the USRP by diviting the
        #   ADC rate by the current decimation_rate
        self.USRP_rate = adc_rate / decimation_rate
        print self.USRP_rate

        # Set up chosen daughterboard
        self.usrp.set_mux(usrp.determine_rx_mux_value(self.usrp,
                                        daughterboard_port))
        self.daughterboard = usrp.selected_subdev(self.usrp,
                                        daughterboard_port)

        g = self.daughterboard.gain_range()
        print "Rx Gain Range: "
        print g

        # Set gain of daughterboard to max
        self.daughterboard_gain = \
        self.daughterboard.gain_range()[1]
        self.daughterboard.set_gain(self.daughterboard_gain)

        # Set center frequency of daughterboard
        result = self.usrp.tune(self.daughterboard._which,
                        self.daughterboard,
                        center_frequency)
        # If the center frequency could not be set, print an error
        if result == False:
            print "Failed to set daughterboard frequency"


    # This has to be called after the FFT block is created,
    #   but the USRP rate has to be created before the block
    #   is instanciated.
```

```python
def add_fft(self, fft_block):
    self.FFT_usrp = fft_block

    # Add FFT to output of USRP
    self.connect(self.usrp, self.FFT_usrp)


# Create a function to be called by the packet receiver.
#   A handle for this function will be passed to the packet
#   receiver block and when a packet is received, its payload
#   will be sent here.
def receive_pkt_callback(ok, payload):
    # Unpack the packet corresponding to the way to transmitter
    #   packed the data
    (pktno,) = struct.unpack('!H', payload[0:2])
    packets_received += 1
    # The packet demodulator performs a CRC check on the
    #   packets, if the check fails the first argument of the
    #   callback function returns false (in this case, the
    #   variable 'ok'), the second variable is the payload.
    if ok:
        # If the CRC check is ok, increment the correct counter
        packets_correct += 1

    # Display information about the transmission
    print "ok = %5s   pktno = %4d   n_rcvd = %4d   n_right = %4d"\
    % (ok, pktno, n_rcvd, n_right)

####### Display class #######
# This is a seperate class creating the GUI for the program.  It is
#   not nessary to put the GUI compoents into a seperate class but
#   was done here for clarity.  To seperate the radio from the GUI,
#   the GUI and or radio can be programed in functions called by
#   the '__init__' function.
class receiver_display(stdgui2.std_top_block):

    # Initialization function to set up the GUI, these arguemnts
    #   will be explained as they are used, excpect for argv which
    #   is never used.
    def __init__(self, frame, panel, vbox, argv):
        stdgui2.std_top_block.__init__(self, frame, panel, vbox,
        argv)

        # Create frame and panel to put GUI elements onto
        self.frame = frame
        self.panel = panel

        # Create an instance of the radio module.
        self.radio = digital_radio()

        # Create FFT (Fast Fourier Transform) block with display,
        #   this block will have to be connected in our radio.
        #   Since the USRP sample rate is not known yet,this will
```

```
#    have to be set after the rate is calcualted by the
#    radio.

# This block is to display a FFT of the output of the usrp
self.FFT_usrp = fftsink2.fft_sink_c(self.panel,
                   title = "Signal from USRP",
                   fft_size = 512,
                   average = False,
                   sample_rate = self.radio.USRP_rate,
                   ref_scale = 32768.0,
                   ref_level = 50,
                   y_divs = 12)

self.radio.add_fft(self.FFT_usrp)

# Add this display to the vertical resize box (vbox)
#    The vbox adds new elements along the vertical axis, the
#    vbox object was added by the __init__ function of the
#    parent class.  The line that is called is'vbox =
#    wx.BoxSizer(wx.VERTICAL)' to create the vbox object.
vbox.Add(self.FFT_usrp.win,4, wx.EXPAND)

# GSDR provides easy ways to connect action listeners (or
#    callback functions) to elements of the GUI.  They have
#    done this with their form framework.  The form module,
#    in gr-wxgui, allows the setup of these GUI elements in
#    one command.  Here we create an edit box and slider to
#    adjust the center frequency.

# Inside of Init functions, there are limits on when, in
#    '__init__' functions,  references to self can be made.
#    We will make two references to the form to add controls
#    to.
self.controls = controls = form.form()

# Create a box for the frequency controls, we will add
#    parts horizontally
freq_box = wx.BoxSizer(wx.HORIZONTAL)


# Provides a callback function for the frequency text box,
#    simply gets the frequency from the key_values and calls
#    the 'set_frequency' function.  This function is defined
#    inside of the '__init__' function and must be defined
#    before it is assigned to the text box GUI object.
def update_frequency_slider(key_values):
    # Unpacks the frequency value from the 'key_values' and
    #    calls the set_frequency function.
    return self.set_frequency(key_values['freq'])

# This is the special text box provided by the 'form'
#    module for inputing floating point numbers.  Since the
#    callback function needs the value the box is set to, we
```

```
#    need to use the callback function provided by the form
#    framework.  The callback function needs to take in an
#    argument that are the key values for the GUI object,
#    the function can then use those values.
controls['freq'] = form.float_field(
    parent = self.panel, sizer = freq_box,
    label = "Frequency", weight = 1,
    callback = \
    controls.check_input_and_call(update_frequency_slider))

# Create a space between the input box and the slider.
freq_box.Add((5,0), 0)

# Create a slider control for the frequency, the forward
#    slash concatinates two lines that would not be combined
#    for other reasons.  Since a flaoting point value is
#    expected, the key_values unpacking is unnessary, as the
#    frequency value will be pass to the callback value by
#    default.  Here the range should be from whatever
#    daughterboard we are using, here we assume FM radio
#    bands.  The range is three values, minimum, maximum, and
#    step size.
controls['freq_slider'] = \
    form.quantized_slider_field(parent = self.panel,
    sizer = freq_box, weight = 3,
    range = (87.9e6, 108.1e6, 0.1e6),
    callback = self.set_frequency)

# Add the frequency controls to the main sizing box.
vbox.Add(freq_box, 0, wx.EXPAND)

# GSDR treats this class as the top level flow graph.  This
#    means we have to connect our radio flow graph to this
#    flow graph, but since the radio flow graph is the
#    entire radio, it will be the only block in the graph.
self.connect(self.radio)

#
# Start the flow graph
#pdb.set_trace()
#radio.start()

# Since python is running the GUI, we do not want to put a
#    wait command here ('radio.wait()'), waiting here would
#    prevent the GUI from exiting its init function and the
#    GUI would not start. The program instead runs
#    'MainLoop()' and waits for action listers attached to
#    elements in the GUI.
```

```python
# Create a callback function for the controls to use.  The self
#   here will refer to _____.
def set_frequency(self, target_frequency):
    """
    Set the center frequency we're interested in.

    @param target_freq: frequency in Hz
    @rypte: bool

    Tuning is a two step process.  First we ask the front end
    to tune as close to the desired frequency as it can.  Then
    we use the result of that operation and our
    target_frequency to determine the value for the digital
    down converter.
    """
    r = usrp.tune(self.radio.usrp, 0, self.radio.daughterboard,
                                            target_frequency)

    if r:
        self.freq = target_frequency
        # update displayed value
        self.controls['freq'].set_value(target_frequency)
        # update displayed value
        self.controls['freq_slider'].set_value(target_frequency)
        #self.update_status_bar()
        #self._set_status_msg("OK", 0)
        return True

    #self._set_status_msg("Failed", 0)
    return False



###### Start of program ######
# This checks to see if this module is being run by itself, or
#   imported by another module
if __name__ == '__main__':
    # If it is being run by itself, try to run the main function
    # Create a new instance of the program
    program = stdgui2.stdapp (receiver_display, "DBPSK Receiver",
                                                nstatus=1)
    # Call the main loop of the program, this function is provided
    #   by the class the 'receiver_display' call inherits from,
    #   stdgui2, this class inherits from 'wx.App' where a default
    #   value for MainLoop can be found.  This default function can
    #   be overwritten to add custom functionallity.
    program.MainLoop ()
```

Adaptive transmitter code

```
#! /usr/bin/env python
# The previous line allows this program to be run from the linux
# command line using: ./DBPSKTx.py

# This program transmits digital data using Differencial Binary Phase
#    Shift Keying (DBPSK)

# Import nessary objects from other modules
from gnuradio import gr, gru, modulation_utils
from gnuradio import blks2, usrp
from gnuradio import eng_notation

# Needed for struct
import struct

# Needed for the timers
import time
import threading

# Needed for System error
import sys

# These are global variables
# Making these parameters global variables is bad programming pratice.
#    These should be set using command line options, for sake of
#    simplicity
#    they have been hard coded. See GSDR examples to see how to use the
#    options parcer.
center_frequency = 2400000000
interpolation_rate = 64
decimation_rate = 32
Tx_amplifier = 32768
daughterboard_port = (1,0)
daughterboard_gain = 0
max_number_packets = 100000
# Specify the size of the packets to be sent, in bytes
packet_size = 1000
repeat_packets = True
# Raised Cosine roll off factor
roll_off = 0.35
# Gray code modulator
gray_code = True
```

```
# Create modulator arguments
modulator_args = {'samples_per_symbol': 2,
                  'excess_bw': roll_off,
                  'verbose': False,
                  'gray_code': True,
                  'log': False}
tx_time = 9.5
noise_time = 0.5
wait_factor = 3

###### Transmitter Class ######
# Here, we inherit the parent class, 'gr.top_block', for our class
#    'digital_radio'
class tranmitter(gr.top_block):

    # Initialization function, sets up the radio
    def __init__(self):
        # Initilizes parent class
        gr.top_block.__init__(self)

        # Setup blocks for the radio

        # Create Transmitter
        # Random data packets will be created in the main function and
        #    inserted into this block to be modulated and transmitted

        # Get a list of the possible modulations
        mods = modulation_utils.type_1_mods()
        # Create a dbpsk modulator object
        modulator = mods['dbpsk']

        print getattr (modulator, "__name__")

        self.packet_modulator = \
        blks2.mod_pkts(modulator(**modulator_args),
                       access_code=None,
                       msgq_limit=4,
                       pad_for_usrp=True,
                       use_whitener_offset=False)

        # Create amplifier
        self.amp = gr.multiply_const_cc(Tx_amplifier)

        # Setup USRP, this function will add self.usrp and
        # self.daughterboard
        #    to this object
        self.setup_USRP()

        # Connect the blocks in a flow graph
        self.connect(self.packet_modulator, self.amp, self.usrp)

    # Set up USRP
    def setup_USRP(self):
```

```python
# Create USRP Block
self.usrp = usrp.sink_c();

# Read the DAC rate from the USRP
dac_rate = self.usrp.dac_rate();

self.usrp.set_interp_rate(interpolation_rate)

# Setup chosen daughterboard
self.usrp.set_mux(usrp.determine_tx_mux_value(self.usrp,
                                daughterboard_port))
self.daughterboard = usrp.selected_subdev(self.usrp,
                                daughterboard_port)

# Set gain of daughterboard to max
self.daughterboard_gain = \
                        self.daughterboard.gain_range()[2]
self.daughterboard.set_gain(self.daughterboard_gain)

# Set center frequency of daughterboard
result = self.usrp.tune(self.daughterboard._which,
                self.daughterboard,
                center_frequency)
# If the center frequency could not be set, print an error
if result == False:
    print "Failed to set daughterboard frequency"

# Turn on Transmitter
self.daughterboard.set_enable(True)

print "Using TX d'board %s"    % (self.daughterboard.
                                    side_and_name(),)
print "Tx amplitude      %s"   % (self.daughterboard_gain)

print self.daughterboard.gain_range()
#print "modulation:       %s"  % (self.usrp._modulator_
                                    class.__name__)
print "Tx Frequency:    %s"    % (eng_notation.num_to_str
                                    (center_frequency))

    def send_pkt(self, payload='', eof=False):
        """
        Calls the transmitter method to send a packet
        """
        return self.packet_modulator.send_pkt(payload, eof)

####### Receiver Class #######
#
class noise_calculator(gr.top_block):
# Create class variables to keep track of the number of packets
#    recevied and the number correct.
    packets_received = 0
    packets_correct = 0
```

```
# Initialization function, sets up the radio
def __init__(self):
    # Initilizes parent class
    gr.top_block.__init__(self)

    # Set up blocks for the radio

    # Set up USRP receiver, this function will add self.usrp
    #   and self.daughterboard to this object
    self.setup_USRP()

    # Design filter to get actual channel we want
    sw_decim = 1
    samples_per_symbol = 2
    chan_coeffs = gr.firdes.low_pass (1.0,  # gain
        sw_decim * samples_per_symbol, # sampling rate
        1.0,                      # midpoint of trans. band
        0.5,                      # width of trans. band
        gr.firdes.WIN_HANN)   # filter type

    # Receive channel filter, reduces out of channel noise and
    # complex in and out, float taps
    self.channel_filter = gr.fft_filter_ccc(sw_decim,
                                          chan_coeffs)


    # Calculate the RMS value of this channel
    self.power_calculator = gr.probe_avg_mag_sqrd_c(-60)



    # Connect the blocks in a flow graph
    self.connect(self.usrp, self.channel_filter,
              self.power_calculator)

# Set up USRP
def setup_USRP(self):
    # Create USRP Block
    self.usrp = usrp.source_c();

    # Read the DAC rate from the USRP
    adc_rate = self.usrp.adc_rate();

    self.usrp.set_decim_rate(decimation_rate)

    # Calcualte the sample rate from the USRP by dividing the
    #   ADC rate by the current decimation_rate
    self.USRP_rate = adc_rate / decimation_rate
    print self.USRP_rate
```

```
                # Set up chosen daughterboard
                self.usrp.set_mux(usrp.determine_rx_mux_value(self.usrp,
                                       daughterboard_port))
                self.daughterboard = usrp.selected_subdev(self.usrp,
                                       daughterboard_port)

                g = self.daughterboard.gain_range()
                print "Rx Gain Range: "
                print g

                # Set gain of daughterboard to max
                self.daughterboard_gain = \
                self.daughterboard.gain_range()[1]
                self.daughterboard.set_gain(self.daughterboard_gain)

                # Set center frequency of daughterboard
                result = self.usrp.tune(self.daughterboard._which,
                              self.daughterboard,
                              center_frequency)
                # If the center frequency could not be set, print an error
                if result == False:
                    print "Failed to set daughterboard frequency"


        # This has to be called after the FFT block is created,
        #   but the USRP rate has to be created before the block
        #   is instanciated.
        def add_fft(self, fft_block):
            self.FFT_usrp = fft_block

            # Add FFT to output of USRP
            self.connect(self.usrp, self.FFT_usrp)

####### Main Transmitter Class #######
#
class adaptive_transmitter():
    def __init__(self):
        # Create the Transmitter
        self.tx_radio = tranmitter()
        # Create the Noise Calculator
        self.noise_radio = noise_calculator()

        # Start by calculating channel noise by calling
        #   the function for the Tx radio finishing.
        self.Tx_Callback()

    # Returns true when the radio is currently transmitting.
    def transmitter_on(self):
        return self.Transmitting

    # This function is called when the transmit timer finishes
    def Tx_Callback(self):
        # Set transmitting flag for transmit polling
```

```python
        self.Transmitting = False
        # Stop the tx radio
        self.tx_radio.stop()
        # Tell the radio to wait for data to finish propagating through
        self.tx_radio.wait()
        print "\n Stopped Tx Radio, starting noise calculation"
        # Create a new noise calculation timer
        self.noise_timer = threading.Timer(noise_time, self.Noise_
                                                         Callback)
        # Start the noise calculation radio
        self.noise_radio.start()
        # Start the noise calculation timer
        self.noise_timer.start()

    # This function is called when the revicer timer finishes
    def Noise_Callback(self):
        print self.noise_radio.power_calculator.level()
        self.noise_radio.stop()
        self.noise_radio.wait()
        print "Stopped noise calcualtion, starting transmission"
        self.tx_timer = threading.Timer(tx_time, self.Tx_Callback)
        self.tx_radio.start()
        self.Transmitting = True
        self.tx_timer.start()

    # Sends packet to the correct funtion in the trasmit radio
    def send_pkt(self, payload='', eof=False):
            return self.tx_radio.send_pkt(payload, eof)

####### Main Function #######
# This function is called when we run this program from the command line
def main():
    # Create an instance of the radio we created, the __init__ function
    #   is called automatically. Here parameters can be passed to the
    #   class, but none are used for this simple example.
    radio = adaptive_transmitter()

    # Create random data to be sent to the adaptive radio.
    while repeat_packets:
        # Reset packet counter
        pkt_num = 0
        # Craete and send packets
        while pkt_num < max_number_packets:
            if radio.transmitter_on():
                # Create data in the size of a packet, leaving room
                #   for the header.
                data = (packet_size - 2) * chr(pkt_num & 0xff)
                # Put together data with a packet header
                payload = struct.pack('!H', pkt_num & 0xffff) + data
                # Send packet to radio to be transmitted
                radio.send_pkt(payload)
                # Increment packet counter
                pkt_num += 1
```

```
                    # Let the user know the radio is running by printing
                    #   something to the command line
                    sys.stderr.write('.')
            else:
                    time.sleep(noise_time / wait_factor)

        ##### TO DO ##### explain this line more
        radio.send_pkt(eof=True)

####### Start of program #######
# This checks to see if this module is being run by itself or
# imported by
#   another module
if __name__ == '__main__':
    # If it is being run by itself, try to run the main function
    try:
        main()
    # If there is a keyboard interrupt, exit main function before
    # it finishes
    except KeyboardInterrupt:
        # Exit the program on the interrupt.
        sys.exit()
```

References

[1] M. McHenry, D. McCloskey, and G. Lane-Roberts, "Spectrum occupancy measurements, location 4 of 6: Republican national convention, New York City, NY, Aug. 30, 2004–Sept. 3, 2004, revision 2," tech. report, Shared Spectrum Company, Aug. 2005.

[2] J. Mitola III, "Software radios: Survey, critical evaluation and future directions," *IEEE Aerospace and Electronic Systems Magazine*, vol. 8, pp. 25–36, Apr. 1993.

[3] R. I. Lackey and D. W. Upmal, "SPEAKeasy: The military software radio," *IEEE Communications Magazine*, vol. 33, pp. 51–61, May 1995.

[4] V. Bose, "A software driven approach to SDR design," *COTS Journal*, Jan. 2004.

[5] J. Chapin and V. Bose, "The Vanu software radio system," Software Defined Radio Technical Conference, San Diego, CA, USA, 2002.

[6] J. Mitola III, "Cognitive radio for flexible mobile multimedia communications," in *Proceedings of the IEEE International Workshop on Mobile Multimedia Communications*, San Diego, CA, USA, vol. 1, pp. 3–10, Nov. 1999.

[7] J. Mitola III, "Cognitive radio: An integrated agent architecture for software defined radio." PhD thesis, Royal Institute of Technology (KTH), Stockholm, Sweden, May 2000.

[8] M. Nekovee, "Dynamic spectrum access—concepts and future architectures," *BT Technology Journal*, vol. 24, pp. 111–116, May 2006.

[9] B. Fette, *Cognitive Radio Technology*. Boston, MA, USA: Elsevier, 2006.

[10] QinetiQ LTD, "An evaluation of software defined radio." [Online]: www.ofcom.org.uk/research/technology/research/emer_tech/sdr.

[11] S. Haykin, "Cognitive radio: Brain-empowered wireless communications," *IEEE Journal on Selected Areas in Communications*, vol. 23, pp. 201–220, Feb. 2005.

[12] Federal Communications Commission, "First report and order and further notice of proposed rulemaking in the matter of unlicesed operation in TV broadcast bands." ET Docket No. 04-186, Oct. 2006.

[13] Federal Communications Commission, "The FCC's Office of Engineering and Technology release report on tests of prototype white space devices." ET Docket No. 04-186, Oct. 2008.

[14] Federal Communications Commission, "FCC adopts rules for unlicenesd use of television white space." Official FCC announcement, Nov. 2009.

[15] Office of Communications, "Digital dividend review, a statement on our approach to awarding the digital dividend." [Online]: www.ofcom.org.uk/condocs/ddr/statement, Dec. 2007.

[16] Office of Communications, "Digital dividend: Cognitive access." [Online]: www.ofcom.org.uk/consult/condocs/cognitive, Feb. 2009.

[17] IEEE 802.22 Working Group on Wireless Regional Area Networks. [Online]: www.iee802.org/22, Feb. 2009.

[18] C. Cordeiro, K. Challapali, D. Birru, and S. Shankar, "IEEE 802.22: An introduction to the first wireless standard based on cognitive radios," *Journal of Communications*, vol. 1, no. 1, pp. 38–47, 2006.

[19] J. Mitola III, "The software radio architecture," *IEEE Communications Magazine*, 1995.

[20] K. E. Nolan, P. D. Sutton, L. E. Doyle, T. W. Rondeau, B. Le, and C. W. Bostian, "Demonstration and analyses of collaboration, coexistence, and interoperability of cognitive radio platforms," in *IEEE Consumer Communications and Networking Conference*, Las Vegas, NV, USA, Jan. 2007.

[21] D. Scaperoth, B. Le, T. Rondeau, D. Maldonado, C. W. Bostian, and S. Harrison, "Cognitive radio platform development for interoperability," in *Military Communications Conference*, Oct. 2006.

[22] I. F. Akyildiz, W. Y. Lee, M. C. Vuran, and S. Mohanty, "NeXt generation/dynamic spectrum access/cognitive radio wireless networks: A survey," *Elsevier Computer Networks Journal*, vol. 50, pp. 2127–2159, Sept. 2006.

[23] Y. Yuan, P. Bahl, R. Chandra, P. A. Chou, J. I. Ferrell, T. Moscibroda, S. Narlanka, and Y. Wu, "KNOWS: Kognitiv networking over white spaces," in *IEEE Symposium on New Frontiers in Dynamic Spectrum Access Networks*, 2007.

[24] T. A. Weiss and F. K. Jondral, "Spectrum pooling: An innovative strategy for the enhancement of spectrum efficiency," *IEEE Communications Magazine*, 2004.

[25] D. Maldonado, B. Le, A. Hugine, T. W. Rondeau, and C. W. Bostian, "Cognitive radio applications to dynamic spectrum allocation: A discussion and an illustrative example," in *IEEE International Symposium on New Frontiers in Dynamic Spectrum Access Networks*, Baltimore, MD, USA, Nov. 2005.

[26] I. K. Ibrahim and D. Taniar, *Mobile Multimedia: Communication Engineering Perspective*, ch. 5. Commack, NY, USA: Nova Publishers, 2006.

[27] Federal Communications Commission, "Spectrum policy task force report." ET Docket No. 02-135, 2002.

[28] A. Petrin, "Maximizing the utility of radio spectrum: Broadband spectrum measurements and occupancy model for use by cognitive radio." PhD thesis, Georgia Institute of Technology, Atlanta, GA, USA, July 2005.

[29] D. A. Roberson, C. S. Hood, J. L. LoCicero, and J. T. MacDonald, "Spectral occupancy and interference studies in support of cognitive radio technology deployment," in *Proceedings of the First IEEE Workshop on Network Technologies for Software-Defined Radio Networks*, Sept. 2006.

[30] R. Chiang, G. Rowe, and K. Sowerby, "A quantitative analysis of spectral occupancy measurements for cognitive radio," in *Proceedings of the IEEE Vehicular Technology Conference*, Apr. 2007.

[31] M. Wellens, A. de Baynast, and P. Mahonen, "Exploiting historical spectrum occupancy information for adaptive spectrum sensing," in *Proceedings of the IEEE Wireless Communications and Network Conference*, Las Vegas, NV, USA, pp. 717–722, Mar. 2008.

[32] M. Nekovee, "Quantifying the availability of TV white spaces for cognitive radio operation in the UK," in *Proceedings of the IEEE International Conference on Communications*, Dresden, Germany, June 2009.

[33] M. J. Marcus, "Unlicensed cognitive sharing of TV spectrum: The controversy at the federal communications commission," *IEEE Communications Magazine*, vol. 43, pp. 24–25, May 2005.

[34] Federal Communications Commission, "Unlicensed operation in the TV broadcast bands." ET Docket No. 04-186, 2004.

[35] L. B. Le and E. Hossain, "Resource allocation for spectrum underlay in cognitive radio networks," *IEEE Transactions on Wireless Communications*, vol. 7, pp. 5306–5315, Dec. 2008.

[36] Q. Qu, L. B. Milstein, and D. R. Vaman, "Cognitive radio based multi-user resource allocation in mobile ad hoc networks using multi-carrier CDMA modulation," *IEEE Journal on Selected Areas in Communications*, vol. 26, pp. 70–82, Jan. 2008.

[37] R. Zhang and Y. Liang, "Exploiting multi-antennas for opportunistic spectrum sharing in cognitive radio networks," *IEEE Journal of Selected Topics in Signal Processing*, vol. 2, pp. 88–102, Feb. 2008.

[38] U. S. Jha and R. Prasad, *OFDM towards Fixed and Mobile Broadband Wireless Access*. London, UK: Artech House, 2007.

[39] A. Attar, O. Holland, M. R. Nakhai, and A. H. Aghvami, "Interference-limited resource allocation for cognitive radio in orthogonal frequency-division multiplexing networks," *IET Communications*, vol. 2, pp. 806–814, July 2008.

[40] "700 MHz wireless spectrum auction." [Online]: http://en.wikipedia.org/wiki/700_MHz_wireless_spectrum_auction.

[41] International Telecommunications Union, "History of the International Telecommunications Union." [Online]: www.itu.int/net/about/history.aspx.

[42] International Telecommunications Union, "Mission statement of the International Telecommunications Union." [Online]: www.itu.int/net/about/mission.aspx.

[43] U.S. Department of Commerce, National Telecommunications and Information Administration, Office of Spectrum Management, "United States frequency allocations—The radio spectrum," Oct. 2003.

[44] Parliament of the United Kingdom. The Communications Act of 2003, 2003 Chapter 21, Part 1, Section 3, July 2003.

[45] S. Woolley, "Dead air," *Forbes Magazine*, Nov. 2002.

[46] F. H. Sanders, "Broadband spectrum surveys in Denver, CO, San Diego, CA, and Los Angeles, CA: Methodology, analysis, and comparative results," in *Proceedings of the IEEE Symposium of Electromagnetic Compatibility*, 1998.

[47] M. McHenry and K. Steadman, "Spectrum occupancy measurements, location 1 of 6: Riverbend Park, Great Falls, VA," tech. rep., Shared Spectrum Company, Aug. 2005.

[48] M. McHenry and K. Steadman, "Spectrum occupancy measurements, location 2 of 6: Tyson's Square Center, Vienna, VA, April 9, 2004," tech. rep., Shared Spectrum Company, Aug. 2005.

[49] M. McHenry and S. Chunduri, "Spectrum occupancy measurements, location 3 of 6: National Science Foundation building roof, April 16, 2004, revision 2," tech. report, Shared Spectrum Company, Aug. 2005.

[50] M. McHenry and K. Steadman, "Spectrum occupancy measurements, location 5 of 6: National Radio Astronomy Observatory (NRAO), Green Bank, WV, October 10-11, 2004, revision 3," tech. rep., Shared Spectrum Company, Aug. 2005.

[51] M. McHenry, D. McCloskey, and J. Bates, "Spectrum occupancy measurements, location 6 of 6: Shared Spectrum building roof, Vienna, VA, December 15-16, 2004," tech. rep., Shared Spectrum Company, Aug. 2005.

[52] A. Petrin and P. Steffes, "Analysis and comparison of spectrum measurements performed in urban and rural areas to determine the total amount of spectrum usage," in *Proceedings of the International Symposium on Advanced Radio Technologies*, pp. 9-12, Mar. 2005.

[53] M. McHenry, D. McCloskey, D. Roberson, and J. MacDonald, "Spectrum occupancy measurements, Chicago, IL, November 16-18, 2005," tech. rep., Shared Spectrum Company, Dec. 2005.

[54] M. McHenry, P. Tenhula, D. McCloskey, D. Roberson, and C. Hood, "Chicago spectrum occupancy measurements and analysis and a long-term studies proposal," in *Proceedings of the First International Workshop on Technology and Policy for Accessing Spectrum*, Aug. 2006.

[55] M. Wellens, J. Wu, and P. Mahonen, "Evaluation of spectrum occupancy in indoor and outdoor scenario in the context of cognitive radio," in *Proceedings of the Second International Conference on Cognitive Radio Oriented Wireless Networks and Communications*, pp. 420-427, Aug. 2007.

[56] D. Roberson, "Structural support for cognitive radio system deployment," in *Proceedings of the Second International Conference on Cognitive Radio Oriented Wireless Networks and Communications*, Aug. 2007.

[57] R. Bacchus, A. Fertner, C. Hood, and D. Roberson, "Long-term, wide-band spectral monitoring in support of dynamic spectrum access networks at the IIT spectrum observatory," in *Proceedings of the Third IEEE Symposium on New Frontiers in Dynamic Spectrum Access Networks*, Oct. 2008.

[58] Federal Communications Commission, "Code of Federal Regulations, title 47 (telecommunication), volume 1 (FCC), part 15 (radio frequency devices), subpart 15, subpart c (intentional radiators), sec. 15.247 operation within the bands 902–928 MHz, 2400–2483.5 MHz, and 5725–5850 MHz." 47CFR15.247, Oct. 2001.

[59] Federal Communications Commission, "Revision of part 15 of the commissions rules regarding ultra-wideband transmissions systems." ET Docket No. 98-153, Feb. 2002.

[60] Federal Communications Commission, "Facilitating opportunities for flexible, efficient and reliable spectrum use employing cognitive radio technology." ET Docket No. 03-108, Mar. 2005.

[61] U.K. Office of Communications, "Spectrum framework review," June 2005.

[62] R. Galvin, J. Schlesinger, and J. Lewis, *Spectrum Management for the 21st Century—A Report on the CSIS Commission on Spectrum Management*. Washington, DC, USA: The CSIS Press, 2003.

[63] Federal Communications Commission, "Second report and order and memorandum and order FCC 08-260: In the matter of unlicensed operation in the TV broadcast bands—ET Docket No. 04-186 and additional spectrum for unlicensed devices below 900 MHz and in the 3 GHz band—ET Docket No. 02-380," Nov. 2008.

[64] U.K. Office of Communications, "Statement on licence-exempting cognitive devices using interleaved spectrum," July 2009.

[65] Institute for Telecommunication Sciences, "Radio spectrum measurement system (RSMS)." [Online]: www.its.bldrdoc.gov/programs/rsms/history/index.php.

[66] R. Thobaben, "Sendora project main page, European Union seventh framework programme." [Online]: www.sendora.eu.

[67] B. Mercier, V. Fodor, R. Thobaben, M. Skoglund, V. Koivunen, S. Lindfors, J. Ryynanen, E. G. Larsson, C. Petrioli, G. Bongiovanni, O. Grondalen, K. Kansanen, G. E. Oien, T. Ekman, A. M. Hayar, R. Knopp, and B. Beferull Lozano, "Sensor networks for cognitive radio: Theory and system design," in *ICT Mobile Summit*, June 2008.

[68] M. Cave, C. Doyle, and W. Webb, *Essentials of Modern Spectrum Management*. Cambridge, UK: Cambridge University Press, 2007.

[69] W. Stallings, *Wireless Communications and Networks*, 2nd ed. Upper Saddle River, NJ, USA: Pearson/Prentice-Hall, 2005.

[70] T. Standage, *The Victorian Internet, The Remarkable Story of the Telegraph and the Nineteenth Century's On-Line Pioneers*. New York, NY, USA: Walker Publishing Company, 1998.

[71] S. Haykin, *Communication System*. New York, NY, USA: John Wiley and Sons, 2000.

[72] A. Papoulis and U. Pillai, *Probability, Random Variables and Stochastic Proccesses*, 4th ed. New York, NY, USA: McGraw-Hill, 2002.

[73] R. Prasad, *Universal Wireless Personal Communications*. Artech House Mobile Communications, Norwood, MA, USA: Artech House, 1998.

[74] R. P. Ramachandran, "Bandwidth efficient filter banks for transmultiplexers." PhD thesis, McGill University, Montréal, QC, Canada, Sept. 1990.

[75] P. P. Vaidyanathan, *Multirate Systems and Filter Banks*. Prentice-Hall Signal Processing Series, Upper Saddle River, NJ, USA: Prentice-Hall, 1993.

[76] N. J. Fliege, *Multirate Digital Signal Processing: Multirate Systems, Filter Banks, Wavelets*. New York, NY, USA: John Wiley and Sons, 1994.

[77] S. D. Sandberg and M. A. Tzannes, "Overlapped discrete multitone modulation for high speed copper wire communications," *IEEE Journal on Selected Areas in Communications*, vol. 13, pp. 1571-1585, Dec. 1995.

[78] S. Colieri, M. Ergen, A. Puri, and A. Bahai, "A study of channel estimation in ofdm systems," in *Proceedings of the IEEE 56th Vehicular Technology Conference*, Sept. 2002.

[79] N. Benvenuto, S. Tomasin, and L. Tomba, "Equalization methods in OFDM and FMT systems for broadband wireless communications," *IEEE Transactions on Communications*, vol. 50, pp. 1413-1418, Sept. 2002.

[80] G. Cherubini, E. Eleftheriou, S. Olcer, and J. M. Cioffi, "Filter bank modulation techniques for very high-speed digital subscriber lines," *IEEE Communications Magazine*, vol. 38, pp. 98-104, May 2000.

[81] C. G, E. Eleftheriou, and S. Olcer, "Filtered multitone modulation for very high-speed digital subscriber lines," *IEEE Journal on Selected Areas in Communications*, vol. 20, pp. 1016-1028, June 2002.

[82] J. A. C. Bingham, "Multicarrier modulation for data transmission: An idea whose time has come," *IEEE Communications Magazine*, pp. 5-14, Apr. 1990.

[83] J. A. C. Bingham, *ADSL, VDSL, and Multicarrier Modulation*. New York, NY, USA: John Wiley and Sons, 2000.

[84] B. R. Saltzberg, "Comparison of single-carrier and multitone digital modulation for ADSL applications," *IEEE Communications Magazine*, pp. 114-121, Nov. 1998.

[85] S. B. Weinstein and P. Ebert, "Data transmission by frequency division multiplexing using the discrete fourier transform," *IEEE Transactions on Communications*, vol. 19, pp. 628-634, Oct. 1971.

[86] B. Hirosaki, "An analysis of automatic equalizers for orthogonally multiplexed QAM systems," *IEEE Transactions on Communications*, vol. 28, pp. 73-83, Jan. 1980.

[87] T. Pollet, M. Peeters, M. Moonen, and L. Vandendorpe, "Equalization for DMT-based broadband modems," *IEEE Communications Magazine*, vol. 38, pp. 106-113, May 2000.

[88] I. Kalet, "The multitone channel," *IEEE Transactions on Communications*, vol. 37, pp. 119-124, Feb. 1989.

[89] B. R. Saltzberg, "Performance of an efficient parallel data transmission system," *IEEE Transactions on Communications*, vol. 15, pp. 805-811, Dec. 1967.

[90] J. G. Proakis, *Digital Signal Processing: Principles, Algorithms, and Applications*, 3rd ed. Upper Saddle River, NJ, USA: Prentice-Hall, 1996.

[91] Z. Wang, X. Ma, and G. B. Giannakis, "OFDM or single-carrier block transmissions?" *IEEE Transactions on Communications*, vol. 52, pp. 380-394, Mar. 2004.

[92] J. Louveaux, "Filter bank based multicarrier modulation for xDSL transmission." PhD thesis, Université Catholique de Louvain, Louvain-la-Neuve, Belgium, May 2000.

[93] A. Viholainen, "Transmultiplexer design for VDSL modems," Master's thesis, Tampere University of Technology, Tampere, Finland, Aug. 1998.

[94] R. P. Ramachanrdan and P. Kabal, "Bandwidth efficient transmultiplexers, part 1: Synthesis," *IEEE Transactions on Signal Processing*, vol. 40, pp. 70-84, Jan. 1992.

[95] R. P. Ramachanrdan and P. Kabal, "Bandwidth efficient transmultiplexers, part 2: Subband complements and performance aspects," *IEEE Transactions on Signal Processing*, vol. 40, pp. 1108-1121, May 1992.

[96] G. W. Wornell, "Emerging applications of multirate signal processing and wavelets in digital communications," *Proceedings of the IEEE*, vol. 84, pp. 586-603, Apr. 1996.

[97] A. Scaglione, G. Giannakis, and S. Barbarossa, "Redundant filterbank precoders and equalizers, part 1: Unification and optimal design," *IEEE Transactions on Signal Processing*, vol. 47, pp. 1988-2006, July 1999.

[98] A. Scaglione, G. Giannakis, and S. Barbarossa, "Redundant filterbank precoders and equalizers, part 2: Blind channel estimation, synchronization and direct equalization," *IEEE Transactions on Signal Processing*, vol. 47, pp. 2007-2022, July 1999.

[99] A. N. Akansu, M. V. Tazebay, and R. A. Haddad, "A new look at digital orthogonal transmultiplexers for CDMA communications," *IEEE Transactions on Signal Processing*, vol. 45, pp. 263-267, Jan. 1997.

[100] A. N. Akansu, P. Duhamel, X. Lin, and M. de Courville, "Orthogonal transmultiplexers in communications: A review," *IEEE Transactions on Signal Processing*, vol. 46, pp. 979-995, Apr. 1998.

[101] T. Karp and N. J. Fliege, "Modified DFT filter banks with perfect reconstruction," *IEEE Transactions on Circuits and Systems—II*, vol. 46, pp. 1404-1414, Nov. 1999.

[102] M. Sablatash and J. Lodge, "Spectrally efficient multiplexing of O-QPSK or VSB signals using wavelet packet-based filter banks," *Digital Signal Processing*, vol. 13, pp. 58–92, Jan. 2003.

[103] X.-G. Xia, "New precoding for intersymbol interference cancellation using non-maximally decimated multirate filterbanks with ideal FIR equalizers," *IEEE Transactions on Signal Processing*, vol. 45, pp. 2431–2441, Oct. 1997.

[104] L. W. Couch II, *Digital and Analog Communication Systems*, 4th ed. New York, NY, USA: Macmillan Publishing Company, 1993.

[105] R. D. Gitlin, J. F. Hayes, and S. B. Weinstein, *Data Communications Principles*. New York, NY, USA: Plenum Press, 1992.

[106] E. A. Lee and D. G. Messerschmitt, *Digital Communications*, 2nd ed. Dordrecht, The Netherlands: Kluwer Academic Publishers, 1994.

[107] R. W. Lucky, J. Salz, and E. J. Weldon, Jr, *Principles of Data Communication*. New York, NY, USA: McGraw-Hill, 1968.

[108] J. G. Proakis, *Digital Communications*, 4th ed. New York, NY, USA: McGraw-Hill, 2001.

[109] Q. Zhao and B. M. Sadler, "A survey of dynamic spectrum access," *IEEE Signal Processing Magazine*, vol. 24, pp. 79–89, May 2007.

[110] Q. Zhao and A. Swami, "A decision-theoretic framework for opportunistic spectrum access," *IEEE Wireless Communications Magazine*, Special Issue on Cognitive Wireless Networks, vol. 14, pp. 14–20, Aug. 2007.

[111] Federal Communications Commission, "Revision of parts 2 and 15 of the commissions rules to permit unlicensed national information infrastructure (U-NII) devices in the 5 GHz band." FCC ET Docket-03-122, Nov. 2003.

[112] C. Cordeiro, K. Challapali, D. Birru, and N. Sai Shankar, "IEEE 802.22: The first worldwide wireless standard based on cognitive radios," in *Proceedings of IEEE Symposium on New Frontiers in Dynamic Spectrum Access Networks*, Baltimore, MD, USA, Nov. 2005.

[113] S. M. Kay, *Fundamentals of Statistical Signal Processing: Detection Theory*, vol. 2. Upper Saddle River, NJ, USA, Prentice-Hall, 1998.

[114] F. F. Digham, M.-S. Alouini, and M. Simon, "On the energy detection of unknown signals over fading channels," in *Proceedings of the IEEE International Conference on Communications*, May 2003.

[115] R. Tandra and A. Sahai, "SNR walls for signal detection," *IEEE Journal of Selected Topics in Signal Processing*, vol. 2, pp. 4–17, Feb. 2008.

[116] M. Ghogho, M. Cardenas-Juarez, A. Swami, and T. Witworth, "Locally optimum detection for spectrum sensing in cognitive radio," in *Proceedings of The 4th International Conference on Cognitive Radio Oriented Wireless Networks and Communications*, Hannover, Germany, June 2009.

[117] W. A. Gardner, "Exploitation of spectral redundancy in cyclostationary signals," *IEEE Signal Processing Magazine*, vol. 8, pp. 14–36, Apr. 1991.

[118] E. Serpedin, F. Panduru, I. Sari, and G. B. Giannakis, "Bibliography on cyclostationarity," *Signal Processing*, vol. 85, pp. 2233–2303, Dec. 2005.

[119] D. Cabric, S. M. Mishra, and R. W. Brodersen, "Implementation issues in spectrum sensing for cognitive radios," in *Proceedings of the Asilomar Conference on Circuits, Systems, and Computers*, vol. 1, pp. 772–776, Nov. 2004.

[120] K. Kim, I. Akbar, K. Bae, J.-S. Um, C. Spooner, and J. Reed, "Cyclostationary approaches to signal detection and classification in cognitive radio," in *Proceedings of IEEE Symposium on New Frontiers in Dynamic Spectrum Access Networks*, Dublin, Ireland, Apr. 2007.

[121] P. Sutton, K. Nolan, and L. Doyle, "Cyclostationary signatures in practical cognitive radio applications," *IEEE Journal on Selected Areas in Communications*, Jan. 2008.

[122] T. Li, W. Mow, V. Lau, M. Siu, R. Cheng, and R. Murch, "Robust joint interference detection and decoding for OFDM-based cognitive radio systems with unknown interference," *IEEE Journal on Selected Areas in Communications*, pp. 566–575, Apr. 2007.

[123] N. Kundargi and A. H. Tewfik, "Sequential pilot sensing of ATSC signals in IEEE 802.22 cognitive radio networks," in *Proceedings of IEEE International Conference on Acoustics, Speech, and Signal Processing*, Mar. 2008.

[124] H. Yu, Y. Sung, and Y. H. Lee, "On optimal operating characteristics of sensing and training for cognitive radios," in *Proceedings of IEEE International Conference on Acoustics, Speech, and Signal Processing*, pp. 2785-2788, Mar. 2008.

[125] S. Mishra, A. Sahai, and R. Brodersen, "Cooperative sensing among cognitive radios," in *Proceedings of the IEEE International Conference on Communications*, June 2006.

[126] A. Kattepur, A. Hoang, Y.-C. Liang, and M. Er, "Data and decision fusion for distributed spectrum sensing in cognitive radio networks," in *Proceedings of the International Conference on Information, Communications, and Signal Processing*, Dec. 2007.

[127] M. Gandetto and C. Regazzoni, "Spectrum sensing: A distributed approach for cognitive terminals," *IEEE Journal on Selected Areas in Communications*, vol. 25, no. 3, pp. 546-557, Apr. 2007.

[128] G. Ganesan and Y. Li, "Cooperative spectrum sensing in cognitive radio, part I: Two user networks; part II: Multiuser networks," *IEEE Transactions on Wireless Communications*, June 2007.

[129] G. Ganesan, Y. Li, B. Bing, and S. Li, "Spatiotemporal sensing in cognitive radio networks," *IEEE Journal on Selected Areas in Communications*, vol. 26, no. 1, pp. 5-12, Jan. 2008.

[130] Z. Quan, S. Cui, and A. Sayed, "Optimal linear cooperation for spectrum sensing in cognitive radio networks," *IEEE Journal of Selected Topics in Signal Processing*, vol. 2, no. 1, pp. 28-40, Feb. 2008.

[131] J. Unnikrishnan and V. Veeravalli, "Cooperative sensing for primary detection in cognitive radio," *IEEE Journal of Selected Topics in Signal Processing*, vol. 2, no. 1, pp. 18-27, Feb. 2008.

[132] A. Pandharipande and J. Linnartz, "Performance analysis of primary user detection in a multiple antenna cognitive radio," in *Proceedings of the IEEE International Conference on Communications*, pp. 6482-6486, June 2007.

[133] Z. Tian and G. B. Giannakis, "A wavelet approach to wideband spectrum sensing for cognitive radios," in *Proceedings of the International Conference on Cognitive Radio Oriented Wireless Networks and Communications*, June 2006.

[134] N. M. Neihart, S. Roy, and D. J. Allstot, "A parallel, multi-resolution sensing technique for multiple antenna cognitive radios," in *Proceedings of the IEEE Symposium on Circuits and Systems*, May 2007.

[135] Z. Tian and G. B. Giannakis, "Compressed sensing for wideband cognitive radios," in *Proceedings of IEEE International Conference on Acoustics, Speech, and Signal Processing*, pp. 1357-1360, Apr. 2007.

[136] Q. Zhao, "Spectrum opportunity and interference constraint in opportunistic spectrum access," in *Proceedings of IEEE International Conference on Acoustics, Speech, and Signal Processing*, Apr. 2007.

[137] Y. Chen, Q. Zhao, and A. Swami, "Joint design and separation principle for opportunistic spectrum access in the presence of sensing errors," *IEEE Transactions on Information Theory*, vol. 54, pp. 2053-2071, May 2008.

[138] J. Kingman, *Poisson Processes*. Oxford, UK: Clarendon Press, 1993.

[139] W. Ren, Q. Zhao, and A. Swami, "Power control in cognitive radio networks: How to cross a multi-lane highway," *IEEE Journal on Selected Areas in Communications*, Special Issue on Stochastic Geometry and Random Graphs for Wireless Networks, vol. 27, no. 7, pp. 1283-1296, Sept. 2009.

[140] Q. Zhao, W. Ren, and A. Swami, "Spectrum opportunity detection: How good is listen-before-talk?" in *Proceedings of the Asilomar Conference on Signals, Systems, and Computers*, Nov. 2007.

[141] T. Sheldon, *Encyclopedia of Networking & Telecommunications*. New York, NY, USA: Osborne/McGraw-Hill, 2001.

[142] P. Wang, L. Xiao, S. Zhou, and J. Wang, "Optimization of detection time for channel efficiency in cognitive radio systems," in *Proceedings of the IEEE Wireless Communications and Networking Conference*, Mar. 2007.

[143] Q. Zhao, B. Krishnamachari, and K. Liu, "On myopic sensing for multi-channel opportunistic access: Structure, optimality, and performance," *IEEE Transactions on Wireless Communications*, vol. 7, no. 12, pp. 5431–5440, Dec. 2008.

[144] K. Liu and Q. Zhao, "Indexability of restless bandit problems and optimality of Whittle's index for dynamic multichannel access," Submitted to *IEEE Transactions on Information Theory*. [Online]: http://arxiv.org/abs/0810.4658

[145] Federal Communications Commission, "Memorandum opinion and order." FCC 07-99, May 2007.

[146] "IEEE standards coordinating committee 41 (dynamic spectrum access networks)." [Online]: www.ieeep1900.org/.

[147] Federal Communications Commission, "Cognitive radio technologies proceeding (CRTP)." [Online]: www.fcc.gov/oet/cognitiveradio/.

[148] Shared Spectrum Company, "Spectrum measurements." [Online]: www.sharedspectrum.com/?section=measurements.

[149] A. Tanenbaum, *Computer Networks*, 4th ed. Upper Saddle River, NJ, USA: Prentice-Hall, 2003.

[150] R. Myerson, *Game Theory: Analysis of Conflict*, 3rd ed. Cambridge, MA, USA: Harvard University Press, 1999.

[151] Office of Communications Consultation, "Spectrum framework review," Nov. 2004.

[152] X. Jing and D. Raychaudhuri, "Spectrum co-existence of IEEE 802.11b and 802.16a networks using the CSCC etiquette protocol," in *Proceedings of the IEEE International Symposium on New Frontiers in Dynamic Spectrum Access Networks*, Baltimore, MD, USA, pp. 243–250, Nov. 8–11, 2005.

[153] D. Raychaudhuri and X. Jing, "A spectrum etiquette protocol for efficient coordination of radio devices in unlicensed bands," in *Proceedings of the IEEE International Symposium on Personal, Indoor, and Mobile Radio Communications*, vol. 1, Beijing, China, pp. 172–176, Sept. 7–10, 2003.

[154] O. Holland, A. Attar, M. Sooriyabandara, T. Farnham, H. Aghvami, M. Muck, V. Ivanov, and K. Nolte, "Architectures and protocols for dynamic Spectrum Sharing in Heterogeneous Wireless Access Networks," in *Heterogeneous Wireless Access Networks: Architectures and Protocols*, Springer, 2008.

[155] M. Buddhikot and K. Ryan, "Spectrum management in coordinated dynamic spectrum access based cellular networks," in *Proceedings of the IEEE International Symposium on New Frontiers in Dynamic Spectrum Access Networks*, Baltimore, MD, USA, pp. 299–307, Nov. 2005.

[156] 3GPP, "Network sharing: Architecture and functional description (release 6)." 3GPP TS 23.251 V6.6.0, Mar. 2006.

[157] J. Perez-Romero, O. Salient, R. Agusti, and L. Giupponi, "A novel on-demand cognitive pilot channel enabling dynamic spectrum allocation," in *Proceedings of the IEEE International Symposium on New Frontiers in Dynamic Spectrum Access Networks*, Dublin, Ireland, pp. 46–54, April 17–20, 2007.

[158] O. Holland, M. Muck, P. Martigne, D. Bourse, P. Cordier, S. B. Jemaa, P. Houze, D. Grandblaise, C. Klock, T. Renk, P. Jianming, P. Slanina, K. Mobner, L. Giupponi, J. P. Romero, R. Agusti, A. Attar, and A. H. Aghvami, "Development of a radio enabler for reconfiguration management within the IEEE P1900.4 working group," in *Proceedings of the IEEE International Symposium on New Frontiers in Dynamic Spectrum Access Networks*, Dublin, Ireland, pp. 232–239, Apr. 17–20, 2007.

[159] WRC-11, "Resolution 956 [com6/18]," Oct. 22-Nov. 16, 2007.

[160] A. P. Hulbert, "Spectrum sharing through beacons," in *Proceedings of the IEEE International Symposium on Personal, Indoor, and Mobile Radio Communications*, vol. 2, Berlin, Germany, pp. 989-993, Sept. 2005.

[161] O. Holland, A. Attar, N. Olaziregi, N. Sattari, and H. Aghvami, "A universal resource awareness channel for cognitive radio," in *Proceedings of the IEEE International Symposium on Personal, Indoor, and Mobile Radio Communications*, Helsinki, Finland, pp. 1-5, Sept. 2006.

[162] A. Attar, M. R. Nakhai, and A. H. Aghvami, "Cognitive radio game for secondary spectrum access problem," *IEEE Transactions on Wireless Communications*, vol. 8, no. 4, pp. 2121-2131, Apr. 2009.

[163] A. Attar, M. R. Nakhai, and A. H. Aghvami, "Cognitive radio game: A framework for efficiency, fairness and QoS guarantee," in *Proceedings of the IEEE International Conference on Communications*, Beijing, China, May 19-23, 2008.

[164] R. Cottle, J. S. Pang, and R.Stone, *The Linear Complimentarity Problem*. New York, NY, USA: Academic Press, Feb. 1992.

[165] R. Etkin, A. Parekh, and D. Tse, "Spectrum sharing for unlicensed bands," *IEEE Journal on Selected Areas in Communications*, vol. 25, pp. 517-528, Apr. 2007.

[166] O. Popescu, D. C. Popescu, and C. Rose, "Simultaneous water filling in mutually interfering systems," *IEEE Transactions on Wireless Communications*, vol. 6, pp. 1102-1113, Mar. 2007.

[167] R. Tandra and A. Sahai, "Noise calibration, delay coherence and SNR walls for signal detection," in *Proceedings of the IEEE International Symposium on New Frontiers in Dynamic Spectrum Access Networks*, Chicago, IL, USA, Oct. 14-17, 2008.

[168] J. D. Poston and W. D. Horne, "Discontiguous OFDM considerations for dynamic spectrum access in idle TV channels," in *Proceedings of the IEEE International Symposium on New Frontiers in Dynamic Spectrum Access Networks*, Baltimore, MD, USA, pp. 607-610, Nov. 8-11, 2005.

[169] S. Kondo and L. Milstein, "Performance of multicarrier DS CDMA systems," *IEEE Transactions on Communications*, vol. 44, pp. 238-246, Feb. 1996.

[170] R. Rajbanshi, Q. Chen, A. M. Wyglinski, G. J. Minden, and J. B. Evans, "Quantitative comparison of agile modulation technique for cognitive radio transceivers," in *Proceedings of the IEEE Consumer Communications and Networking Conference*, Las Vegas, NV, USA, pp. 1144-1148, Jan. 2007.

[171] A. Attar, M. R. Nakhai, and A. H. Aghvami, "Cognitive radio transmission based on direct-sequence MC-CDMA," *IEEE Transactions on Wireless Communications*, vol. 7, no. 4, pp. 1157-1162, Apr. 2008.

[172] A. Attar, M. R. Nakhai, and A. H. Aghvami, "Sharing with legacy RANs using cognitive MC-CDMA," in *Proceedings of the IEEE International Symposium on Personal, Indoor, and Mobile Radio Communications*, Cannes, France, Sept. 14-18, 2008.

[173] G. Fischer, "RF-MEMS for frequency agile software defined rf systems," in *Proceedings of the ESSDERC Workshop on Invisible Electronics for Ambient Intelligence Applications*, Grenoble, France, Sept. 2005.

[174] M. Costa, "Writing on dirty paper," *IEEE Transactions on Information Theory*, vol. 29, pp. 439-441, May 1983.

[175] N. Devroye, P. Mitran, and V. Tarokh, "Achievable rates in cognitive radio channels," *IEEE Transactions on Information Theory*, vol. 52, no. 5, pp. 1813-1827, May 2006.

[176] R. Zhang, "On peak versus average interference power constraints for protecting primary users in cognitive radio networks," *IEEE Transactions on Wireless Communications*, vol. 8, no. 4, pp. 2112-2120, Apr. 2009.

[177] A. Ghasemi and E. S. Sousa, "Fundamental limits of spectrum-sharing in fading environments," *IEEE Transactions on Wireless Communications*, vol. 6, pp. 649-658, Feb. 2007.

[178] L. Musavian and S. Aissa, "Ergodic and outage capacities of spectrum-sharing systems in fading channels," in *Proceedings of the IEEE Global Telecommunications Conference*, New Orleans, LA, USA, Nov. 30–Dec. 4, 2008.

[179] A. Goldsmith, *Wireless Communications*. Cambridge, UK: Cambridge University Press, Aug. 2005.

[180] J. E. Dennis and R. B. Schnabel, "Numerical methods for unconstrained optimization and nonlinear equations," *SIAM Classics in Applied Mathematics*, Jan. 1987.

[181] R. Rajbanshi and A. M. Wyglinski, and G. J. Minden, "An efficient implementation of the NC-OFDM transceivers for cognitive radios," in *Proceedings of the International Conference on Cognitive Radio Oriented Wireless Networks and Communications*, Mykonos, Greece, June 2006.

[182] J. Mitola III, *Cognitive Radio Architecture*. Hoboken, NJ, USA: John Wiley and Sons, Inc., 2006.

[183] R. Rajbanshi, A. M. Wyglinski, and G. J. Minden, *Cognitive Radio Communication Networks*, ch. 5. New York, NY, USA: Springer-Verlag, 2007.

[184] M. Z. Win and R. A. Scholtz, "Impulse radio: How it works," *IEEE Communication Letters*, vol. 2, pp. 36–38, Feb. 1998.

[185] Federal Communications Commission, "FCC first report and order: Revision of part 15 of the commissions's rules regarding ultra-wideband transmission systems." ET Docket No. 98-153, Apr. 2002.

[186] K. S. Gilhousen, I. M. Jacobs, R. Padovani, A. J. Viterbi, L. A. Weaver Jr., and C. E. Wheatley III, "On the capacity of a cellular CDMA system," *IEEE Transactions on Vehicular Technology*, vol. 40, pp. 303–312, May 1991.

[187] R. Menon, R. M. Buehrer, and J. H. Reed, "Outage probability based comparison of underlay and overlay spectrum sharing techniques," in *Proceedings of the IEEE International Symposium on New Frontiers in Dynamic Spectrum Access Networks*, vol. 1, Baltimore, MD, USA, pp. 101–109, Nov. 2005.

[188] L. Wang and C. Tellambura, "Narrowband interference suppression in time-hopping impulse radio ultra-wideband communications," *IEEE Transactions on Communications*, vol. 54, June 2006.

[189] C. Rose, S. Ulukus, and R. D. Yates, "Wireless systems and interference avoidance," *IEEE Transactions on Wireless Communications*, vol. 1, pp. 415–428, July 2002.

[190] R. Etkin, A. Parekh, and D. Tse, "Spectrum sharing for unlicensed bands," in *Proceedings of the IEEE International Symposium on New Frontiers in Dynamic Spectrum Access Networks*, vol. 1, Baltimore, MD, USA, pp. 251–258, Nov. 2005.

[191] U. Berthold and F. K. Jondral, "Guidelines for designing OFDM overlay systems," in *Proceedings of the IEEE International Symposium on New Frontiers in Dynamic Spectrum Access Networks*, vol. 1, Baltimore, MD, pp. 626–629, Nov. 2005.

[192] T. Weiss, J. Hillenbrand, A. Krohn, and F. Jondral, "Mutual interference in OFDM-based spectrum pooling systems," in *Proceedings of the IEEE Vehicular Technology Conference*, vol. 4, Milan, Italy, pp. 1873–1877, May 2004.

[193] S. Kapoor and S. Nedic, "Interference suppression in DMT receivers using windowing," in *Proceedings of the IEEE International Conference on Communications*, vol. 2, New Orleans, LA, USA, pp. 778–782, June 2000.

[194] I. Cosovic, S. Brandes, and M. Schnell, "A technique for sidelobe suppression in OFDM systems," in *Proceedings of the IEEE Global Telecommunications Conference*, vol. 1, St. Louis, MO, USA, pp. 204–208, Nov. 2005.

[195] S. Brandes, I. Cosovic, and M. Schnell, "Sidelobe suppression in OFDM systems by insertion of cancellation carriers," in *Proceedings of the IEEE Vehicular Technology Conference*, vol. 1, Dallas, TX, USA, pp. 152–156, Sept. 2005.

[196] R. Rajbanshi, "OFDM-based cognitive radio for DSA networks." Phd dissertation, University of Kansas, Lawrence, KS, USA, May 2007.

[197] H. Tang, "Some physical layer issues of wide-band cognitive radio systems," in *Proceedings of the IEEE International Symposium on New Frontiers in Dynamic Spectrum Access Networks*, vol. 1, Baltimore, MD, USA, pp. 151–159, Nov. 2005.

[198] IEEE802.11a, "Wireless LAN medium access control (MAC) and physical layer (PHY) specifications: High-speed physical layer in the 5 GHz band." IEEE, tech. rep., 1999.

[199] ETSI-TS-101475, "Broadband radio access networks (BRAN); HIPERLAN Type 2: Physical (PHY) layer." ETSI, tech. rep., 2001.

[200] S. Pagadarai, A. M. Wyglinski, and R. Rajbanshi, "A novel sidelobe suppression technique for OFDM-based cognitive radio transmission," in *Proceedings of the IEEE International Symposium on New Frontiers in Dynamic Spectrum Access Networks*, Chicago, IL, USA, Oct. 2008.

[201] S. Pagadarai, R. Rajbanshi, and A. M. Wyglinski, "Sidelobe suppression for OFDM-based cognitive radios using constellation expansion," in *Proceedings of the IEEE Wireless Commnunications and Networking Conference*, Las Vegas, NV, USA, Apr. 2008.

[202] M. Senst, M. Jordan, M. Dörpinghaus, M. Färber, G. Ascheid, and H. Meyr, "Joint reduction of peak-to-average power ratio and out-of-band power in OFDM systems," in *Proceedings of the IEEE Global Telecommunications Conference*, Washington DC, USA, Nov. 2007.

[203] I. Cosovic and T. Mazzoni, "Suppression of sidelobes in OFDM systems by multiple choice sequences," *European Transactions on Communications*, vol. 1, pp. 623–630, Dec. 2006.

[204] R. G. Alves, P. L. Osorio, and M. N. S. Swamy, "General FFT pruning algorithm," in *IEEE Transactions on Circuits and Systems*, vol. 3, pp. 1192–1195, Aug 2000.

[205] J. W. Cooley and J. W. Tukey, "An algorithm for the machine calculation of complex Fourier series," *Mathematical Computation*, vol. 19, pp. 297–301, Apr. 1965.

[206] T. A. Wilkison and A. E. Jones, "Minimization of the peak to mean envelope power ratio of multicarrier transmission schemes by block coding," in *Proceedings of the IEEE Vehicular Technology Conference*, vol. 2, Chicago, IL, USA, pp. 825–829, July 1995.

[207] H. Ahn, Y. M. Shin, and S. Im, "A block coding scheme for peak to average power ratio reduction in an orthogonal frequency division multiplexing system," in *Proceedings of the 51st IEEE Vehicular Technology Conference—Spring*, vol. 1, Tokyo, Japan, pp. 56–60, May 2000.

[208] Y. Zhang, A. Yongacoglu, J. Chouinard, and L. Zhang, "OFDM peak power reduction by subblock coding and its extended versions," in *Proceedings of the IEEE Vehicular Technology Conference*, vol. 1, Houston, TX, USA, pp. 695–699, May 1999.

[209] C. Tellambura, "A coding technique for reducing peak to average power ratio in OFDM," in *Proceedings of the IEEE Global Telecommunications Conference*, vol. 5, Sydney, Australia, pp. 2783–2787, Nov. 1998.

[210] R. W. Bauml, R. F. H. Fischer, and J. B. Huber, "Reducing the peak to average power ratio of multicarrier modulation by selective mapping," *Electronic Letters*, vol. 32, pp. 2056–2057, Oct. 1996.

[211] S. H. Muller and J. B. Huber, "OFDM with reduced peak to average power ratio by optimum combination of partial transmit sequences," *Electronic Letters*, vol. 33, pp. 368–369, Feb. 1997.

[212] H. Nikookar and K. S. Lidsheim, "Random phase updating algorithm for OFDM transmission with low PAPR," *IEEE Transactions on Broadcasting*, vol. 48, pp. 123–128, June 2002.

[213] J. Tellado, *Multicarrier modulation with low PAR: Applications to DSL and wireless.* Boston, MA, USA: Kluwer Academic Publishers, 2000.

[214] H. Yu and G. Wei, "Computation of the continuous time PAR of an OFDM signal," in *Proceedings of the IEEE International Conference on Acoustic, Speech, and Signal Processing*, vol. 2003, Hong Kong, pp. iv, 529–531, Apr. 2003.

[215] N. Y. Ermolova and P. Vainikainen, "On the relationship between peak factor of a multicarrier signal and aperiodic correlation of the generating sequence," *IEEE Communication Letters*, vol. 7, pp. 107–108, Mar. 2003.

[216] I. Kalet, "The multitone channel," *IEEE Transactions on Communications*, vol. 37, pp. 119–124, Feb. 1989.

[217] R. F. H. Fischer and J. B. Huber, "A new loading algorithm for discrete multitone transmission," in *Proceedings of the IEEE Global Telecommunications Conference*, vol. 1, London, UK, pp. 724–728, Nov. 1996.

[218] A. M. Wyglinski, "Physical layer loading algorithms for indoor wireless multicarrier systems." PhD thesis, McGill University, Montreal, Quebec, Canada, Nov. 2004.

[219] Federal Communications Commission, "Part 15—Radio frequency devices, subpart E—Unlicenced national information infrastructure devices, section 403—Definitions." Code of Federal Regulations, FCC 47CFR15.403, Oct. 2000.

[220] R. Rajbanshi, A. M. Wyglinski, and G. J. Minden, "Subcarrier power adjustment technique for peak-to-average power ratio reduction of OFDM systems," in *Proceedings of the Military Communications Conference*, Washington, DC, USA, May 2006.

[221] Y. H. You, I. T. Hwang, C. K. Song, and H. K. Song, "PAPR analysis for multi-band OFDM signals," *Electronic Letters*, vol. 41, pp. 261–262, Mar. 2005.

[222] R. van Nee and A. de Wild, "Reducing the peak-to-average power ratio of OFDM," in *Proceedings of the IEEE Vehicular Technology Conference*, Ottawa, Ontario, Canada, pp. 2072–2076, May 1998.

[223] J. Armstrong, "Peak-to-average power reduction for OFDM by repeated clipping and frequency domain filtering," *Electronic Letters*, vol. 38, pp. 246–247, Feb. 2002.

[224] A. D. S. Jayalath and C. Tellambura, "The use of interleaving to reduce the peak to average power ratio of an OFDM signal," in *Proceedings of the IEEE Global Telecommunications Conference*, vol. 1, San Francisco, CA, USA, pp. 82–86, Nov. 2000.

[225] M. A. Tzannes, M. C. Tzannes, and H. Resnikoff, "The DWMT: A multicarrier transceiver for ADSL using *M*-band wavelet transforms," ANSI Contribution T1E1.4/93-067, Mar. 1993.

[226] A. Lindsey, "Wavelet packet modulation for orthogonally multiplexed communication," *IEEE Transactions on Signal Processing*, vol. 45, pp. 1336–1339, May 1997.

[227] A. Lindsey and J. C. Dill, "Wavelet packet modulation: A generalized method for orthogonally multiplexed communications," in *Proceedings of the 27th Southeastern Symposium on Systems Theory*, pp. 392–396, Mar. 1995.

[228] M. Hawryluck, A. Yongacoglu, and M. Kavehrad, "Efficient equalization of discrete wavelet multitone over twisted pair," in *Proceedings of the International Zurich Seminar on Broadband Communications*, Zurich, Switzerland, pp. 185–191, 1998.

[229] L. Vandendorpe, L. Cuvelier, F. Deryck, J. Louveaux, and O. van de Wiel, "Fractionally spaced linear and decision-feedback detectors for transmultiplexers," *IEEE Transactions on Signal Processing*, vol. 46, pp. 996–1011, Apr. 1998.

[230] A. N. Akansu, P. Duhamel, X. Lin, and M. de Courville, "Orthogonal transmultiplexers in communication: A review," *IEEE Transactions on Signal Processing*, vol. 46, pp. 979–995, Apr. 1998.

[231] A. D. Rizos, J. G. Proakis, and T. Q. Nguyen, "Comparison of DFT and cosine modulated filter banks in multicarrier modulation," in *Proceedings of the IEEE Global Telecommunications Conference*, vol. 2, Zurich, Switzerland, pp. 687–691, 1994.

[232] W. Rhee, C. J. Chuang, and L. J. Cimini, "Performance comparison of OFDM and multitone with polyphase filterbank for wireless communications," in *Proceedings of the IEEE Vehciular Technology Conference*, pp. 768–772, May 1998.

[233] B. Farhang-Boroujeny, "Multicarrier modulation with blind detection capability using cosine modulated filter banks," *IEEE Transactions on Communications*, vol. 51, pp. 2057–2070, Dec. 2003.

[234] Joint Tactical Radio Systems, "Software communications architecture specification," Nov. 2002.

[235] S. M. Mishra, D. Cabric, and C. Chang, "A real time cognitive radio testbed for physical and link layer experiments," in *Proceedings of the IEEE International Symposium on New Frontiers in Dynamic Spectrum Access*, pp. 562–567, 2005.

[236] J. Chapin and V. Bose, "The vanu software radio system," in *Proceedings of the Software Defined Radio Technical Conference*, Nov. 2002.

[237] C. J. Rieser, "Biologically inspired cognitive radio engine model utilizing distributed genetic algorithms for secure and robust wireless communications and networking." PhD thesis, Virginia Polytechnic Institute and State University, April 2004.

[238] "Shared spectrum company." [Online]: www.sharedspectrum.com.

[239] J. H. Reed et al., "Development of a cognitive engine and analysis of WRAN cognitive radio algorithms," report submitted to etri, Mobile and Portable Radio Research Group (MPRG), Wireless@Virginia Tech, Virginia Polytechnic Institute and State University, Blacksburg, VA, USA, Dec. 2006.

[240] C. Rieser, T. Rondeau, C. Bostian, and T. Gallagher, "Cognitive radio testbed: Further details and testing of a distributed genetic algorithm based cognitive engine for programmable radios," in *Proceedings of the Military Communications Conference*, 2004.

[241] J. H. Reed et al., "Applying artificial intelligence to the development of a cognitive radio engine," report submitted to aro, Mobile and Portable Radio Research Group (MPRG), Wireless@Virginia Tech, Virginia Polytechnic Institute and State University, Blacksburg, VA, USA, July 2006.

[242] DARPA XG program web site. [Online]: www.darpa.mil/ato/programs/XG/.

[243] T. Newman, R. Rajbanshi, A. Wyglinski, J. Evans, and G. Minden, "Population adaptation for genetic algorithm-based cognitive radios," *ACM/Springer Mobile Ad Hoc Networks—Special Issue on Cognitive Radio Oriented Wireless Networks and Communications*, 2008.

[244] S. Yarkan and H. Arslan, "Exploiting location awareness toward improved wireless system design in cognitive radio," *IEEE Communications Magazine*, vol. 46, no. 1, pp. 128–136, Jan. 2008.

[245] T. Weingart, D. C. Sicker, and D. Grunwald, "A statistical method for reconfiguration of cognitive radios," *IEEE Wireless Communications*, vol. 14, no. 4, pp. 34–40, Aug. 2007.

[246] M. Pursley and T. Royster, "Low-complexity adaptive transmission for cognitive radios in dynamic spectrum access networks," *IEEE Journal on Selected Areas in Communications*, vol. 26, no. 1, pp. 83–94, Jan. 2008.

[247] H. Arslan and S. Yarkan, *Enabling Cognitive Radio through Sensing, Awareness, and Measurements*. Dordrecht, The Netherlands: Springer, 2007.

[248] R. Rajbanshi, Q. Chen, A. M. Wyglinski, J. B. Evans, and G. J. Minden, "Comparative study of frequency agile data transmission schemes for cognitive radio transceivers," in *Proceedings of the First International Workshop on Technology and Policy for Accessing Spectrum*, Boston, MA, USA, Aug. 2006.

[249] C. M. Fonseca and P. J. Fleming, "An overview of evolutionary algorithms in multiobjective optimization," *Evolutionary Computation*, vol. 3, no. 1, pp. 1–16, 1995.

[250] P. J. Fleming, "Designing control systems with multiple objectives," in *Proceedings of the IEE Colloquium Advances in Control Technology*, pp. 4/1–4/4, 1999.

[251] E. Zitzler and L. Thiele, "Multiobjective evolutionary algorithms: A comparative case study and the strength pareto approach," *IEEE Transactions on Evolutionary Computation*, vol. 3, no. 4, pp. 257–271, 1999.

[252] L. A. Zadeh, "Optimality and non-scalar-valued performance criteria," *IEEE Transactions on Automatic Control*, vol. 8, pp. 59–60, 1963.

[253] A. Goicoechea, D. Hansen, and L. Duckstein, *Multiobjective Decision Analysis with Engineering and Business Applications*. New York, NY, USA: John Wiley and Sons, 1982.

[254] A. E. Hans, "Multicriteria optimization for highly accurate systems," in *Multicriteria Optimization in Engineering and Sciences*, pp. 309–352, 1988.

[255] M. Schniederjans, *Goal Programming: Methodology and Applications*. Norwell, MA, USA: Kluwer Academic Publishers, 1995.

[256] J. Ignizio, *Goal Programming and Extensions*. Lanham, MD, USA: Lexington Books, 1976.

[257] S. Forrest, "Genetic algorithms," *ACM Computing*, vol. 28, pp. 77–80, March 1996.

[258] D. E. Goldberg, *Genetic Algorithms in Search, Optimization, and Machine Learning*. Reading, MA, USA: Addison-Wesley, 1989.

[259] J. P. Ignizio, *Introduction to Expert Systems: The Development and Implementation of Rule-Based Expert Systems*. New York, NY, USA: McGraw-Hill, 1991.

[260] B. G. Buchanan and E. H. Shortliffe, *Rule-Based Expert Systems: The MYCIN Experiments of the Stanford Heuristic Programming Project*. Reading, MA, USA: Addison-Wesley, 1985.

[261] CLIPS expert system web site. [Online]: www.ghg.net/clips/CLIPS.html.

[262] K. DeJong and W. Spears, "An analysis of the interacting roles of population size and crossover in genetic algorithms," in *Proceedings of the First Workshop Parallel Problem Solving from Nature*, 1990.

[263] J. W. Myers, K. B. Laskey, and K. A. DeJong, "Learning Bayesian networks from incomplete data using evolutionary algorithms," in *Proceedings of the Genetic and Evolutionary Computation Conference*, vol. 1, pp. 458–465, 1999.

[264] D. G. Li and A. C. Watson, "Genetic algorithms in optical thin film optimization design," in *Proceedings of the 3rd International Conference on Computational Intelligence and Multimedia Applications*, p. 86, 1999.

[265] M. Srinivas and L. M. Patnaik, "Genetic algorithms: A survey," *Computer Magazine*, vol. 27, no. 6, pp. 17–26, 1994.

[266] J. L. Kolodner and D. Leake, "A tutorial introduction to case-based reasoning," in *Case-Based Reasoning: Experiences, Lessons and Future Directions* (D. Leake, ed.), pp. 31–65, Cambridge, MA, USA: MIT Press, 1996.

[267] S. Shiu and S. K. Pal, *Foundations of Soft Case-Based Reasoning*. Wiley Series on Intelligent Systems, Hoboken, NJ: Wiley-Interscience, 2004.

[268] T. R. Newman, B. A. Barker, A. M. Wyglinski, A. Agah, J. B. Evans, and G. J. Minden, "Cognitive engine implementation for wireless multicarrier transceivers," *Wireless Communications and Mobile Computing*, vol. 7, no. 9, pp. 1129–1142, 2007.

[269] P. Gupta and P. Kumar, "The capacity of wireless networks," *IEEE Transactions on Information Theory*, vol. 46, pp. 388–404, Mar. 2000.

[270] C. E. Perkins, "IP mobility support for IPv4." IETF RFC 3344, Oct. 2002.

[271] M. Grossglauser and D. Tse, "Mobility increases the capacity of ad hoc wireless networks," *IEEE/ACM Transactions on Networking*, vol. 10, pp. 477–486, Aug. 2002.

[272] A. B. McDonald and T. F. Znati, "A mobility-based framework for adaptive clustering in wireless adhoc networks," *IEEE Journal on Selected Areas in Communications*, vol. 17, pp. 1466–1487, Aug. 1999.

[273] W. Su, S.-J. Lee, and M. Gerla, "Mobility prediction and routing in ad hoc wireless networks," *International Journal of Network Management*, vol. 11, pp. 3–30, Jan./Feb. 2001.

[274] J. Zhang, Q. Zhang, B. Li, X. Luo, and W. Zhu, "Energy-efficient routing in mobile ad hoc networks: Mobility-assisted case," *IEEE Transactions on Vehicular Technology*, vol. 55, pp. 369–379, Jan. 2006.

[275] IEEE, "Wireless LAN media access control (MAC) and physical layer (PHY) specifications," 1999.

[276] W. Ye, J. Heidemann, and D. Estrin, "Medium access control with coordinated, adaptive sleeping for wireless sensor networks," *IEEE Transactions on Networking*, vol. 12, pp. 493–506, June 2004.

[277] A. Kansal and M. B. Srivastava, "An environmental energy harvesting framework for sensor networks," in *Proceedings of the 2003 International Symposium on Low Power Electronics and Design*, Seoul, Korea, pp. 481–486, Aug. 2003.

[278] M. Kim, D. Kotz, and S. Kim, "Extracting a mobility model from real user traces," in *Proceedings of IEEE INFOCOM 2006*, Barcelona, Spain, pp. 1-13, Apr. 2006.

[279] C. Tuduce and T. Gross, "A mobility model based on WLAN traces and its validation," in *Proceedings of IEEE INFOCOM 2005*, Miami, FL, USA, pp. 664-674, Apr. 2005.

[280] J. Yoon, B. D. Noble, M. Liu, and M. Kim, "Building realistic mobility models from coarse-grained traces," in *Proceedings of ACM MOBISYS 2006*, Uppsala, Sweden, pp. 177-190, June 2006.

[281] T. Camp, J. Boleng, and V. Davies, "A survey of mobility models for ad hoc network research," *Wireless Communications and Mobile Computing*, vol. 2, pp. 483-502, Sept. 2002.

[282] D. Johnson and D. Maltz, "Dynamic source routing in ad hoc wireless networks," in *Mobile Computing* (T. Imelinsky and H. Korth, eds.), pp. 153-181, Norwood, MA, USA: Kluwer Adacemmic Publishers, 1996.

[283] J. Broch, D. Maltz, D. Johnson, Y. Hu, and J. Jetcheva, "Multi-hop wireless ad hoc network routing protocols," in *Proceedings of ACM MOBICOM'98*, Dallas, TX, USA, pp. 85-97, Oct. 1998.

[284] W. Navidi and T. Camp, "Stationary distribution for the random waypoint mobility model," *IEEE Transactions on Mobile Computing*, vol. 3, pp. 99-108, Jan./Mar. 2004.

[285] J. Yoon, M. Liu, and B. Noble, "Random waypoint considered harmful," in *Proceedings of IEEE INFOCOM 2003*, San Francisco, CA, USA, pp. 1312-1321, Mar. 2003.

[286] J. Yoon, M. Liu, and B. Noble, "Sound mobility models," in *Proceedings of ACM MOBICOM 2003*, San Diego, CA, USA, pp. 205-216, Sept. 2003.

[287] J.-Y. L. Boudec and M. Vojnovic, "The random trip model: Stability, stationary regime, and perfect simulation," *IEEE Transactions on Networking*, vol. 14, pp. 1153-1166, Dec. 2006.

[288] J.-Y. L. Boudec, "On the stationary distribution of speed and location of random waypoint," *IEEE Transactions on Mobile Computing*, vol. 4, pp. 404-405, July/Aug. 2005.

[289] G. Lin, G. Noubir, and R. Rajamaran, "Mobility models for ad hoc network simulation," in *Proceedings of IEEE INFOCOM 2004*, Hong Kong, China, pp. 454-463, Apr. 2004.

[290] F. Baccelli and P. Bremaud, *Elements of Queueing Theory: Palm Martingale Calculus and Stochastic Recurrences*, 2nd ed. Berlin, Germany: Springer-Verlag, 2003.

[291] D. Tse and P. Viswanath, *Fundamentals of Wireless Communication*. Cambridge, UK: Cambridge University Press, 2005.

[292] P. Viswanath, D. N. C. Tse, and R. Laroia, "Opportunistic beamforming using dumb antennas," *IEEE Transactions on Information Theory*, vol. 48, pp. 1277-1294, June 2002.

[293] R. Knopp and P. A. Humblet, "Information capacity and power control in single-cell multiuser communications," in *Proceedings of IEEE ICC 1995*, Seattle, WA, USA, pp. 331-335, June 1995.

[294] D. N. C. Tse, "Optimal power allocation over parallel gaussian channels," in *Proceedings of the IEEE International Symposium on Information Theory*, Ulm, Germany, p. 27, June 1997.

[295] S. Sanayei and A. Nosratinia, "Opportunistic downlink transmission with limited feedback," *IEEE Transactions on Information Theory*, vol. 53, pp. 4363-4372, Nov. 2007.

[296] M. Sharif and B. Hassibi, "A comparison of time-sharing, DPC, and beamforming for MIMO broadcast channels with many users," *IEEE Transactions on Communications*, vol. 55, pp. 11-15, Jan. 2007.

[297] T. M. Cover and J. Thomas, *Elements of Information Theory*, 2nd ed. Hoboken, NJ, USA: Wiley & Sons, 2006.

[298] R. Knopp, "Achieving multiuser diversity under hard fairness constraints," in *Proceedings of IEEE ISIT 2002*, Lausanne, Switzerland, p. 451, June/July 2002.

[299] L. Yang, M. Kang, and A. Mohamed-Slim, "On the capacity-fairness tradeoff in multiuser diversity systems," *IEEE Transactions on Vehicular Technology*, vol. 56, pp. 1901–1907, July 2007.

[300] D. Bertsekas and R. Gallager, *Data Networks* 2nd ed. Upper Saddle River, NJ, USA: Prentice-Hall, 1992.

[301] H. Levy, "Non-uniform structures and synchronization patterns in shared-channel communication networks." PhD thesis, University of California, Los Angeles, USA, Aug. 1984.

[302] H. Takagi and L. Kleinrock, "A tutorial on the analysis of polling systems," tech. rep., Report No. 850005, Computer Science Department, University of California, Los Angeles, USA, Feb. 1985.

[303] S.-M. Hur, S. Mao, Y. Hou, K. Nam, and J. Reed, "On exploiting location information for concurrent transmission in multi-hop wireless networks," *IEEE Transactions on Vehicular Technology*, vol. 58, Jan. 2009.

[304] Y. Xiao and J. Rosdahl, "Throughput and delay limits of IEEE 802.11," *IEEE Communication Letters*, vol. 6, pp. 355–357, Aug. 2002.

[305] C. E. Jones, K. M. Sivalingam, P. Agrawal, and J.-C. Chen, "A survey of energy efficient network protocols for wireless networks," *Springer Wireless Networks Journal*, vol. 7, pp. 343–358, Aug. 2001.

[306] Y. Hou, Y. Shi, and H. Sherali, "Rate allocation in wireless sensor networks with network lifetime requirement," in *Proceedings of ACM MOBIHOC 2004*, Tokyo, Japan, pp. 67–77, May 2004.

[307] Y. Hou, Y. Shi, H. Sherali, and J. Wieselthier, "Online lifetime-centric multicast routing for ad hoc networks with directional antennas," in *Proceedings of IEEE INFOCOM 2005*, Miami, FL, USA, pp. 761–772, Mar. 2005.

[308] J.-H. Chang and L. Tassiulas, "Maximum lifetime routing in wireless sensor networks," *IEEE Transactions on Networking*, vol. 12, pp. 609–619, Aug. 2004.

[309] H. Balakrishnan, V. Padmanabhan, S. Seshan, and R. Katz, "A comparison of mechanisms for improving TCP performance over wireless links," *IEEE Transactions on Networking*, vol. 5, pp. 756–769, Dec. 1997.

[310] N. Vaidya, "TCP for mobile and wireless hosts." [Online]: www.crhc.uiuc.edu/wireless/tutorials.html, Oct. 1998.

[311] A. A. Hanbali, E. Altman, and P. Nain, "A survey of TCP over ad hoc networks," *IEEE Communications Surveys and Tutorials*, vol. 7, pp. 22–36, Aug. 2005.

[312] J. Li, C. Blake, D. S. J. De Couto, H.-I. L, and R. Morris, "Capacity of ad hoc wireless networks," in *Proceedings of ACM MOBICOM 2001*, Rome, Italy, pp. 61–69, July 2001.

[313] Z. Fu, P. Zerfos, H. Luo, S. Lu, L. Zhang, and M. Gerla, "The impact of multihop wireless channel on TCP throughput and loss," in *Proceedings of IEEE INFOCOM 2003*, San Francisco, CA, USA, pp. 1744–1753, Apr. 2003.

[314] S. Xu and T. Saadawi, "Does the IEEE 802.11 MAC protocol work well in multihop wireless ad hoc networks?" *IEEE Communications Magazine*, vol. 39, pp. 130–137, June 2001.

[315] M. Chiang, "Balancing transport and physical layers in wireless multihop networks: Jointly optimal congestion control and power control," *IEEE Journal of Selected Areas in Communications*, vol. 23, pp. 104–116, Jan. 2005.

[316] X. Lin, N. B. Shroff, and R. Srikant, "A tutorial on cross-layer optimization in wireless networks," *IEEE Journal on Selected Areas in Communications*, vol. 24, pp. 1452–1463, Aug. 2006.

[317] M. J. Neely, E. Modiano, and C. E. Rohrs, "Dynamic power allocation and routing for time varying wireless networks," *IEEE Journal on Selected Areas in Communications*, vol. 23, pp. 89–103, Jan. 2005.

[318] S. Kompella, S. Mao, Y. T. Hou, and H. D. Sherali, "On path selection and rate allocation for video in wireless mesh networks," *IEEE/ACM Transactions on Networking*, vol. 17, pp. 212–224, Feb. 2009.

[319] Y. Shi, Y. T. Hou, and H. D. Sherali, "Cross-layer optimization for data rate utility problem in UWB-based ad hoc networks," *IEEE Transactions on Mobile Computing*, vol. 7, pp. 764–777, June 2008.

[320] M. van der Schaar and S. Shankar N, "Cross-layer wireless multimedia transmission: Challenges, principles, and new paradigms," *IEEE Wireless Communications*, vol. 12, pp. 50–58, Aug. 2005.

[321] Q. Zhang and Y.-Q. Zhang, "Cross-layer design for QoS support in multihop wireless networks," *Proceeding of the IEEE*, vol. 96, pp. 64–76, Jan. 2008.

[322] V. Kawadia and P. R. Kumar, "A cautionary perspective on cross-layer design," *IEEE Wireless Communications*, vol. 12, pp. 3–11, Feb. 2005.

[323] D. Clark, G. Partridge, J. C. Ramming, and J. T. Wroclawski, "A knowledge plane for the Internet," in *Proceedings of ACM SIGCOMM 2003*, Karlsruhe, Germany, Aug. 2003.

[324] P. Mähönen, "Cognitive trends in making: Future of networks," in *Proceedings of the 15th IEEE International Symposium on Personal, Indoor and Mobile Radio Communications*, vol. 2, Barcelona, Spain, pp. 1449–1454, 2004.

[325] Q. H. Mahmoud, *Cognitive Networks: Towards Self-Aware Networks*. New York, NY, USA: Wiley, 2007.

[326] R. W. Thomas, D. H. Friend, L. A. DaSilva, and A. B. MacKenzie, "Cognitive networks: Adaptation and learning to achieve end-to-end performance objectives," *IEEE Communications Magazine*, vol. 44, pp. 51–57, Dec. 2006.

[327] R. W. Thomas, L. A. DaSilva, and A. B. MacKenzie, "Cognitive networks," in *Proceedings of the IEEE International Symposium on New Frontiers in Dynamic Spectrum Access Networks*, pp. 352–360, Nov. 2005.

[328] P. Mähönen, M. Petrova, J. Riihijärvi, and M. Wellens, "Cognitive wireless networks," in *Proceedings of IEEE INFOCOM*, Barcelona, Spain, 2006.

[329] J. Mitola and G. Q. Maguire, "Cognitive radio: making software radios more personal," *IEEE Personal Communications*, vol. 6, no. 4, pp. 13–18, Aug. 1999.

[330] M. Sherman, A. N. Mody, R. Martinez, C. Rodriguez, and R. Reddy, "IEEE standards supporting cognitive radio and networks, dynamic spectrum access, and coexistence," *IEEE Communications Magazine*, vol. 46, no. 7, pp. 72–79, 2008.

[331] M. Petrova and P. Mähönen, "Cognitive resource manager: A cross-layer architecture for implementing cognitive radio networks," in *Cognitive Wireless Networks* (F. Fitzek and M. Katz, eds.). Dordrecht, The Netherlands: Springer, 2007.

[332] D. Wilkins, G. Denker, M. O. Stehr, D. Elenius, R. Senanayake, and C. Talcott, "Policy-based cognitive radios," *IEEE Wireless Communications*, vol. 14, pp. 41–46, Aug. 2007.

[333] D. Lewis, K. Feeney, K. Foley, L. Doyle, T. Forde, P. Argyroudis, J. Keeney, and D. O'Sullivan, "Managing policies for dynamic spectrum access," *Lecture Notes in Computer Science*, vol. 4195, p. 285, 2006.

[334] D. D. Clark, J. Wroclawski, K. R. Sollins, and R. Braden, "Tussle in cyberspace: Defining tomorrow's Internet," *IEEE/ACM Transactions on Networking*, vol. 13, pp. 462–475, June 2005.

[335] M. Sooriyabandara, T. Farnham, C. Efthymiou, M. Wellens, J. Riihijärvi, P. Mähönen, A. Gefflaut, J. Galache, D. Melpignano, and A. van Rooijen, "Unified link layer API: A generic and open API to manage wireless media access," *Computer Communications*, vol. 31, no. 5, pp. 962–979, 2008.

[336] D. Wilkins, G. Denker, M. O. Stehr, D. Elenius, R. Senanayake, and C. Talcott, "Policy-based cognitive radios," *IEEE Wireless Communications*, vol. 14, pp. 41–46, Aug. 2007.

[337] H. Urkowitz, "Energy detection of unknown deterministic signals," *Proceedings of the IEEE*, vol. 55, pp. 523–531, Apr. 1967.

[338] A. E. Leu, K. Steadman, M. McHenry, and J. Bates, "Ultra sensitive TV detector measurements," in *Proceedings of the IEEE International Symposium on New Frontiers in Dynamic Spectrum Access Networks*, Baltimore, MD, USA, Nov. 2005.

[339] C. Cordeiro, M. Ghosh, D. Cavalcanti, and K. Challapali, "Spectrum sensing for dynamic spectrum access of TV bands," in *Proceedings of the International Conference on Cognitive Radio Oriented Wireless Networks and Communications*, Orlando, FL, USA, Aug. 2007.

[340] W. A. Gardner, "Signal interception: A unifying theoretical framework for feature detection," *IEEE Transactions on Communications*, vol. 36, pp. 897-906, Aug. 1988.

[341] A. Tkachenko, D. Cabric, and R. W. Brodersen, "Cyclostationary feature detector experiments using reconfigurable BEE2," in *Proceedings of the IEEE International Symposium on New Frontiers in Dynamic Spectrum Access Networks*, Dublin, Ireland, Apr. 2007.

[342] S. M. Mishra, A. Sahai, and R. W. Brodersen, "Cooperative Sensing among Cognitive Radios," in *Proceedings of the IEEE International Conference on Communications*, Istanbul, Turkey, June 2006.

[343] E. Visotsky, S. Kuffner, and R. Peterson, "On collaborative detection of TV transmissions in support of dynamic spectrum sharing," in *Proceedings of the IEEE International Symposium on New Frontiers in Dynamic Spectrum Access Networks*, Baltimore, MD, USA, Nov. 2005.

[344] A. Ghasemi and E. S. Sousa, "Collaborative spectrum sensing for opportunistic access in fading environments," in *Proceedings of the IEEE International Symposium on New Frontiers in Dynamic Spectrum Access Networks*, vol. 1, Baltimore, MD, USA, pp. 131-136, Nov. 2005.

[345] A. Taherpour, Y. Norouzi, M. Nasiri-Kenari, A. Jamshidi, and Z. Zeinalpour-Yazdi, "Asymptotically optimum detection of primary user in cognitive radio networks," *IET Communications*, vol. 1, pp. 1138-1145, Dec. 2007.

[346] F. E. Visser, G. J. M. Janssen, and P. Pawełczak, "Multinode spectrum sensing based on energy detection for dynamic spectrum access," in *Proceedings of the IEEE Vehicular Technology Conference*, Singapore, China, pp. 1394-1398, May 2008.

[347] Y. Zhao, B. Le, and J. H. Reed, "Network support—The radio environment map," *Cognitive Radio Technology*, 2006.

[348] Y. Zhao, L. Morales, J. Gaeddert, K. K. Bae, J. S. Um, and J. H. Reed, "Applying radio environment maps to cognitive wireless regional area networks," in *Proceedings of the IEEE International Symposium on New Frontiers in Dynamic Spectrum Access Networks*, pp. 115-118, 2007.

[349] D. E. Knuth, "Two notes on notation," *American Mathematical Monthly*, vol. 99, no. 5, pp. 403-422, 1992.

[350] M. Chiang, S. H. Low, A. R. Calderbank, and J. C. Doyle, "Layering as optimization decomposition: A mathematical theory of network architectures," *Proceedings of the IEEE*, vol. 95, no. 1, pp. 255-312, 2007.

[351] F. Kelly, A. Maulloo, and D. Tan, "Rate control for communication networks: Shadow prices, proportional fairness and stability," *Journal of the Operational Research Society*, vol. 49, no. 3, pp. 237-252, 1998.

[352] S. H. Low and D. E. Lapsely, "Optimization flow control. I. Basic algorithm and convergence," *IEEE/ACM Transactions on Networking*, vol. 7, no. 6, pp. 861-874, 1999.

[353] J. Mo and J. Walrand, "Fair end-to-end window-based congestion control," *IEEE/ACM Transactions on Networking*, vol. 8, no. 5, pp. 556-567, 2000.

[354] M. Petrova, P. Mähönen, and J. Riihijärvi, "Evolution of radio resource management: A case for cognitive resource manager with VPI," in *Proceedings of the IEEE International Conference on Communications*, pp. 6471-6475, 2007.

[355] R. W. Thomas, L. A. DaSilva, M. V. Marathe, and K. N. Wood, "Critical design decisions for cognitive networks," in *Proceedings of the IEEE International Conference on Communications*, pp. 3993-3998, 2007.

[356] J. W. Stewart III, *BGP4: Inter-Domain Routing in the Internet*. Boston, MA, USA: Addison-Wesley Longman Publishing Co., 1998.

[357] M. M. Buddhikot, P. Kolodzy, S. Miller, K. Ryan, and J. Evans, "DIMSUMNet: New directions in wireless networking using coordinated dynamic spectrum access," in *Proceedings of the IEEE International Symposium on a World of Wireless Mobile and Multimedia Networks*, 2005.

[358] M. Buddhikot, "Understanding dynamic spectrum access: Taxonomy, models and challenges," in *Proceedings of the IEEE International Symposium on New Frontiers in Dynamic Spectrum Access Networks*, pp. 649-663, 2007.

[359] T. Finin, R. Fritzson, D. McKay, and R. McEntire, "KQML—A language and protocol for knowledge and information exchange," in *Proceedings of the Thirteenth International Workshop on Distributed Artificial Intelligence*, pp. 126-136, 1994.

[360] M. I. Jordan, "Graphical models," *Statistical Science*, vol. 19, no. 1, pp. 140-155, 2004.

[361] E. Meshkova, K. Rerkrai, J. Riihijärvi, and P. Mähönen, "Optimizing component-oriented systems: A case study in wireless sensor networks," in *Demonstration in ACM SIGCOMM 2008*, Seattle, Washington, USA, Aug. 2008.

[362] H. Celebi and H. Arslan, "Utilization of location information in cognitive wireless networks," *IEEE Wireless Communications*, vol. 14, pp. 6-13, Aug. 2007.

[363] A. F. Karr, *Point Processes and Their Statistical Inference* 2nd ed. New York, NY, USA: Marcel Dekker, 1991.

[364] D. Stoyan, W. S. Kendall, and J. Mecke, *Stochastic Geometry and Its Applications*. New York, NY, USA: John Wiley & Sons, 1995.

[365] J. Riihijärvi and P. Mähönen, "Exploiting spatial statistics of primary and secondary users towards improved cognitive radio networks," in *Proceedings of the International Conference on Cognitive Radio Oriented Wireless Networks and Communications*, Singapore China, May 2008.

[366] J. Riihijärvi, P. Mähönen, and M. Rübsamen, "Characterizing wireless networks by spatial correlations," *IEEE Communications Letters*, vol. 11, pp. 37-39, January 2007.

[367] P. Mähönen, M. Petrova, and J. Riihijärvi, "Applications of topology information for cognitive radios and networks," in *Proceedings of the IEEE International Symposium on New Frontiers in Dynamic Spectrum Access Networks*, Dublin, Ireland, April 2007.

[368] M. N. M. van Lieshout and A. J. Baddeley, "A nonparametric measure of spatial interaction in point patterns," *Statistica Neerlandica*, vol. 3, pp. 344-361, 1996.

[369] J. Riihijärvi, P. Mähönen, M. Wellens, and M. Gordziel, "Characterization and modelling of spectrum for dynamic spectrum access with spatial statistics and random fields," in *Proceedings of the First IEEE International Workshop on Cognitive Radios and Networks (CRNETS)*, Cannes, France, Sept. 2008.

[370] M. Wellens, J. Riihijärvi, M. Gordziel, and P. Mähönen, "Evaluation of cooperative spectrum sensing based on large scale measurements," in *Proceedings of the IEEE Symposium on New Frontiers in Dynamic Spectrum Access Networks*, Chicago, IL, USA, Oct. 2008.

[371] N. A. C. Cressie, *Statistics for Spatial Data*. New York, NY, USA: John Wiley & Sons, 1991.

[372] P. Eugster, P. Felber, R. Guerraoui, and A. Kermarrec, "The many faces of publish/subscribe," *ACM Computing Surveys*, vol. 35, no. 2, pp. 114-131, 2003.

[373] D. Tse and P. Viswanath, *Fundamentals of Wireless Communication*. Cambridge, UK: Cambridge University Press, 2005.

[374] E. C. van der Meulen, "Transmission of information in a T-terminal discrete memoryless channel." PhD thesis, Berkeley, CA, USA: Department of Statistics, University of California, 1968.

[375] E. C. van der Meulen, "Three-terminal communication channels," *Advances in Applied Probability*, vol. 3, pp. 120-154, 1971.

[376] T. M. Cover and A. A. El Gamal, "Capacity theorems for the relay channel," *IEEE Transactions on Information Theory*, vol. 25, pp. 572-584, Sept. 1979.

[377] M. R. Aref, "Information flow in relay networks." PhD thesis, Stanford, CA, USA: Stanford University, 1980.

[378] M. Aref and A. A. El Gamal, "The capacity of the semideterministic relay channel," *IEEE Transactions on Information Theory*, vol. 28, p. 536, May 1982.

[379] Z. Zhang, "Partial converse for a relay channel," *IEEE Transactions on Information Theory*, vol. 34, pp. 1106–1110, Sept. 1988.

[380] E. C. van der Meulen and P. Vanroose, "Uniquely decodable code for deterministic relay channels," *IEEE Transactions on Information Theory*, vol. 38, pp. 1203–1212, July 1992.

[381] R. Ahlswede and A. H. Kaspi, "Optimal coding strategies for certain permuting channels," *IEEE Transactions on Information Theory*, vol. 33, pp. 310–314, May 1987.

[382] B. Schein and R. Gallager, "The Gaussian parallel relay network," *Proceedings of the IEEE International Symposium on Information Theory*, 2000.

[383] B. Schein, "Distributed coordination in network information theory." PhD thesis, Cambridge, MA, USA: MIT, 2001.

[384] P. Gupta and P. R. Kumar, "Toward and information theory of large networks: An achievable rate region," *Proceedings of the IEEE International Symposium on Information Theory*, p. 150, June 2001.

[385] P. Gupta and P. R. Kumar, "Toward an information theory of large networks: An achievable rate region," *IEEE Transactions on Information Theory*, vol. 49, pp. 1877–1894, Aug. 2001.

[386] M. Gastpar, G. Kramer, and P. Gupta, "Cooperative strategies and capacity theorems for relay networks," *IEEE Transactions on Information Theory*, May 2004.

[387] M. Gastpar and M. Vetterli, "On the capacity of wireless networks: The relay case," *Proceedings of IEEE INFOCOM*, June 2002.

[388] M. Gastpar and M. Vetterli, "On asymptotic capacity of gaussian relay networks," *Proceedings of the IEEE International Symposium on Information Theory*, p. 195, July 2002.

[389] M. Gastpar, G. Kramer, and P. Gupta, "The multiple-relay channel: Coding and antenna clustering capacity," *Proceedings of the IEEE International Symposium on Information Theory*, p. 136, July 2002.

[390] Z. Yang and A. Høst-Madsen, "Routing and power allocation in asynchronous gaussian multiple-relay channels," *EURASIP Journal on Wireless Communications and Networking*, vol. 2006, no. 6, 2006.

[391] K. G. Seddik, A. K. Sadek, W. Su, and K. J. R. Liu, "Outage analysis and optimal power allocation for multinode relay networks," *IEEE Signal Processing Letters*, vol. 14, pp. 377–380, June 2007.

[392] Y. Fan and J. Thompson, "Decode-and-forward cooperative diversity with power allocation in wireless networks," *IEEE Transactions on Communications*, vol. 6, pp. 793–799, Mar. 2007.

[393] A. Høst-Madsen and J. Zhang, "Capacity bounds and power allocation for the wireless relay channel," *IEEE Transactions on Information Theory*, 2003.

[394] K. K. Wong and E. Elsheikh, "Optimizing time and power allocation for cooperation diversity in a decode-and-forward three-node relay channel," *Journal of Communications*, vol. 3, pp. 43–52, Apr. 2008.

[395] R. Annavajjala, P. C. Cosman, and L. B. Milstein, "Statistical channel knowledge-based optimum power allocation for relaying protocols in the high SNR regime," *IEEE Journal on Selected Areas in Communications*, vol. 25, pp. 292–305, Feb. 2007.

[396] G. Kramer and A. J. van Wijngaarden, "On the white Gaussian multiple-access relay channel," *IEEE Transactions on Information Theory*, vol. 46, p. 40, June 2000.

[397] A. Nosratinia and T. E. Hunter, "Grouping and partner selection in cooperative wireless networks," *IEEE Journal on Selected Areas in Communications*, vol. 25, pp. 369–378, Feb. 2007.

[398] B. Zhao and M. C. Valenti, "Practical relay networks: A generalization of hybrid-ARQ," *IEEE Journal on Selected Areas in Communications*, vol. 23, pp. 7–18, Jan. 2005.

[399] K. Ching Liang, X. Wang, and I. Berenguer, "Minimum error-rate linear dispersion codes for cooperative relays," *IEEE Transactions on Vehicular Technology*, vol. 6, pp. 2143–2157, July 2007.

[400] Y. Li and X.-G. Xia, "A family of distributed space-time trellis codes with asynchronous cooperative diversity," *IEEE Transactions on Communications*, vol. 55, pp. 790–800, Apr. 2007.

[401] J. Castura and Y. Mao, "Rateless coding for wireless relay channels," *IEEE Transactions on Wireless Communications*, vol. 6, pp. 1638–1642, May 2007.

[402] P. Liu, Z. Tao, Z. Lin, E. Erkip, and S. Panwar, "Cooperative wireless communications: A cross-layer approach," *IEEE Wireless Communications*, vol. 13, pp. 84–92, Aug. 2006.

[403] R. C. King, "Multiple-access channels with generalized feedback." PhD thesis, Stanford, CA, USA: Stanford University, 1978.

[404] A. B. Carleial, "Multiple-access channels with different generalized feedback signals," *IEEE Transactions on Information Theory*, vol. 28, pp. 841–850, Nov. 1982.

[405] F. M. J. Willems, "Information theoretical results for the discrete memoryless multiple-access channel." PhD thesis, Leuven, Belgium: Katholieke University, 1982.

[406] F. J. Willems, E. C. van der Meulen, and J. P. M. Schalkwijk, "An achievable rate region for the multiple-access channel with generalized feedback," *Proceedings of the Allerton Conference on Communications, Control and Computing*, pp. 284–292, Oct. 1983.

[407] F. J. Willems, "An achievable rate region for the multiple-access channel with generalized feedback," *IEEE Transactions on Information Theory*, vol. 28, pp. 841–850, Nov. 1982.

[408] F. J. Willems and E. C. van der Meulen, "The discrete memoryless multiple-access channel with cribbing encoders," *IEEE Transactions on Information Theory*, vol. 31, pp. 313–327, May 1985.

[409] J. N. Laneman and G. W. Wornell, "Distributed space-time coded protocols for exploiting cooperative diversity in wireless networks," *IEEE Transactions on Information Theory*, vol. 49, pp. 2415–2425, Oct. 2003.

[410] A. Sendonaris, E. Erkip, and B. Aazhang, "User cooperation diversity—Part I: System description," *IEEE Transactions on Communications*, vol. 51, pp. 1927–1938, Nov. 2003.

[411] A. Sendonaris, E. Erkip, and B. Aazhang, "User cooperation diversity—Part II: Implementation aspects and performance analysis," *IEEE Transactions on Communications*, vol. 51, pp. 1939–1948, Nov. 2003.

[412] J. N. Laneman, D. Tse, and G. W. Wornell, "Cooperative diversity in wireless networks: Efficient protocols and outage behavior," *IEEE Transactions on Information Theory*, vol. 50, pp. 3062–3080, Dec. 2004.

[413] A. Nosratinia and T. E. Hunter, "Diversity through coded cooperation," *IEEE Transactions on Wireless Communications*, vol. 5, pp. 283–289, Feb. 2006.

[414] T. E. Hunter, S. Sanayei, and A. Nosratinia, "Outage analysis of coded cooperation," *IEEE Transactions on Information Theory*, vol. 52, pp. 375–391, Feb. 2006.

[415] R. U. Nabar, H. Bolcskei, and F. W. Kneubuhler, "Fading relay channels: Performance limits and space-time signal design," *IEEE Journal on Selected Areas in Communications*, vol. 22, pp. 1099–1109, Aug. 2004.

[416] A. Nosratinia, T. E. Hunter, and A. Hedayat, "Cooperative communication in wireless networks," *IEEE Communications Magazine*, vol. 42, pp. 74–80, Oct. 2004.

[417] K. K. Wong and E. Elsheikh, "Optimized cooperative diversity for a three-node decode-and-forward relay channel," *Proceedings of the IEEE International Symposium on Wireless Pervasive Computing*, Feb. 2007.

[418] K. K. Wong and E. Elsheikh, "Unleashing the full potential of relaying," *Proceedings of the London Communications Symposium*, Sept. 2007.

[419] E. Elsheikh and K. K. Wong, "Wireless cooperative networks: Partnership selection and fairness," *Proceedings of the IFIP Wireless Days Conference*, Nov. 2008.

[420] A. S. Ibrahim, A. K. Sadek, W. Su, and K. J. R. Liu, "Cooperative communications with relay-selection: When to cooperate and whom to cooperate with," *IEEE Transactions on Wireless Communications*, vol. 7, pp. 2814-2827, July 2008.

[421] A. S. Ibrahim, A. K. Sadek, W. Su, and K. J. R. Liu, "Cooperative communications with partial channel state information: When to cooperate," *Proceedings of the IEEE Global Telecommunications Conference*, pp. 3068-3072, Nov. 2005.

[422] L. Zheng and D. N. C. Tse, "Diversity and multiplexing: A fundamental tradeoff in multiple-antenna channels," *IEEE Transactions on Information Theory*, vol. 49, pp. 1073-1096, May 2003.

[423] M. Yuksel and E. Erkip, "Diversity-multiplexing tradeoff in cooperative wireless systems," *Conference on Information Sciences and Systems*, pp. 1062-1067, Mar. 2006.

[424] Q. Zhang, J. Jia, and J. Zhang, "Cooperative relay to improve diversity in cognitive radio networks," *IEEE Communications Magazine*, vol. 47, pp. 111-117, Feb. 2009.

[425] S. Zahedi, "On reliable communication over the relay channel." PhD thesis, Stanford, USA: Stanford University, 2005.

[426] C. E. Shannon, "A mathematical theory of communication," *Bell Systems Technical Journal*, vol. 27, pp. 379-423, 623-656, July-Oct. 1948.

[427] A. J. Goldsmith and P. P. Varaiya, "Capacity of fading channels with channel side information," *IEEE Transactions on Information Theory*, vol. 43, pp. 1986-1992, Nov. 1997.

[428] M. Veluppillai, "Cooperative diversity and partner Selection in wireless networks." PhD thesis, Waterloo, ON, Canada: University of Waterloo, 2007.

[429] R. Y. Rubinstein and D. P. Kroese, *The Cross-Entropy Method: A Unified Approach to Combinatorial Optimization, Monte-Carlo Simulation and Machine Learning*. New York, NY, USA: Springer-Verlag, 2004.

[430] P.-T. de Boer, D. Kroese, S. Mannor, and R. Rubinstein, "A tutorial on the cross-entropy method," *Annals of Operations Research*, vol. 134, pp. 19-67, Jan. 2005.

[431] J. Kennedy and R. C. Eberhart, *Swarm Intelligence*. San Mateo, CA, USA: Morgan Kaufmann, 2001.

[432] E. Khandani, E. Modiano, J. Abounadi, and L. Zheng, "Reliability and route diversity in wireless networks," *Proceedings of the Conference on Information Sciences and Systems*, Mar. 2005.

[433] L. L. Xie and P. R. Kumar, "A network information theory for wireless communication: Scaling laws and optimal operation," *IEEE Transactions on Information Theory*, vol. 50, pp. 748-767, May 2004.

[434] L. L. Xie and P. R. Kumar, "An achievable rate for the multiple level relay channel," *IEEE Transactions on Information Theory*, vol. 51, pp. 1348-1358, Apr. 2005.

[435] G. Kramer, I. Maric, and R. Yates, "Cooperative communications," *Foundations and Trends in Networking*, vol. 1, no. 3-4, pp. 271-425, 2006.

[436] I. E. Telatar, "Capacity of multi-antenna Gaussian channels," *European Transactions on Telecommunications*, vol. 10, pp. 585-595, Nov.-Dec. 1999.

[437] A. Narula, M. D. Trott, and G. W. Wornell, "Performance limits of coded diversity methods for transmitter antenna arrays," *IEEE Transactions on Information Theory*, vol. 45, pp. 2418-2433, 1999.

[438] S. Alamouti, "A simple transmit diversity technique for wireless communications," *IEEE Journal on Selected Areas in Communications*, vol. 16, pp. 1451-1458, Oct. 1998.

[439] G. Foschini and M. Gans, "On limits of wireless communication in a fading environment when using multiple antennas," *Wireless Personal Communications*, vol. 6, pp. 311-335, Mar. 1998.

[440] V. Tarokh, N. Seshadri, and A. R. Calderbank, "Space-time codes for high data rate wireless communication: Performance criterion and code construction," *IEEE Transactions on Information Theory*, vol. 6, pp. 744-765, Mar. 1998.

[441] M. Debbah and R. R. Muller, "MIMO channel modeling and the principle of maximum entropy," *IEEE Transactions on Information Theory*, vol. 51, no. 5, pp. 1667-1690, 2005.

[442] Y. Fan and J. Thompson, "MIMO configurations for relay channels: Theory and practice," *IEEE Transactions on Wireless Communications*, vol. 6, pp. 1774-1786, May 2007.

[443] S. Cui, A. J. Goldsmith, and A. Bahai, "Energy-efficiency of MIMO and cooperative MIMO techniques in sensor networks," *IEEE Journal on Selected Areas in Communications*, vol. 22, pp. 1089-1098, Aug. 2004.

[444] B. Wang and J. Zhang, "MIMO relay channel and its application for cooperative communication in ad hoc networks," *Proceedings of the Allerton Conference on Communication, Control and Computing*, vol. 41, no. 3, pp. 1556-1565, 2003.

[445] Federal Communications Commission. [Online]: www.fcc.gov/oet/cognitiveradio/.

[446] Federal Communications Commission, "Secondary markets initiative." [Online]: wireless. fcc.gov/licensing/secondarymarkets/.

[447] N. Devroye, P. Mitran, and V. Tarokh, "Cognitive decomposition of wireless networks," in *Proceedings of the International Conference on Cognitive Radio Oriented Wireless Networks and Communications*, Mykonos Island, Greece, Mar. 2006.

[448] N. Devroye, P. Mitran, M. Sharif, S. S. Ghassemzadeh, and V. Tarokh, "Information theoretic analysis of cognitive radio systems," in *Cognitive Wireless Communication Networks* (V. Bhargava and E. Hossain, eds.). New York, NY, USA: Springer, 2007.

[449] N. Devroye and V. Tarokh, "Fundamental limits of cognitive radio networks," in *Cognitive Wireless Networks: Concepts, Methodologies and Vision* (F. H. P. Fitzek and M. Katz, eds.). Dordrecht, The Netherlands: Springer, 2007.

[450] A. Goldsmith, S. A. Jafar, I. Maric, and S. Srinivasa, "Breaking spectrum gridlock with cognitive radios: An information theoretic perspective," *Proceedings of the IEEE, 2008*, vol. 97, no. 5, May 2009.

[451] A. F. Molisch, *Wireless Communications*. Hoboken, NJ, USA: John Wiley & Sons, 2005.

[452] R. W. Yeung, *A First Course in Information Theory*. New York, NY, USA: Springer, 2002.

[453] I. Csiszár and J. Körner, *Information Theory: Coding Theorems for Discrete Memoryless Systems*. New York, NY, USA: Academic Press, 1981.

[454] A. B. Carleial, "Interference channels," *IEEE Transactions on Information Theory*, vol. IT-24, pp. 60-70, Jan. 1978.

[455] H. Sato, "Two user communication channels," *IEEE Transactions on Information Theory*, vol. IT-23, pp. 295-304, May 1977.

[456] R. Etkin, D. Tse, and H. Wang, "Gaussian interference channel capacity to within one bit." [Online]: www.citebase.org/abstract?id=oai:arXiv.org:cs/0702045, 2007.

[457] G. Kramer, "Outer bounds on the capacity of Gaussian interference channels," *IEEE Transactions on Information Theory*, vol. 50, Mar. 2004.

[458] X. Shang, G. Kramer, and B. Chen, "A new outer bound and the noisy-interference sum-rate capacity for gaussian interference channels," *IEEE Transactions on Information Theory*, vol. 55, no. 2, pp. 689-699, Feb. 2009.

[459] H. Chong, M. Motani, H. Garg, and H. El Gamal, "On the Han-Kobayashi region for the interference channel," in *IEEE Transactions on Information Theory*, vol. 54, no. 7, pp. 3188-3195, July 2008.

[460] M. Grossglauser and D. Tse, "Mobility increases the capacity of ad hoc wireless networks," *IEEE/ACM Transactions on Networking*, vol. 10, pp. 477-486, Aug. 2002.

[461] L.-L. Xie and P. R. Kumar, "On the path-loss attenuation regime for positive cost and linear scaling of transport capacity in wireless networks," *IEEE Transactions on Information Theory*, vol. 52, pp. 2313-2328, June 2006.

[462] S. R. Kulkarni and P. Viswanath, "A deterministic approach to throughput scaling in wireless networks," *IEEE Transactions on Information Theory*, vol. 50, pp. 1041–1049, June 2004.

[463] A. Agarwal and P. R. Kumar, "Capacity bounds for ad hoc and hybrid wireless networks," *ACM SIGCOMM Computer Communication Review*, vol. 34, no. 3, 2004.

[464] O. Leveque and I. E. Telatar, "Information-theoretic upper bounds on the capacity of large extended ad hoc wireless networks," *IEEE Transactions on Information Theory*, vol. 51, pp. 858–865, Mar. 2005.

[465] S. Ahmad, A. Jovicic, and P. Viswanath, "On outer bounds to the capacity region of wireless networks," *IEEE Transactions on Information Theory*, vol. 52, pp. 2770–2776, June 2006.

[466] A. Jovicic, S. R. Kulkarni, and P. Viswanath, "Upper bounds to transport capacity of wireless networks," *IEEE Transactions on Information Theory*, vol. 50, pp. 2555–2565, Nov. 2004.

[467] F. Xue and P. R. Kumar, *Scaling Laws for Ad Hoc Wireless Networks: An Information Theoretic Approach*. Hanover, MA, USA: NOW Publishers, 2006.

[468] M. Franceschetti, O. Dousse, D. Tse, and P. Thiran, "Closing the gap in the capacity of wireless networks via percolation theory," *IEEE Transactions on Information Theory*, vol. 53, pp. 1009–1018, Mar. 2007.

[469] A. Ozgur, O. Leveque, and E. Preissmann, "Scaling laws for one- and two-dimensional random wireless networks in the low attenuation regime," *IEEE Transactions on Information Theory*, vol. 53, no. 10, pp. 3573–3585, Oct. 2007.

[470] A. Ozgur, O. Leveque, and D. Tse, "Hierarchical cooperation achieves optimal capacity scaling in ad hoc networks," in *IEEE Transactions on Information Theory*, vol. 53, no. 10, pp. 3549–3572, Oct. 2007.

[471] A. S. Avestimehr, S. N. Diggavi, and D. N. C. Tse, "Approximate capacity of Gaussian relay networks." [Online]: http://arxiv.org/abs/0802.3535.

[472] T. Cover, A. El Gamal, and M. Salehi, "Multiple access channels with arbitrarily correlated sources," *IEEE Transactions on Information Theory*, vol. IT-26, pp. 648–657, Nov. 1980.

[473] B. Rimoldi and R. Urbanke, "A rate-splitting approach to the Gaussian multiple-access channel," *IEEE Transactions on Information Theory*, vol. 42, no. 2, pp. 364–375, 1996.

[474] K. Marton, "A coding theorem for the discrete memoryless broadcast channel," *IEEE Transactions on Information Theory*, vol. 25, pp. 306–311, May 1979.

[475] T. Cover, "Broadcast channels," *IEEE Transactions on Information Theory*, vol. 18, pp. 2–14, Jan. 1972.

[476] H. Weingarten, Y. Steinberg, and S. Shamai, "The capacity region of the Gaussian MIMO broadcast channel," *IEEE Transactions on Information Theory*, vol. 52, no. 9, pp. 3936–3964, Sept. 2006.

[477] M. Mohseni and J. M. Cioffi, "A proof of the converse for the capacity of Gaussian MIMO broadcast channels," *Proceedings of the IEEE International Sympoisum on Information Theory*, July 2006.

[478] G. Caire and S. Shamai, "On the achievable throughput of a multi-antenna Gaussian broadcast channel," *IEEE Transactions on Information Theory*, vol. 49, pp. 1691–1705, July 2003.

[479] H. Sato, "The capacity of Gaussian interference channel under strong interference," *IEEE Transactions on Information Theory*, vol. 27, Nov. 1981.

[480] M. Costa and A. E. Gamal, "The capacity region of the discrete memoryless interference channel with strong interference," *IEEE Transactions on Information Theory*, vol. 33, pp. 710–711, Sept. 1987.

[481] T. S. Han and K. Kobayashi, "A new achievable rate region for the interference channel," *IEEE Transactions on Information Theory*, vol. 27, pp. 49–60, Jan. 1981.

[482] C. Nair and H. El Gamal, "An outer bound to the capacity region of the broadcast channel," *IEEE Transactions on Information Theory*, vol. 53, pp. 350–355, Jan. 2007.

[483] I. Sason, "On achievable rate regions for the Gaussian interference channel," *IEEE Transactions on Information Theory*, vol. 50, pp. 1345–1356, June 2004.

[484] S. A. Jafar and M. Fakhereddin, "Degrees of freedom for the MIMO interference channel," in *Proceedings of the IEEE International Symposium on Information Theory*, Seattle, WA, USA, July 2006.

[485] P. A. Parker, D. W. Bliss, and V. Tarokh, "On the degrees-of-freedom of the mimo interference channel," in *Proceedings of the Annual Conference on Information Sciences and Systems*, Princeton, NJ, USA, Mar. 2008.

[486] S. A. Jafar and S. Srinivasa, "Capacity limits of cognitive radio with distributed and dynamic spectral activity," *IEEE Journal on Selected Areas in Communications*, vol. 25, pp. 529–537, Apr. 2007.

[487] S. Srinivasa, S. A. Jafar, and N. Jindal, "On the capacity of the cognitive tracking channel," *Proceedings of the IEEE International Symposium on Information Theory*, July 2006.

[488] O. Simeone, Y. Bar-Ness, and U. Spagnolini, "Stable throughput of cognitive radios with and without relaying capability," *IEEE Transactions on Communications*, vol. 55, pp. 2351–2360, Dec. 2007.

[489] A. Tkachenko, "Testbed design for cognitive radio spectrum sensing experiments." PhD thesis, University of California Berkeley, USA, 2007.

[490] S. M. Mishra, A. Sahai, and R. W. Brodersen, "Cooperative sensing among cognitive radios," in *Proceedings of the IEEE International Conference on Communications*, Istanbul, Turkey, June 2006.

[491] G. Ganesan and Y. Li, "Cooperative spectrum sensing in cognitive radio networks," in *Proceedings of the IEEE International Symposium on New Frontiers in Dynamic Spectrum Access Networks*, Baltimore, MD, USA, Nov. 2005.

[492] C. R. C. da Silva, B. Choi, and K. Kim, "Distributed spectrum sensing for cognitive radio systems," in *Proceedings of the Information Theory and Applications Workshop*, La Jolla, CA, USA, Feb. 2007.

[493] R. Zhang and Y. Liang, "Exploiting multi-antennas for opportunistic spectrum sharing in cognitive radio networks." [Online]: http://front.math.ucdavis.edu/0711.4414.

[494] L. Zhang, Y. Xin, and Y.-C. Liang, "Power allocation for multi-antenna multiple access channels in cognitive radio networks," in *Proceedings of the Annual Conference on Information Sciences and Systems*, Princeton, NJ, USA, Mar. 2007.

[495] M. H. Islam, Y.-C. Liang, and A. T. Hoang, "Joint power control and beamforming for secondary spectrum sharing," in *Proceedings of the 66th IEEE Vehicular Technology Conference*, Oct. 2007.

[496] S. Yiu, M. Vu, and V. Tarokh, "Interference reduction by beamforming in cognitive networks," in *Proceedings of the IEEE Global Telecommunications Conference*, Nov. 2008.

[497] L. Zhang, Y.-C. Liang, Y. Xin, and H. V. Poor, "Robust cognitive beamforming with partial channel state information." [Online]: http://front.math.ucdavis.edu/0711.4414.

[498] P. J. Kolodzy, "Interference temperature: A metric for dynamic spectrum utilization," *International Journal of Network Management*, pp. 103–113, Mar. 2006.

[499] M. C. Gastpar, "On capacity under receive and spatial spectrum-sharing constraints," *IEEE Transactions on Information Theory*, vol. 53, pp. 471–487, Feb. 2007.

[500] A. Ghasemi and E. S. Sousa, "Capacity of fading channels under spectrum-sharing constraints," in *Proceedings of the IEEE International Conference on Communications*, Istanbul, Turkey, June 2006.

[501] W. Weng, T. Peng, and W. Wang, "Optimal power control under interference temperature constraints in cognitive radio network," in *Proceedings of the IEEE Wireless Communications and Networking Conference*, Hong Kong, China, Mar. 2007.

[502] Y. Xing, C. N. Marthur, M. A. Haleem, R. Chandramouli, and K. P. Subbalakshmi, "Dynamic spectrum access with QoS and interference temperature constraints," *IEEE Transactions on Mobile Computing*, vol. 6, pp. 423–433, Apr. 2007.

[503] S. Verdu, *Multiuser Detection*, 3rd ed. Cambridge, UK: Cambridge University Press, 2003.

[504] M. Vu and V. Tarokh, "Scaling laws of single-hop cognitive networks," *IEEE Transactions on Wireless Communications*, vol. 8, no. 8, pp. 4089–4097, Aug. 2009.

[505] M. Vu, N. Devroye, and V. Tarokh, "The primary exclusive regions in cognitive networks," *IEEE Transactions on Wireless Communications*, vol. 8, no. 7, pp. 3380–3385, July 2009.

[506] N. Hoven and A. Sahai, "Power scaling for cognitive radio," *Proceedings of the International Conference on Wireless Networks, Communications and Mobile Computing*, vol. 1, pp. 250–255, June 2005.

[507] S.-W. Jeon, N. Devroye, M. Vu, S.-Y. Chung, and V. Tarokh, "Cognitive networks achieve throughput scaling of a homogeneous network," Submitted to *IEEE Transactions on Information Theory*, Mar. 2008.

[508] P. Popovski, H. Yomo, K. Nishimori, R. D. Taranto, and R. Prasad, "Opportunistic interference cancellation in cognitive radio systems," in *Proceedings of the IEEE International Symposium on New Frontiers in Dynamic Spectrum Access Networks*, Dublin, Ireland, Apr. 2007.

[509] N. Devroye, P. Mitran, and V. Tarokh, "Achievable rates in cognitive radio channels," in *Proceedings of the Annual Conference on Information Sciences and Systems*, Princeton, NJ, USA, Mar. 2005.

[510] W. Wu, S. Vishwanath, and A. Arapostathis, "Capacity of a class of cognitive radio channels: Interference channels with degraded message sets," *IEEE Transactions on Information Theory*, vol. 53, pp. 4391–4399, June 2007.

[511] S. Gel'fand and M. Pinsker, "Coding for channels with random parameters," *Probability, Control and Information Theory*, vol. 9, no. 1, pp. 19–31, 1980.

[512] A. Kusnetsov and B. Tsybakov, "Coding in a memory with defective cells," *Problemy Peredachi Informatsii*, vol. 10, pp. 52–60, Apr.–June 1974.

[513] C. Heegard and A. El Gamal, "On the capacity of computer memories with defects," *IEEE Transactions on Information Theory*, vol. 29, pp. 731–739, Sept. 1983.

[514] Y. Steinberg and N. Merhav, "Identification in the presence of side information with application to watermarking," *IEEE Transactions on Information Theory*, vol. 47, pp. 1410–1422, May 2001.

[515] I. Maric, R. D. Yates, and G. Kramer, "Capacity of interference channels with partial transmitter cooperation," *IEEE Transactions on Information Theory*, vol. 53, pp. 3536–3548, Oct. 2007.

[516] J. Jiang and X. Xin, "On the achievable rate regions for interference channels with degraded message sets," *IEEE Transactions on Information Theory*, vol. 54, pp. 4707–4712, Oct. 2008.

[517] A. Jovicic and P. Viswanath, "Cognitive radio: An information-theoretic perspective," *IEEE International Symposium on Information Theory*, Seattle, WA, USA, July 2006.

[518] J. Jiang, X. Xin, and H. Garg, "Interference channels with common information," *IEEE Transactions on Information Theory*, vol. 54, pp. 171–187, Jan. 2008.

[519] Y. Liang, A. Somekh-Baruch, H. V. Poor, S. Shamai (Shitz), and S. Verdu, "Capacity of cognitive interference channels with and without secrecy," *IEEE Transactions on Information Theory*, vol. 55, pp. 604–619, Feb. 2009.

[520] W. Wu, S. Vishwanath, and A. Ariposthathis, "On the capacity of interference channel with degraded message sets," *IEEE Transactions on Information Theory*, vol. 53, pp. 4391–4399, Nov. 2007.

[521] N. Devroye, "Information theoretic limits of cooperation and cognition in wireless networks." PhD thesis, Harvard University, Cambridge, MA, USA, 2007.

[522] S. Yang and D. Tuninetti, "A new achievable region for interference channel with generalized feedback," in *Proceedings of the Annual Conference on Information Sciences and Systems*, Princeton, NJ, USA, Mar. 2008.

[523] O. O. Koyluoglu and H. El Gamal, "On power control and frequency re-use in the two-user cognitive channel," *IEEE Transactions on Wireless Communications*, vol. 8, no. 7, pp. 3546–3553, July 2009.

[524] S. H. Seyedmehdi, Y. Xin, and Y. Lian, "An achievable rate region for the causal cognitive radio," in *Proceedings of the Allerton Conference on Communications, Control and Computing*, Monticello, IL, USA, Sept. 2007.

[525] P. Mitran, N. Devroye, and V. Tarokh, "On compound channels with side-information at the transmitter," *IEEE Transactions on Information Theory*, vol. 52, pp. 1745–1755, Apr. 2006.

[526] P. Grover and A. Sahai, "What is needed to exploit knowledge of primary transmissions?" [Online]: www.citebase.org/abstract?id=oai:arXiv.org:cs/0702071, 2007.

[527] N. Devroye, P. Mitran, and V. Tarokh, "Achievable rates in cognitive networks," in *Proceedings of the IEEE International Symposium on Information Theory*, Adelaide, Australia, Sept. 2005.

[528] P. Cheng, G. Yu, Z. Zhang, H.-H. Chen, and P. Qiu, "On the maximum sum-rate capacity of cognitive multiple access channel." [Online]: http://arxiv.org/abs/cs.IT/0612024.

[529] M. Maddah-Ali, A. S. Motahari, and A. K. Khandani, "Signaling over MIMO multi-base systems: Combination of multi-access and broadcast schemes," in *Proceedings of the IEEE International Symposium on Information Theory*, Seattle, WA, USA, July 2006.

[530] N. Devroye and M. Sharif, "The multiplexing gain of MIMO X-channels with partial transmit side-information," in *Proceedings of the IEEE International Symposium on Information Theory*, Nice, France, June 2007.

[531] S. A. Jafar and S. Shamai, "Degrees of freedom region for the MIMO X channel," *IEEE Transactions on Information Theory*, vol. 54, pp. 151–170, Jan. 2008.

[532] A. Host-Madsen and A. Nosratinia, "The multiplexing gain of wireless networks," in *Proceedings of the IEEE International Symposium on Information Theory*, Adelaide, Australia, Sept. 2005.

[533] C. Huang and S. A. Jafar, "Degrees of freedom on the MIMO interference channel with cooperation and cognition." [Online]: http://newport.eecs.uci.edu/syed/newpage/resume.html#publicationslistl.

[534] M. Alicherry, R. Bhatia, and L. Li, "Joint channel assignment and routing for throughput optimization in multi-radio wireless mesh networks," in *Proceedings of ACM MobiCom*, Cologne, Germany, pp. 58–72, Aug. 2005.

[535] R. Draves, J. Padhye, and B. Zill, "Routing in multi-radio, multi-hop wireless mesh networks," in *Proceedings of ACM MobiCom*, Philadelphia, PA, USA, pp. 114–126, Sept. 2004.

[536] M. Kodialam and T. Nandagopal, "Characterizing the capacity region in multi-radio multi-channel wireless mesh networks," in *Proceedings of ACM MobiCom*, Cologne, Germany, pp. 73–87, Aug. 2005.

[537] P. Kyasanur and N. H. Vaidya, "Capacity of multi-channel wireless networks: Impact of number of channels and interfaces," in *Proceedings of ACM MobiCom*, Cologne, Germany, pp. 43–57, Aug. 2005.

[538] A. Raniwala and T. Chiueh, "Architecture and algorithms for an IEEE 802.11-based multi-channel wireless mesh network," in *Proceedings of IEEE INFOCOM*, Miami, FL, USA, pp. 2223–2224, Mar. 2005.

[539] Y. Shi and Y. T. Hou, "Optimal power control for multi-hop software defined radio networks," in *Proceedings of IEEE INFOCOM*, Anchorage, AK, USA, pp. 1694–1702, May 2007.

[540] H. D. Sherali and W. P. Adams, *A Reformulation-Linearization Technique for Solving Discrete and Continuous Nonconvex Problems*. Norwood, MA, USA: Kluwer Academic Publishers, 1999.

[541] M. R. Garey and D. S. Johnson, *Computers and Intractability: A Guide to the Theory of NP-Completeness*. New York, NY, USA: W. H. Freeman and Company, 1979.

[542] G. L. Nemhauser and L. A. Wolsey, *Integer and Combinatorial Optimization*. New York, NY, USA: John Wiley & Sons, 1999.

[543] A. J. Goldsmith and S.-G. Chua, "Adaptive coded modulation for fading channels," *IEEE Transactions on Communications*, vol. 46, pp. 595–602, May 1998.

[544] P. Pawełczak, S. Pollin, H.-S. W. So, A. Motamedi, A. Bahai, R. V. Prasad, and R. Hekmat, "State of the art in opportunistic spectrum access MAC design: Research challenges," in *Proceedings of the International Conference on Cognitive Radio Oriented Wireless Networks and Communications*, Singapore, China, May 2008.

[545] R. V. Prasad, P. Pawełczak, J. Hoffmeyer, and S. Berger, "Cognitive functionality in next generation wireless networks: Standardization efforts," *IEEE Communications Maagazine*, vol. 46, pp. 72–78, Apr. 2007.

[546] E. Noam, "Taking the next step beyond spectrum auctions: Open spectrum access," *IEEE Communications Magazine*, vol. 33, pp. 66–73, Dec. 1995.

[547] C. Jackson, "Dynamic sharing of radio spectrum: A brief history," in *Proceedings of the IEEE International Symposium on New Frontiers in Dynamic Spectrum Access Networks*, Baltimore, MA, USA, Nov. 2005.

[548] P. Pawełczak, R. Venkatesha Prasad, L. Xia, and I. Niemegeers, "Cognitive radio emergency networks-requirements and design," in *Proceedings of the IEEE International Symposium on New Frontiers in Dynamic Spectrum Access Networks*, Baltimore, MA, USA, Nov. 2005.

[549] Institute of Electrical and Electronics Engineering, Inc., "IEEE Communications Society—Technical Committee on Cognitive Networks." [Online]: www.eecs.ucf.edu/tccn/, 2008.

[550] J. Hoffmeyer, D. Stewart, S. Berger, B. Eydt, F. Frantz, F. Granelli, K. Kontson, K. Nolan, P. Pawełczak, R. V. Prasad, R. Roy, D. Swain, and P. Tenhula, "Standard definitions and concepts for dynamic spectrum access—Terminology relating to emerging wireless networks, system functionality, and spectrum management," tech. rep., IEEE 1900.1 Standard, Oct. 2008.

[551] Scientific American, "Cognitive radio: Overview/intelligent radios." [Online]: www.sciam.com/article.cfm?id=0005A161-2620-13F6-A62083414B7F0101, 2006.

[552] T. W. Rondeau and C. W. Bostian, "Cognitive techniques: physical and link layers," in *Cognitive Radio Technology* (B. A. Fette, ed.), New York, NY, USA: Elsevier Science and Technology Books, 2006.

[553] S. Pollin, "Coexistence and dynamic sharing in cognitive radio networks," in *Cognitive Wireless Communication Networks* (E. Hossain and V. K. Bhargava, eds.), New York, NY, USA: Springer, 2007.

[554] P. Pawełczak, S. Pollin, H.-S. W. So, A. Bahai, R. V. Prasad, and R. Hekmat, "Quality of service of opportunistic spectrum access: A medium access control approach," *IEEE Wireless Communications*, vol. 15, Oct. 2008.

[555] P. G. Cook and W. Bonser, "Architectural overview of the SPEAKeasy system," *IEEE Journal on Selected Areas in Communications*, vol. 17, pp. 650–661, Apr. 1999.

[556] J. Place, D. Kerr, and D. Schaefer, "Joint tactical radio system," in *Proceedings of the Military Communications Conference*, Los Angeles, CA, USA, Oct. 2000.

[557] M. McHenry, E. Livsics, T. Nguyen, and N. Majumdar, "XG dynamic spectrum access field test results," *IEEE Communications Magazine*, vol. 45, pp. 51–57, June 2007.

[558] P. Camana, "Integrated communications, navigation, identification avionics (ICNIA)—The next generation," *IEEE Aerospace and Electronic Systems Magazine*, vol. 3, pp. 23–26, Aug. 1988.

[559] F. Granelli, P. Pawełczak, R. Venkatesha Prasad, K. P. Subbalakshmi, and R. Chandramouli, "Open issues in research and IEEE SCC41 standardization activities in the framework of cognitive communications," *IEEE Communications Magazine*, July 2008. Submitted.

[560] Institute of Electrical and Electronic Engineerings, Inc., "Functional requirements for IEEE 802.22 WRAN standard." 802.22/05−0007r46, Sept. 2006.

[561] Internation Telecommunication Union, "Method for point-to-area prediction for terrestrial services in the frequency range 30 MHz to 3000 MHz." ITU-R P.1546-1, Oct. 2005.

[562] Federal Communications Commission, "Federal Communications Commission (1996) telecommunications: Glossary of telecommunications terms." [Online]: www.its.bldrdoc.gov/fs-1037/dir-017/_2541.htm, June 2004.

[563] O. Ileri and N. B. Mandayam, "Dynamic spectrum access models: Toward an engineering perspective in the spectrum debate," *IEEE Communications Magazine*, Jan. 2008.

[564] International Telecommunications Union, "New broadband statistics." [Online]: www.itu.int/osg/spu/newslog/ITUs+New+Broadband+Statistics+For+1+January+2005.aspx, Jan. 2005.

[565] D. Cavalcanti and M. Ghosh, "Cognitive radio networks: Enabling new wireless broadband opportunities," *Proceedings of the International Conference on Cognitive Radio Oriented Wireless Networks and Communications*, May 2008.

[566] Institute of Electrical and Electronic Engineerings, Inc., "IEEE 802.16 working group on broadband wireless standards." [Online]: http://wirelessman.org.

[567] A. Mody, "Protocol reference model enhancements in 802.22." IEEE 22-08-0121-06, July 2008.

[568] S. Shellhammer, V. Tawil, G. Chouinard, M. Muterspaugh, and M. Ghosh, "Spectrum sensing simulation model." IEEE 802.22-06/0028r6, June 2006.

[569] G. Chouinard, "WRAN reference model spreadsheet." IEEE 802.22-04-0002r12, June 2006.

[570] Advanced Television Systems Committee, "ATSC A74 recommended practice guideline document entitled: ATSC recommended practice: Receiver performance guidelines." Sections 4.5.2 & 4.5.3, ATSC A74, June 2004.

[571] S. Kay, *Fundamentals of Statistical Signal Processing: Detection Theory*. Upper Saddle River, NJ, USA: Prentice-Hall, 1998.

[572] S. Shellhammer, "Performance of the power detector." IEEE 802.22-06/0075r0, May 2006.

[573] S. Shellhammer, S. Shankar N, R. Tandra, and J. Tomcik, "Performance of power detector sensors of DTV signals in IEEE 802.22 WRANs," *Proceedings of the ACM International Workshop on Technology and Policy for Accessing Spectrum*, Aug. 2007.

[574] Institute of Electrical and Electronic Engineerings, Inc., "IEEE 802.16h task group." [Online]: http://wirelessman.org/le/.

[575] Institute of Electrical and Electronic Engineerings, Inc., "IEEE standards coordinating committee 41." [Online]: http://www.scc41.org/.

[576] S. Shankar, C. Cordeiro, and K. Challapali, "Spectrum agile radios: Utilization and sensing architectures," in *Proceedings of the IEEE International Symposium on New Frontiers in Dynamic Spectrum Access Networks*, pp. 160–169, Nov. 2005.

[577] H. Chan, A. Perrig, and D. Song, "Secure hierarchical in-network aggregation in sensor networks," in *Proceedings of the 13th ACM Conference on Computer and Communications Security*, pp. 278–287, Oct.–Nov. 2006.

[578] W. Du, J. Deng, Y. Han, and P. Varshney, "A witness-based approach for data fusion assurance in wireless sensor networks," in *Proceedings of the IEEE Global Telecommunications Conference*, pp. 1435–1439, Dec. 2003.

[579] Y. Yang, X. Wang, S. Zhu, and G. Cao, "SDAP: A secure hop-by-hop data aggregation protocol for sensor networks," in *Proceedings of the ACM International Symposium on Mobile Ad Hoc Networking and Computing*, pp. 356–367, 2006.

[580] B. Przydatek, D. Song, and A. Perrig, "SIA: Secure information aggregation in sensor networks," in *Proceedings of the First International Conference on Embedded Networked Sensor Systems*, 2003.

[581] L. Hu and D. Evans, "Secure aggregation for wireless networks," in *Proceedings of the Workshop on Security and Assurance in Ad Hoc Networks*, 2003.

[582] IEEE 802.22 Working Group, "IEEE P802.22/D0.1 draft standard for wireless regional area networks part 22: Cognitive wireless ran medium access control (MAC) and physical layer (PHY) specifications: Policies and procedures for operation in the TV bands." IEEE Document: 22-06-0067-00-0000_P802-22_D0.1, May 2006.

[583] D. Grandblaise and W. Hu, "Inter base stations adaptive on demand channel contention for IEEE 802.22 WRAN self-coexistence." IEEE Document: IEEE 802.22-07/0024r0, Jan. 2007.

[584] K. Challapali, S. Mangold, and Z. Zhong, "Spectrum agile radio: Detecting spectrum opportunities," in *Proceedings of the Sixth Annual International Symposium on Advanced Radio Technologies*, Mar. 2004.

[585] M. P. Olivieri, G. Barnett, A. Lackpour, A. Davis, and P. Ngo, "A scalable dynamic spectrum allocation system with interference mitigation for teams of spectrally agile software defined radios," in *Proceedings of the IEEE International Symposium on New Frontiers in Dynamic Spectrum Access Networks*, pp. 170–179, Nov. 2005.

[586] K. Dogancay and D. A. Gray, "Closed-form estimators for multi-pulse TDOA localization," in *Proceedings of the Eighth International Symposium on Signal Processing and Its Applications*, pp. 543–546, Aug. 2005.

[587] T. He, C. Huang, B. M. Blum, J. A. Stankovic, and T. F. Abdelzaher, "Range-free localization schemes in large scale sensor networks," in *Proceedings of ACM MobiCom*, pp. 81–94, Sept. 2003.

[588] T. Locher, R. Wattenhofer, and A. Zollinger, "Received-signal-strength-based logical positioning resilient to signal fluctuation," in *Proceedings of the First ACIS International Workshop on Self-Assembling Wireless Sensor Networks*, May 2005.

[589] D. Niculescu and B. Nath, "Ad hoc positioning system (APS)," in *Proceedings of the IEEE Global Telecommunications Conference*, pp. 2926–2931, Nov. 2001.

[590] B. H. Wellenhoff, H. Lichtenegger, and J. Collins, *Global Positioning System: Theory and Practice*, 4th ed. Dordrecht, The Netherlands: Springer Verlag, 1997.

[591] B. Wild and K. Ramchandran, "Detecting primary receivers for cognitive radio applications," in *Proceedings of the IEEE International Symposium on New Frontiers in Dynamic Spectrum Access Networks*, pp. 124–130, Nov. 2005.

[592] R. Chen and J. Park, "Ensuring trustworthy spectrum sensing in cognitive radio networks," in *Proceedings of the 1st IEEE Workshop on Networking Technologies for Software Defined Radio Networks*, pp. 110–119, Sept. 2006.

[593] T. S. Rappaport, *Wireless Communications: Principles and Practice*. Prentice Hall Communications Engineering and Emerging Technology, Upper Saddle River, NJ, USA: Prentice-Hall, 1996.

[594] Federal Communications Commission, "E911 requirements for IP-enabled service providers." ET Docket No. 05-196, May 2005.

[595] T. Roos, P. Myllymaki, and H. Tirri, "A statistical modeling approach to location estimation," *IEEE Transactions on Mobile Computing*, vol. 1, pp. 59–69, Jan.–Mar. 2002.

[596] R. Chen, J. Park, and J. Reed, "Defense against primary user emulation attacks in cognitive radio networks," *IEEE Journal on Selected Areas in Communications*, vol. 26, pp. 25–37, Jan. 2008.

[597] K. Jain, J. Padhye, V. N. Padmanabha, and L. Qiu, "Impact of interference on multi-hop wireless network performance," in *Proceedings of ACM MobiCom*, pp. 66–80, Sept. 2003.

[598] S. Capkun, M. Cagalj, and M. Srivastava, "Secure localization with hidden and mobile base stations," in *Proceedings of IEEE INFOCOM*, pp. 1–10, Apr. 2006.

[599] S. Capkun and J.-P. Hubaux, "Secure positioning in wireless networks," *IEEE Journal on Selected Areas in Communications*, vol. 24, pp. 221–232, Feb. 2006.

[600] A. Pandharipande, J.-M. Kim, D. Mazzarese, and B. Ji, "IEEE P802.22 Wireless RANs: Technology proposal package for IEEE 802.22." [Online]: www.ieee802.org/22/, Nov. 2005.

[601] E. Visotsky, S. Kuffner, and R. Peterson, "On collaborative detection of TV transmissions in support of dynamic spectrum sharing," in *Proceedings of the IEEE International Symposium on New Frontiers in Dynamic Spectrum Access Networks*, pp. 338–345, Nov. 2005.

[602] L. Lu, S.-Y. Chang, J. Zhang, L. Qian, J. Wen, V. K. N. Lau, R. S. Cheng, R. D. Murch, W. H. Mow, and K. B. Letaief, "Technology proposal clarifications for IEEE 802.22 WRAN systems." [Online]: www.ieee802.org/22/, Mar. 2006.

[603] J. Hillenbrand, T. A. Weiss, and F. K. Jondral, "Calculation of detection and false alarm probabilities in spectrum pooling systems," *IEEE Communications Letters*, vol. 9, pp. 349–351, Apr. 2005.

[604] C. Bettstetter, G. Resta, and P. Santi, "The node distribution of the random waypoint mobility model for wireless ad hoc networks," *IEEE Transactions on Mobile Computing*, vol. 2, pp. 257–269, Mar. 2003.

[605] G. Chouinard, "IEEE P802.22 Wireless RANs: Minutes of the channel model sub-group teleconference." [Online]: www.ieee802.org/22/, July 2005.

[606] K. Bian and J. Park, "Security vulnerabilities in IEEE 802.22," in *Proceedings of the Fourth International Wireless Internet Conference*, Nov. 2008.

[607] A. N. Mody, R. Reddy, M. J. Sherman, T. Kiernan, and D. J. Shyy, "Security and the protocol reference model enhancements in IEEE 802.22." IEEE Document: IEEE 802.22-08/0083r04, June 2008.

[608] J. Chaplin, "Software engineering for software radios: Experiences at MIT and Vanu, Inc.," in *Software Defined Radio—Enabling Technologies* (W. Tuttlebee, ed.). New York, NY, USA: John Wiley & Sons, 2002.

[609] Free Software Foundation, "GNU Radio." [Online]: www.gnuradio.org.

[610] L. B. Michael, M. J. Mihaljevic, S. Haruvama, and R. Kohno, "A framework for secure download for software-defined radio," *IEEE Communications Magazine*, vol. 40, pp. 88–96, July 2002.

[611] A. Brawerman and J. A. Copeland, "Towards a fraud-prevention framework for software defined radio mobile devices," *EURASIP Journal on Wireless Communications and Networking*, vol. 5, pp. 401–412, Aug. 2005.

[612] SAFECOM, "Public safety statement of requirements." [Online]: www.safecomprogram.gov/SAFECOM/library/, 2006.

[613] MESA, "Project MESA: Statement of requirements." [Online]: www.projectmesa.org/ftp/, 2008.

[614] R. I. Desourdis, D. R. Smith, W. D. Speights, R. J. Dewey, and J. R. DiSalvo, *Emerging Public Safety Wireless Communication Systems*. Norwood, MA, USA: Artech House, 2001.

[615] Zeeburg Nieuws, "Tweede kamer wil opheldering over UMTS-veiling." [Online]: www.zeeburgnieuws.nl/rubriek/internet/umtsveiling.html, 2006.

[616] P. Anker, "UMTS in Nederland." [Online]: www.frequentieland.nl/umts/umts_nl.htm, 2009.

[617] Project 25 Technology Interest Group, "P25 homepage." [Online]: www.project25.org/, 2009.

[618] TETRAPOL Forum, "TETRAPOL Homepage." [Online]: www.tetrapol.net/, 2009.

[619] The TETRA MoU Association Ltd., "Terrestial trunked radio, radio communication standards, tetra private digital mobile radio pmr." [Online]: www.tetra-association.com/, 2009.

[620] ETSI, "Terrestrial trunked radio; voice plus data; part 2: Air interface," tech. rep. EN 300 392-2, European Telecommunications Standards Institute, 2006.

[621] Commissie Oosting, "De vuurwerkramp—eindrapport," 2001.

[622] R. Schiphorst, *A T-DAB Field Trial Using a Low-Mast Infrastructure*. Enschede, The Netherlands: Signals and Systems Group, University of Twente, Dec. 2006.

[623] S. Olafsson, B. Glower, and M. Nekovee, "Future management of spectrum," *BT Technology Journal*, vol. 25, no. 2, pp. 52–63, 2007.

[624] T. K. Forde and L. E. Doyle, "A combinatorial clock auction for OFDMA-based cognitive wireless networks," in *Proceedings of the Third International Conference on Wireless Pervasive Computing*, May 2008.

[625] X. Zhou, S. Gandhi, S. Suri, and H. Zheng, "EBay in the sky: Strategy-proof wireless spectrum auctions," in *Proceedings of ACM MobiCom*, Sept. 2008.

[626] X. Zhou and H. Zheng, "TRUST: A general framework for truthful double spectrum auctions," in *Proceedings of IEEE INFOCOM*, Apr. 2009.

[627] V. Krishna, *Auction Theory*. San Diego, CA, USA, Academic Press, Mar. 2002.

[628] S. Gandhi, C. Buragohain, L. Cao, H. Zheng, and S. Suri, "A general framework for wireless spectrum auctions," in *Proceedings of the IEEE International Symposium on New Frontiers in Dynamic Spectrum Access Networks*, 2007.

[629] F. Hillier and G. Lieberman, *Introduction to Operations Research*. New York, NY, USA: McGraw-Hill Science, 2004.

[630] W. Vickery, "Counterspeculation, auctions and competitive sealed tenders," *Journal of Finance*, vol. 16, pp. 8–37, 1961.

[631] D. Lehmann, L. Oallaghan, and Y. Shoham, "Truth revelation in approximately efficient combinatorial auctions," *Journal of the ACM*, vol. 49, no. 5, pp. 577–602, 2002.

[632] A. Mu'alem and N. Nisan, "Truthful approximation mechanisms for restricted combinatorial auctions: Extended abstract," in *Proceedings of the 18th National Conference on Artificial Intelligence*, pp. 379–384, 2002.

[633] E. Clarke, "Multipart pricing of public goods," *Public Choice*, vol. XI, pp. 17–33, 1971.

[634] T. Groves, "Incentive in terms," *Econometrica*, vol. 41, pp. 617–631, 1973.

[635] K. Jain, J. Padhye, V. N. Padmanabhan, and L. Qiu, "Impact of interference on multi-hop wireless network performance," in *Proceedings of ACM MobiCom*, 2003.

[636] S. Sakai, M. Togasaki, and K. Yamazaki, "A note on greedy algorithms for the maximum weighted independent set problem," *Discrete Applied Mathematics*, vol. 126, no. 2–3, pp. 313–322, 2003.

[637] M. Babaioff and N. Nisan, "Concurrent auctions across the supply chain," in *Proceedings of ACM Conference on Electronic Commerce*, 2001.

[638] R. P. McAfee, "A dominant strategy double auction," *Journal of Economic Theory*, vol. 56, pp. 434–450, Apr. 1992.

[639] S. Ramanathan, "A unified framework and algorithm for channel assignment in wireless networks," *Wireless Networks*, vol. 5, no. 2, pp. 81–94, 1999.

[640] A. P. Subramanian, H. Gupta, S. R. Das, and M. M. Buddhikot, "Fast spectrum allocation in coordinated dynamic spectrum access based cellular networks," in *Proceedings of the IEEE International Symposium on New Frontiers in Dynamic Spectrum Access Networks*, Nov. 2007.

[641] C. Reis, R. Mahajan, M. Rodrig, D. Wetherall, and J. Zahorjan, "Measurement-based models of delivery and interference in static wireless networks," in *Proceedings of ACM SIGCOMM*, Sept. 2006.

[642] G. Brar, D. M. Blough, and P. Santi, "Computationally efficient scheduling with the physical interference model for throughput improvement in wireless mesh networks," in *Proceedings of ACM MobiCom*, 2006.

[643] Spectrum Bridge, Inc., "Spectrum Bridge Inc.—spectrum value, spectrum experts." [Online]: www.spectrumbridge.com.

[644] A. P. Subramanian, M. Al-Ayyoub, H. Gupta, S. Das, and M. M. Buddhikot, "Near optimal dynamic spectrum allocation in cellular networks," in *Proceedings of the IEEE International Symposium on New Frontiers in Dynamic Spectrum Access Networks*, 2008.

[645] Y. Xing, R. Chandramouli, and C. Cordeiro, "Price dynamics in competetive agile spectrum access markets," *IEEE Journal on Selected Areas in Communications*, vol. 25, no. 3, pp. 613–621, 2007.

[646] D. Niyato, E. Hossein, and Z. Han, "Dynamics of multiple-seller and multiple-buyer spectrum trading in cognitive radio networks: A game theoretic modelling approach," *IEEE Transactions on Mobile Computing*, vol. 8, no. 8, pp. 1009–1021, 2009.

[647] V. Rodriguez, K. Mossner, and R. Tafazoli, "Auction-based optimal bidding, pricing and service priorities for multi-rate, multi-class cdma," in *Proceedings of the IEEE International Symposium on Personal, Indoor and Mobile Radio Communications*, pp. 1850–1854, Sept. 2005.

[648] J. Huang, R. Berry, and M. L. Honig, "Auction-based spectrum sharing," *ACM Mobile Networks and Applications*, vol. 11, no. 3, pp. 405–618, 2006.

[649] D. Porter, S. Rassenti, A. Roopnarine, and V. Smith, "Combinatorial auction design," *Proceedings of the National Academy of Science of the United States of America*, vol. 100, pp. 11153–11157, Sept. 2003.

[650] F. K. Jondral, "Software-defined radio—Basics and evolution to cognitive radio," *EURASIP Journal on Wireless Communications and Networking*, vol. 3, pp. 275–283, 2005.

[651] Wikipedia, "Software-defined radio." [Online]: http://en.wikipedia.org/wiki/Software_defined_radio.

[652] Wikipedia, "GNU radio." [Online]: http://en.wikipedia.org/wiki/Gnu_radio.

[653] C. Chang, J. Wawrzynek, and R. W. Brodersen, "BEE2: A high-end reconfigurable computing system," *IEEE Design and Test of Computers*, vol. 22, June 2005.

[654] Xilinx, "FPGA and CPLD solutions from Xilinx, Inc." [Online]: www.xilinx.com/.

[655] The MathWorks, "Simulink—simulation and model-based design." [Online]: www.mathworks.com/products/simulink/.

[656] Berkeley Wireless Research Center, University of California Berkeley, "Berkeley emulation engine 2 homepage." [Online]: http://bee2.eecs.berkeley.edu/.

[657] H. So, A. Tkachenko, and R. W. Brodersen, "A unified hardware/software runtime environment for FPGA based reconfigurable computers using BORPH," in *Proceedings of the International Conference on Hardware-Software Codesign and System Synthesis*, Oct. 2006.

[658] M. Steyaert, B. D. Muer, P. Leroux, M. Borremans, , and K. Mertens, "Low-voltage low-power CMOS-RF transceiver design," *IEEE Transactions on Microwave Theory and Technology*, vol. 50, Jan. 2002.

[659] P. Zhang et al., "A single-chip dual-band direct-conversion IEEE 802.11a/b/g WLAN transceiver in 0.18 micron CMOS," *IEEE Journal of Solid-State Circuits*, vol. 40, Sept. 2005.

[660] T. Maeda et al., "Low-power-consumption direct-conversion CMOS transceiver for multi-standard 5-GHz wireless LAN systems with channel bandwidths of 5–20 MHz," *IEEE Journal of Solid-State Circuits*, vol. 41, Feb. 2006.

[661] R. Bagheri, A. Mirzaei, M. E. Heidari, S. Chehrazi, M. Lee, M. Mikhemar, W. K. Tang, and A. A. Abidi, "Software-defined radio receiver: Dream to reality," in *IEEE Communications Magazine*, Aug. 2006.

[662] E. Bautista, B. Bastani, , and J. Heck, "A high IIP2 downconversion mixer using dynamic matching," *IEEE Journal of Solid-State Circuits*, vol. 35, Dec. 2000.

[663] O. Charlon et al., "A low-power high-performance SiGe BiCMOS 802.11a/b/g transceiver IC for cellular and bluetooth co-existence applications," *IEEE Journal of Solid-State Circuits*, vol. 44, July 2006.

[664] J. Juan, R. Stengel, F. Martin, and D. Bockelman, U.S. patent 6,891,420: "Cascaded delay locked loop circuit," 2006.

[665] D. Bockelman and J. Juan, U.S. patent 6,353,649: "Time interpolating direct digital synthesizer," 2002.

[666] F. Martin, R. Stengel, and J. Juan, U.S. patent 7,154,978: "Method and apparatus for digital frequency synthesis," 2005.

[667] T. Gradishar and R. Stengel, U.S. patent 7,143,125: "Method and apparatus for noise shaping in direct digital synthesis circuits," 2006.

[668] A. Fehske, J. Gaeddert, and J. H. Reed, "A new approach to signal classification using spectral correlation and neural networks," in *First IEEE International Symposium on New Frontiers in Dynamic Spectrum Access Networks*, Nov. 2005.

[669] W. Gardner, W. Brown, and C. Chen, "Spectral correlation of modulated signals: Part II—Digital modulation," *IEEE Transactions on Communications*, vol. 35, July 1987.

[670] P. Sutton, L. E. Doyle, and K. E. Nolan, "A reconfigurable platform for cognitive networks," in *Proceedings of the International Conference on Cognitive Radio Oriented Wireless Networks and Communications*, 2006.

[671] D. Taubenheim, W. Chiou, N. Correal, P. Gorday, S. Kyperountas, S. Machan, M. Pham, Q. Shi, E. Callaway, and R. Rachwalski, "Implementing an experimental cognitive radio system for DySPAN," in *Proceedings of the IEEE Global Telecommunications Conference*, 2007.

[672] F. Adachi, H. Wakana, H. Morikawa, M. Kuroda, H. Harada, S. Isobe, R. Miura, and H. Ogawa, "Network and access technologies for new generation mobile communications overview of national R&D project in NICT: Research articles," *Wireless Communications & Mobile Computing*, vol. 7, pp. 937–950, 2007.

[673] D. Naughton, G. Baldwin, and R. Farrell, "Performance requirements for analog-to-digital converters in wideband reconfigurable radios," in *Proceedings of the SPIE Microtechnologies for the New Millennium Symposium: VLSI Circuits and Systems*, 2005.

[674] F. Agnelli, G. Albasini, I. Bietti, A. Gnudi, A. Lacaita, D. Manstretta, R. Rovatti, E. Sacchi, P. Savazzi, and F. Svelto et al., "Wireless multi-standard terminals: System analysis and design of a reconfigurable RF front-end," *IEEE Circuits and Systems Magazine*, vol. 6, pp. 38–59, 2006.

[675] G. Piazza, P. J. Stephanou, and A. P. Pisano, "Single-chip multiple-frequency ALN MEMS filters based on contour-mode piezoelectric resonators," *Journal of Microelectromechanical Systems*, vol. 16, 2007.

[676] M. Vidojkovic, V. Vidojkovic, M. A. T. Sanduleanu, J. van der Tang, P. Baltus, and A. van Roermund, "A 1.2 V inductorless receiver front-end for multi-standard wireless applications," in *Proceedings of the IEEE Radio and Wireless Symposium*, 2008.

[677] F. Ge, Q. Chen, Y. Wang, C. W. Bostian, T. W. Rondeau, and B. Le, "Cognitive radio: From spectrum sharing to adaptive learning and reconfiguration," in *Proceedings of the IEEE Aerospace Conference*, 2008.

[678] J. Mitola III, "Cognitive radio." Licentiate dissertation, The Royal Institute of Technology, Stockholm, Sweden, Sept. 1999.

[679] F. Fitzek and M. Katz, eds., *Cooperation in Wireless Networks: Principles and Applications—Real Egoistic Behavior Is to Cooperate!* Dordrecht, The Netherlands: Springer, 2006.

[680] *Proceedings of the IEEE International Symposium on New Frontiers in Dynamic Spectrum Access Networks*. Piscatawy, NJ, USA: Institute of Electrical and Electronics Engineers, Inc., 2005 and 2007.

[681] *Proceedings of the International Conference on Cognitive Radio Oriented Wireless Networks and Communications*, Institute for Computer Sciences, Social-Informatics and Telecommunications Engineering, 2007.

[682] SDR Forum, "Cognitive radio working group notes." [Online]: www.sdrforum.org/CRWG.

[683] J. Strassner, "IEEE P1900.5 Policy language backgrounder, part 1," Mar. 2008.

[684] E. Meshkova, J. Riihijarvi, P. Mahonen, and C. Kavadias, "Modeling the home environment using ontology with applications in software configuration management," in *Proceedings of the 15th International Conference on Telecommunications*, St. Petersburg, Russia, June 2008.

[685] H. I. Christensen, "Cognitive vision," *AAAI Magazine*, vol. 25, pp. 8–9, Oct. 2004.

[686] *Proceedings of the Human Language Technologies Conference*, North American Chapter of the Association for Computational Linguistics, Apr. 2007.

[687] J. M. III, "Software radio architecture: A mathematical perspective," *IEEE Journal on Selected Areas in Communications*, vol. 17, pp. 514–538, Apr. 1999.

[688] J. Mitola III, ed., *Software Radio Architecture*. New York, NY, USA: John Wiley & Sons, 2000.

[689] H. Huang, Z. Zhang, P. Cheng, G. Yu, and O. Qui, "Throughput analysis of cognitive MIMO system," in *Proceedings of the International Workshop on Cross Layer Design*, Jinan, China, Sept. 2007.

[690] D. Novo, B. Bougard, P. Raghavan, T. Schuster, L. Van der Perre, H.-S. Kim, and H. Yang, "Energy-performance exploration of a CGA-based SDR processor," in *Proceedings of the SDR Forum Technical Conference*, 2006.

[691] C. Linn and E. Koski, "The JTRS program: Software-defined radios as a software product line," in *Proceedings of the 10th International Software Product Line Conference*, Aug. 2006.

[692] United States Department of the Navy, "Joint program executive office—joint tactical radio system website." [Online]: http://jtrsjpeo.mil.

[693] Object Management Group, Inc., "OMG corba website." [Online]: www.omg.org/corba.

[694] B. S. Manoj, M. Zorzi, and R. Rao, "A new paradigm for cognitive networking," in *Proceedings of the IEEE International Symposium on New Frontiers in Dynamic Spectrum Access Networks*, 2007.

[695] L. Kovacs and A. Vidacs, "Spectrum auction and pricing in dynamic spectrum allocation networks," in *Proceedings of the IEEE International Symposium on New Frontiers in Dynamic Spectrum Access Networks*, 2007.

[696] S. Gandhi, C. Buragohain, L. Cao, H. Zheng, and S. Suri, "Towards real-time dynamic spectrum auctions," *Computer Networks*, vol. 52, Mar. 2008.

[697] D. Grandblaise, C. Kloeck, K. Moessner, V. Rodriguez, E. Mohyeldin, M. K. Pereirasamy, J. Luo, and I. Martoyo, "Techno-economic of collaborative based secondary spectrum usage—E2R research: Project outcomes overview," in *Proceedings of the IEEE International Symposium on New Frontiers in Dynamic Spectrum Access Networks*, 2005.

[698] Google, Inc., "Letter to the US Federal Communications Commission," 2007.

[699] Cantor Fitzgerald, "Cantor spectrum and tower exchange." [Online]: www.cantor.com/sales_trading.

[700] H. Harada, "Software defined radio prototype toward cognitive radio communication systems," in *Proceedings of the IEEE International Symposium on New Frontiers in Dynamic Spectrum Access Networks*, 2005.

[701] H. Harada, "Regulatory perspective of Japan," in *Proceedings of the VCE Regulatory Workshop*, London, UK, Apr. 2007.

[702] H. Harada, "Advances in flexible radio technology to support cognitive radio," in *Proceedings of the VCE International Research Workshop on Intelligent Spectrum Usage for Personal Communications*, London, UK, Apr. 2007.

[703] W. Webb, "Keynote address." IEEE International Symposium on New Frontiers in *Dynamic Spectrum Access Networks 2007 Keynote Presentation*.

[704] K. E. Nolan, E. Ambrose, and D. O'Mahony, "Cognitive radio: Value creation and value migration," *Proceedings of the SDR Forum Technical Conference 2006*, 2006.

[705] International Telecommunication Union, "Radio-spectrum management for a converging world." [Online]: www.itu.int/osg/spu/ni/spectrum/.

[706] R. Ercole, "Innovation, spectrum regulation, and DySPAN technologies access to markets," in *Proceedings of the IEEE International Symposium on New Frontiers in Dynamic Spectrum Access Networks*, 2005.

[707] S. Forum, "Academic challenge." [Online]: www.sdrforum.org.

[708] P. Mannion, "Smart radios stretch spectrum," *EE Times*, Dec. 2005.

[709] F. Seelig, "A description of the August 2006 XG demonstrations at Fort A.P. Hill," in *Proceedings of the IEEE International Symposium on New Frontiers in Dynamic Spectrum Access Networks*, Apr. 2007.

[710] F. Perich, "Policy-based network management for NeXt generation spectrum access control," in *Proceedings of the IEEE International Symposium on New Frontiers in Dynamic Spectrum Access Networks*, Apr. 2007.

[711] Y. Chen, H. Huang, Z. Zhang, V. Lau, and P. Qiu, "Spectrum access for cognitive radio network employing rateless code," in *Proceedings of the IEEE International Communications Conference*, 2008.

[712] N. Nilsson, ed., *Principles of Artificial Intelligence*. Palo Alto, CA, USA: Springer-Verlag, 1980.

[713] H.-M. Huang, "Toward a generic model for autonomy levels for unmanned systems (ALFUS)," in *Proceedings of the Performance Metrics for Intelligent Systems Workshop*, Gaithersburg, MD, USA, Sept. 2003.

[714] G. Liu, X. Chen, X. Wang, and B. Zhang, "On multi-agent collaborative planning and its application in RoboCup," in *Proceedings of the Third World Congress on Intelligent Control and Automation*, 2000.

[715] P. Stone, "Reinforcement learning for RoboCup Soccer keepaway, Sage journals on line." [Online]. www.adb.sagepub.com, 2005.

[716] R. H. Bordini, M. Dastani, J. Dix, and A. E. F. Seghrouchni, eds., *Multi-Agent Programming Languages, Platforms and Applications*. New York, NY, USA: Springer-Verlag, 2005.

[717] B. Bougard, L. Hollevoet, F. Naessens, T. Schuster, C. H. A. Ng, and L. Van der Perre, "A low power signal detection and pre-synchronization engine for energy-aware software defined radio," in *Proceedings of the SDR Forum Technical Symposium*, 2006.

[718] *Proceedings of the First Dagstuhl Workshop on Cognitive Radio*, Dagstuhl, Germany, 2003.

[719] T. Kanter, "A service architecture, test bed and application for extensible and adaptive mobile communication," in *Proceedings of the IEEE Professional Communication Society Conference*, 2001.

[720] F. Fitzek, P. Kyritsi, and M. Katz, "Power consumption and spectrum usage paradigms in cooperative wireless networks," in *Cooperation in Wireless Networks: Principles and Applications—Real Egoistic Behavior Is to Cooperate!* (F. Fitzek and M. Katz, eds.). Dordrecht, The Netherlands: Springer, 2006.

[721] A. B. Olsen and P. Koch, "Energy aware task allocation in cooperative wireless networks," in *Cooperation in Wireless Networks: Principles and Applications—Real Egoistic Behavior Is to Cooperate!* (F. Fitzek and M. Katz, eds.). Dordrecht, The Netherlands: Springer, 2006.

[722] A. Chakrabarti, A. Sabharwal, and B. Aazhang, "Cooperative communications," in *Cooperation in Wireless Networks: Principles and Applications—Real Egoistic Behavior Is to Cooperate!* (F. Fitzek and M. Katz, eds.). Dordrecht, The Netherlands: Springer, 2006.

[723] P. Mahonen, M. Petrova, and J. Riihijarvi, "Cooperation in ad-hoc networks," in *Cooperation in Wireless Networks: Principles and Applications—Real Egoistic Behavior Is to Cooperate!* (F. Fitzek and M. Katz, eds.). Dordrecht, The Netherlands: Springer, 2006.

[724] J. N. Laneman, "Cooperative diversity," in *Cooperation in Wireless Networks: Principles and Applications—Real Egoistic Behavior Is to Cooperate!* (F. Fitzek and M. Katz, eds.). Dordrecht, The Netherlands: Springer, 2006.

[725] S. Cui and A. J. Goldsmith, "Cooperation techniques in cross-layer design," in *Cooperation in Wireless Networks: Principles and Applications—Real Egoistic Behavior Is to Cooperate!* (F. Fitzek and M. Katz, eds.). Dordrecht, The Netherlands: Springer, 2006.

[726] K. Wrona and P. Mahonen, "Stability and security in wireless cooperative networks," in *Cooperation in Wireless Networks: Principles and Applications—Real Egoistic Behavior Is to Cooperate!* (F. Fitzek and M. Katz, eds.). Dordrecht, The Netherlands: Springer, 2006.

[727] C. Mathur and K. P. Subbalakshmi, "Security issues in cognitive radio networks," in *Cognitive Networks* (Q. H. Mahmoud, ed.). New York, NY, USA: John Wiley & Sons, 2007.

[728] O. Shin, N. Devroye, P. Mitran, H. Ochiai, S. S. Ghassemzadeh, H. T. Kung, and V. Tarokh, "Cooperation, competition and cognition in wireless networks," in *Cooperation in Wireless Networks: Principles and Applications—Real Egoistic Behavior Is to Cooperate!* (F. Fitzek and M. Katz, eds.). Dordrecht, The Netherlands: Springer, 2006.

[729] P. Eggers, P. Kyrirtsi, and I. Z. Kovacs, "Cooperative antenna systems," in *Cooperation in Wireless Networks: Principles and Applications—Real Egoistic Behavior Is to Cooperate!* (F. Fitzek and M. Katz, eds.). Dordrecht, The Netherlands: Springer, 2006.

[730] T. Masden, "Cooperative header compression," in *Cooperation in Wireless Networks: Principles and Applications—Real Egoistic Behavior Is to Cooperate!* (F. Fitzek and M. Katz, eds.). Dordrecht, The Netherlands: Springer, 2006.

[731] J. C. H. Lin and A. Stefanov, "Cooperative coding and its application to OFDM systems," in *Cooperation in Wireless Networks: Principles and Applications—Real Egoistic Behavior Is to Cooperate!* (F. Fitzek and M. Katz, eds.). Dordrecht, The Netherlands: Springer, 2006.

[732] Y. Takatori, "Cooperative methods for spatial channel control," in *Cooperation in Wireless Networks: Principles and Applications—Real Egoistic Behavior Is to Cooperate!* (F. Fitzek and M. Katz, eds.). Dordrecht, The Netherlands: Springer, 2006.

[733] Q. H. Mahmoud, ed., *Cognitive Networks*. New York, NY, USA: John Wiley & Sons, 2007.

[734] K. Liebnitz, N. Wakamiya, and M. Murata, "Biologically inspired networking," in *Cognitive Networks* (Q. H. Mahmoud, ed.), New York, NY, USA: John Wiley & Sons, 2007.

[735] J. Strassner, "The role of autonomic networking in cognitive networks," in *Cognitive Networks* (Q. H. Mahmoud, ed.), New York, NY, USA: John Wiley & Sons, 2007.

[736] Y. Rekhter and P. Gross, "Application of the border gateway protocol in the Internet." [Online]: www.ietf.org/rfc/rfc1772.txt.

[737] J. Strassner, "An architectural blueprint for autonomic computing," tech. rep., IBM, 2003.

[738] A. Ginsberg, W. Horne, and J. Poston, "The semantic side of cognitive radio," in *Cognitive Networks* (Q. H. Mahmoud, ed.), New York, NY, USA: John Wiley & Sons, 2007.

[739] M. Kokar, D. Brady, and K. Baclawski, "Roles of ontologies in cognitive radios," in *Cognitive Radio Technology* (B. A. Fette, ed.), Boston, MA, USA: Elsevier, 2006.

[740] M. Cummings, D. Bourse, T. Cooklev, S. Das, M. Kokar, B. Lyles, R. Normoyle, J.Strassner, and P. Subrahmanyam, "IEEE 802.21: The leading edge of a larger challenge," in *IEEE Communications Magazine*, May 2008.

[741] J. O. Neel, J. H. Reed, and R. P. Gilles, "Convergence of cognitive radio networks," in *Proceedings of the IEEE Wireless Communications and Networking Conference*, 2004.

[742] T. Russell, ed., *Signaling System #7*, 5th edition ed. New York, NY, USA: McGraw-Hill, 2006.

[743] M. Sexton and A. Reid, eds., *Transmission Networking: Sonet and the Synchronous Digital Hierarchy*. Norwood, MA, USA: Artech House, 1992.

[744] "Session initiation protocol, Internet Engineering Task Force (IETF) RFC 3261 and SIP enumservice, RFC 3764," Apr. 2004.

[745] J. Davies, ed., *Understanding IPv6*. Redmond, WA, USA: Microsoft Press, 2003.

[746] G. Carenini and R. Sharma, "Exploring more realistic evaluation measures for collaborative filtering," in *Proceedings of the American Association for Artificial Intelligence*, 2004.

[747] A.-B. Garcia, M. Alvarez-Campana, E. Vazquez, G. GuCnon, and J. Berrocal, "ATM transport between UMTS base stations and controllers: Supporting topology and dimensioning decisions," in *Proceedings of the IEEE International Symposium on Personal, Indoor and Mobile Radio Communications*, 2004.

[748] D. Bourse, M. Muck, D. Bateman, W. Koenig, K. Nolte, P. Martigne, I. Gaspard, E. Bogenfeld, N. Kiukkonen, and E. Buracchini, "FP7 E3 project key challenges," in *Proceedings of the SDR Forum Technical Conference*, Nov. 2007.

[749] S. Zhong, "Performance evaluation of the functional description language in a SDR environment," in *Proceedings of the SDR Forum Technical Conference*, Nov. 2007.

[750] R. C. Meier, U.S. patent 20030112767, "communication network providing wireless and hard-wired dynamic routing," 2003.

[751] A. Bravo, "Advanced positioning and location based services in 4G mobile-IP radio access networks," in *Proceedings of the IEEE International Symposium on Personal, Indoor and Mobile Radio Communications*, 2004.

702 References

[752] N. Bhatia, "Policy management in context-aware networks," M.S. dissertation, Royal Institute of Technology, Stockholm, Sweden, 2007.

[753] F. Capar and F. Jondral, "Spectrum pricing for excess bandwidth in radio networks," in *Proceedings of the IEEE International Symposium on Personal, Indoor and Mobile Radio Communications*, Sept. 2004.

[754] K. E. Nolan, P. D. Sutton, L. E. Doyle, T. W. Rondeau, B. Le, and C. W. Bostian, "Dynamic spectrum access and coexistence experiences involving two independently developed cognitive radio testbeds," in *Proceedings of the IEEE International Symposium on New Frontiers in Dynamic Spectrum Access Networks*, Apr. 2007.

[755] K. Nolan, Personal communications, Mar. 2008.

[756] D. Maldonado, B. Le, A. Hugine, T. W. Rondeau, and C. W. Bostian, "Cognitive radio applications to dynamic spectrum allocation," in *Proceedings of the IEEE International Symposium on New Frontiers in Dynamic Spectrum Access Networks*, 2005.

[757] U.S. Department of Homeland Security, "Statement of requirements for public safety wireless communications and interoperability," Mar. 2004.

[758] S. Li, "Contribution for IEEE 802.22 WRAN systems: Sensing scheme for DVB-T," tech. rep., University of Electronic Science & Technology of China, Nov. 2007.

[759] S.-G. Jung, S.-S. Lee, Y.-B. Park, and J.-K. Kim, "Development of URC testing and certification system," in *Proceedings of the IEEE International Symposium on Computational Intelligence in Robotics and Automation*, June 2005.

[760] T. Kotoku, "OMG Working Group meeting," Jan. 2006.

[761] Object Management Group, OMG home page. [Online]: www.omg.org.

[762] ISC West, ISC west home page. [Online]: www.iscwest.com.

[763] Nedstat, "Nedstat—stream sense video analytics." [Online]: www.nedstat.co.uk/web/nedstatuk.nsf/pages/analytics_stream_sense.

[764] Agenti, "AgentVi.com—futureproofed vedio analytics." [Online]: www.agentvi.com/.

[765] Truveo, "Truveo developer center." [Online]: http://developer.truveo.com/.

[766] J. Mitola III, "Crowncom 2007 keynote presentation." International Conference on Cognitive Radio Oriented Wireless Networks and Communications, 2007.

[767] J. S. Olsson, "Combining evidence for improved speech retrieval," in *Proceedings of the NAACL Human Language Technologies Conference*, Apr. 2007.

[768] J. Campbell, "Speaker recognition: A tutorial," in *Proceedings of the IEEE*, Sept. 1997.

[769] KMWorld, "Text analytics: On the trail of business intelligence." [Online]: www.kmworld.com, Nov. 2007.

[770] IBM, "Omnifind enterprise edition." [Online]: www-306.ibm.com/software/data/enterprise-search/omnifind-enterprise/uima.html, Jan. 2008.

[771] Google, Inc., "Android—An open handset alliance project." [Online]: www.code.google.com/android/.

[772] B. Abu Shawar and E. Atwell, "Different measurements metrics to evaluate a chatbot system," in *Proceedings of the NAACL Human Language Technologies Conference*, Apr. 2007.

[773] J. Stroessner, "Autonomic networks," in *Cooperation in Wireless Networks: Principles and Applications—Real Egoistic Behavior Is to Cooperate!* (F. Fitzek and M. Katz, eds.). Dordrecht, The Netherlands: Springer, 2006.

[774] M. Cummings, T. Cooklev, B. Lyles, and P. A. Subrahmanyam, "Commercial wireless metalanguage scenario," in *Proceedings of the SDR Forum Technical Conference*, Nov. 2007.

[775] Y. Gao, "Speech-to-speech translation, IBM user interface technologies." [Online]: http://domino.watson.ibm.com, 2007.

[776] F. Och, "Statistical machine translation." [Online]: www.googleresearch.blogspot.com, Apr. 2006.

[777] L. Yongquan, "Current natural language computing in China." [Online]: http://llc. oxfordjournals.org.

[778] K. Deschacht and M.-F. Moens, "Text analysis for automatic image annotation," in *Proceedings of the 45th Annual Meeting of the Association of Computational Linguistics*, June 2007.

[779] R. Gopal, "Model based framework for implementing situation management infrastructure," in *Proceedings of the Military Communications Conference*, 2007.

[780] K. Chrisler, "A user-centered approach to the wireless world," in *Technologies for the Wireless Future* (R. Tafazolli, ed.), New York, NY, USA: John Wiley & Sons, 2005.

[781] R. Krishnan, "Towards policy-defined cognitive radio," tech. rep., National Science Foundation, 2004.

[782] L. Berlemann, S. Mangold, and B. H. Walke, "Policy-based spectrum sharing in cognitive radio networks," in *Proceedings of the IEEE International Symposium on New Frontiers in Dynamic Spectrum Access Networks*, 2005.

[783] D. Elenius, G. Denker, M.-O. Stehr, R. Senanayake, C. Talcott, and D. Wilkins, "CoRaL— Policy language and reasoning techniques for spectrum policies," tech. rep., SRI International, USA, 2007.

[784] Next Generation Communication, "Maude XG policy reasoner." [Online]: http://xg.csl. sri.com/maude.php.

[785] J. Strassner, "DEN," tech. rep., Motorola Inc., 2007.

[786] "E2R-E3 program plan." SDR Forum, 2008.

[787] Z. Boufidis, M. Stamatelatos, and N. Alonistioti, "Actors, management plane, and policy provision challenges for end-to-end reconfiguration," in *Proceedings of the E2R Workshop on Reconfigurable Mobile Systems and Networks Beyond 3G*, 2004.

[788] L. Kagal, T. Berners-Lee, D. Connolly, and D. Weitzner, "Promoting interoperability between heterogeneous policy domains," tech. rep., National Science Foundation, 2008.

[789] M. Hondo, T. Nadalin, R. Nagaratnam, M. Kudoh, G. Karjoth, B. Pfitzmann, and M. Schunter, "Position paper: Privacy policies as a component of policy-enabled governance," tech. rep., IBM, Sept. 2006.

[790] C. Gunter, "Ensuring privacy conformance in inter-domain systems," in *Proceedings of the W3C Workshop on Languages for Privacy Policy Negotiation and Semantics-Driven Enforcement*, Sept. 2006.

[791] L. Qiang and Z. Ping, "A novel mobility management architecture based on GLL in configurable networks," in *Proceedings of the Sixth International Conference on ITS Telecommunications*, 2006.

[792] "RFC IETF GeoPriv standard."

[793] I. Horrocks, P. F. Patel-Schneider, H. Boley, S. Tabet, B. Grosof, and M. Dean, "SWRL: A semantic web rule language combining OWL and RuleML." [Online]: www.w3.org/ Submission/SWRL/.

[794] J. Hendler, "AAAI 2004 keynote address inter alia," 2004.

[795] G. Lakoff, *Women, Fire and Dangerous Things*. Upper Saddle River, NJ, USA: Prentice-Hall, 1987.

[796] T. Ziemke, J. Zlatev, and R. M. Frank, eds., *Body, Language, Mind: Embodiment*. Hawthorn, NY, USA: Mouton de Gruyter, 2007.

[797] J. Mitola, "Orthogonalization of language," tech. rep., Stevens Institute of Technology, 2008.

[798] M. Johnson and T. Rohrer, "We are live creatures: Embodiment, american pragmatism," in *Body, Language, Mind: Embodiment* (T. Ziemke, J. Zlatev, and R. M. Frank, eds.), Hawthorn, NY, USA: Mouton de Gruyter, 2007.

[799] Mobile VCE, "Welcome to the mobile VCE." [Online]: http://www.mobilevce.com.

[800] M. A. Beach, C. M. Tan, and A. R. Nix, "Indoor dynamic double directional measurements," in *Proceedings of the International Conference on Electromagnetics in Advanced Applications*, Turin, Italy, 2003.

[801] Dartmouth University, "CRAWDAD: A community resource for archiving wireless data at dartmouth." [Online]: http://crawdad.cs.dartmouth.edu.

[802] J. Stine, "A location-based method for specifying RF spectrum rights," in *Proceedings of the IEEE International Symposium on New Frontiers in Dynamic Spectrum Access Networks*, 2007.

[803] S. Gandhi, C. Buragohain, L. Cao, H. Zheng, and S. Suri, "A general framework for wireless spectrum auctions," in *Proceedings of the IEEE International Symposium on New Frontiers in Dynamic Spectrum Access Networks*, 2007.

[804] S. Sengupta, M. Chatterjee, and S. Ganguly, "An economic framework for spectrum allocation and service pricing with competitive wireless service providers," in *Proceedings of the IEEE International Symposium on New Frontiers in Dynamic Spectrum Access Networks*, 2007.

[805] S. D. Jones, E. Jung, X. Liu, N. Merheb, and I.-J. Wang, "Characterization of spectrum activities in the U.S. public safety band for opportunistic spectrum access," in *Proceedings of the IEEE International Symposium on New Frontiers in Dynamic Spectrum Access Networks*, 2007.

[806] Altera, "EP1C12 datasheets." [Online]: www.datasheetsite.com/datasheet/EP1C12.

[807] L. K. Patton, "A GNU radio based software-defined radar," Master's thesis, Wright State University, Dayton, OH, USA, Apr. 2007.

[808] D. Shen, "GNU radio tutorials." [Online]: www.nd.edu/dshen/GNU/, 2005.

[809] F. A. Hamza, "The USRP under 1.5X magnifying lens." [Online]: http://gnuradio.org/trac/wiki/UsrpFAQ, 2008.

Index

A

Adaptable radio devices, 378–379
Adaptation engine, 178–179
Agile transmission techniques
  dynamic spectrum access
    spectrum pooling, 151
    underlay and overlay
      transmission, 151–154
  NC-OFDM-based cognitive radio
    challenges and solutions, 155
  FFT pruning, 165–167
  interference mitigation,
    156–167
  peak-to-average power ratio,
    167–174
  noncontiguous orthogonal
    frequency division
    multiplexing, 154–155
Ambient Gaussian noise density,
  340–341
Analog/baseband processing, 544
Analog receiver
  first line, 511
  graph construction
    flow graph, 515
    FM receiver block diagram,
      514
    graph creation, C++ and
      Python, 513
    usrp block, 514–515
  initialization function, 512–513
  modules importation, 511–512
  Python programs, 511
Analog-to-digital converters
  (ADCs), 1
Antijamming, 485–486
APCO Project 25 systems, 473
Application-specific integrated
  circuits (ASICs), 3
Applications programmer interfaces
  (APIs), 604
ATSC cyclostationary sensing
  technique
  feature, 424–425
  received signal processing, 425
  statistic test, 425–426
  TV signal sensing procedure, 426
  wireless microphone signal
    sensing procedure,
    426–427
Automatic gain control (AGC), 525

B

Backbone network, 468
Base station (BS), 434, 469
Berkeley emmulation engine (BEE2)
  computing devices, 552–553

cyclostationary feature detector
  computation, spectral
    correlation function
    (SCF), 550
  QPSK signal, spectral
    correlation function,
    551–552
  robustness, sampling clock
    offsets, 550–551
  signal detection, 550
  signal sampling, 551–552
  vs. energy detector, 551
energy detector
  FFT-based architecture,
    547–548
  FPGA, 549
  narrowband architecture, 547,
    549
  sensing time *vs.* SNR,
    549–550
  wideband architecture,
    548–549
lab equipment, 553
multigigabit transceivers (MGTs),
  541
multiple antenna processing, 547
networking capabilities
  BWRC test bed, 546–547
  Ethernet interface, 546
  XAUI interface, 545–546
reconfigurable radio frequency
  front end, 543–544
software design flow, 544–545
system architecture and
  implementation, 541–542
Bit error probability
  derivation
    dynamic spectrum access,
      56–57
    lower bound, 56
    upper bound, 55
  M-ary phase shift keying, 57
Border gateway protocol (BGP),
  251
BPSK-OFDM transceiver system,
  169
Branch-and-bound solution
  procedure, 345–347
Broadband wireless access, IEEE
  802.22 standards
  dynamic spectrum access
    models, 389
  IEEE 802.16h, SCC 41 standard,
    427–428
  IEEE 802.22.1 standard, 427
  medium-access control (MAC)
    layer
    channel estimation, 399
    coexistence beacon protocol,
      406

    IEEE 802.16 standard, 399
    incumbent database support,
      411–412
    incumbent detection, 402–403
    multichannel operation,
      403–404
    network discovery and
      coordination, 406
    orthogonal frequency division
      multiplexing (OFDM),
      402
    quality of service (QoS)
      support, 408–409
    resource sharing mechanisms,
      406–407
    self-coexistence window
      (SCW), 400
    spectrum automaton (SA),
      410–411
    spectrum management model,
      409–410
    spectrum sensing function,
      411
    synchronization, 404–405
    synchronous timing structure,
      399
    time-frequency structure, 401
    unlicensed spectrum access
      model, 405
  overview
    applications, 391
    effective isotropic radiated
      power (EIRP), 390
    reference architecture,
      391–393
    WRAN standard, 390
  physical layer
    channel coding and
      modulation schemes,
      397–398
    coexistence beacon protocol
      (CBP), 396–397
    control header and map
      definition, 395–396
    orthogonal frequency division
      multiple access
      (OFDMA), 393–394
    preamble definition, 395
    RF mask, 398
    symbol and TTG/RTG values,
      395
    transmit power control, 398
  spectrum sensing
    ATSC cyclostationary sensing
      technique, 424–427
    base station keep-out region,
      414–415
    CPE keep-out region, 415–416
    DTV broadcasting station, 412
    DTV power *vs.* distance, 413

Printed and bound by CPI Group (UK) Ltd, Croydon, CR0 4YY

03/10/2024

01040317-0007